# HORMESIS IN HEALTH AND DISEASE

# OXIDATIVE STRESS AND DISEASE

Series Editors

## LESTER PACKER, PhD
## ENRIQUE CADENAS, MD, PhD

UNIVERSITY OF SOUTHERN CALIFORNIA SCHOOL OF PHARMACY
LOS ANGELES, CALIFORNIA

1. Oxidative Stress in Cancer, AIDS, and Neurodegenerative Diseases, *edited by Luc Montagnier, René Olivier, and Catherine Pasquier*
2. Understanding the Process of Aging: The Roles of Mitochondria, Free Radicals, and Antioxidants, *edited by Enrique Cadenas and Lester Packer*
3. Redox Regulation of Cell Signaling and Its Clinical Application, *edited by Lester Packer and Junji Yodoi*
4. Antioxidants in Diabetes Management, *edited by Lester Packer, Peter Rösen, Hans J. Tritschler, George L. King, and Angelo Azzi*
5. Free Radicals in Brain Pathophysiology, *edited by Giuseppe Poli, Enrique Cadenas, and Lester Packer*
6. Nutraceuticals in Health and Disease Prevention, *edited by Klaus Krämer, Peter-Paul Hoppe, and Lester Packer*
7. Environmental Stressors in Health and Disease, *edited by Jürgen Fuchs and Lester Packer*
8. Handbook of Antioxidants: Second Edition, Revised and Expanded, *edited by Enrique Cadenas and Lester Packer*
9. Flavonoids in Health and Disease: Second Edition, Revised and Expanded, *edited by Catherine A. Rice-Evans and Lester Packer*
10. Redox–Genome Interactions in Health and Disease, *edited by Jürgen Fuchs, Maurizio Podda, and Lester Packer*
11. Thiamine: Catalytic Mechanisms in Normal and Disease States, *edited by Frank Jordan and Mulchand S. Patel*
12. Phytochemicals in Health and Disease, *edited by Yongping Bao and Roger Fenwick*
13. Carotenoids in Health and Disease, *edited by Norman I. Krinsky, Susan T. Mayne, and Helmut Sies*
14. Herbal and Traditional Medicine: Molecular Aspects of Health, *edited by Lester Packer, Choon Nam Ong, and Barry Halliwell*
15. Nutrients and Cell Signaling, *edited by Janos Zempleni and Krishnamurti Dakshinamurti*
16. Mitochondria in Health and Disease, *edited by Carolyn D. Berdanier*
17. Nutrigenomics, *edited by Gerald Rimbach, Jürgen Fuchs, and Lester Packer*
18. Oxidative Stress, Inflammation, and Health, *edited by Young-Joon Surh and Lester Packer*

19. Nitric Oxide, Cell Signaling, and Gene Expression, *edited by Santiago Lamas and Enrique Cadenas*
20. Resveratrol in Health and Disease, *edited by Bharat B. Aggarwal and Shishir Shishodia*
21. Oxidative Stress and Age-Related Neurodegeneration, *edited by Yuan Luo and Lester Packer*
22. Molecular Interventions in Lifestyle-Related Diseases, *edited by Midori Hiramatsu, Toshikazu Yoshikawa, and Lester Packer*
23. Oxidative Stress and Inflammatory Mechanisms in Obesity, Diabetes, and the Metabolic Syndrome, *edited by Lester Packer and Helmut Sies*
24. Lipoic Acid: Energy Production, Antioxidant Activity and Health Effects, *edited by Mulchand S. Patel and Lester Packer*
25. Dietary Modulation of Cell Signaling Pathways, *edited by Young-Joon Surh, Zigang Dong, Enrique Cadenas, and Lester Packer*
26. Micronutrients and Brain Health, *edited by Lester Packer, Helmut Sies, Manfred Eggersdorfer, and Enrique Cadenas*
27. Adipose Tissue and Inflammation, *edited by Atif B. Awad and Peter G. Bradford*
28. Herbal Medicine: Biomolecular and Clinical Aspects, Second Edition, *edited by Iris F. F. Benzie and Sissi Wachtel-Galor*
29. Inflammation, Lifestyle and Chronic Diseases: The Silent Link, *edited by Bharat B. Aggarwal, Sunil Krishnan, and Sushovan Guha*
30. Flavonoids and Related Compounds: Bioavailability and Function, *edited by Jeremy P. E. Spencer and Alan Crozier*
31. Mitochondrial Signaling in Health and Disease, *edited by Sten Orrenius, Lester Packer, and Enrique Cadenas*
32. Vitamin D: Oxidative Stress, Immunity, and Aging, *edited by Adrian F. Gombart*
33. Carotenoids and Vitamin A in Translational Medicine, *edited by Olaf Sommerburg, Werner Siems, and Klaus Kraemer*
34. Hormesis in Health and Disease, *edited by Suresh I. S. Rattan and Éric Le Bourg*

# HORMESIS IN HEALTH AND DISEASE

EDITED BY
SURESH I. S. RATTAN • ÉRIC LE BOURG

CRC Press
Taylor & Francis Group
Boca Raton London New York

CRC Press is an imprint of the
Taylor & Francis Group, an **informa** business

First published in paperback 2024

First published 2014 by CRC Press
2385 NW Executive Center Drive, Suite 320, Boca Raton FL 33431

and by CRC Press
4 Park Square, Milton Park, Abingdon, Oxon, OX14 4RN

*CRC Press is an imprint of Taylor & Francis Group, LLC*

© 2014, 2024 Taylor & Francis Group, LLC

Reasonable efforts have been made to publish reliable data and information, but the author and publisher cannot assume responsibility for the validity of all materials or the consequences of their use. The authors and publishers have attempted to trace the copyright holders of all material reproduced in this publication and apologize to copyright holders if permission to publish in this form has not been obtained. If any copyright material has not been acknowledged please write and let us know so we may rectify in any future reprint.

Except as permitted under U.S. Copyright Law, no part of this book may be reprinted, reproduced, transmitted, or utilized in any form by any electronic, mechanical, or other means, now known or hereafter invented, including photocopying, microfilming, and recording, or in any information storage or retrieval system, without written permission from the publishers.

For permission to photocopy or use material electronically from this work, access www.copyright.com or contact the Copyright Clearance Center, Inc. (CCC), 222 Rosewood Drive, Danvers, MA 01923, 978-750-8400. For works that are not available on CCC please contact mpkbookspermissions@tandf.co.uk

*Trademark notice*: Product or corporate names may be trademarks or registered trademarks and are used only for identification and explanation without intent to infringe.

Publisher's Note
The publisher has gone to great lengths to ensure the quality of this reprint but points out that some imperfections in the original copies may be apparent.

**Library of Congress Cataloging-in-Publication Data**

Hormesis in health and disease / editors, Suresh I.S. Rattan and Éric Le Bourg.
   p. ; cm.
Includes bibliographical references and index.
ISBN 978-1-4822-0545-9 (alk. paper)
   I. Rattan, Suresh I. S., editor of compilation. II. Le Bourg, Éric, editor of compilation.
   [DNLM: 1. Hormesis--physiology. 2. Stress, Physiological. 3. Longevity--physiology.
QT 162.S8]

QP82.2.S8
616.9'8--dc23                                                                                                  2013045240

ISBN : 978-1-48-220545-9 (hbk)
ISBN : 978-1-03-292083-2 (pbk)
ISBN : 978-0-42-917107-9 (ebk)

DOI: 10.1201/b17042

**Visit the Taylor & Francis Web site at**
**http://www.taylorandfrancis.com**

**and the CRC Press Web site at**
**http://www.crcpress.com**

# Contents

Preface ......................................................................................................................... ix
Editors ...................................................................................................................... xiii
Contributors ............................................................................................................. xv

## SECTION I  History and Terminology

**Chapter 1**  Brief History of Hormesis and Its Terminology ................................... 3
*Edward J. Calabrese*

**Chapter 2**  Pre- and Postconditioning Hormesis .................................................. 13
*Fred A.C. Wiegant*

## SECTION II  Evidence for Hormesis in Human Beings

**Chapter 3**  Exercise and Hormesis: Shaping the Dose–Response Curve ............ 37
*Zsolt Radak*

**Chapter 4**  Nutritional Components: How They Enhance the Ability
to Adapt ............................................................................................... 45
*Antje R. Weseler and Aalt Bast*

**Chapter 5**  Periodic Fasting and Hormesis ........................................................... 79
*Yan Y. Lam and Eric Ravussin*

**Chapter 6**  Iron, Metabolic Syndrome, and Hormesis ......................................... 93
*Kupper A. Wintergerst and Lu Cai*

**Chapter 7**  Radiation Exposure .......................................................................... 107
*Alexander Vaiserman*

**Chapter 8**  Thermal Hydrotherapy as Adaptive Stress Response: Hormetic
Significance, Mechanisms, and Therapeutic Implications ............... 153
*Giovanni Scapagnini, Sergio Davinelli, Nicola Angelo
Fortunati, Davide Zella, and Marco Vitale*

**Chapter 9** Cardiac Ischemic Preconditioning and the Ischemia/
Reperfusion Injury ........................................................................... 167

*Andreas Simm and Rüdiger Horstkorte*

**Chapter 10** Cerebral Ischemia................................................................. 185

*Yannick Béjot and Philippe Garnier*

**Chapter 11** Optimal Stress, Psychological Resilience, and
the Sandpile Model........................................................................ 201

*Martha Stark*

## SECTION III  Molecular Mechanisms of Hormesis

**Chapter 12** Molecular Stress Response Pathways as the Basis of Hormesis ...... 227

*Dino Demirovic, Irene Martinez de Toda, and Suresh I.S. Rattan*

**Chapter 13** Inflammatory Pathways...................................................... 243

*Salvatore Chirumbolo*

**Chapter 14** Oxidative Stress Response Pathways: Role of Redox Signaling
in Hormesis ................................................................................... 281

*Li Li Ji*

## SECTION IV  Hormesis in Risk Assessment

**Chapter 15** Relating Hormesis to Ethics and Policy: Conceptual Issues and
Scientific Uncertainty.................................................................... 307

*George R. Hoffmann*

**Chapter 16** Hormesis and Risk Assessment......................................... 339

*Edward J. Calabrese*

**Index** ............................................................................................................ 357

# Preface

There is now a general accordance that some mild stresses have positive effects on the survival, health, aging, and longevity in animal models, particularly in the fly *Drosophila melanogaster* and nematode *Caenorhabditis elegans* (reviews in, e.g., Rattan [2008] and Le Bourg [2009]). Studies have also been performed in mammalian models, including human cells in culture, on such hormetic effects, especially with respect to aging and longevity (e.g., Demirovic and Rattan 2013). The idea of using stress to improve health is, at first sight, paradoxical. Indeed, being subjected to stress is not what we usually wish for because, as defined by Selye (1970), "in a strictly medical sense, stress is the rate of wear and tear to which a living being is exposed at any one moment." Furthermore, according to Selye (1975), stress is "a syndrome accompanied by objectively measurable somatic manifestations, and elicited by a variety of emotional and physical agents." This definition does not bear any notion of beneficial and life-supporting effects of stress, and Selye (1975) thus made a distinction between two kinds of stress: "demands for adaptation, experienced as agreeable or beneficial ... are designated as eustress, in opposition to distress."

Later on, Sacher (1977) reviewed the life-prolonging effects of "the class of phenomena in which small quantities of a depressive or toxic agent are stimulatory," that is, stresses with hormetic effects on life span. However, he considered that hormetic effects "are unlikely to occur in the healthy active animal, and are more likely to be significant in the ill or depressed animal," and that "hormesis is in one sense an obstacle in the path of gerontological research, and efforts to understand and annul it would be well justified." In that goal, it was necessary "to breed vigorous animal genotypes and ... to develop living environments that are optimal for their behavioral, physiological, and immunological health" (Sacher 1977). Thus, anyone observing any positive effects of a mild stress on life span could have simply concluded that the results were not of interest and that rearing methods in the lab were probably not optimal.

However, in contrast to Sacher (1977), Frolkis (1993) wrote that "in 1970 we proposed ... that a variety of repeated stress exposures of a mild magnitude trains the defensive mechanisms of the organism ... and may increase the lifespan" and reported a successful experience in rats (see Frolkis [1982], pp. 9, 92). Because Sacher's (1977) article was in English and Frolkis et al. (1976) reported their results in an article written in Russian, it is not surprising that Sacher had a greater impact on biogerontologists for many years and that studying the positive effects of mild stress was then not considered as a top priority.

This was, however, not the end of the story, because during the last decade of the twentieth century, some biogerontologists reported positive effects of mild stress on longevity and stress resistance, mainly in invertebrates (review of these early results in Minois [2000]). After this review was published, many more new results showing that mild stresses can have positive effects on aging and longevity were published, and these positive results, not explained by the use of suboptimal living conditions, revivified the interest in hormesis. Then, the first book on the effects of mild stress on aging was published (Le Bourg and Rattan 2008), quickly followed by a more general book (Mattson and Calabrese 2010) and by a debate among experts on the use of mild stress in human beings (Le Bourg and Rattan 2010). Thus, there was now not only a consensus on the idea that mild stresses can have positive effects on aging, longevity, and resistance to other stresses in model organisms, but also that it could be of use in human beings, even if there was not enough clarity on the precise ways to use mild stress.

Interestingly, studies have been or are being performed on the effects of mild stress in human beings, even if the authors did or do not always interpret their results as hormetic effects. As a matter of fact, it seems that there is a large body of results showing hormetic effects on aging, health, and resistance to severe stresses and diseases in human beings. However, these data are dispersed in the literature and are not always interpreted as hormetic effects, restricting their full apprehension, appreciation, and application. Thus, the paradox is that, even if many authors ignored the existence of hormetic effects, they have gathered results showing such effects. For instance, many studies have shown that moderate and repeated exercise can have positive effects on health and life expectancy (e.g., Paffenbarger et al. 1986), but the authors did not conclude that exercise is a mild stress with hormetic effects.

We thus feel that the time is now ripe to devote an entire book reviewing the evidence for hormesis in humans, as achieved through exercise, radiation, temperature, nutrition, ischemia, fasting, and mental challenge, along with discussing mechanisms of hormesis, and its ethical and legal issues. Our motivation is to convince clinicians, researchers, health-care personnel, the health-care industry, and social and political policy makers to initiate and implement new strategies to learn the strengths and limits of the dual nature of stress, especially with respect to mild stress-induced hormesis in maintaining health, in the prevention of diseases and even as a therapeutic approach for certain diseases. However, we must emphasize that as editors our aim was to gather contributions from top experts who were absolutely free to present their views in whatever way they chose within the framework of the subject. Thus, this book is an attempt to present the current state of research, questions, debates, doubts, and controversies in hormesis and is composed of four sections: history and terminology, evidence for hormesis in human beings, molecular mechanisms of hormesis, and hormesis in risk assessment.

Section I, which gives some information on the history and terminology of hormesis and describes its main features, is a necessary step before presenting the evidence that various mild stresses or stimuli can be useful to improve health span. It may be added that apart from the established terminology for hormesis as proposed by a consensus paper written by more than 50 authors and published in 2007 (Calabrese et al. 2007), some additional terms have also been proposed and used, such as *hormetin* for any condition that is potentially hormetic (Rattan 2008), and *hormetics* for the field of study or the science of hormesis (Rattan 2012).

Section II, composed of nine chapters, shows that hormetic effects can rely on various stresses and can be observed in various organs or at the organism level. These stresses are versatile and include physical exercise, nutritional components, fasting, micronutrients, irradiation, heat, ischemia, and even mental challenge. Thus, various mild stresses, challenges, or stimuli can improve health and health span. This is clearly the case not only for exercise, which is now routinely recommended by physicians as the simplest and cheapest means to improve health, but also for ischemic and cardiac preconditioning when facing the often deadly consequences of ischemic attack and myocardial infarction. Regarding fasting, even a single fasting episode could be of help, for instance, to increase the effect of chemotherapy against cancer (Lee and Longo 2011), and there is a debate on lengthening the fast before surgery to improve postoperative recovery (Mitchell et al. 2010). However, not all hormetic effects are due to environmental challenges such as temperature, because some nutritional components appear to have beneficial effects by enhancing the capacity of the organism to resist stressful events when used at a low dose, despite toxicity at a higher dose.

Section III is composed of three chapters that review the possible mechanisms of hormesis as elucidated so far, mostly through experiments performed on animal models. Deciphering the mechanisms of the hormetic effects is probably a bigger challenge than discovering new mild stresses with such hormetic effects. Firm answers are not still available but it seems clear that signaling pathways and immediate and delayed stress response pathways involving receptors and postreceptor transcription factors and downstream effectors are at play.

Section IV discusses how hormesis may have wider consequences for everybody, as there is a debate regarding the implementation of this concept in risk assessment regulations as well as other ethical aspects. The editors support the idea that when there is a controversy in science, it is appropriate to favor the expression of the divergent opinions to help other scientists reach their own conclusions. However, the debate on risk assessment is not the exclusive domain of scientists because, for instance, modifying or not modifying safety regulations is of concern to everybody.

All the chapters in this book show that health beneficial hormetic effects do exist, not only in animal models or in cells cultured outside the body but also in human beings. Questions however remain: for example, could it be possible to use mild stress at any age to enhance health span with no negative effects? This seems to be possible with exercise, which is not contraindicated at old age, but could sauna, for example, be used at any age in most people or should it be used only in a therapeutic approach? Will mild stresses become commonplace in future therapeutic settings, such as a longer period of fasting before surgery (Mitchell et al. 2010), or hyperbaric hyperoxia or carbon monoxide to improve recovery from cardiac surgery (Lavitrano et al. 2004; Li et al. 2011; Liu et al. 2012)?

Relying on hormesis as a preventive and/or therapeutic strategy is of topical interest and significance. This is because modern health-care practice not only has to treat diseases for which there are clear causative biological and chemical agents but also deal with the problems emerging as a result of lifestyle and long life span. Almost all age-related diseases, such as Alzheimer's, Parkinson's, type 2 diabetes, osteoporosis, sarcopenia, and even cancers, have complex origins and cannot be *treated* in a traditional sense (Rattan 2013). Furthermore, the long-term management of and attempts at treating such diseases are costly. Therefore, it is necessary to discover, develop, and promote new ways for maintaining health, preventing or delaying the onset of diseases, and extending the human health span. Hormetics, the science of hormesis, appears to be a promising area for further research, development, and application. This book is a step in that direction.

## REFERENCES

Calabrese, E.J., Bachman, K.A., Bailer, J. et al. 2007. Biological stress response terminology: Integrating the concepts of adaptive response and preconditioning stress within a hormetic dose-response framework. *Toxicology and Applied Pharmacology* 222: 122–128.

Demirovic, D., and Rattan, S.I.S. 2013. Establishing cellular stress response profiles as biomarkers of homeodynamics, health and hormesis. *Experimental Gerontology* 48: 94–98.

Frolkis, V.V. 1982. *Aging and Life-Prolonging Processes*. Wien, Austria: Springer-Verlag.

Frolkis, V.V. 1993. Stress-age syndrome. *Mechanisms of Ageing and Development* 69: 93–107.

Frolkis, V.V., Bogatskaya, L.N., Stupina, A.S. et al. 1976. Comparative characterization of some influences on life prolongation. In: D.F. Chebotarev and V.V. Frolkis, eds. *Biological Abilities of Increasing the Animal's Life Span*. Kiev, Ukraine: Institute of Gerontology, pp. 138–150 (in Russian).

Lavitrano, M., Smolenski, R.T., Musumeci, A. et al. 2004. Carbon monoxide improves cardiac energetics and safeguards the heart during reperfusion after cardiopulmonary bypass in pigs. *FASEB Journal* 18: 1093–1095.

Le Bourg, E. 2009. Hormesis, aging, and longevity. *Biochimica et Biophysica Acta* 1790: 1030–1039.

Le Bourg, E., and Rattan, S.I.S., eds. 2008. *Mild Stress and Healthy Aging. Applying Hormesis in Aging Research and Interventions*. Berlin, Germany: Springer.

Le Bourg, E., and Rattan, S.I.S. 2010. Is hormesis applicable as a pro-healthy aging intervention in mammals and human beings, and how? Introduction to a special issue of *Dose-Response*. *Dose-Response* 8: 1–3.

Lee, C., and Longo, V.D. 2011. Fasting vs. dietary restriction in cellular protection and cancer treatment: From model organisms to patients. *Oncogene* 30: 3305–3316.

Li, Y., Dong, H., Chen, M. et al. 2011. Preconditioning with repeated hyperbaric oxygen induces myocardial and cerebral protection in patients undergoing coronary artery bypass graft surgery: A prospective, randomized, controlled clinical trial. *Journal of Cardiothoracic and Vascular Anesthesia* 25: 908–916.

Liu, W., Liu, K., Tao, H. et al. 2012. Hyperoxia preconditioning: The next frontier in neurology? *Neurological Research* 34: 415–421.

Mattson, M.P., and Calabrese, E.J., eds. 2010. *Hormesis*. Dordrecht, the Netherlands: Springer.

Minois, N. 2000. Longevity and aging: Beneficial effects of exposure to mild stress. *Biogerontology* 1: 15–29.

Mitchell, J.R., Verweij, M., Brand, K. et al. 2010. Short-term dietary restriction and fasting precondition against ischemia reperfusion injury in mice. *Aging Cell* 9: 40–53.

Paffenbarger, R.S., Hyde, R.T., Hsieh, C.C. et al. 1986. Physical activity, other life-style patterns, cardiovascular disease and longevity. *Acta Medica Scandinavica* 711 (suppl): 85–91.

Rattan, S.I.S. 2008. Hormesis in aging. *Ageing Research Reviews* 7: 63–78.

Rattan, S.I.S. 2012. Rationale and methods of discovering hormetins as drugs for healthy aging. *Expert Opinion on Drug Discovery* 7: 439–448.

Rattan, S.I.S. 2013. Healthy ageing, but what is health? *Biogerontology* 14: 673–677.

Sacher, G.A. 1977. Life table modification and life prolongation. In: C.E. Finch and L. Hayflick, eds. *Handbook of the Biology of Aging*. New York: Van Nostrand Reinhold, pp. 582–638.

Selye, H. 1970. Stress and aging. *Journal of the American Geriatrics Society* 18: 669–680.

Selye, H. 1975. Confusion and controversy in the stress field. *Journal of Human Stress* 1: 37–44.

# Editors

**Suresh I.S. Rattan, PhD, DSc**, is an internationally renowned biogerontologist at the Laboratory of Cellular Ageing, Department of Molecular Biology and Genetics, Aarhus University, Denmark. He is one of the pioneers of the testing and application of hormesis in aging research and interventions. His research expertise includes elucidating the molecular mechanisms of human cellular aging, especially with respect to protein synthesis, modifications, and turnover. He discovered the aging-modulatory effects of kinetin and zeatin, which are now used in several skin care products. He has published numerous original research papers and reviews in international journals and books and has edited/coedited several books for various international publishers. Dr. Rattan is also the author of popular science books for children in several languages, including English, Punjabi, Hindi, and Danish.

**Éric Le Bourg, PhD, DSc**, is an internationally renowned biogerontologist at the French National Center for Scientific Research (CNRS), University Paul-Sabatier, Toulouse, France. He is well known for his studies on the effects of mild stress in *Drosophila* but is also working on learning, behavior, and demographic matters. He has published numerous papers in academic journals, edited books, newspapers etc., and has written several books on aging in French for the lay public and academics on biology, demographic, and social matters.

Both editors have previously collaborated and edited the book *Mild Stress and Healthy Aging* (Springer, 2008) and brought out two special issues of the international peer-reviewed journals *Biogerontology* (2006) and *Dose Response* (2010).

# Contributors

**Aalt Bast**
Department of Toxicology
Maastricht University
Maastricht, the Netherlands

**Yannick Béjot**
Dijon Stroke Registry
University Hospital and Medical School of Dijon
University of Burgundy
Dijon, France

**Lu Cai**
Departments of Pediatrics, Radiation Oncology, and Pharmacology and Toxicology
Kosair Children's Hospital Research Institute
University of Louisville
Louisville, Kentucky

**Edward J. Calabrese**
Department of Public Health
Environmental Health Sciences
University of Massachusetts
Amherst, Massachusetts

**Salvatore Chirumbolo**
Laboratory of Physiopathology of Obesity-LURM Est
Department of Medicine
University of Verona
Verona, Italy

**Sergio Davinelli**
Department of Medicine and Health Sciences
University of Molise
Campobasso, Italy

and

Department of Biochemistry and Molecular Biology
Institute of Human Virology
University of Maryland–School of Medicine
Baltimore, Maryland

**Dino Demirovic**
Department of Molecular Biology and Genetics
Laboratory of Cellular Ageing
Aarhus University
Aarhus, Denmark

**Nicola Angelo Fortunati**
Spa Centre of Grotta Giusti Natural Spa Resort
Monsummano Terme
Pistoia, Italy

**Philippe Garnier**
INSERM U1093 Cognition, Action and Sensorimotor Plasticity
University of Burgundy
Dijon, France

**George R. Hoffmann**
Department of Biology
College of the Holy Cross
Worcester, Massachusetts

**Rüdiger Horstkorte**
Institute for Physiological Chemistry
Martin-Luther University
 Halle-Wittenberg
Halle (Saale), Germany

**Li Li Ji**
Laboratory of Physiological Hygiene
 and Exercise Science
School of Kinesiology
University of Minnesota Twin Cities
Minneapolis, Minnesota

**Yan Y. Lam**
John S. McIlhenny Skeletal Muscle
 Physiology Laboratory
Pennington Biomedical Research Center
Baton Rouge, Louisiana

**Éric Le Bourg**
Research Center on Animal Cognition
UMR CNRS 5169
University Paul-Sabatier
Toulouse, France

**Irene Martinez de Toda**
Department of Molecular Biology and
 Genetics
Laboratory of Cellular Ageing
Aarhus University
Aarhus, Denmark

**Zsolt Radak**
Faculty of Physical Education and Sport
 Science
Institute of Sport Science
Semmelweis University
Budapest, Hungary

**Suresh I.S. Rattan**
Department of Molecular Biology and
 Genetics
Laboratory of Cellular Ageing
Aarhus University
Aarhus, Denmark

**Eric Ravussin**
John S. McIlhenny Skeletal Muscle
 Physiology Laboratory
Pennington Biomedical Research
 Center
Baton Rouge, Louisiana

**Giovanni Scapagnini**
Department of Medicine and Health
 Sciences
University of Molise
Campobasso Italy and Sannio Tech
Apollosa (BN), Italy

**Andreas Simm**
Heart Centre
University Hospital Halle (Saale)
Halle (Saale), Germany

**Martha Stark**
Department of Psychiatry
Beth Israel Deaconess Medical
 Center
Harvard Medical School
Boston, Massachusetts

**Alexander Vaiserman**
Laboratory of Epigenetics
Institute of Gerontology
Kiev, Ukraine

**Marco Vitale**
Department of Biomedical,
 Biotechnological and Translational
 Sciences (S.Bi.Bi.T.)
University of Parma
Parma, Italy

**Antje R. Weseler**
Department of Toxicology
Maastricht University
Maastricht, the Netherlands

**Fred A.C. Wiegant**
Department of Biology
Institute of Education
Utrecht University
Utrecht, the Netherlands

**Kupper A. Wintergerst**
Wendy L. Novak Diabetes Care Center
Kosair Children's Hospital
and
Children's Metabolic Bone Center
Department of Pediatrics
University of Louisville School of
   Medicine
Louisville, Kentucky

**Davide Zella**
Department of Biochemistry and
   Molecular Biology
Institute of Human Virology
University of Maryland–School of
   Medicine
Baltimore, Maryland

# Section I

*History and Terminology*

# 1 Brief History of Hormesis and Its Terminology

*Edward J. Calabrese*

## CONTENTS

1.1 Brief History ........................................................................................................ 3
1.2 Biomedical Sciences and Hormesis ................................................................... 9
References .................................................................................................................. 10

## 1.1 BRIEF HISTORY

The concept of hormesis started with the findings of Hugo Schulz (1887, 1888) that a number of chemical disinfectants stimulated the metabolism of yeast at low concentrations, although they were inhibitory at higher concentrations. Schulz created a major controversy because he used these findings to assert that he had discovered the explanatory principle of homeopathy, a medical practice then in serious conflict with what is today called traditional medicine. Thus, before Schulz's biphasic dose-response could receive a fair assessment by the scientific and medical communities, it was rejected and dismissed. These actions were based on issues other than science. Schulz and his dose-response would become quickly marginalized, ridiculed, and treated as a scientific outcast for the next 50 years (Calabrese 2005, 2011).

The reason why Schulz linked the biphasic dose-response to homeopathy was based on a series of interrelated events (Calabrese 2011). First, Schulz had earlier read a report in which a homeopathic drug named veratrine was used to treat patients with gastroenteritis successfully. Soon thereafter, the bacterial cause of this condition was identified and cultured. Schulz subsequently obtained the bacterium and assessed whether the veratrine could kill the disease-causing organism. Regardless of the dose used, the veratrine was not able to kill the bacterium. This failure of the veratrine did not affect Schulz's belief that the treatment was effective. He simply could not provide an explanation. In the meantime, Schulz conducted experiments with yeasts during which multiple chemical disinfectants affected a low-dose stimulation and high-dose inhibition on metabolism, responses he was able to reliably replicate. Following subsequent discussion with colleague Rudolf Arndt, Schulz sought to integrate this series of observations, that is, the veratrine findings and the chemical disinfectant–induced biphasic dose-responses. He concluded that the veratrine did cure the patients with the gastroenteritis, that it did so not by killing the disease-causing organism, but by enhancing the adaptive capacity of the individual to resist the infection. He based this conclusion on the findings with the yeast, with the low-dose stimulation representing the capacity of the body to affect an adaptive response. Within this context, the biphasic dose research was to provide the theoretical framework for how low (not infinitesimally low) doses of homeopathic drugs might act. Thus, by linking the biphasic dose-response concept with the practice of homeopathy, Schulz made it politically impossible for those in the medical sciences to explore the scientific foundations of this biphasic dose-response model. In fact, by linking the biphasic dose-response to homeopathy and its political conflict with traditional medicine, he created the conditions by which just the opposite would occur. His actions created more hostility than interest in his new biphasic dose-response. The area of low dose-responses and biphasic dose-responses would not be systemically studied; when biphasic dose-responses were observed, they often tended to be discounted as trivial or simple variability. Thus, the biphasic dose-response concept got off to a bad start in the biomedical community, a start from which it would never fully recover over successive generations of biomedical scientists. With respect to the biphasic dose-response, the reach of history has been insidious and controlling, with very well-disguised actions impeding its fair assessment.

The biphasic dose-response would often be seen by others in experiments and reported in leading journals during the remainder of Schulz's life (Calabrese and Baldwin 2000a–e). Although one might think this would have led to at least grudging acceptance of Schulz's biphasic dose-response concept by the scientific and medical communities, it failed to do so, at least in his lifetime, which ended during his 79th year in 1932. The principal reason was that traditional medicine, which had gained considerable influence, power, and prestige during the early decades of the twentieth century, continued to marginalize Schulz while also deliberately misrepresenting his association with homeopathy, by falsely aligning him with the high-dilution wing of homeopathy (Calabrese 2005, 2011). By the end of the third decade of the twentieth century, homeopathy was no longer a serious rival of medicine, Schulz would soon die, and medicine would continue to ensure that the biphasic dose-response would never rise to prominence by its exclusion from textbooks, professional society activities, research funding, and governmental advisory activities.

Because homeopathy scooped traditional medicine on adopting a dose-response model, traditional medicine would respond, not only by challenging the biphasic dose-response but also by proposing a dose-response model of its own. Unable to consider the biphasic dose-response relationship of its rival, traditional medicine selected the threshold dose-response for its default model, and made it the centerpiece for drug evaluation and risk assessment. The threshold dose-response was not an unreasonable model to be selected. There were articles in the published literature that supported a threshold dose-response concept (Shackell 1923, 1925; Shackell et al. 1924/1925). The concept of a threshold was easy to understand and resonated in the common experience of people who might reflect on their own dose-related experiences, such as with consumption of alcohol, the ingestion of certain medicinal drugs, or other dose-related activities. However, medicine neglected to do due diligence when it gave this model precedence in the risk assessment domain. It failed to validate the capacity of the threshold dose-response model to make accurate predictions in the low-dose zone, where people typically live. This lack of validation would only become recognized in the first decade of the twenty-first century, highlighting a critical failure of pharmacology and toxicology, which let the validation requirement for the dose-response slip past their collective oversight.

A significant problem with the acceptance of hormesis by the scientific community, beyond that of powerful political conflicts between medicine and homeopathy, was that this biphasic dose-response concept was not well understood, even by its supporters, including Schulz, with respect to the quantitative characteristics of the biphasic dose-response, its temporal features, generalizability, conditions for optimization, and mechanisms. Of particular importance was the fact that the hormetic stimulation was consistently modest, almost always less than a factor of twofold greater than the control group. The maximum stimulation would typically be only some 30%–60% greater than the control group. This modest increase made it difficult to detect a hormetic response and to differentiate it from background variation. Second, the low-dose stimulation would sometimes be seen as an apparent direct stimulatory response. However, with other biological models and end points it would be observed only after an initial disruption in homeostasis or after modest toxicity (Calabrese 1999, 2001; Calabrese and Baldwin 2002). In this latter case, the experiment would require that there be a dose–time response. It was not usually obvious that both types of low-dose stimulatory responses occurred or how they related to each other, if they actually did, or what their respective significance was, if any. In the case of radiation biology experts of the early twentieth century, the direct stimulatory response was seen as potentially biologically significant, whereas compensatory stimulatory responses were often dismissed as a *trivial* reparative process. Yet, subsequent detailed evaluations of the dose-response literature would show that both types of responses could occur and would have the same quantitative features of the dose-response. Thus, during the early decades of the twentieth century, researchers struggled with understanding the nature of the dose-response, especially in the low-dose zone, and with the concept of its replication, although not actually realizing that the low-dose stimulatory response was inherently modest. Hypothesis testing and the statistical power concept, which were expanding their influence in the scientific community from the early 1920s, did not inform the early evolution of the hormetic dose-response concept as they do today. Such a circumstance would come to place greater requirements on the strength of the study design, statistical power, replication of findings, and mechanistic understandings. Another factor affecting the lack of appreciation of the general nature of the hormetic dose-response is that research in the area of radiation hormesis rarely recognized or cited those working in the area of chemical hormesis. This failure to integrate across biological disciplines profoundly weakened the strength of both the radiation and chemical hormesis perspectives (Table 1.1).

From approximately 1930 to 1980, the concept of hormesis continued to marginally expand under the guise of different terms with a slow but steady flow of publications, yet never achieving much visibility or significance. However, on occasion the concept of hormesis would have the opportunity to gain special

## TABLE 1.1
## Assessment of Factors Leading to the Historical Marginalization of Chemical and Radiation Hormesis

### Limitations More Unique to Either Chemical or Radiation Hormesis

Chemical Hormesis
- Close historical relationship with homeopathy.
- Strong criticisms by leading pharmacologists.
- Agricultural, industrial, and medical applications were not recognized.
- Clinical or epidemiological data supporting the hormesis concept were lacking.
- Failure of applications as seen with the USDA studies.

Radiation Hormesis
- Despite the efforts of Luckey to organize the published radiation literature, radiation hormesis had a much weaker database than chemical hormesis. His efforts were so strong that it created just the opposite impression.
- Technical criticism of the hormesis concept by the U.S. National Research Council (NRC) in 1936 in the Edna Johnson (1936) paper affected acceptance of this concept at the highest levels of scientific expertise/reputation in the United States.
- Fear of health risks for mutation induced cancer even at very low doses based on the work of Harman J. Muller and other radiation geneticists.
- Discoveries concerning adaptive response and DNA repair occurred after dose-response concepts were formulated. This had a profound effect on the historical foundations of the dose-response.

### Limitations Held in Common by Chemical and Radiation Hormesis

- Failed to understand the quantitative features of the hormetic dose-response.
- Did not appreciate that the low-dose stimulation was inherently modest.
- Failed to appreciate the requirement for strong study designs, with numerous and appropriately spaced doses below the toxic and/or pharmacological threshold.
- Failed to appreciate the extra resources needed to assess hormetic concepts, especially with respect to number of subjects and sampling over time and the need for replication.
- The presence of only a relatively small number of scientists researching in the area of dose-response and from diverse disciplines prevented the development of a critical pool/size of scientists.
- The historical timing was not good for the broad acceptance of hormesis; higher priorities were diverted to assess the nature of the dose-response above the threshold rather than below it.
- There was a far greater governmental need to define community and worker exposure standards. This led to a marginalizing of the low-dose hormetic effect.
- The failure to have an organized presence on the nature of the dose-response in the low dose zone affected funding, publications, and other professional/educational outlets.

recognition, but it would fail because it was not well-understood and this lack of understanding affected how it could or should have been assessed. For example, the U.S. Department of Agriculture (USDA) attempted to assess the capacity of ionizing radiation to enhance vegetable growth and productivity for commercial purposes (Alexander 1950). They created an elaborate and multisite research program during the mid to late 1940s using a variety of vegetables. However, this research incorporated only one dose and used this same dose for each vegetable tested. In addition, there was no preliminary testing to identify a possible threshold response below which a possible hormetic treatment should be applied. This attempt to test the hormetic concept for a commercial application reflected a profound lack of understanding of the hormetic dose-response and was in effect bound to fail to detect general hormetic responses, which it did. Despite the extensive efforts of this U.S. federal agency, the research strategy and study design were inadequate and inappropriate to test a hormesis hypothesis. When looked at with hindsight, it is obvious that those in charge still did not understand the fundamental aspects of the hormesis dose-response relationship even 15 years after Schulz's death. Nonetheless, the negative findings of this major effort by the USDA were important as they delivered a crushing blow to the hormesis concept within this agency, along with influencing possible parallel activities in other agencies, the university system, and private sector organizations. It would take several decades (i.e., a new generation of scientists) for the hormesis concept to generate new interest.

By the late 1970s, scientific change was occurring in the dose-response domain. Numerous researchers from very divergent biological subdisciplines had begun to report biphasic dose-responses as a reasonably common occurrence. This was enhanced by marked improvements in the capacity to measure lower doses as well as by a greater interest in cell culture research, which permitted the testing of a larger number of concentrations/doses than used in whole animal studies. Of further importance is that some investigators had begun to study the hormesis phenomenon in considerable depth. Of particular note were three individuals from different biological fields: Thomas Luckey, who published the first substantial book on hormesis in 1980 dealing with ionizing radiation. Later, he provided a detailed update of his earlier book (Luckey 1991). Next, Elmer Szabadi, a neuropharmacologist, summarized the frequency of biphasic dose-response across the discipline of pharmacology. He also provided a receptor-based mechanism that could account for such responses (Szabadi 1977). The third person was Tony Stebbing of the Plymouth Research Station in the United Kingdom. Stebbing, a marine toxicologist, unexpectedly discovered the phenomenon of hormesis in an assessment of heavy metals on marine microorganisms (Stebbing 1976, 1981, 1998). Besides his copious documentation of this phenomenon, he proposed a cybernetic feedback model mechanism to account for his observations. Although there were other contemporary researchers (Sacher 1963; Frolkis 1982) with contributions to the reawakening of the hormesis concept, these three provided the initial scientific leadership that would stimulate the scientific community to give the hormesis concept a new consideration.

A major event in the history of hormesis was that a conference was held on Radiation Hormesis in Oakland, California in August 1985. According to Sadao Hattori of the Japanese Electric Power Institute, he was challenged and inspired by the 1980 book of Luckey and needed to learn more about hormesis. If Luckey was correct, he said, on the nature of the dose-response, then it would have significant implications for radiation risk assessment and his industry, which included nuclear power. Hattori contacted the U.S. Electric Power Research Institute in Palo Alto, California, and through their interactions, the conference in 1985 took place. The significance of the conference was severalfold. It provided a professionally unifying element to the topic of hormesis; it was international in scope; the conference led to the development of a radiation hormesis research program funded by the Japanese electric power industry; the peer-reviewed proceedings were published in a respected journal, *Health Physics*; it led to a debate on hormesis in the journal *Science* in 1989; and it led to the formation of Biological Effects of Low-Level Exposure at the University of Massachusetts that created a long-term research and information exchange (e.g., conferences, workshops, and publication vehicles) including the journal *Dose-Response* (formerly *Non-Linearity in Biology, Toxicology and Medicine*) on hormesis. This effort, therefore, helped to encourage others to research the hormesis topic and to explore funding opportunities by U.S. federal agencies such as the Department of Energy, National Institutes of Health, National Science Foundation, Environmental Protection Agency, and the military, as well as various private sector organizations and foundations.

Despite the book of Luckey (1980) and the publications of Stebbing and the conference proceedings in *Health Physics*, the number of citations in the 1980s for hormesis or hormetic averaged only 10–15 per year. However, by 2012, the number of citations had reached 4,500 (Figure 1.1), with a collective total approaching 30,000. The major increase that started in the late 1990s was associated with funding support from U.S. federal agencies such as the U.S. Nuclear Regulatory Commission (NRC) and long-term funding support of the Air Force.

One of the subtle but important scientific and technical developments that enhanced the assessment of hormesis, as noted earlier, was the dramatic shift to the use of cell culture during the 1980s. This was important because it provided a cost-effective means to assess a larger number of concentrations (doses) than would

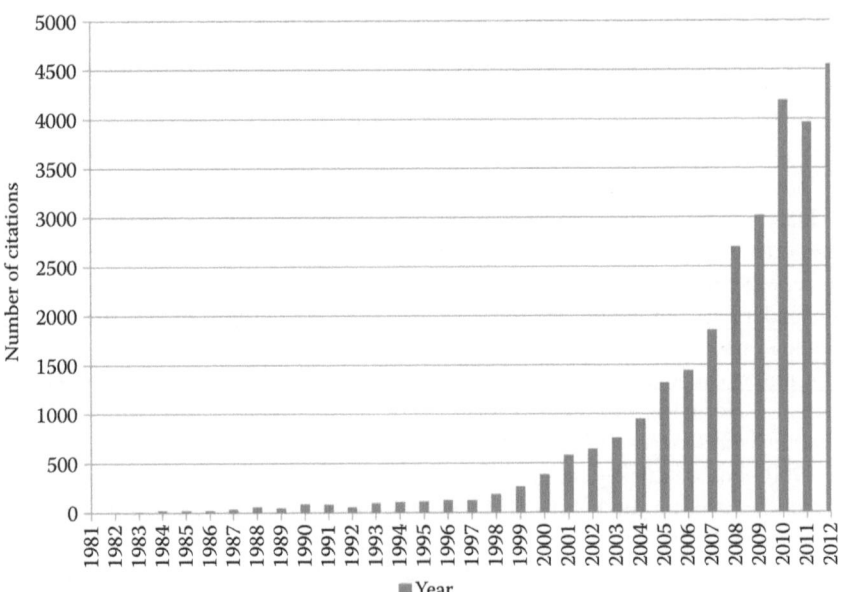

**FIGURE 1.1** Citations of "hormesis" or "hormetic" in the Web of Knowledge database.

be typically done via the use of whole animal studies. Over the past 15 years, the proportion of studies showing hormetic dose-responses with cell cultures has far exceeded that with whole animals.

Another scientific development enhancing the study of hormesis has been progress in the understanding of the underlying mechanism(s) of hormetic-biphasic dose-response relationships. The documentation of hormetic mechanisms has been shown at the level of receptor and/or cell-signaling pathways. To date, between 400 and 500 highly reproducible hormetic dose-responses have the low-dose stimulation mediated by a specific receptor or cell-signaling pathway (Calabrese 2013). Use of inhibitors of the receptor/signaling pathway has been able to block the hormetic dose-response stimulation. Thus, the phenomenon of hormesis has progressed from a descriptive phenomenon to vastly improved understanding of underlying mechanisms.

## 1.2 BIOMEDICAL SCIENCES AND HORMESIS

Despite the fact that the pharmaceutical industry has not used the term *hormesis* in their research publications, regulatory agency submissions, and advertising, many pharmaceutical drugs are based on preclinical research that is fundamentally hormetic in nature. For example, numerous anxiolytic (Calabrese 2008a) and antiseizure (Calabrese 2008b) drugs display hormetic-like biphasic dose-response relationships, with quantitative features fully consistent with the hormetic dose-response. Most drugs that enhance memory do so in a biphasic manner, reflecting a hormetic dose-response. The preclinical findings of Alzheimer's disease drugs approved by the U.S. Food and Drug Administration display hormetic-like biphasic dose-responses (Calabrese 2008c). Similar dose-response profiles have been reported for stroke-related research and in studies dealing with the prevention and treatment of brain trauma injury (Calabrese 2008d; Zhao 2013). Of particular importance is that the concepts of pre- and postconditioning that are being applied to prevent neurodegeneration, heart disease, and stroke, and to enhance surgical successes are based on the hormetic concept (Calabrese et al. 2007). That is, when the pre- or postconditioning treatment is presented in a dose-response framework, the results are hormetic, with a low-dose stimulation and a high-dose inhibition.

The biomedical field is slowly recognizing the term hormesis and its generalizing influence. Nonetheless, the hormetic-biphasic dose-response concept has been a mainstay guiding the development of drugs that are designed to enhance biological performance. The hormesis concept should be central to the drug development process, as it defines the ceiling of the response for pharmaceutical agents and the width of the stimulatory response. In fact, it is the modest stimulatory nature of the hormetic dose-response that places significant challenges to the detection of drug efficacy during clinical trials with human subjects that display considerable heterogeneity of responsiveness.

The fact that low-dose stimulatory responses can result in both beneficial and harmful effects is expected to become better integrated within the biological and biomedical sciences over the next decade. For example, biofilms produced by *Staphylococcus epidermidis* affecting prosthetic device infections can be enhanced in a hormetic manner by low doses of antibiotics (Kaplan et al. 2011). Likewise, the hormetic response has recently been recognized for its potential threat to agriculture based on the concerns with low-dose enhancement of fungal growths (Garzon and Flores 2013). The prolonged debate over endocrine-disrupting chemicals is centered on the nature of the dose-response in the low-dose zone, which again conforms to the quantitative features of the hormetic dose-response, whether one uses the term nonmonotonic or hormetic (Calabrese 2008e). Thus, the biological strategies are the same: all reflect the hormetic dose-response.

Recent developments in the dose-response area are leading to a convergence of biological concepts, such as preconditioning, postconditioning, adaptive response, nonmonotonic responses, Arndt–Schulz law, Yerkes–Dodson law, U-shaped dose-responses, J-shaped responses, and rebound responses (Calabrese 2007). All of these apparently diverse biological concepts have a central component in common: they all are manifestations of the hormetic dose-response. The hormetic dose-response therefore represents a previously unrecognized general biological principle that integrates responses following a basic and highly conserved adaptive strategy. This dose-response concept has profound implications for drug development, environmental assessment, cause–benefit analysis, and lifestyle choices, including optimal aging.

## REFERENCES

Alexander, L.T. 1950. Radioactive materials as plant stimulants—Field results. *Agronomy Journal* 42: 252–255.
Calabrese, E.J. 1999. Evidence that hormesis represents an "overcompensation" response to a disruption in homeostasis. *Ecotoxicology and Environmental Safety* 42: 135–137.
Calabrese, E.J. 2001. Overcompensation stimulation: A mechanism for hormetic effects. *Critical Reviews in Toxicology* 31(4–5): 425–470.
Calabrese, E.J. 2005. Historical blunders: How toxicology got the dose-response relationship half right. *Cellular and Molecular Biology* 51(7): 643–654.
Calabrese, E.J. 2008. Converging concepts: Adaptive response, preconditioning, and the Yerkes–Dodson law are manifestations of hormesis. *Aging Research Reviews* 7: 8–20.
Calabrese, E.J. 2008a. An assessment of anxiolytic drug screening tests: Hormetic dose responses predominate. *Critical Reviews in Toxicology* 38(6): 489–542.
Calabrese, E.J. 2008b. Modulation of the epileptic seizure threshold: Implications of biphasic dose responses. *Critical Reviews in Toxicology* 38(6): 543–556.
Calabrese, E.J. 2008c. Alzheimer's disease drugs: An application of the hormetic dose-response model. *Critical Reviews in Toxicology* 38(5): 419–451.
Calabrese, E.J. 2008d. Drug therapies for stroke and traumatic brain injury often display U-shaped dose responses: Occurrence, mechanisms, and clinical implications. *Critical Reviews in Toxicology* 38(6): 557–577.
Calabrese, E.J. 2008e. Hormesis: Why it is important to toxicology and toxicologists. *Environmental Toxicology and Chemistry* 27(7): 1451–1474.
Calabrese, E.J. 2011. Toxicology rewrites its history and rethinks its future: Giving equal focus to both harmful and beneficial effects. *Environmental Toxicology and Chemistry* 30(12): 2658–2673.
Calabrese, E.J. 2013. Hormetic mechanisms. *Critical Reviews in Toxicology* 43(7): 580–606.
Calabrese, E.J. and Baldwin, L.A. 2000a. Chemical hormesis: Its historical foundations as a biological hypothesis. *Human and Experimental Toxicology* 19(1): 2–31.

Calabrese, E.J. and Baldwin, L.A. 2000b. The marginalization of hormesis. *Human and Experimental Toxicology* 19(1): 32–40.

Calabrese, E.J. and Baldwin, L.A. 2000c. Radiation hormesis: Its historical foundations as a biological hypothesis. *Human and Experimental Toxicology* 19(1): 41–75.

Calabrese, E.J. and Baldwin, L.A. 2000d. Radiation hormesis: The demise of a legitimate hypothesis. *Human and Experimental Toxicology* 19(1): 76–84.

Calabrese, E.J. and Baldwin, L.A. 2000e. Tales of two similar hypotheses: The rise and fall of chemical and radiation hormesis. *Human and Experimental Toxicology* 19(1): 85–97.

Calabrese, E.J. and Baldwin, L.A. 2002. Defining hormesis. *Human and Experimental Toxicology* 21: 91–97.

Calabrese, E.J., Bachmann, K.A., Bailer, A.J., Bolger, P.M., Borak, J., Cai, L., Cedergreen, N. et al. 2007. Biological stress response terminology: Integrating the concepts of adaptive response and preconditioning stress within a hormetic dose-response framework. *Toxicology and Applied Pharmacology* 222(1): 122–128.

Frolkis, V.V. 1982. *Aging and Life-Prolonging Processes*. Springer, Heidelberg.

Garzon, C.D. and Flores, F.J. 2013. Hormesis: Biphasic dose-responses to fungicides in plant pathogens and their potential threat to agriculture. In: *Fungicides—Showcases of Integrated Plant Disease Management from Around the World*, M. Nita (ed.). InTech, Rijeka, Croatia. Chapter 12. ISBN: 980-953-307-398-4, DOI: 10.5772/55359. Available from: http://www.intechopen.com/books/fungicides-showcases-of-integrated-plant-disease-management-from-around-the-world/hormesis-biphasic-dose-responses-to-fungicides-in-plant-pathogens-and-their-potential-threat-to-agri.

Johnson E. (1936). Effects of X-rays upon green plants. In: Duggar BM (ed). *Biological Effects of Radiation*, Vol. II. McGraw-Hill Book Co., Inc. New York, pp. 961–985.

Kaplan, J.B., Jabbouri, S., and Sadovskaya, I. 2011. Extracellular DNA-dependent biofilm formation by *Staphylococcus epidermidis* RP62A in response to subminimal inhibitory concentrations of antibiotics. *Research in Microbiology* 162: 535–541.

Luckey, T.D. 1980. *Ionizing Radiation and Hormesis*. CRC Press, Boca Raton, FL.

Luckey, T.D. 1991. *Radiation Hormesis*. CRC Press, Boca Raton, FL.

Sacher, G.A. 1963. Effects of X-rays on the survival of *Drosophila* imagoes. *Physiological Zoology* 36: 295–311.

Schulz, H. 1887. Zur Lehre von der Arzneiwirdung. *Virchows Archiv fur Pathologische Anatomie und Physiologie und fur Klinische Medizin* 108: 423–445.

Schulz, H. 1888. Uber Hefegifte. *Pflugers Archiv fur die Gesamte Physiologie des Menschen und der Tiere* 42: 517–541.

Shackell, L.F. 1923. Studies in protoplasm poisoning. I. Phenols. *Journal of General Physiology* 5: 783–805.

Shackell, L.F. 1925. The relation of dosage to effect. II. *Journal of Pharmacology and Experimental Therapeutics* 25: 275–288.

Shackell, L.F., Williamson, W., Deitchmann, M.M., Katzman, G.M., and Kleinman, B.S. 1924/1925. The relation of dosage to effect. *Journal of Pharmacology and Experimental Therapeutics* 23–24: 53–5.

Stebbing, A.R.D. 1976. Effects of low metal levels on a clonal hydroid. *Journal of the Marine Biological Association of the United Kingdom* 56(4): 977–994.

Stebbing, A.R.D. 1981. Hormesis—Stimulation of colony growth in *Campanularia flexuosa* (hydrozoa) by copper, cadmium and other toxicants. *Aquatic Toxicology* 1(3–4): 227–238.

Stebbing, A.R.D. 1998. A theory for growth hormesis. *Mutation Research* 403(1–2): 249–258.

Szabadi, E. 1977. A model of two functionally antagonistic receptor populations activated by same agonist. *Journal of Theoretical Biology* 69(1): 101–112.

Zhao, H. 2013. Hurdles to clear before clinical translation of ischemic postconditioning against stroke. *Translational Stroke Research* 4(1): 63–70.

# 2 Pre- and Postconditioning Hormesis

*Fred A.C. Wiegant*

## CONTENTS

2.1 Introduction ........................................................................................... 14
    2.1.1 Preconditioning Hormesis ............................................................. 15
    2.1.2 Postconditioning Hormesis ............................................................ 16
    2.1.3 Terminology and Mechanism of Action ........................................ 16
2.2 Sensitivity and (De)sensitization: The Time Window in Which Mild Stress Is Beneficial .............................................................................. 18
    2.2.1 Preconditioning and the Issue of Desensitization .......................... 18
    2.2.2 Postconditioning and the Issue of Sensitization ............................ 19
        2.2.2.1 Homologous Postconditioning Hormesis with Heat Shock ............................................................... 20
        2.2.2.2 Homologous Postconditioning Hormesis with Arsenite or Cadmium ........................................... 21
        2.2.2.3 Homologous Postconditioning Hormesis with Ischemia/Reperfusion .......................................... 21
    2.2.3 Sensitivity and Time Window: Discussion and Conclusion ........... 22
2.3 Issue of Specificity in Pre- and Postconditioning ................................. 22
    2.3.1 Cross-Tolerance: The Issue of Specificity in Preconditioning Hormesis ........................................................................................ 23
    2.3.2 Cross-Sensitization: The Issue of Specificity in Heterologous Postconditioning Hormesis ............................................................ 23
        2.3.2.1 Heterologous Postconditioning Hormesis: The Issue of Similarity ........................................................ 24
        2.3.2.2 Heterologous Postconditioning Hormesis: The Similia Principle Put to the Test ....................................... 25
        2.3.2.3 Molecular Aspects of Heterologous Postconditioning Hormesis ............................................................. 25
2.4 Underlying Mechanism of Action in Pre- and Postconditioning .......... 26
    2.4.1 Preconditioning Hormesis and the Induction of Protective Mechanisms ................................................................................... 26

2.4.2 Postconditioning Hormesis and the Enhancement of
Recuperative Mechanisms ................................................................. 27
2.4.2.1 Studies on Protein Denaturation and Renaturation:
Luciferase as a Model Protein ........................................... 28
2.5 Discussion and Conclusions ........................................................................ 28
References ............................................................................................................. 29

## 2.1 INTRODUCTION

In the biomedical sciences, mild stress conditions applied in the framework of preconditioning and postconditioning strategies have received increased attention because of their observed beneficial effects in experimental and clinical settings. In general, beneficial effects of mild stress have been linked to hormesis, which refers to the phenomenon in which exposure to a low dose of an agent that is toxic at higher doses induces a beneficial effect on cells or organisms (Luckey 1980; Calabrese and Baldwin 2002a; Rattan 2004, 2008; Calabrese et al. 2007; Calabrese 2008a). Beneficial effects of mild stress, observed either before or after a more severe damaging condition, are defined in terms of a protective effect in the case of preconditioning hormesis and a curative effect in the case of postconditioning hormesis.

The phenomenon of hormesis has been described with a variety of terms, including the widely observed inverted U-shaped dose-response curve; the Yerkes–Dodson law in experimental psychology; subsidy gradients in ecological analysis; functional antagonism in enzymology; adaptive response in molecular biology and physiology; preconditioning in medicine; biphasic response in pharmacology; and hormesis in toxicology (Calabrese 2008a). Originally, hormesis has been defined in the field of toxicology as "a stimulatory effect of subinhibitory concentrations of any toxic substance on any organism" (Southam and Ehrlich 1943). Owing to Calabrese and others, the term *hormesis*, which reflects an adaptive response of cells or organisms to a moderate (usually intermittent) stress exposure, has gained wider general acceptance and, importantly, has increasingly come into use in the different scientific disciplines mentioned previously (Calabrese and Baldwin 1998, 2002b, 2003; Calabrese et al. 2007). In a consensus paper, researchers from various scientific fields have agreed on the universality of the phenomena they are studying (i.e., adaptive response, resilience, induced (thermo)tolerance, dynamic buffering capacity, preconditioning) and agreed to use the term *hormesis* to indicate this phenomenon (Calabrese et al. 2007).

What is particularly interesting about hormesis is that all the major groups of chemicals and even some physical agents can induce a hormetic effect. In this respect, the various types of environmental stress that have been reported to induce an adaptive or hormetic response include radiation, temperature (heat shock), allergens, pathogens, ischemia, exercise, pharmaceuticals, toxins and heavy metals, dietary energy restriction, or psychologically induced stress. Also, exposure to low doses of specific plant components (phytochemicals) that are toxic at high doses, and which are increasingly held responsible for the beneficial health effects in our normal diet as well as in the framework of naturopathic strategies, are considered inducers of hormesis (Mattson et al. 2007; Mattson 2008; Son et al. 2008; Frenkel and

Wiegant 2011; Bast and Hanekamp 2012). This implies that there is no single shared property of the chemical or physical agents that can be responsible for the stimulatory effect at low doses. Therefore, it is more likely that not the *agent* itself, but the biological *response* to the agent can be held accountable for the effect (Calabrese and Baldwin 2001; Stebbing 2003a,b). In this respect, it has been suggested to refer to all agents and conditions that are able to induce a hormetic effect with the general term *hormetin* (Rattan et al. 2009; Rattan 2012).

This chapter focuses on a number of limitations, challenges, and opportunities in the field of conditioning, especially in relation to the following three aspects:

1. The period of sensitivity in which stimulation of an adaptive and/or recuperative response can take place. This is related to the time window in which conditioning results in an optimal effect.
2. The issue of specificity relates to the question whether any mild stress condition can be used for conditioning purposes (cross-tolerance and cross-sensitization) or whether there are limitations with respect to the type of stress condition that can be applied to achieve effective conditioning.
3. Mechanism of action of pre- and postconditioning. A better understanding of the underlying mechanism may improve the application and design of conditioning strategies.

For most of these aspects, examples are described and discussed from the field of hyperthermia research or from the field of ischemic reperfusions studies. Special attention is paid to results from our research program, which has focused on postconditioning hormesis (Wiegant and van Wijk 2010; van Wijk and Wiegant 2011).

### 2.1.1 PRECONDITIONING HORMESIS

Within the framework of hormesis, different conditioning approaches are used to elicit an adaptive or hormetic response. In preconditioning hormesis, cells or organisms are exposed to low doses of stressors (heat shock, oxidative stress, ischemia/reperfusion), which raise the resistance against subsequent exposure to high doses of the same agent. The result of preconditioning is known by a variety of terms, including adaptive response, tolerance, desensitization, and reprogramming. Examples of preconditioning hormesis are ischemic preconditioning and heat-shock-induced thermotolerance.

In clinical settings, ischemic preconditioning is relevant in cases of transplantation of organs, such as liver (Song et al. 2012), kidney (Bon et al. 2012), and heart (Liem et al. 2007; Hausenloy and Yellon 2011), as well as of implanting stem cells in damaged heart tissue (Haider and Ashraf 2012). The main aim of preconditioning is to induce an adaptive response that enhances resilience and survival of cells or donor organs following transplantation against prolonged shortage of oxygen and nutrients.

In experimental settings, various strategies have been used to trigger an adaptive response, including chronic exposure to mild stress conditions, a single conditioning exposure, or a sequence of intermittent exposures. In the field of ischemia, the clinical possibilities and limitations of conditioning are subjects of intensive research and discussions (Schmidt et al. 2012; Zhao et al. 2013), where strategies such as repeated brief ischemia/reperfusion cycles, which recruit endogenous cytoprotective and recovery pathways and which limit reperfusion injury, have developed into a range of modalities. These modalities can be applied before (preconditioning), during (perconditioning), or after the injurious ischemic insult (postconditioning) either directly to the heart or on a distant tissue (remote preconditioning). Interestingly, remote ischemic conditioning by brief nonlethal episodes of ischemia and reperfusion to an organ or tissue remote from the heart also seems to activate innate cardioprotective mechanisms (Pilcher et al. 2012). Preconditioning, however, seems to be of little practical use in patients as the onset of infarction is usually unpredictable. In contrast, postconditioning strategies are more promising in this respect and are currently an intensive focus of studies in the field of ischemia (Zhao et al. 2013).

### 2.1.2 Postconditioning Hormesis

In postconditioning hormesis a low dose of stress is administered *after* exposure to a more severe stress with the aim to enhance repair and recovery processes (Calabrese et al. 2007; Calabrese 2008b). Promising discoveries indicating the beneficial effects of exposing a biological system to a low adapting dose after a severe stress condition are emerging from different scientific disciplines. Recently, a variety of examples were described in clinical and experimental settings in which mild stress, applied according to a postconditioning protocol, is shown to be beneficial (Wiegant et al. 2011). These include examples from immunology, from cardiovascular pathologies where ischemic postconditioning shows consistent and clear cardioprotective effects, in cancer where mild stress was shown to suppress metastasis and delay tumor growth, in neurological and psychological diseases, as well as in the field of intoxication where exposure to subtoxic levels of toxic agents stimulates a protective response for both prevention and treatment of the effects resulting from exposure to higher doses of toxin. A research program in the field of molecular biology reported on beneficial effects of mild doses of stress (including toxic compounds) within the postconditioning framework (van Wijk et al. 1994a; Wiegant et al. 1997, 1998, 1999, 2011). This research program has also been discussed in relation to the similia principle, being one of the pillars of homeopathy in which remedies are applied within a strictly defined postconditioning approach (Wiegant and van Wijk 2010; van Wijk and Wiegant 2011).

### 2.1.3 Terminology and Mechanism of Action

With respect to the terminology, the term *conditioning* may be preceded by the type of inducing agent. For example, *ischemic* postconditioning hormesis, which involves low doses of hypoxic stress following a myocardial infarction, or *chemical* preconditioning hormesis, which involves exposure to low levels of a chemical toxicant before a subsequent exposure to a severe stress condition such as a heat shock or a large amount of toxic agent. A further distinction in conditioning can be made depending on whether the low-dose stress is of the same type of stress or is different from the high-dose stress. Conditioning can be classified as *homologous* (i.e., the low-dose stress is the same as the high-dose stress) or *heterologous* (i.e., the low-dose stress is not the same as the high-dose stress). Currently, many research efforts are focused on optimizing protocols for preconditioning and postconditioning hormesis, as well as improving our understanding of the underlying protective and recuperative mechanisms that can explain the beneficial effect of using mild stress in conditioning.

A fascinating capacity of cells, being the basic building blocks of organismal life, is their innate ability to withstand and recover from various forms of damage, using a variety of defense, maintenance, and repair mechanisms. Activation of these adaptive self-recovery mechanisms with mild stress conditions confers beneficial effects to biological systems, including the activation of the adaptive capacity of biological systems in response to stress (van Wijk et al. 1994a; Calabrese and Baldwin 1998, 2001; Stebbing 2003a, b; Rattan 2004; Agutter 2008; Mattson 2008). In the literature, different, and sometimes overlapping, parts of this general capacity of cells have been referred to with a wide variety of terms, such as *cell vitality* (and vitagenes) (Calabrese et al. 2010), *allostasis* (McEwen and Wingfield 2003), *homeodynamics* (Rattan 2012), *resilience* (Cuesta and Singer 2012), *programmed cell life*, and *adaptive response* (Calabrese et al. 2007). In a recent paper, the term *cellular quality control mechanisms* was proposed to indicate the term encompassing all beneficial mechanisms involved in cellular defense, maintenance, and repair (Wiegant et al. 2012).

Because many examples are used in this chapter from the heat-shock studies, it is of relevance to shortly elaborate on the general effect of an exposure to an enhanced temperature during conditioning. Heat shock is best known for its effect on the denaturation of cellular proteins or proteotoxic action and the subsequent induction of heat-shock proteins (HSPs). Proteotoxicity, originally defined by Hightower (1991), is used to indicate the detrimental action of denatured proteins in cells. It is a phenomenon of increasing interest in biomedical disciplines. Damage to cellular proteins occurs (1) after heat shock, (2) after ingestion of environmental pollutants such as heavy metals, and (3) following ischemia. The damage incurred is frequently due to reactive oxygen species (oxidative stress). Denatured proteins are increasingly recognized as crucial factors in the development of various chronic diseases including neurodegenerative, atherosclerotic, and diabetic, and in the process of aging (Gregersen et al. 2006; Liu et al. 2007; Cohen and Dillin 2008; Morimoto 2008; van Wijk et al. 2008; Calderwood et al. 2009). To limit proteotoxicity, a set of HSPs (chaperones) are produced that are involved in cellular repair, recovery, and defense mechanisms. Bacteria and eukaryotes defend against misfolded (toxic) protein aggregation utilizing two protein types: molecular chaperones (typically HSP27, HSP60, HSP70, HSP90, and HSP100) and the ATP-dependent proteases (including the 26S proteasome) (Ellis 2007). It has since been demonstrated that chaperones possess many active functions: they repair structural damages by forcefully disentangling aggregated proteins, and unfolding and refolding them into *re-educated and born again* functional proteins (Csermely and Vigh 2007; Ellis 2007). The various HSPs appear to be differentially induced depending on the stress condition (Wiegant et al. 1994; Ryan and Hightower 1996).

## 2.2 SENSITIVITY AND (DE)SENSITIZATION: THE TIME WINDOW IN WHICH MILD STRESS IS BENEFICIAL

Both in pre- and in postconditioning, it has been observed that there is a limited time frame in which the low-dose application (mild stressor condition) can exert a beneficial effect. Biological systems will react to mild stress conditions if they are in a state of (enhanced) sensitivity. In this part, the nature and molecular background will be described as well as the time frame in which sensitivity is expressed. Below, this aspect is described for both pre- and postconditioning using hyperthermia/heat shock as well as ischemia/reperfusion as examples.

### 2.2.1 Preconditioning and the Issue of Desensitization

Preconditioning with heat shock will induce a state of thermotolerance that depends on the temperature and the length of the preconditioning heat shock used, as well as on the half-life of the HSPs that have been synthesized. The occurrence of thermotolerance and its relation with stress proteins have been the subject of numerous excellent reviews (Hahn and Li 1990; Nover 1991; Parsell and Lindquist 1993). During and after the period of heat-induced increased synthesis of HSPs, a state of refractoriness (or tolerance) to further induction of HSP synthesis was observed by a second heat shock (Laszlo 1988; Mizzen and Welch 1988; Li and Mak 1989; Tuijl et al. 1993).

In our program, desensitization of HSP induction has been verified, employing high doses of heat shock, sodium arsenite, and cadmium chloride as primary and secondary inducers of HSPs (Wiegant et al. 1996). Normally, the development of tolerance is observed when a fractionated treatment protocol is used. In this protocol, an initial exposure to heat is followed by incubation at a normal temperature (37°C) and subsequently by an exposure to high temperatures again to determine the degree of developed self-tolerance. Preconditioned cells survive this high temperature in comparison with non-pretreated cells. Examination of the kinetics of induction and decay of thermotolerance showed that an interval of about 4–8 hours at normal temperature was required before the maximum thermotolerance manifested itself. The tolerance persists for at least 24 hours, which is related to the half-life of the stress proteins.

The same fractionated conditioning protocol was also applied using chemical stressors (arsenite and cadmium). Stimulation of HSP synthesis following arsenite as the secondary stressor is severely inhibited in arsenite-pretreated cells, whereas reinduction following cadmium is specifically inhibited in cadmium-pretreated cells. Therefore, during the period of tolerance further conditioning has no additional stimulatory effects. In conclusion, the time frame in which the tolerant state persists appears to depend on the severity of preconditioning, the degree/fullness to which the adaptive response was induced, and the concentration of HSPs that were synthesized by preconditioning treatments and their half-lives.

Ischemic preconditioning consists of short periods of ischemia that increase myocardial tolerance to a subsequent long-lasting ischemic period. Total occlusion of a coronary artery will already cause irreversible cellular damage and cell death within 20 minutes. A sublethal stress stimulus used in terms of preconditioning is able to induce an adaptive response with a protective effect against such a lethal insult. In ischemic preconditioning an early and late phase can be discerned (also indicated as primary and secondary windows of protection). The early phase lasts about 2–3 hours and does not provide full protection, whereas the late phase lasts 3–4 days and provides full protection and thus has greater clinical relevance. The late phase requires a simultaneous activation of multiple stress response responsive genes including NO synthase and antioxidant genes.

### 2.2.2 Postconditioning and the Issue of Sensitization

In postconditioning, the time between severe stress exposure and mild postconditioning treatment should neither be too short nor too long. Following a severe and damage-inducing stress exposure, a state of enhanced sensitivity is observed. For instance, it has been demonstrated that the period of enhanced sensitivity depends on both the time and the temperature of the initial heat treatment. A lower initial heat treatment reduces both the degree of thermosensitization being developed and the period during which thermosensitization is present (Hahn and Li 1990). The period of enhanced sensitivity is followed by a period of desensitization or development of thermotolerance. This state of enhanced sensitivity is transient in nature and appears to be crucial for the outcome of mild stress conditioning. The effect of mild stress conditions can be beneficial or detrimental. However, when the mild dose is too high and is applied immediately following the high-dose application, more damage may occur. When the mild stressor dose is well-attuned to the initial damaging condition and when the period of time between the high dose and mild dose used in postconditioning is not too long, a clear beneficial effect may be obtained due to a development of adaptive and recuperative mechanisms leading to higher resilience and resistance. When the mild stressor is applied beyond the sensitized state and during the tolerant phase, the mild stress condition may not be *sensed* anymore and will therefore be without effect.

In postconditioning it is therefore relevant to focus on the state of enhanced sensitivity, which has been used for opposite purposes: to inflict damage or to stimulate self-recovery processes, depending on the dose of the mild stress condition used in postconditioning. In therapeutic strategies focused on hyperthermia-induced tumor cell death, sensitization was originally applied using the so-called step-down protocol in which cells are exposed to an initial severe heat treatment, immediately followed by a second treatment at a lower hyperthermic temperature (Henle 1980; for reviews see Lindegaard and Overgaard [1990], Hahn and Li [1990], and Nover 1991). Here the sensitized state is used to increase cell damage and cell death in tumor cells by applying a moderate but lethal stress condition, which will mainly affect the heat-treated tumor but will leave the surrounding tissue undisturbed. In case a beneficial effect of mild stress is aimed for, the impact of the stress condition used in postconditioning should be mild. However, the enhanced state of sensitivity also offers the opportunity to stimulate self-recovery processes by mild stressor conditions, which under normal conditions do not have any effect on cellular physiology.

With respect to the origin of the sensitive state, it has been postulated that the cause of the change in sensitivity, which occurs after the application of harmful circumstances to cells, is related to the pool of free HSPs. Between the time the damage induced by a stress condition has consumed most of the free HSPs and the time that the subsequent production of these proteins has taken place, the cell is in a period of increased sensitivity for damaging agents. When the increased production of newly produced HSPs has replenished the free pool, a reduced sensitivity (i.e., a tolerance) to the damaging circumstances develops (van Wijk and Wiegant 2006). It can be postulated that during the period of sensitivity the cell will react to low doses, which under normal conditions do not have any effect on cellular physiology.

An exposure to damaging conditions other than heat shock also results in (biphasic) alterations in sensitivity to a second exposure with the same agent, which has been shown in a variety of stress conditions, including chemical stress conditions as well as in ischemia/reperfusion protocols (Wiegant et al. 1993, 1997; van Wijk et al. 1994b). Next, beneficial effects of postconditioning are described using heat, chemical stress conditions, or ischemia/reperfusion as mild stress conditions following a more severe stress exposure. As many examples of our research program on postconditioning hormesis are used to illustrate these different aspects mentioned earlier, this program is described first. A more elaborate description can be found in studies by van Wijk and Wiegant (2006, 2011) and Wiegant and van Wijk (2010). In our research on postconditioning hormesis, a harmful physical stressor (heat shock) and several chemical stressors were used. Cells were exposed to different strengths of stress by varying doses and time to select the harmful (high-dose) and subsequent postconditioning hormetic (low-dose) effects. The stress condition was carefully selected to prevent the occurrence of much irreversible damage or compromised resilience ability. For an experimental protocol, stressed cells should be able to express their recovery and survival capacities as well to allow a further increase by changes in environmental circumstances. Only then, can an addition in survival capacity in response to a subsequent exposure to a low-dose postconditioning treatment can be achieved. A final harmful test stress exposure is used to record the possible increase in cell's survival capacity (Wiegant et al. 1999; Wiegant and van Wijk 2010). The studies focused on (1) the proteotoxic response of cells and (2) the regulation of HSP (stress) synthesis. The question was raised regarding the possible stimulating effect of the postconditioning hormetic treatment on the stress protein response. Both homologous and heterologous postexposure conditionings were studied using chemical (arsenite and cadmium) and physical (heat) stress conditions. In the first instance, postexposure studies were carried out using the homologous strategy. Cell cultures that were first exposed to a harmful (high-dose) stress condition were subsequently incubated with lower doses of the initial stressor.

### 2.2.2.1 Homologous Postconditioning Hormesis with Heat Shock

To study the effect of homologous heat postconditioning, we first exposed cells to a heat shock at 42°C or 43.5°C and subsequently incubated at mild *fever-like* temperatures. An enhanced survival capacity and an enhanced synthesis of stress proteins were observed when the initial heat shock was followed by a postexposure to a lower hyperthermic (fever-like) temperature (Schamhart et al. 1992; van Wijk et al. 1994b). Similar observations were reported by Delpino et al. (1992). In these heat postconditioning studies, the effect of mild postexposure appeared to be related to the severity of the preexposure temperature. Whereas the fever-like temperature enhanced survival capacity and HSP synthesis when applied *postconditionally* following 42°C, the same fever-like temperature depressed survival capacity and inhibited HSP synthesis when applied following exposure of cells to 43.5°C (van Wijk et al. 1994b). A further increase in HSP synthesis is interpreted as a beneficial phenomenon, because more molecular chaperones will be available to assist in poststress recovery (van Wijk and Wiegant 2006; Wiegant and van Wijk 2010).

### 2.2.2.2 Homologous Postconditioning Hormesis with Arsenite or Cadmium

When cells were initially exposed to a damaging concentration of chemical compounds such as arsenite (100 or 300 µM) or cadmium (10 or 30 µM) and subsequently exposed to low-dose conditions of arsenite (1–10 µM) or cadmium (0.1–1.0 µM), both HSP synthesis and survival capacity reflecting a tolerant state were enhanced due to postconditioning (Ovelgönne et al. 1995; Wiegant et al. 1997).

### 2.2.2.3 Homologous Postconditioning Hormesis with Ischemia/Reperfusion

Currently, a major issue in the management of acute myocardial infarction is to identify the time frame in which application of preconditioning and postconditioning at reperfusion is still profitable (Zhao 2007, 2009; Zhao et al. 2013). Postconditioning, a transient period of brief ischemia following prolonged severe ischemia, applied at the onset of reperfusion, reduces myocardial infarction in both animals and humans (Staat et al. 2005; Skyschally et al. 2009). Most studies have focused on applying postconditioning immediately following severe ischemia. An important question is whether the cardioprotective effect can also be achieved when postconditioning is initiated later. The objective of a number of recent studies is to identify the length of the cardioprotective window of low-dose application. It is of relevance in identifying the time delay that postconditioning at reperfusion can still be applied.

Several researchers using various organs or tissues have reported on the beneficial effect of delayed postconditioning ischemic stimulus, for example:

- Delayed postconditioning applied up to 30 minutes after the onset of reperfusion was shown decrease infarct size in the mouse myocardium (Barrère-Lemaire et al. 2012).
- Delayed postconditioning applied 1 hour after retinal ischemia effectively prevented damage in the rat retina. Interestingly, postconditioning (during 5 minutes) 24 hours (but not 48 hours) after prolonged ischemia significantly improved functional recovery and decreased histological damage induced by prolonged ischemia (Dreixler et al. 2011). Within a defined time window, delayed postischemic conditioning ameliorated postischemic injury in rat retinas. Delayed stimuli are thus still able to provide protection in the retina following ischemia.
- In an in vivo mouse model of myocardial ischemia/reperfusion, Roubille et al. (2011) showed that delaying the intervention of postconditioning to 30 minutes does not abrogate the cardioprotective effect of postconditioning.

### 2.2.3 SENSITIVITY AND TIME WINDOW: DISCUSSION AND CONCLUSION

In summary, using a variety of stress conditions, the biphasic action of a substance has been well-demonstrated in a variety of experimental and clinical fields. These observations led to the conclusion that the capacity of low doses of homologous substances applied during postconditioning is effective only during the early period of recovery. A small dose can exert a stimulatory effect on the recovery and the development of the survival capacity of cells that have been previously disturbed by a high dose of the same substance or a more severe condition. It is of interest that this stimulatory effect of mild stress is dependent on the initial exposure condition. The more severe the initial stress conditions, the milder the postconditioning stress condition can be to stimulate recovery mechanisms and HSP induction effectively. Small doses can, however, unexpectedly merge into a harmful range, especially when the initial pathology-inducing stress condition has been severe.

These observations are in agreement with the *Law of Initial Values* formulated by Wilder (1962). This *law* states that the response of a (cellular) function to any (outside) agent depends to a large degree on the *initial level* of that function at the start of the experiment. Therefore, the higher the initial stimulus, the smaller the response to a *function-raising* substance and the greater the response to *function-depressing* agents. Conversely, the lower the *initial level*, the greater the response to function-raising agents and the lesser to function-depressing ones.

From the studies presented here, it can be predicted that a beneficial effect using mild stress in postconditioning can only be observed when (1) the mild stress condition is applied during the limited period of increased sensitivity following the disturbance and (2) the stress applied in low doses is carefully attuned to the severity of the initial disturbance. Relevant follow-up questions are: what determines the time window of protection, what is the time window in which meaningful result can be achieved, and how should the strength of the mild stress condition that is used for postconditioning be determined and attuned to the pathological state or to the initial severe stress condition. An additional intriguing question relates to the specificity of the stimulation of survival capacity and of HSP synthesis by postconditioning. In other words: does an initial damaging condition only sensitizes a cell population to stimulation by small doses of the same stress condition, or are cells also stimulated by small doses of heterologous stressors? This will be dealt with in Section 2.3.

## 2.3 ISSUE OF SPECIFICITY IN PRE- AND POSTCONDITIONING

In preconditioning, homologous and heterologous strategies can be used in which the same or a different stressor is used as the initial prestress condition in comparison with the subsequent severe stress condition to which protection is initiated. Cross-conditioning is the capability of one stressor to induce tolerance against another. In this respect, various types of pharmacological interventions as well as different types of preconditioning, such as remote preconditioning, the use of heat shock, and hyperbaric oxygen, have been developed to attenuate the functional impairment accompanying ischemia/reperfusion injury (Theodoraki et al. 2011). The question is whether mild cross-conditioning is able to confer the same degree of effectiveness in protecting against any severe or damaging stressor in case of heterologous conditioning.

### 2.3.1 CROSS-TOLERANCE: THE ISSUE OF SPECIFICITY IN PRECONDITIONING HORMESIS

In cross-tolerance, a primary exposure to a stressful stimulus results in an adaptive response whereby the cell or organism is resistant to a subsequent stress that is different from the initial stress (i.e., exposure to heat stress leading to resistance to oxidant stress). The heat-shock response is one of the most commonly described examples of stress adaptation and is characterized by the rapid expression of a unique group of proteins collectively known as HSPs, which have important cytoprotective roles during severe stress as well as during pathologies. Interestingly, different stress conditions induce a different gene expression profile (Wiegant et al. 1994; Ryan and Hightower 1996). At the cellular level, the pattern of induced stress proteins (both HSPs and glucose-regulated proteins [GRPs]) is stressor specific (Wiegant et al. 1994). It has been suggested that the type of damage involved by a heat shock, heavy metal, and oxidative stress condition or ischemia/reperfusion might be different and require specific response patterns represented by a combination of a unique set of HSPs and focused on recuperative mechanisms that are well-adapted to the type of damage inflicted. Considering the stressor-specific induction of HSPs, the question of relevance is to what degree heterologous stress combinations provide protection to a specific stressor.

This is especially relevant since it has been observed that cross-tolerance is relatively modest in comparison with self-tolerance in which the same stressor is used as primary and secondary stress exposure. In this respect, Kirino (2002) described that metabolic and physical stress can induce cross-tolerance to cerebral ischemia, but that the protection by cross-tolerance is relatively modest. It is of interest to determine whether mild stress conditions applied within the framework of heterologous postconditioning hormesis also show lower efficiency in comparison with homologous postconditioning.

### 2.3.2 CROSS-SENSITIZATION: THE ISSUE OF SPECIFICITY IN HETEROLOGOUS POSTCONDITIONING HORMESIS

An intriguing question relates to the specificity of the stimulation of survival capacity and of HSP synthesis by postconditioning. Does an initial damaging condition only sensitize a cell population to stimulation by small doses of the same stress condition, or are they also stimulated by small doses of heterologous stressors? A number of experiments were performed in which cells were pretreated with a heat shock (42°C for 30 minutes), sodium arsenite (100 µM for 1 hour), or cadmium chloride (10 µM for 1 hour). Each of these cell cultures were then exposed to the postconditioning treatment of a lower hyperthermic temperature (39.5°C), a lower concentration of arsenite (1 µM), a lower concentration of cadmium (0.3 µM), or to control conditions (37°C) (Wiegant et al. 1998). To determine the amounts of the various HSPs that were specifically produced by the small-dose application, we quantified the amount of each stress protein that had been produced superimposed on the specific synthesis in cells which were not posttreated. Only this extra synthesis of the indicated stress proteins was important and reflected the effect of the small-dose treatment. Since the small-dose conditions used were without effect on stress protein synthesis in naive cells, the observed patterns of stress proteins were due solely to sensitization of cell cultures by a pretreatment condition. In most conditions, subsequent incubation with a small dose of any stress condition was able to induce an extra synthesis of various stress proteins. However, the degree of enhanced stimulation of the various stress proteins appeared to vary according to the combination of stressors used (Wiegant et al. 1996, 1998).

To identify the possible relationship between high- and low-dose stress in heterologous conditions more extensively, we decided to analyze the approach prescribed by the similia principle. This principle, a fundamental pillar of the homeopathic therapeutic strategy, predicts the effectiveness of low-dose stimulation by stress conditions based on the degree of similarity in their ability to induce *symptoms*. A more detailed elaboration of this research program with respect to the evaluation of the similia principle at the cellular level using the degree of similarity in the ability of stress conditions to induce the various HSPs has been described elsewhere (van Wijk and Wiegant 2006, 2011; Wiegant and van Wijk 2010). In short, a damaging stress condition (heat shock) is followed by exposure to a smaller dose of various stress conditions (Wiegant et al. 1998, 1999). The stress conditions used for postexposure purposes included arsenite, several heavy metal ions (cadmium, mercury, lead, and copper), two oxidative stress conditions (menadione and diethyldithiocarbamate), and a mild (fever-like) hyperthermic temperature. These stress conditions differ both in the extent and the pattern in which various stress proteins are stimulated (Wiegant et al. 1999).

### 2.3.2.1 Heterologous Postconditioning Hormesis: The Issue of Similarity

On the basis of the differences between stress-induced protein patterns, one particular condition corresponding with a particular pattern can be taken as standard; all the other conditions and patterns can then be compared to this. In the study, heat shock was selected as the initial damaging condition and the similarity between response patterns on exposure to the stress condition used in postconditioning was estimated based on the major HSPs (HSP28, 32, 60, 68, 70, 84, and 100) and on the major GRPs (GRP78 and 94). The ability to quantify the *overlap* between the HSP pattern induced by the initial heat shock and the HSPs that the stress conditions are potentially able to induce is a crucial prerequisite to study the similia principle. The specificity of low-dose stimulation on cell survival capacity was evaluated by analyzing the effect of a range of mild stress conditions that were used postconditionally following a heat shock (Wiegant et al. 1999). A heat shock was selected (42°C for 30 minutes) that did not substantially affect actual cell survival but was able to stimulate suboptimal development of cell survival capacity.

### 2.3.2.2 Heterologous Postconditioning Hormesis: The Similia Principle Put to the Test

To select the small-dose treatment, we established conditions that were just below the limit of having an effect on survival capacity as well as on the synthesis of stress proteins during an 8-hour period. When the various small-dose stressors were applied after an initial heat shock, an increase in survival capacity was observed. This increase was labeled *survival stimulation factor* (the ratio of relative survival with low-dose postconditioning vs. the effect on survival due to the initial heat shock only). It represents the degree of stimulation of survival capacity as exerted by postconditioning with low doses. However, the increase appeared to depend on the nature of the low-dose stressor. Heterologous postconditioning showed an increase of the *survival stimulation factor* but not as much as homologous postconditioning. The stimulation of survival capacity was correlated with the degree of similarity between the stress protein pattern induced by heat shock and the pattern of stress protein induction characteristic (specific) for the compounds that were applied in small doses. The survival stimulation factor plotted against the degree of similarity showed a highly significant correlation. It was concluded that, in general, a higher percentage of similarity in stress protein induction pattern predicts a higher stimulation of survival capacity (Wiegant et al. 1999).

### 2.3.2.3 Molecular Aspects of Heterologous Postconditioning Hormesis

To explain the variability of small-dose stimulatory action on heat-shocked cells, it was hypothesized that such conditions induce an increased survival only if they are able to stimulate the specific endogenous defense and recuperative mechanisms required by damaged cells. Since stress proteins are viewed as a reflection of the initiation of endogenous defense at the molecular level, the research evaluated whether observed differences in stimulation by small doses were related to the specificity in the overall pattern of stress proteins. Indeed, a significant correlation was observed for most stress proteins between enhanced additional synthesis due to low-dose stress and the degree of similarity between the inductions of individual HSPs by high- and low-dose stress conditions (Wiegant et al. 1999). It was also concluded that during the period of enhanced sensitivity, cells react to substances applied in low doses, to which they would normally not react, and that the stimulation of recovery processes depends on the similarity in effect between high- and low-dose stress conditions.

In summary, the degree of stimulation appears to be stressor specific. The specificity is present not only in the development of the survival capacity, but also in the subsequent enhancement of the heat-shock-induced synthesis of stress proteins. The degree of stimulation of survival capacity by sequential postconditioning exposure to low doses of the mentioned stressors is determined by the degree of stress protein pattern similarity between the stress condition used as a low-dose and the initial high-dose preexposure conditions. This observation supports the validity of the similia principle at the cellular level.

## 2.4 UNDERLYING MECHANISM OF ACTION IN PRE- AND POSTCONDITIONING

The search for an underlying mechanism of action to explain the beneficial effect observed in pre- and postconditioning hormesis is ongoing. A brief summary of the proposed mechanisms is given in the following sections.

### 2.4.1 Preconditioning Hormesis and the Induction of Protective Mechanisms

One theory that has been proposed assumes that exposure of an organism to mild stress results in a disturbance in homeostasis (Calabrese and Baldwin 2002a; Stebbing 2003b). In response, the expression of specific genes will be triggered, which supports the development of the adaptive response at the molecular and cellular levels, contributing to an explanation of the phenomenon of hormesis (van Wijk et al. 1994a; Calabrese and Baldwin 2000; Mattson 2008; Rattan 2008, 2012; Calabrese et al. 2010). These stress resistance genes include HSPs, chaperones, and antioxidant enzymes. Moreover, it has been suggested that in its attempt to restore homeostasis the cell overactivates the maintenance and repair mechanism, leading to a so-called overcorrection, which can account for the benefits of hormesis (Stebbing 2003a,b). A number of authors have focused on the involvement of HSPs in hormesis and their role in quality control. Mild stress that causes minor cellular damage activates HSFs leading to the recruitment and production of extra HSPs that confer increased stress resistance to the cell, which enhances cell viability and lifespan (Landry et al. 1982; Laszlo 1988; van Wijk et al. 1994b; Wiegant et al. 2008, 2009; Frenkel and Wiegant 2011).

Exposure to mild stress is suggested to activate the quality control system, which induces the upregulation of HSPs in moderation, leading to increased tolerance to future stress and accounting for the benefits of hormesis. Increased resistance against subsequent, more severe stress is caused by an increased concentration of chaperones after the initial exposure to mild stress. Enhanced resistance even after homeostasis has been reinstated can be explained because chaperones will remain functional in the cell as long as their half-life permits them. Consequently, by activating the heat-shock response through mild stress the cell exhibits an increased capacity to sense and clear damaged proteins, leading to increased viability of cells and organisms (Morley and Morimoto 2004; Wiegant et al. 2009). In this respect, it has been shown that repeated mild heat-shock (RMHS) treatment increases the basal level of a number of HSPs (Rattan 1998; Verbeke et al. 2001, 2002; Fonager et al. 2002). A significant increase in the basal level of HSP27 and HSP70 throughout aging was observed in RMHS-treated cells, and was linked to improved functionality and survival of the cells. In this respect, HSP27 has also been shown to offer enhanced resistance to harmful cytotoxic damage induced by heat shock and oxidative stress (Arrigo 2007). The augmented basal levels of HSPs are suggestive of an adaptive response of the RMHS-exposed cells to increasing intracellular stress in the course of aging. RMHS-treated cells thus reveal hormesis-like effects. The enhanced expression of HSPs under these conditions may explain an increased ability to cope with oxidative stress (Rattan 1998; Verbeke et al. 2001, 2002; Fonager et al. 2002). Further research should aim to elucidate whether an increase in tolerance results from a higher capacity or a higher need for cellular chaperones.

The damage-eliminating side of the quality control system, the proteasome-mediated degradation of denatured proteins, and autophagy are also believed to be stimulated in the hormetic response to allow the cell to cope with increased amounts of damaged proteins. As a result, the proteasome may recognize and degrade not only the damaged proteins that resulted from the mild stress, but may also clear dysfunctional proteins that were already present before the stress exposure. This will obviously explain a net beneficial effect of mild stress following recovery. Results that confirm this hypothesis show that proteasome activity is enhanced after exposure to mild stress. For instance, an increase has been observed in the expression of ubiquitin, which is necessary for targeting proteins for degradation (Kimura et al. 2009). Moreover, exposure to RMHS has been shown to upregulate the 20S and 11S proteasomal subunits, contributing to an increased degradation capacity (Beedholm et al. 2004). Finally, it has been suggested that some HSPs synthesized in response to stress can enhance proteasomal degradation by accelerating the peptidase activity of the 20S proteasome when the 11S activator is bound to it, because the 11S activator acts as an adaptor molecule between the HSP and the proteasome (Beedholm et al. 2004).

Because of this overactivation of repair mechanisms, not only the damage inflicted by mild stress but also other, previously unnoticed damage or gradually accumulated damage might be removed (Sørensen et al. 2005). Hormesis is an adaptive response to low levels of stress or damage resulting in improved fitness for some physiological systems for a finite period. The duration of enhanced resistance is dependent on the period of the enhanced expression of stress resistance genes, as well as on the half-life of proteins responsible for stress resistance and quality control. In more specific terms, hormesis is defined as a modest overcompensation to a disruption in homeostasis. Therefore, it has been proposed that the key conceptual features of hormesis are the disruption of homeostasis, activation of corrective cellular pathways that lead to modest overcompensation, the reestablishment of homeostasis, and the adaptive nature of the process on future exposure to stressors (Calabrese and Baldwin 2002a; Stebbing 2003b). The various mechanisms involved in detection and repair of damage are increasingly indicated with the overall term *cellular quality control*. Because hormesis is a response to mild stressors and the cell responds to stressful conditions by activating the *cellular quality control* systems, it has been suggested that the signaling pathways activated by the *quality control systems* represent the molecular mechanisms underlying the beneficial effects of hormesis (Wiegant et al. 2012).

### 2.4.2 Postconditioning Hormesis and the Enhancement of Recuperative Mechanisms

In our research program we focused on unraveling the molecular events in postconditioning hormesis. More specifically, the question was whether the molecular events in HSP induction, including the presence of damaged proteins; changes in the HSP induction due to changes in the regulation of binding of the heat-shock factor (HSF) to the heat-shock element (HSE), and subsequent synthesis of HSP-mRNA could explain the prolonged synthesis of HSPs, the stronger... etc. until postconditioning. According to the generally accepted model of HSP induction, heat-denatured proteins are the molecular signal for the induction of heat-shock gene transcription. An increase in denatured proteins causes activation of the HSF, which binds to the HSE, the regulatory element of heat-shock genes, initiating their transcription and resulting in HSP-mRNA production and subsequent synthesis of stress proteins. As chaperones, these stress proteins have the capacity to partly bind to denatured proteins and assist in their renaturation, thereby decreasing the amount of denatured proteins. This action terminates the signal for HSP induction. Interference with the process of renaturation, due to postconditioning, might result in a prolonged binding of HSPs to denatured proteins, a continuation of the activated state of HSF, and a continuation of the synthesis of HSPs.

#### 2.4.2.1 Studies on Protein Denaturation and Renaturation: Luciferase as a Model Protein

Previously, it had been shown that mild hyperthermic conditions lead to a prolonged existence of an activated form of HSF in cells that were previously heat shocked (Ovelgönne and van Wijk 1995). Souren et al. (1999) tested the hypothesis that protein renaturation is inhibited when heat-shocked cells are postconditioned using mild hyperthermic conditions. In mammalian cells, firefly luciferase (Luc) provides one of the best nontoxic and sensitive enzymes to study characteristics of the process of denaturation and renaturation of thermolabile proteins in the same temperature range as the onset of heat-induced cellular protein denaturation. Using the method of monitoring intracellular Luc activity, Souren et al. (1999) were able to show that postconditioning inhibits or delays the renaturation of heat-damaged Luc. A longer presence of damaged proteins due to postconditioning implies a prolonged signal to continue HSP induction and may explain the longer time that HSPs are required for binding to stress-damaged proteins. This, in turn, may explain the continuation of the activated state of HSF and thus the prolonged induction and synthesis of HSP-mRNAs and of HSPs observed during postconditioning (Souren et al. 1999). In the end, a higher level of HSPs in the cell is considered to be more beneficial and therefore may explain the beneficial effect of postconditioning.

## 2.5 DISCUSSION AND CONCLUSIONS

Preconditioning and postconditioning hormesis are promising clinical strategies as reflected by an increasing number of studies in a variety of fields. Some limitations, challenges, and opportunities have been addressed in this chapter. These are as follows:

- The sensitive state, which determines the time window in which mild stress is able to achieve its beneficial and stimulatory action, needs to be identified in more detail. A better understanding of the sensitive state may improve therapeutic applications. For this, pre- and postconditioning stressors may require improved attunement to each other within both the homologous and heterologous conditioning strategies.

- Specificity of stress conditions, which is of relevance for cross-protective or cross-recuperative use in which stress conditions (hormetins) can be applied. Postexposure of heat-shocked cells to different chemical low-dose stress conditions demonstrated a differential stimulation of both survival capacity development and HSP synthesis. The degree of stimulation appeared to depend on the similarity of both the molecular and cellular effects of the stress conditions used as initial disturbance and the postexposure treatment.
- Mechanism of action: Although the underlying mechanisms in which recovery mechanisms are activated by low doses require more research, promising results indicating the beneficial effects of postexposure conditioning are emerging from different scientific disciplines. Further research in this field will be an interesting challenge for both fundamental and clinical measurements. A better understanding will allow more precise use of hormetic conditioning in a variety of situations.
- It may be stated that regulation of HSPs levels, overexpression of components of the ubiquitin-proteasome system, and overexpression of autophagy are a credible molecular foundation for the beneficial effects of preconditioning hormesis. Although mechanistic explanations of hormesis have been suggested during the past few years (Stebbing 2003b; Agutter 2008; Mattson 2008), the beneficial effect of low-dose stress is experimentally best explained by the proteotoxicity and stress protein response defense system (van Wijk and Wiegant 2010; Wiegant et al. 2012). The various molecular elements of cellular quality control mechanisms may form a theoretical basis to explain the phenomenon of hormesis. To support this hypothesis, specific studies need to be designed that explore the role of the quality control system in the generation of beneficial effects in response to mild stress. A clear mechanism of hormesis will support the application of hormesis in various fields of science.

## REFERENCES

Agutter, P.S. 2008. Elucidating the mechanism(s) of hormesis at the cellular level: The universal cell response. *Am J Pharmacol Toxicol* 3: 100–110.

Arrigo, A.P. 2007. The cellular networking of mammalian Hsp27 and its functions in the control of protein folding, redox state and apoptosis. *Adv Exp Med Biol* 594: 13–26.

Barrère-Lemaire, S., J. Nargeot, and C. Piot. 2012. Delayed post-conditioning: Not too late? *Trends Cardiovasc Med* 22(7): 173–179.

Bast, A. and J.C. Hanekamp. 2012. Chemicals and health—Thought for food. *Dose Response* 11(3): 295–300.

Beedholm, R., B.F.C. Clark, and S.I.S. Rattan. 2004. Mild heat stress stimulates 20S proteasome and its 11S activator in human fibroblasts undergoing aging in vitro. *Cell Stress Chaperones* 9: 49–57.

Bon, D., N. Chatauret, S. Giraud, R. Thuillier, F. Favreau, and T. Hauet. 2012. New strategies to optimize kidney recovery and preservation in transplantation. *Nat Rev Nephrol* 8(6): 339–347.

Calabrese, E.J. 2008a. Converging concepts: Adaptive response, preconditioning, and the Yerkes–Dodson law are manifestations of hormesis. *Ageing Res Rev* 7: 8–20.

Calabrese, E.J. 2008b. Hormesis: Why is it important to toxicology and toxicologists? *Environ Toxicol Chem* 27: 1451–1474.

Calabrese, E.J. and L.A. Baldwin. 1998. Hormesis as a biological hypothesis. *Environ Health Perspect* 106: 357–362.

Calabrese, E.J. and L.A. Baldwin. 2000. Radiation hormesis: The demise of a legitimate hypothesis. *Hum Exp Toxicol* 19: 76–84.

Calabrese, E.J. and L.A. Baldwin. 2001. Hormesis: A generalizable and unifying hypothesis. *Crit Rev Toxicol* 31(4–5): 353–424.

Calabrese, E.J. and L.A. Baldwin. 2002a. Defining hormesis. *Hum Exp Toxicol* 21: 91–97.

Calabrese, E.J. and L.A. Baldwin. 2002b. Applications of hormesis in toxicology, risk assessment and chemotherapeutics. *Trends Pharmcol Sci* 23: 331–337.

Calabrese, E.J. and L.A. Baldwin. 2003. Hormesis: The dose–response revolution. *Annu Rev Pharmacol Toxicol* 43: 175–197.

Calabrese, E.J., K.A. Bachmann, A.J. Bailer, P.M. Bolger, J. Borak, L. Cai, N. Cedergreen et al. 2007. Biological stress response terminology: Integrating the concepts of adaptive response and pre-conditioning stress within a hormetic dose–response framework. *Toxicol Appl Pharmacol* 222: 122–128.

Calabrese, V., C. Cornelius, A.M. Giuffrida Stella, and E.J. Calabrese. 2010. Cellular stress responses, mitostress and carnitine insufficiencies as critical determinants in aging and neurodegenerative disorders: Role of hormesis and vitagenes. *Neurochem Res* 35: 1880–1915.

Calderwood, S.K., A. Murshid, and T. Prince. 2009. The shock of aging: Molecular chaperones and the heat shock response in longevity and aging. *Gerontology* 55: 550–558.

Cohen, E. and A. Dillin. 2008. The insulin paradox: Aging, proteotoxicity and neurodegeneration. *Nat Rev Neurosci* 9: 759–767.

Csermely, P. and L. Vigh. (eds.). 2007. *Molecular Aspects of the Stress Response: Chaperones, Membranes and Networks.* Landes Bioscience, New York.

Cuesta, J.M. and M. Singer. 2012. The stress response and critical illness: A review. *Crit Care Med* 40(12): 3283–3289.

Delpino, A., F.P. Gentile, F. Di Modugno, M. Benassi, A.M. Mileo, and E. Mattei. 1992. Thermosensitization, heat shock protein synthesis and development of thermotolerance in M-14 human tumor cells subjected to step-down heating. *Radiat Environ Biophys* 31: 323–332.

Dreixler, J.C., J.N. Poston, A.R. Shaikh, M. Alexander, K.Y. Tupper, M.M. Marcet, M. Bernaudin, and S. Roth. 2011. Delayed post-ischemic conditioning significantly improves the outcome after retinal ischemia. *Exp Eye Res* 92(6): 521–527.

Ellis, R.J. 2007. Protein misassembly: Macromolecular crowding and molecular chaperones. In: *Molecular Aspects of the Stress Response: Chaperones, Membranes and Networks*, Csermely P and Vigh L (eds.). Landes Bioscience, New York. pp. 1–13.

Fonager, J.R., R. Beedholm, B.F. Clark, and S.I.S. Rattan. 2002. Mild stress-induced stimulation of heat-shock protein synthesis and improved functional ability of human fibroblasts undergoing aging in vitro. *Exp Gerontol* 37: 1223–1228.

Frenkel, N.C. and F.A.C. Wiegant. 2011. The role of cytoprotective HO-1 and its induction by plant adaptogens and phytochemicals in the development of an enhanced cellular resistance. In: *Adaptation Biology and Medicine, Volume 6: Adaptations and Challenges*, Wang P, Kuo C-H, Takeda N, and Singal PK (eds.). Narosa Publishers, New Delhi, India. pp. 409–422.

Gregersen, N., P. Bross, S. Vang, and J.H. Christensen. 2006. Protein misfolding and human disease. *Annu Rev Genomics Hum Genet* 7: 103–124.

Hahn, G.M. and G.C. Li. 1990. Thermotolerance, thermoresistance and thermosensitization. In: *Stress Proteins in Biology and Medicine*, Morimoto RI, Tissières A, and Georgopoulos C (eds.). Cold Spring Harbor Laboratory Press, New York. pp. 79–100.

Haider, K.H. and M. Ashraf. 2012. Pre-conditioning approach in stem cell therapy for the treatment of the infarcted heart. *Prog Mol Biol Transl Sci* 111: 323–356.

Hausenloy, D.J. and D.M. Yellon. 2011. The therapeutic potential of ischemic conditioning: An update. *Nat Rev Cardiol* 8(11): 619–629.

Henle, K.J. 1980. Sensitization to hyperthermia below 43°C induced in Chinese hamster ovary cells by step-down heating. *J Natl Cancer Inst* 64: 1479–1483.

Hightower, L.E. 1991. Heat shock, stress proteins, chaperones, and proteotoxicity. *Cell* 66: 191–197.

Kimura, Y., H. Yashiroda, T. Kudo, S. Koitabashi, S. Murata, A. Kakizuka, and K. Tanaka. 2009. An inhibitor of a deubiquitinating enzyme regulates ubiquitin homeostasis. *Cell* 137: 549–559.

Kirino, T. 2002. Ischemic tolerance. *J Cereb Blood Flow Metab* 22(11): 1283–1296.

Landry, J., D. Bernier, P. Chretien, L.M. Nicole, R.M. Tanguay, and N. Marceau. 1982. Synthesis and degradation of heat shock proteins during development and decay of thermotolerance. *Cancer Res* 42: 2457–2461.

Laszlo, A. 1988. The relationship of heat shock proteins, thermotolerance, and protein synthesis. *Exp Cell Res* 178: 401–414.

Li, G.C. and J.Y. Mak. 1989. Re-induction of HSP70 synthesis: An assay for thermotolerance. *Int J Hyperthermia* 5: 389–403.

Liem, D.A., H.M. Honda, J. Zhang, D. Woo, and P. Ping. 2007. Past and present course of cardioprotection against ischemia-reperfusion injury. *J Appl Physiol* 103: 2129–2136.

Lindegaard, J.C. and J. Overgaard. 1990. Step-down heating in a C3H mammary carcinoma in vivo: Effects of varying the time and temperature of the sensitizing treatment. *Int J Hyperthermia* 6: 607–617.

Liu, M., I. Hodish, C.J. Rhodes, and P. Arvan. 2007. Proinsulin maturation, misfolding, and proteotoxicity. *Proc Natl Acad Sci USA* 104: 15841–15846.

Luckey, T.D. 1980. *Hormesis with Ionizing Radiation*. CRC Press, Boca Raton, FL.

Mattson, M.P. 2008. Hormesis and disease resistance: Activation of cellular stress response pathways. *Hum Exp Toxicol* 27: 155–162.

Mattson, M.P., T.G. Son, and S. Camandola. 2007. Viewpoint: Mechanisms of action and therapeutic potential of neurohormetic phytochemicals. *Dose Response* 5(3): 174–186.

McEwen, B.S. and J.C. Wingfield. 2003. The concept of allostasis in biology and biomedicine. *Horm Behav* 43: 2–15.

Mizzen, L.A. and W.J. Welch. 1988. Characterization of the thermotolerant cell. I. Effects of protein synthesis activity and the regulation of heat shock protein 70 expression. *J Cell Biol* 106: 1105–1116.

Morimoto, R.I. 2008. Proteotoxic stress and inducible chaperone networks in neurodegenerative disease and aging. *Genes Dev* 22: 1427–1438.

Morley, J.F. and R.I. Morimoto. 2004. Regulation of longevity in *Caenorhabditis elegans* by heat shock factor and molecular chaperones. *Mol Biol Cell* 15: 657–664.

Nover, L. 1991. *Heat Shock Response*. CRC Press, Boca Raton, FL.

Ovelgönne, J.H. and R. van Wijk. 1995. Modulation of *HSP68* gene expression after heat shock in thermosensitized and thermotolerant cells is not solely regulated by binding of HSF to HSE. *Int J Hyperthermia* 11(5): 719–732.

Ovelgönne, H.H., F.A.C. Wiegant, J.E.M. Souren, H. van Rijn, and R. van Wijk. 1995. Enhancement of the stress response by low concentrations of arsenite in arsenite-pretreated H35 hepatoma cells. *Toxicol Appl Pharmacol* 132: 146–155.

Parsell, D.A. and S. Lindquist. 1993. The function of heat-shock proteins in stress tolerance: Degradation and reactivation of damaged proteins. *Annu Rev Genet* 27: 437–496.

Pilcher, J.M., P. Young, M. Weatherall, I. Rahman, R.S. Bonser, and R.W. Beasley. 2012. A systematic review and meta-analysis of the cardioprotective effects of remote ischaemic preconditioning in open cardiac surgery. *J R Soc Med* 105(10): 436–445.

Rattan, S.I. 2004. Mechanisms of hormesis through mild heat stress in human cells. *Ann N Y Acad Sci* 1019: 554–558.

Rattan, S.I. 2012. Rationale and methods of discovering hormetins as drugs for healthy ageing. *Expert Opin Drug Discov* 7(5): 439–448.

Rattan, S.I.S. 1998. Repeated mild heat shock delays ageing in cultured human skin fibroblasts. *Biochem Mol Biol Int* 45: 753–759.

Rattan, S.I.S. 2008. Hormesis in aging. *Ageing Res Rev* 7: 63–78.

Rattan. S.I., R.A. Fernandes, D. Demirovic, B. Dymek, and C.F. Lima. 2009. Heat stress and hormetin-induced hormesis in human cells: Effects on aging, wound healing, angiogenesis, and differentiation. *Dose Response* 7(1): 90–103.

Roubille, F., A. Franck-Miclo, A. Covinhes, C. Lafont, F. Cransac, S. Combes, A. Vincent et al. 2011. Delayed postconditioning in the mouse heart in vivo. *Circulation* 124(12): 1330–1336.

Ryan, J.A. and L.E. Hightower. 1996. Stress proteins as molecular biomarkers for environmental toxicology. *EXS* 77: 411–424.

Schamhart, D.H.J., G. Zoutewelle, H. van Aken, and R. van Wijk. 1992. Effects on the expression of heat shock proteins by step-down heating and hypothermia in rat hepatoma cells with a different degree of heat sensitivity. *Int J Hyperthermia* 8: 701–716.

Schmidt, M.R., A.D. Sloth, J. Johnsen, and H.E. Botker. 2012. Remote ischemic conditioning: The cardiologist's perspective. *J Cardiovasc Med (Hagerstown)* 13(11): 667–674.

Skyschally, A., P. van Caster, E.K. Iliodromitis, R. Schulz, D.T. Kremastinos, and G. Heusch. 2009. Ischemic postconditioning: Experimental models and protocol algorithms. *Basic Res Cardiol* 104: 469–483.

Son, T.G., S. Camandola, and M.P. Mattson. 2008. Hormetic dietary phytochemicals. *Neuromolecular Med* 10(4): 236–246.

Song, X., N. Zhang, H. Xu, L. Cao, and H. Zhang. 2012. Combined preconditioning and postconditioning provides synergistic protection against liver ischemic reperfusion injury. *Int J Biol Sci* 8(5): 707–718.

Sørensen, J.G., M.M. Nielsen, M. Kruhoffer, J. Justesen, and V. Loeschcke. 2005. Full genome gene expression analysis of the heat stress response in *Drosophila melanogaster*. *Cell Stress Chaperones* 10: 312–328.

Souren, J.E.M., F.A.C. Wiegant, P. van Hof, J.M. van Aken, and R. van Wijk. 1999. The effect of temperature and protein synthesis on the renaturation of firefly luciferase in intact H9c2 cells. *Cell Mol Life Sci* 55: 1473–1481.

Southam, C.M. and J. Ehrlich. 1943. Effects of extract of western red-cedar heartwood on certain wood-decaying fungi in culture. *Phytopathology* 33: 517–524.

Staat, P., G. Rioufol, C. Piot, Y. Cottin, T.T. Cung, I. L'Huillier, J.F. Aupetit et al. 2005. Postconditioning the human heart. *Circulation* 112: 2143–2148.

Stebbing, A.R.D. 2003a. Adaptive responses account for the β-curve—Hormesis is linked to acquired tolerance. *Nonlinearity Biol Toxicol Med* 1: 493–511.

Stebbing, A.R.D. 2003b. A mechanism for hormesis—A problem in the wrong discipline. *Crit Rev Toxicol* 33: 463–467.

Theodoraki, K., A. Tympa, I. Karmaniolou, A. Tsaroucha, N. Arkadopoulos, and V. Smyrniotis. 2011. Ischemia/reperfusion injury in liver resection: A review of preconditioning methods. *Surg Today* 41(5): 620–629.

Tuijl, M.J., S. Cluistra, C.M. van der Kruijssen, and R. van Wijk. 1993. Heat-induced unresponsiveness of heat shock gene expression is regulated at the transcriptional level. *Int J Hyperthermia* 9(1): 125–136.

van Wijk, R. and F.A.C. Wiegant. 2006. *The Simila Principle. An Experimental Approach on the Cornerstone of Homeopathy*. Karl und Veronica Carstens-Stiftung, Essen, Germany.

van Wijk, R. and F.A.C. Wiegant. 2010. Postconditioning hormesis and the homeopathic similia principle: Molecular aspects. *Hum Exp Toxicol* 29: 561–565.

van Wijk, R. and F.A.C. Wiegant. 2011. Postconditioning hormesis and the similia principle. *Front Biosci (Elite Ed)* 3: 1128–1138.

van Wijk, R., E.P.A. van Wijk, F.A.C. Wiegant, and J. Ives. 2008. Free radicals and low-level photon emission in human pathogenesis: State of the art. *Indian J Exp Biol* 46: 273–309.

van Wijk, R., H. Ooms, F.A.C. Wiegant, J.E.M. Souren, J.H. Ovelgönne, J.M. van Aken, and A.W.J.M. Bol. 1994a. A molecular basis for understanding the benefits from subharmful doses of toxicants. *Environ Manage Health* 5: 13–25.

van Wijk, R., J.H. Ovelgönne, E. de Koning, K. Jaarsveld, J. van Rijn, and F.A.C. Wiegant. 1994b. Mild step-down heating causes increased transcription levels of hsp68 and hsp84 mRNA and enhances thermotolerance development in Reuber H35 hepatoma cells. *Int J Hyperthermia* 10: 115–125.

Verbeke, P., B.F. Clark, and S.I. Rattan. 2001. Reduced levels of oxidized and glycoxidized proteins in human fibroblasts exposed to repeated mild heat shock during serial passaging in vitro. *Free Radic Biol Med* 31: 1593–1602.

Verbeke, P., M. Deries, B.F. Clark, and S.I. Rattan. 2002. Hormetic action of mild heat stress decreases the inducibility of protein oxidation and glycoxidation in human fibroblasts. *Biogerontology* 3: 117–120.

Wiegant, F.A.C. and R. van Wijk. 2010. The similia principle: Results obtained in a cellular model system. *Homeopathy* 99: 3–14.

Wiegant, F.A.C., G. Limandjaja, S.A.H. de Poot, L.A. Bayda, O.N. Vorontsova, T.A. Zenina, M. Langelaar-Makkinje, J.A. Post, and G. Wikman. 2008. Plant adaptogens activate cellular adaptive mechanisms by causing mild damage. In: *Adaptation Biology and Medicine, Volume 5: Health Potentials*, Lukyanova L, Takeda N, and Singal PK (eds.). Narosa Publishers, New Delhi, India. pp. 319–332.

Wiegant, F.A.C., H.A.B. Prins, and R. van Wijk. 2011. Postconditioning hormesis put in perspective: An overview of experimental and clinical studies. *Dose Response* 9: 209–224.

Wiegant, F.A.C., J. van Rijn, and R. van Wijk. 1997. Enhancement of the stress response by minute amounts of cadmium in sensitized Reuber H35 hepatoma cells. *Toxicology* 116: 27–37.

Wiegant, F.A.C., J.E.M. Souren, and R. van Wijk. 1999. Stimulation of survival capacity in heat shocked cells by subsequent exposure to minute amounts of chemical stressors: Role of similarity in hsp-inducing effects. *Hum Exp Toxicol* 18: 460–470.

Wiegant, F.A.C., J.E.M. Souren, H. van Rijn, and R. van Wijk. 1993. Arsenite induced sensitization and self-tolerance of Reuber H35 hepatoma cells. *Cell Biol Toxicol* 9: 49–59.

Wiegant, F.A.C., J.E.M. Souren, J. van Rijn, and R. van Wijk. 1994. Stressor-specific induction of heat shock proteins in rat hepatoma cells. *Toxicology* 94: 143–159.

Wiegant, F.A.C., N. Spieker, and R. van Wijk. 1998. Stressor-specific enhancement of hsp induction by low doses of stressors in conditions of self- and cross-sensitization. *Toxicology* 127: 107–119.

Wiegant, F.A.C., N. Spieker, C.A. van der Mast, and R. van Wijk. 1996. Is heat shock protein re-induction during tolerance related to the stressor-specific induction of heat shock proteins? *J Cell Physiol* 169: 364–372.

Wiegant, F.A.C., S. Surinova, E. Ytsma, M. Langelaar-Makkinje, G. Wikman, and J.A. Post. 2009. Plant adaptogens increase lifespan and stress resistance in *C. elegans*. *Biogerontology* 10: 27–42.

Wiegant, F.A.C., S.A.H. de Poot, V.E. Boers-Trilles, and A.M.A. Schreij. 2012. Hormesis and cellular quality control: A possible explanation for the molecular mechanism that underlie the benefits of mild stress. *Dose Response* 11(3): 413–430.

Wilder, J. 1962. Basimetric approach (law of initial value) to biological rhythms. *Ann NY Acad Sci* 98: 1211–1220.

Zhao, H. 2007. The protective effect of ischemic postconditioning against ischemic injury: From the heart to the brain. *J Neuroimmune Pharmacol* 2: 313–318.

Zhao, H. 2009. Ischemic postconditioning as a novel avenue to protect against brain injury after stroke. *J Cereb Blood Flow Metab* 29: 873–885.

Zhao, H., S. Joo, W. Xie, and X. Ji. 2013. Using hormetic strategies to improve ischemic preconditioning and postconditioning against stroke. *Int J Physiol Pathophysiol Pharmacol* 5(2): 61–72.

# Section II

*Evidence for Hormesis in Human Beings*

# 3 Exercise and Hormesis
## *Shaping the Dose-Response Curve*

*Zsolt Radak*

### CONTENTS

3.1 Introduction .................................................................................................. 37
3.2 Exercise Hormesis and Functional Endpoints ............................................. 38
3.3 Antioxidants and Reactive Oxygen Species–Dependent Hormesis
    with Exercise ................................................................................................ 40
3.4 Extension of the Peak of the Hormetic Curve with Exercise ...................... 41
3.5 Conclusions .................................................................................................. 42
Acknowledgment ................................................................................................. 43
References ............................................................................................................ 43

### 3.1 INTRODUCTION

Physical inactivity is a major contributor to the development of lifestyle-related diseases, including cardiovascular diseases (Lee et al. 2013; Oppewal et al. 2013), certain tumors, type 2 diabetes (Kasuga et al. 2013), osteoporosis (Janssen 2012), and even neurodegenerative diseases (Radak et al. 2010; Janssen 2012). This could be partly because the development of the genome at the genesis of humans was associated with a physically active lifestyle. It has been suggested that Stone Age man used 4000 kcal for physical activity on a daily basis (Radak et al. 2013c). Adaptive changes on the DNA sequence are very slow. Therefore, it is not surprising that physical inactivity is a risk factor for a wide range of diseases.

In accordance with these findings, there is a large body of evidence showing that regular exercise reduces the risk of lifestyle-related diseases (Bassil and Gougeon 2013; Lacza and Radak 2013; Mellett and Bousquet 2013; Strasser 2013; Yilmaz et al. 2013), increases mean life span (Carter et al. 2007), and can significantly increase the quality of life, especially in the elderly (Brovold et al. 2013; Frazzitta et al. 2013). It is important to note that the effects of exercise are systemic and complex (Radak et al. 2008). Regular exercise can beneficially affect brain function, resulting in neurogenesis, elevated production of neurotrophic and growth factors, and improved capillarization (Radak et al. 2013a). Regular exercise increases the content of glucose transporter GLUT4, enhances insulin sensitivity, and reduces the risk of the metabolic syndrome (Sato et al. 2007). In addition, regular aerobic exercise beneficially affects the function of kidney, liver (Peeri et al. 2013), testes (Di Luigi et al. 2012), and, of course, heart and skeletal muscle (Okita et al. 2013).

## 3.2 EXERCISE HORMESIS AND FUNCTIONAL ENDPOINTS

The hormesis thesis is that biological systems respond to exposure to chemicals, toxins, and radiation and that hormesis is denoted by a bell-shaped curve. In toxicology, hormesis is a dose-response phenomenon characterized by a low dose of stimulation and a high dose of inhibition, resulting in either a bell-shaped or an inverted U-shaped dose-response curve, which is a nonmonotonic response (Calabrese and Baldwin 2001, 2002; Cook and Calabrese 2006). Recently, we have extended the hormesis theory to reactive oxygen species (ROS), which appear to plateau when modulated by aging or physical exercise (Radak et al. 2005). We have proposed that exercise modulates ROS and the effects can be described by the hormetic curve. We believe that regular exercise results in a typical bell-shaped hormetic curve, because of the regulation of adaptive systems. We have proposed that adaptation is dependent on the modulation of homeostasis. It is clear that homeostasis is a dynamic system with biological and functional/actual endpoints. Biological endpoints are signified by the point at which the system collapses.

Functional/actual endpoints demark the limits of individual tolerance, which are naturally below biological endpoints and are dynamic, variable values. The distance between the optimal zone and biological endpoints represents the zone that can be targeted to induce adaptations to extend functional/actual endpoints. In the case of a high degree of adaptation, the distance between the biological endpoints and the functional endpoints can be narrowed. In other words, the distance between the optimal zone and functional/actual endpoints can be increased (Figure 3.1). It can be assumed that a larger range between optimal zone and functional/actual endpoints represents greater adaptive capability and better tolerance against stressors.

Two examples will suffice to exemplify this point. The resting heart rate of untrained individuals is around 70 beats/min, whereas the maximal heart rate is approximately 220 minus the age of the individual. Therefore, the adaptive range is 130 beats for a 20-year-old and just 50 beats for a centenarian. Well-trained endurance runners, on the other hand, have a significant decrease in resting heart rate, which could be as low as 35 beats/min. If an individual has the same maximal heart rate as the 20-year-old, trained and untrained, the adaptive range increases from 130 to 165 with the extension of the functional endpoints. Another example is lactic acid tolerance. Both trained and untrained individuals have a resting blood lactic acid level around 1.5 mM/L. If they run until exhaustion on a treadmill, untrained individuals stop when their lactic acid levels reach approximately 6–10 mM/L, whereas elite athletes can still continue until their lactic acid levels are over 20 mM/L.

Exercise-associated adaptive response is specific and depends on the intensity and duration of exercise. High-intensity acute exercise produces different metabolic intermediates, such as lactic acid, ammonia, and adenosine monophosphate, than exercise of low intensity and high duration. Single bouts of aerobic exercise, on the

# Exercise and Hormesis

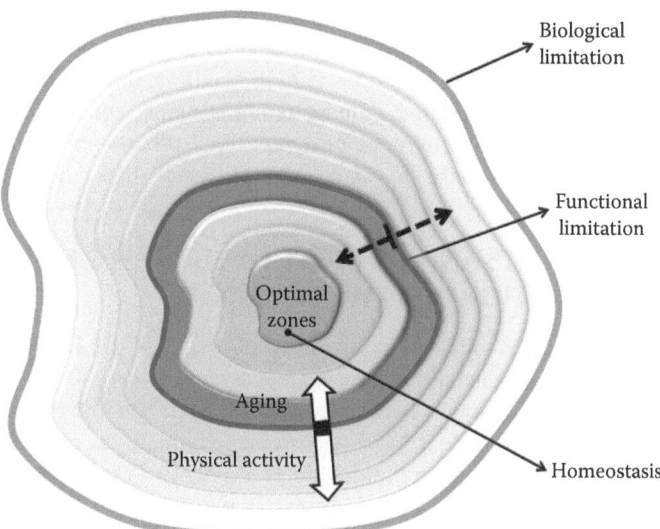

**FIGURE 3.1** The hypothetical adaptive range. The middle of the graph represents the optimal zone of the dynamic homeostasis, whereas the outer line indicates the biological limitations, which cannot be reached without extreme risk of death. The line, called functional limitation, shows the capacity of each individual and it is a mobile value. The functional/actual limit can be readily altered by exercise training. Aging decreases the rate of adaptive response, and the capacity to maintain homeostasis decreases, as shown by the white arrows.

other hand, could result in hypoglycemia and a significant loss of glycogen content in skeletal muscle. Hence, the adaptive response to high-intensity exercise would cause increased tolerance to lactic acid, a higher speed at the anaerobic threshold, and better elimination of ammonia, whereas regular aerobic exercise results in improved carbohydrate handling of the body and increased levels of glycogen in skeletal muscle. These adaptations are brought about by regular exercise without which these adaptive responses would not occur. However, exercise above a certain threshold can cause maladaptations, which decrease the range of the biological and functional/actual adaptive zones. This is termed overtraining. Therefore, the two minimal endpoints of exercise-related dose-response are physical inactivity and overtraining.

A similar type of adaptive response is found for exercise-induced ROS production. It is known that a single bout of exhaustive exercise results in elevated levels of lipid peroxidation, carbonylation of amino acid residues, and 8-oxo-7,8 dihydroguanine (8-oxoG) in DNA. On the other hand, when a single bout of exhaustive exercise is given to well-trained subjects, the body responds without a large elevation in oxidative damage (Radak et al. 2001). In addition, regular exercise–associated adaptation is a precondition against treatment with hydrogen peroxide ($H_2O_2$), which causes a significant degree of damage in untrained subjects (Radak et al. 2000). Moreover, when heart attack or stroke is simulated in untrained and trained animals, the infarct size is significantly smaller in trained groups (Bolli and Marban 1999; Ding et al. 2005), showing that regular exercise acts as a preconditioning tool (Radak et al. 2000) by enhancing the adaptive zone and by narrowing the theoretical distance between functional and biological endpoints. A great deal of evidence exists, that suggests that regular exercise–induced adaptations to ROS handling, through redox signaling, including antioxidant and oxidative damage repair systems, significantly contribute to the health-promoting effects of regular exercise. We further propose here that exercise-mediated adaptive response can shape the typical bell-shaped hormetic curve.

## 3.3 ANTIOXIDANTS AND REACTIVE OXYGEN SPECIES–DEPENDENT HORMESIS WITH EXERCISE

It is clear that moderate amounts of oxidizing agents (e.g., 150 µM $H_2O_2$) can increase the force generation of skeletal muscle, whereas larger doses (300 µM $H_2O_2$) result in a decline in force production (Andrade et al. 1998). It has also been shown that administration of the antioxidant *N*-acetylcysteine to humans decreased the fatigue-associated decline in force generation (Reid et al. 1994). This shows that elevated amounts of ROS cause fatigue (this is in association with the fact that high levels of ROS suppress force generation). On the other hand, there are a few recent reports suggesting that antioxidant supplementation is not always good, because it can suppress the adaptive response to exercise training (Gomez-Cabrera et al. 2005, 2008; Ristow et al. 2009). However, it must be mentioned that this view is not fully accepted (Higashida et al. 2011). If the exercise-generated ROS can be described by a bell-shaped curve, the opposing effects of antioxidant supplementation on ROS-induced adaptation can be easily understood. Figure 3.2 displays our hypothesis that if the antioxidants are supplemented before the ROS reach levels for maximum adaptive response, the antioxidant would depress the physiological response. On the other hand, if the antioxidants are supplemented when the concentration of exercise-generated ROS is associated with a declining physiological response, the supplementation would result in enhanced performance and delayed fatigue. Indeed, the scientific database includes results for both situations, when the antioxidants were

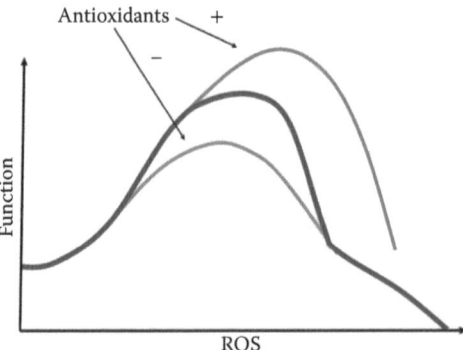

**FIGURE 3.2** Supplementation of antioxidants before (−) the reactive oxygen species (ROS) levels reach the value associated with peak physiological function can attenuate the beneficial effects of exercise. On the other hand, antioxidant treatment after (+) the period of maximum ROS-associated function can result in decreased appearance of fatigue and/or improved function.

stimulating or suppressing. Our concept could be one possible explanation for these divergent results. Moreover, it must be mentioned that the so-called maximal or optimal levels of ROS are dependent on many factors, such as age, history of the exposure to oxidative stress, the effectiveness of the endogenous antioxidant system, and the level of physical fitness. Indeed, in our opinion the level of physical fitness is crucial, and we propose that the adaptive plasticity is a regulator of dose-response; hence, it gives shape to the hormetic curve.

## 3.4 EXTENSION OF THE PEAK OF THE HORMETIC CURVE WITH EXERCISE

As we showed earlier, exercise-associated adaptation can extend the functional endpoints of adaptation, which means that the distance between the functional/actual endpoints and the biological endpoints (limitations) is significantly narrowed. If we reflect these changes into the bell-shaped hormetic curve, it could mean a significant extension of the peak and/or the *optimal* zones (Figure 3.3). This extension would mean that a greater dose could be tolerated by the body with high levels of physiological performance.

Untrained individuals also have a bell-shaped dose-response curve to exercise, which means that during moderate levels of intensity and duration their physiological responses would be better. On the other hand, high intensity or exercise of long duration would cause fatigue and decreased performance. The so-called optimal intensity or/and duration would comprise a very narrow zone. Superbly trained individual would endure much larger and wider optimal exercise loading with enhanced functional parameters of exercise. The same situation is true for the dosage of ROS.

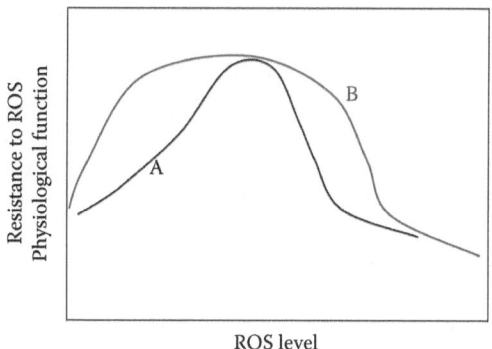

**FIGURE 3.3** The "A" curve is a typical dose-response curve of physical exercise. Moderate exercise increases the physiological function of different organs, increases the rate of prevention against diseases, and improves quality of life. Physical inactivity, strenuous exercise, and overtraining increase the risk of diseases and decrease physiological function. The "B" curve indicates that regular exercise can extend or stretch the levels of reactive oxygen species (ROS) that are associated with high levels of physical function. This means that exercise can increase tolerance against high levels of ROS and can be preventive against oxidative stress–associated diseases.

Well-trained individuals, due to the adaptive response, would endure higher levels of ROS without significant damage of macromolecules, and higher levels of oxidative damage would be tolerated without significant loss of function. This is the effect of exercise-mediated precondition via ROS and is the dose-response phenomenon by which exercise attenuates the incidence of ROS-associated diseases (Radak et al. 2010; Sturek 2011; Hamer 2012; Kovacic et al. 2012; Lacza and Radak 2013).

The bell-shape curve type of dose-response of ROS can be extendented to the accumulation of oxidative damage (Radak et al. 2011). We used different markers of oxidative damage to evaluate the interaction between ROS and macromolecules. Interestingly, the most common oxidative damage markers, such as malondialdehyde (MDA) for lipid peroxidation, carbonyls for oxidative protein damage, or 8-oxoG for DNA damage, are always present in cells. Although the oxidative stress theory is one of the most accepted theories of aging (Harman 1956), in the very young organism there is a measurable amount of lipid peroxidation, carbonyl levels, or even 8-oxoG levels. Moreover, the accumulation of MDA, carbonyls, or 8-oxoG with aging is not a linear process; the actual increase is in the last quarter of the life span (Radak et al. 2013c). It appears that large doses of antioxidants cannot eliminate these damage markers, as they have definite physiological effects on individual cells. Indeed, base levels of oxidative modification of lipids can be important for cell signaling and membrane remodeling. In addition, the ROS-mediated posttranslational modifications of proteins could be important to the homeostasis of protein turnover (Radak et al. 2011), whereas low levels of 8-oxoG might be necessary for transcription (Radak and Boldogh 2010). We suggest that a moderate level of oxidative damage could be not just a consequence of metabolism, but also even important and necessary for cells. There is no question that high levels of accumulation of lipid peroxidation, oxidative protein damage, or 8-oxoG accelerate the progress of aging and neurodegenerative diseases. Therefore, agents, including moderate levels of oxidative damage, that induce the activity of repair enzymes, such as $Ca^{2+}$-independent phospholipase A(2) (iPLA(2)β) for lipids, methionine sulfoxide reductase for proteins, and 8-oxoguanine DNA glycosylase for 8-oxoG, are important for the maintenance and viability of cells. Acute or severe bouts of exercise can lead to moderate increases in oxidative damage, whereas regular exercise–associated adaptive responses result in increased activity of repair enzymes and moderate levels of oxidative damage (Radak et al. 2013a–c).

## 3.5 CONCLUSIONS

The response of biological systems to stressors can be described by a bell-shaped curve. Physical exercise also evokes this hormetic curve response in the organism. The two endpoints of the hormetic curve are inactivity and overtraining, and both of these result in decreased physiological function (Figure 3.1). Antioxidant supplementation, depending on the timing, could suppress, enhance, or prolong high levels of physiological function. Adaptive capacity could extend, that is, stretch the top of the bell-shaped curve, resulting in a greater tolerance for ROS, and possibly other metabolic products, with high performance and loss of function. Oxidative damage markers, such as MDA, carbonyls, or 8-oxoG, are necessary factors of hormesis.

## ACKNOWLEDGMENT

This work was supported through the laboratories of Dr. Zsolt Radak.

## REFERENCES

Andrade, F. H., Reid, M. B., Allen, D. G. and Westerblad, H. 1998. Effect of hydrogen peroxide and dithiothreitol on contractile function of single skeletal muscle fibres from the mouse. *J Physiol.* 509(Pt 2): 565–575.
Bassil, M. S. and Gougeon, R. 2013. Muscle protein anabolism in type 2 diabetes. *Curr Opin Clin Nutr Metab Care.* 16(1): 83–88.
Bolli, R. and Marban, E. 1999. Molecular and cellular mechanisms of myocardial stunning. *Physiol Rev.* 79(2): 609–634.
Brovold, T., Skelton, D. A. and Bergland, A. 2013. Older adults recently discharged from the hospital: effect of aerobic interval exercise on health-related quality of life, physical fitness, and physical activity. *J Am Geriatr Soc.* 61(9):1580–1585.
Calabrese, E. J. and Baldwin, L. A. 2001. U-shaped dose-responses in biology, toxicology, and public health. *Annu Rev Public Health.* 22: 15–33.
Calabrese, E. J. and Baldwin, L. A. 2002. Defining hormesis. *Hum Exp Toxicol.* 21(2): 91–97.
Carter, C. S., Hofer, T., Seo, A. Y. and Leeuwenburgh, C. 2007. Molecular mechanisms of life- and health-span extension: Role of calorie restriction and exercise intervention. *Appl Physiol Nutr Metab.* 32(5): 954–966.
Cook, R. R. and Calabrese, E. J. 2006. Hormesis is biology, not religion. *Environ Health Perspect.* 114(12): A688.
Di Luigi, L., Romanelli, F., Sgro, P. and Lenzi, A. 2012. Andrological aspects of physical exercise and sport medicine. *Endocrine.* 42(2): 278–284.
Ding, Y. H., Young, C. N., Luan, X., Li, J., Rafols, J. A., Clark, J. C., McAllister, J. P., 2nd and Ding, Y. 2005. Exercise preconditioning ameliorates inflammatory injury in ischemic rats during reperfusion. *Acta Neuropathol.* 109(3): 237–246.
Frazzitta, G., Balbi, P., Maestri, R., Bertotti, G., Boveri, N. and Pezzoli, G. 2013. The beneficial role of intensive exercise on Parkinson disease progression. *Am J Phys Med Rehabil.* 92(6): 523–532.
Gomez-Cabrera, M. C., Borras, C., Pallardo, F. V., Sastre, J., Ji, L. L. and Vina, J. 2005. Decreasing xanthine oxidase-mediated oxidative stress prevents useful cellular adaptations to exercise in rats. *J Physiol.* 567(Pt 1): 113–120.
Gomez-Cabrera, M. C., Domenech, E., Romagnoli, M., Arduini, A., Borras, C., Pallardo, F. V., Sastre, J. and Vina, J. 2008. Oral administration of vitamin C decreases muscle mitochondrial biogenesis and hampers training-induced adaptations in endurance performance. *Am J Clin Nutr.* 87(1): 142–149.
Hamer, M. 2012. Psychosocial stress and cardiovascular disease risk: the role of physical activity. *Psychosom Med.* 74(9): 896–903.
Harman, D. 1956. Aging: A theory based on free radical and radiation chemistry. *J Gerontol.* 11(3): 298–300.
Higashida, K., Kim, S. H., Higuchi, M., Holloszy, J. O. and Han, D. H. 2011. Normal adaptations to exercise despite protection against oxidative stress. *Am J Physiol Endocrinol Metab.* 301(5): E779–E784.
Janssen, I. 2012. Health care costs of physical inactivity in Canadian adults. *Appl Physiol Nutr Metab.* 37(4): 803–806.
Kasuga, M., Ueki, K., Tajima, N., Noda, M., Ohashi, K., Noto, H., Goto, A. et al. 2013. Report of the Japan Diabetes Society/Japanese Cancer Association joint committee on diabetes and cancer. *Cancer Sci.* 104(7): 965–976.

Kovacic, J. C., Castellano, J. M. and Fuster, V. 2012. Cardiovascular defense challenges at the basic, clinical, and population levels. *Ann N Y Acad Sci.* 1254: 1–6.

Lacza, G. and Radak, Z. 2013. [Is physical activity an elixir?]. *Orv Hetil.* 154(20): 764–768.

Lee, I. M., Bauman, A. E., Blair, S. N., Heath, G. W., Kohl, H. W., 3rd, Pratt, M. and Hallal, P. C. 2013. Annual deaths attributable to physical inactivity: whither the missing 2 million? *Lancet.* 381(9871): 992–993.

Mellett, L. H. and Bousquet, G. 2013. Cardiology patient page. Heart-healthy exercise. *Circulation.* 127(17): e571–e572.

Okita, K., Kinugawa, S. and Tsutsui, H. 2013. Exercise intolerance in chronic heart failure—Skeletal muscle dysfunction and potential therapies. *Circ J.* 77(2): 293–300.

Oppewal, A., Hilgenkamp, T. I., van Wijck, R. and Evenhuis, H. M. 2013. Cardiorespiratory fitness in individuals with intellectual disabilities—A review. *Res Dev Disabil.* 34(10): 3301–3316.

Peeri, M., Habibian, M., Azarbayjani, M. A. and Hedayati, M. 2013. Protective effect of aerobic exercise against L-NAME-induced kidney damage in rats. *Arh Hig Rada Toksikol.* 64(2): 43–49.

Radak, Z. and Boldogh, I. 2010. 8-Oxo-7,8-dihydroguanine: links to gene expression, aging, and defense against oxidative stress. *Free Radic Biol Med.* 49(4): 587–596.

Radak, Z., Chung, H. Y. and Goto, S. 2005. Exercise and hormesis: oxidative stress-related adaptation for successful aging. *Biogerontology.* 6(1): 71–75.

Radak, Z., Chung, H. Y. and Goto, S. 2008. Systemic adaptation to oxidative challenge induced by regular exercise. *Free Radic Biol Med.* 44(2): 153–159.

Radak, Z., Hart, N., Sarga, L., Koltai, E., Atalay, M., Ohno, H. and Boldogh, I. 2010. Exercise plays a preventive role against Alzheimer's disease. *J Alzheimers Dis.* 20(3): 777–783.

Radak, Z., Ihasz, F., Koltai, E., Goto, S., Taylor, A. W. and Boldogh, I. 2013a. The redox-associated adaptive response of brain to physical exercise. *Free Radic Res.* 12 (10): 856-863.

Radak, Z., Koltai, E., Taylor, A. W., Higuchi, M., Kumagai, S., Ohno, H., Goto, S. and Boldogh, I. 2013b. Redox-regulating sirtuins in aging, caloric restriction, and exercise. *Free Radic Biol Med.* 58: 87–97.

Radak, Z., Sasvari, M., Nyakas, C., Pucsok, J., Nakamoto, H. and Goto, S. 2000. Exercise preconditioning against hydrogen peroxide-induced oxidative damage in proteins of rat myocardium. *Arch Biochem Biophys.* 376(2): 248–251.

Radak, Z., Taylor, A. W., Ohno, H. and Goto, S. 2001. Adaptation to exercise-induced oxidative stress: From muscle to brain. *Exerc Immunol Rev.* 7: 90–107.

Radak, Z., Zhao, Z., Goto, S. and Koltai, E. 2011. Age-associated neurodegeneration and oxidative damage to lipids, proteins and DNA. *Mol Aspects Med.* 32(4–6): 305–315.

Radak, Z., Zhao, Z., Koltai, E., Ohno, H. and Atalay, M. 2013c. Oxygen consumption and usage during physical exercise: The balance between oxidative stress and ROS-dependent adaptive signaling. *Antioxid Redox Signal.* 18(10): 1208–1246.

Reid, M. B., Stokic, D. S., Koch, S. M., Khawli, F. A. and Leis, A. A. 1994. N-acetylcysteine inhibits muscle fatigue in humans. *J Clin Invest.* 94(6): 2468–2474.

Ristow, M., Zarse, K., Oberbach, A., Kloting, N., Birringer, M., Kiehntopf, M., Stumvoll, M., Kahn, C. R. and Bluher, M. 2009. Antioxidants prevent health-promoting effects of physical exercise in humans. *Proc Natl Acad Sci U S A.* 106(21): 8665–8670.

Sato, Y., Nagasaki, M., Kubota, M., Uno, T. and Nakai, N. 2007. Clinical aspects of physical exercise for diabetes/metabolic syndrome. *Diabetes Res Clin Pract.* 77(Suppl 1): S87–S91.

Strasser, B. 2013. Physical activity in obesity and metabolic syndrome. *Ann N Y Acad Sci.* 1281: 141–159.

Sturek, M. 2011. $Ca^{2+}$ regulatory mechanisms of exercise protection against coronary artery disease in metabolic syndrome and diabetes. *J Appl Physiol.* 111(2): 573–586.

Yilmaz, Y., Colak, Y., Kurt, R., Senates, E. and Eren, F. 2013. Linking nonalcoholic fatty liver disease to hepatocellular carcinoma: From bedside to bench and back. *Tumori.* 99(1): 10–16.

# 4 Nutritional Components
## *How They Enhance the Ability to Adapt*

Antje R. Weseler and Aalt Bast

## CONTENTS

4.1 Introduction .................................................................................................. 45
4.2 Molecular Mechanisms and Their Modulation by Nutritional
    Components Mediating the Ability to Adapt ............................................... 47
    4.2.1 Nuclear Factor (Erythroid-Derived 2)-Like 2 .................................. 47
    4.2.2 Heat Shock Transcription Factors ..................................................... 51
    4.2.3 Nuclear Factor-κB ............................................................................. 53
    4.2.4 Sirtuin 1 ............................................................................................. 57
    4.2.5 Forkhead Box O Transcription Factors ............................................. 61
    4.2.6 Metabolic Processes .......................................................................... 62
        4.2.6.1 Phase I Metabolism ............................................................ 63
        4.2.6.2 Phase II Metabolism ........................................................... 63
        4.2.6.3 Phase III Transporters and Cross Talk ............................... 64
    4.2.7 Epigenetic Mechanisms ..................................................................... 64
4.3 Conclusions ................................................................................................... 67
References ............................................................................................................ 67

## 4.1 INTRODUCTION

Drug development is driven by the notion that pharmacotherapeutic advantage is reached through selective toxicity (Albert 1985). Drugs work preferably via interaction with a single target, that is, a receptor, an enzyme, or a transporter. On interaction, a clearly defined strong response is elicited. This response can be regarded as a disturbance of homeostasis and thus is in itself a toxic response. The uniqueness of the target involved renders selectivity to the response. The basis for pharmacotherapy is selective toxicity. In past decades, this approach has led to the discovery of many successful drugs. The paradigm of selective toxicity was nourished by visual techniques (Guner 2000). The term *pharmacophore* was introduced in the early 1900s indicating "a molecular framework that carries (*phoros*) the essential features responsible for a drug's (*pharmacon*) biological activity" (Ehrlich 1909).

Selective and strong responses were also in harmony with the ruling notion on health. The WHO (World Health Organization) recently defined health as "a state of complete physical, mental, and social well-being and not merely the absence of disease or infirmity" (WHO 2006). The first part of this definition is part of the founding constitution of the WHO, which dates from 1946. A *complete physical well-being* indeed requires strong curative actions of drugs, that is, compounds with a selective toxicity profile. It is now increasingly recognized that health is not a static situation but should rather be seen as a dynamic condition (Anonymous 2009) that may differ under various circumstances (Weseler and Bast 2012). Furthermore, health is currently not regarded as a state of perfection but as the ability to adapt. Of course, this changing view also has consequences for the prevention or treatment of diseases. Instead of trying to attain perfection, our preventive and therapeutic approaches should rather aim for increasing the capacity to adapt. In a healthy situation, the physiology should be such that changes can be overcome. Many phrases have been used to describe that phenomenon, such as tuning up metabolism, resilience, elasticity, or heterostasis (Bast and Haenen 2013). Health-promoting compounds should have an added character to provide a better adaptability. Not a strong selective response is needed in this case, but rather a strengthening of physiology should be pursued. This can be attained by consolidating diverse physiological processes. It has been suggested that such a situation can be achieved by combining several drugs all in a relatively low dose, the so-called polypill (Wald and Law 2003).

In combining several pharmacological active substances in one pill, physiology can be boosted, thus increasing protection against changing conditions. The polypill is intended to be consumed by healthy people as means of preventive medicine or might be used in treating illnesses. A mixture of pharmacologically active ingredients in the form of a polypill was first used in the setting of cardiovascular diseases. Originally, the effects were analyzed of a daily administration of six components: a statin, three blood-pressure-lowering drugs (a thiazide, a β-blocker, and an angiotensin-converting enzyme inhibitor) each in half of the standard dose, completed by folic acid and aspirin to maximize cardiovascular disease prevention and minimize adverse effects. The daily pill was aimed to simultaneously reduce four cardiovascular risk factors, that is, low-density lipoprotein cholesterol, blood pressure, serum homocysteine, and platelet function. Impressive cost-effective prevention of cardiovascular diseases has been suggested by the polypill approach (Lonn et al. 2010).

The tactic has also been coined for other age-related disorders such as type 2 diabetes (Kuehn 2006). The realization that chronic age-related diseases might be optimally prevented and/or treated by a polypharmacy methodology led to the recent suggestion that drug discovery should shift the current paradigm from a single to a multitarget approach (Medina-Franco et al. 2013). In this new paradigm, it was advocated that drugs should not be developed as suitable ligands for one specific target but rather through interaction with multiple targets.

We realize that the paradigm shift was already discerned in the study of the effect of food-derived ingredients. Food ingredients and food supplements have a multitude of frequently mild activities (Weseler et al. 2011). It is therefore difficult to describe the clinical effects of food in a reductionist single-target manner. In addition to that, it appears that food-derived compounds can be regarded as specific inherent toxic compounds (Boots et al. 2008; Hanekamp et al. 2012). Polyphenols, for example, interact with other intracellular antioxidants (Jacobs et al. 2010) and more particularly have thiol reactivity (Weseler and Bast 2012). When this activity is displayed too forcefully, toxicity may occur. Mild toxicity, however, leads to mild, nonselective disruptions that can be counteracted by an adaptive response (Hanekamp and Bast 2007). This adaptive reaction might be regarded as a strengthening response, fortifying physiology. We hypothesize that this is, from an evolutionary point of view, an excellent way for organisms to defend themselves against continuous, diet-inflicted, toxic insults. Mild damage leads to adaptation, which causes an invigorated physiology, resulting in an increased ability to adapt—that is, health.

## 4.2 MOLECULAR MECHANISMS AND THEIR MODULATION BY NUTRITIONAL COMPONENTS MEDIATING THE ABILITY TO ADAPT

On a cellular level, the organism's ability to adapt is made up by the cells' ability to cope with stimuli that damage or corrupt their frictionless function. In our aerobic environment, the handling of reactive, oxidizing molecules poses a continuous challenge to cells and requires specific molecular systems that effectively dispose them before they cause any harm. In addition, continuous exposure to xenobiotics and microorganisms requires a battery of protective molecular measures to ensure cellular survival and proper function. As long as the molecular systems of cellular protection, damage removal, and repair are functioning well, cells and eventually the organism are able to adequately adapt to continuously changing environmental stimuli without experiencing any hindrance in their performance. Defects in cellular defense mechanisms or their exhaustion by persistent or abundant exogenous stressors may drive cells into a crisis that eventually may lead to their death. Progressing insights in the regulation of cell function via molecular signaling pathways, transcriptional, posttranslational, and epigenetic mechanisms, have uncovered various pathways that play a key role in facilitating cellular defense and resistance to exogenous triggers. Figure 4.1 provides an overview on the most relevant pathways and mechanisms. Enhancement of these cellular defense mechanisms, either directly or indirectly, can be shown for many nutritional components (Figure 4.2), which thereby increases the capacity of cells and the organism to cope with stressful situations.

### 4.2.1 NUCLEAR FACTOR (ERYTHROID-DERIVED 2)-LIKE 2

A variety of molecular pathways have been identified to directly increase the cells' endogenous defense to oxidants. The most prominent pathway is regulated by the transcription factor Nrf2 (nuclear factor (erythroid-derived 2)-like 2). Nrf2 is a 605 amino acid containing leucine zipper/cap 'n' collar protein that resides in the cytosol bound to its suppressor protein kelch-like ECH-associated protein (Keap) 1. Keap1 facilitates the binding of a Cullin 3 (Cul3)-dependent ubiquitin E3 ligase that ubiquitinates Nrf2 under normal conditions leading to its proteasomal degradation. Human Keap1 possesses 27 cysteine residues of which 10 are particularly reactive because of the proximity of basic amino acids. This collocation favors deprotonation of the cysteine thiol groups, which in turn become stabilized by the positively charged basic amino acids in their surroundings (Dinkova-Kostova et al. 2002; Bindoli and

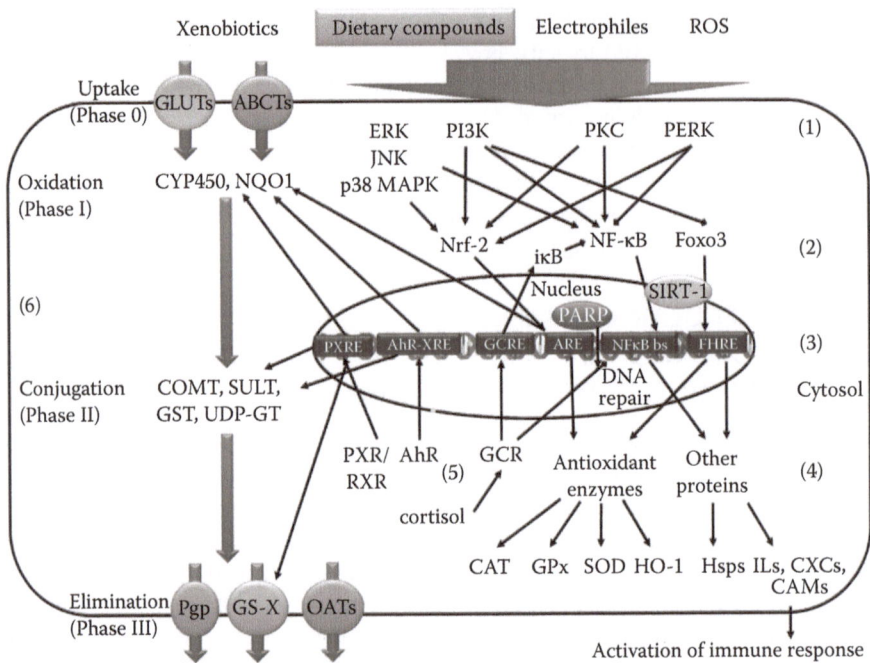

**FIGURE 4.1** Key molecular pathways and mediators that are involved in facilitating cellular protection and prevention against damage induced by xenobiotics (including dietary compounds), electrophiles, and reactive oxygen species (ROS). Signaling cascades (1) have been reported to activate transcription factors (2) that translocate into the nucleus, bind to their respective DNA-binding site, and induce the expression of their target genes (4). Depending on their molecular size and chemical properties, xenobiotics enter cells in various ways (phase 0), for example, via passive diffusion or active transport. Cellular uptake and efflux transporters, and phase I and II detoxification enzymes are transcriptionally regulated via intracellular receptors (5). Phase I drug metabolization renders xenobiotics to substrates of phase II enzymes. Metabolites may leave the cell via various transporters (6).

Rigobello 2013). Thereby, these cysteines become attractive targets for all kinds of electrophiles and reactive oxygen species (ROS). On oxidation, the Nrf2-Keap1-Cul3-ubiquitin E3-ligase complex is destabilized, leading to activation and release of Nrf2. In the nucleus, Nrf2 dimerizes with proteins of the small Maf family and binds to antioxidant or electrophile response elements (ARE/EpRE). A large battery of genes coding for proteins that maintain cellular redox homeostasis and detoxification of xenobiotics are regulated by Nrf2-ARE, including antioxidant and phase II metabolizing enzymes such as superoxide dismutases 1-3 (SOD), catalase (CAT), glutathione peroxidases (GPx), glutathione reductase (GR), hemoxygenase (HO)-1, NAD(P)H:quinone oxidoreductase 1, glutathione S-transferases (GSTs), and uridine diphosphate (UDP)-glucuronosyltransferases (UGTs), all contributing to cellular survival in stressful situations (Birringer 2011; Niture et al. 2013). In addition to the oxidative activation of Nrf2, phosphorylation at a specific serine residue by, among others, mitogen-activated protein kinases (MAPKs), phosphatidylinositol-3-kinase

# Nutritional Components

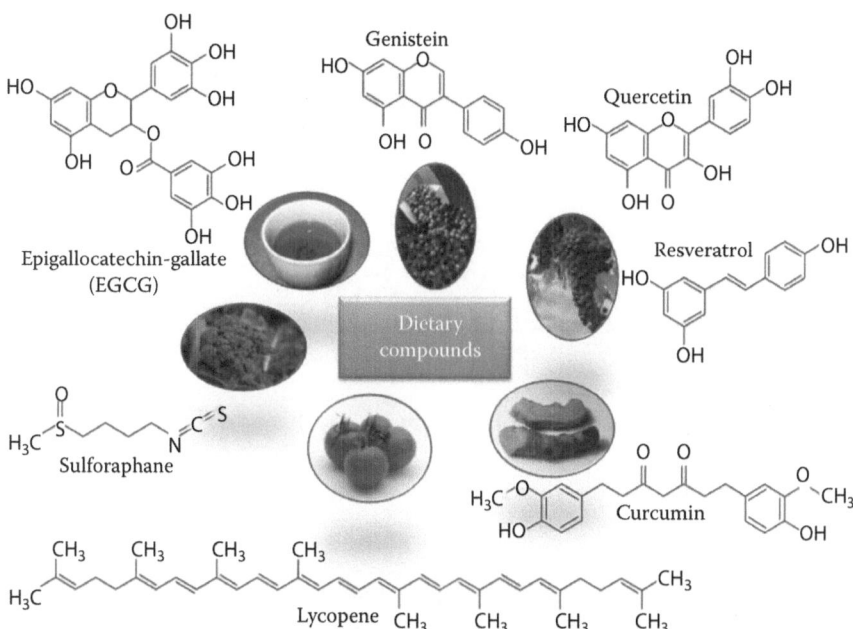

**FIGURE 4.2** Molecular structures and dietary sources of selected compounds that have been reported to modulate stress response and cellular resistance pathways, thereby contributing to an enhanced ability to adapt.

(PI3K), or atypical protein kinase (PK) C (Yu et al. 1999; Huang et al. 2002; Kang et al. 2002) was found to initiate destabilization of the Nrf2–Keap1 complex and Nrf2 migration into the nucleus (Figure 4.3). Because of its central regulatory role in cell defense and survival, Nrf2 has been recently designated as a *master regulator* (Hybertson et al. 2011).

Hydrogen peroxide, hypochlorous acid, peroxynitrite, and enzymatic and nonenzymatic lipid oxidation products such as isoprostanes and 4-hydroxynonenal are examples of endogenous Nrf2 activators (Levonen et al. 2004). Moreover, many environmental and plant-derived compounds such as heavy metals, selenium compounds, α,β-unsaturated carbonyls, quinones, phenolic acids, polyphenols, chalcones, carotenoids, and isothiocyanates are electrophiles that have been shown to oxidize Keap1, thereby activating Nrf2 (excellently reviewed by, among others, Surh [2003], Zhao et al. [2010], Uruno and Motohashi [2011], and Bindoli and Rigobello [2013]). These agents can either directly or indirectly modify sensitive Keap1 cysteines by formation of oxidative metabolites through auto-oxidation (Brigelius-Flohe and Banning 2006; Na and Surh 2008). Polyphenols, for example, as shown for epigallocatechin gallate (EGCG), can activate Nrf2-mediated gene transcription by three mechanisms: (1) EGCG is auto-oxidized, resulting in the production of ROS-oxidizing Keap1 cysteines; (2) oxidized EGCG metabolites bind directly to Keap1 cysteines; or (3) oxidized EGCG metabolites react with glutathione, leading to a transient shift in the cellular redox state, which in turn activates kinase signaling and induces the phosphorylation and subsequent liberation of Nrf2 (Sang et al. 2005; Na and Surh 2008).

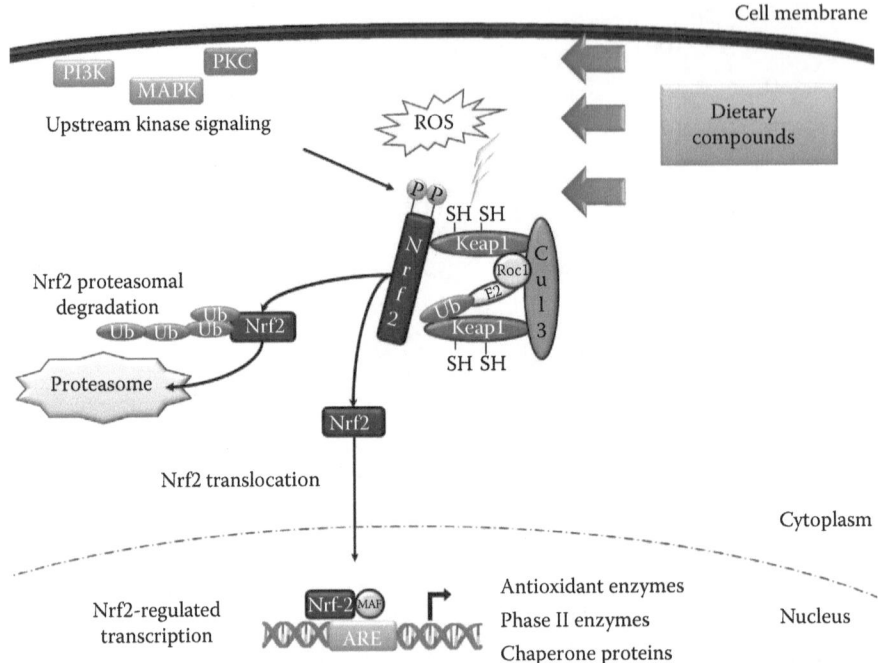

**FIGURE 4.3** The transcription factor Nrf2 (nuclear factor (erythroid-derived 2)-like 2) is bound in the cytosol to its suppressor protein kelch-like ECH-associated protein (Keap) 1 and a Cul3-dependent ubiquitin E3 ligase complex. Ubiquitination of Nrf2 leads to its proteasomal degradation. Reactive oxygen species (ROS) and other electrophiles can oxidize Keap1, destabilizing the Nrf2–Keap1–Cul3 ubiquitin ligase complex. Also, phosphorylation of Nrf2 by various kinases (e.g., mitogen-activated protein kinases [MAPKs], PI3K, or atypical protein kinase [PK] C) fosters diverging of the complex. On liberation, Nrf2 translocates into the nucleus, binds to the antioxidant responsive element (ARE), and initiates the transcription of a large battery of genes, for example, antioxidant defense and phase II detoxification enzymes. Dietary compounds can activate the Nrf2 pathway at different levels by interfering with upstream kinase signaling pathways, oxidizing critical cysteines of Keap1, or by scavenging Nrf2-activating ROS.

Contrary to that, the isothiocyanate group of the glucosinolates from cruciferous, for example, sulforaphane, has a high electrophilicity and thus easily reacts with thiols to a dithiocarbamate (Baird and Dinkova-Kostova 2011). The same is observed for curcumin, which is derived from turmeric, because it possesses two α,β-unsaturated ketone groups that in fact are two Michael acceptor centers (Baird and Dinkova-Kostova 2011). Interestingly, lipophilic carotenoids, such as lycopene and β-carotene, which lack any electrophilicity, have been reported to activate Nrf2 through their more hydrophilic oxidation products that are formed by chemical or enzymatic oxidation and frequently contain an aldehyde moiety (Linnewiel et al. 2009).

Remarkably, all dietary compounds that are described as potent Nrf2 activators are known as good antioxidants. Currently, a vivid debate has emerged in the scientific community, regarding how antioxidants act under physiological conditions

(Carnelio et al. 2007, 2008; Forman et al. 2013; Halliwell 2013). A lot of arguments, in particular from failing clinical efficacy studies (Biesalski et al. 2010), have been accumulated against the relevance of the capability of those compounds to scavenge free radicals under certain physiological conditions (Hollman et al. 2011), and it was suggested to replace the term *antioxidants* with *bioactives* (Sies 2010). Today, it is advocated that antioxidants exclusively exert their biological effects by modulating Nrf2 activity (Forman et al. 2013). Attempts to introduce a new paradigm by defining those effects as *nucleophilic tone* rather than as *antioxidant* have just been made (Forman et al. 2013). Not surprisingly, nature is much more complex. This is nicely reflected by very recent literature that identifies direct intracellular ROS reduction by cocoa flavanols and their metabolites (Ruijters et al. 2013) as well as by hydroxyflavones in nanomolar concentrations (Lombardo et al. 2013). As always, there will be no single exclusive mechanism of those compounds by which they affect biological systems. It is the nature of phytochemicals to target a plethora of cellular and subcellular processes and molecules. This phenomenon is vividly illustrated in the subsequent paragraphs, where we reintroduce many of the compounds that were briefly addressed here as Nrf2 activators, and see that they also modulate, for example, the activity of transcription factors such as nuclear factor-κB (NF-κB) and forkhead box O (FOXO) and enzymes such as sirtuin 1 (SIRT1) and cytochrome P450. Therefore, it is certainly time to clean up such lopsided conceptions of antioxidants (Bast and Haenen 2013) and acknowledge their system-wide activities.

Regardless of the designation, it is evident that the intake of Nrf2-inducing substances as part of the daily diet, directly or indirectly, generates mild electrophilic stress that enhances the cellular capacity to cope with exogenous stressors and improves their resistance against oxidizing environmental stimuli (Figure 4.4a). At the same time, the expression of metabolic phase I and II enzymes increases, enhancing metabolic turnover and thus the speed of xenobiotics' detoxification (*vide infra*). However, if a state of unopposed, chronic ROS generation or unrestricted exposure to electrophiles develops, the Keap1–Nrf2 axes may become disrupted, which prompts deterioration of cell and body functions (Ohta et al. 2008; Taguchi et al. 2011) (Figure 4.4a).

In general, it becomes clear that modulation of the Keap1–Nrf2–ARE pathway is a crucial mechanism by which nutritional components can strengthen cellular stress response, that is, its ability to adapt. The search for Nrf2 modulators and the investigation of their therapeutic capabilities are underway (Suzuki et al. 2013).

### 4.2.2 Heat Shock Transcription Factors

Virtually all organisms, from bacteria to humans, respond to extreme toxic insults such as heat, oxidative stress, heavy metals, toxins from bacterial and other origins, as well as inflammation by the expression of heat shock protein (Hsp) genes to promote the organism's survival under those rigorous circumstances. Hsps are highly conserved proteins that are constitutively expressed under normal growth conditions in all cell types at relatively high levels (i.e., 5%–10% of the total protein content accounts for Hsps), though their level increases significantly under thermal

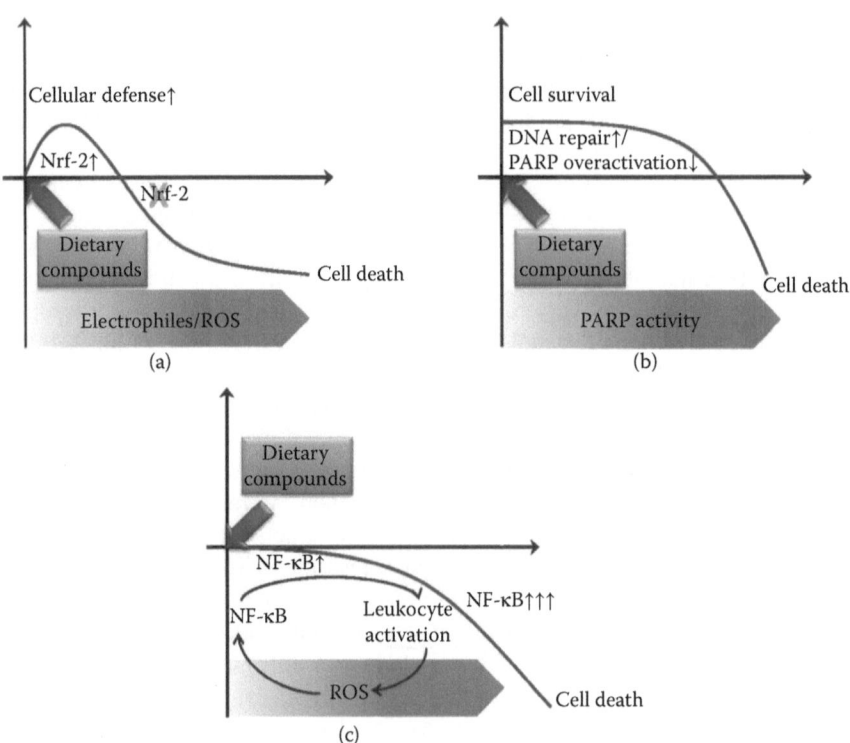

**FIGURE 4.4** Electrophiles including reactive oxygen species (ROS) activate the transcription factor Nrf2 (nuclear factor (erythroid-derived 2)-like 2). In low concentrations, increased Nrf2 activation elicits the transcription of antioxidant defense and phase II enzymes, contributing to an enhanced cellular stress response, that is, ability to adapt. On persisting and increasing levels of oxidative stressors, Nrf2 becomes overactivated and eventually disrupted, inevitably leading to cell death. Dietary compounds mimic Nrf2 response on low stressor levels, thereby enhancing Nrf2-mediated cell defense. The biological effects of dietary compounds being associated with Nrf2 activity frequently follow a bell-shaped dose-response curve (a). The transcription factor NF-κB can be activated by ROS and inflammatory mediators from activated leukocytes. Activated NF-κB leads to the expression of inflammatory mediators, such as cytokines and chemokines, that further attract and activate leukocytes to the site of inflammation. This fuels the production of ROS and activation of NF-κB in a self-amplifying manner. Unopposed NF-κB activity leads inevitably to cell death. In the early stages of this *circulus vitiosus* many dietary compounds can inhibit NF-κB activation either directly or indirectly, by reducing ROS levels (b). Poly(ADP-ribose)-polymerase (PARP) is a nuclear enzyme transferring ADP-ribose units from $NAD^+$ to nuclear proteins such as histones, topoisomerase I and II, and to itself, thereby altering their structure and activities. Normal PARP activity is required to maintain DNA repair capacity, whereas increasing PARP activity activates NF-κB and, if unrestricted, leads to necrotic cell death. Dietary compounds have been discovered as mild PARP inhibitors, attenuating PARP overactivation without compromising DNA repair capacity (c).

or other stress conditions (Jindal 1996). Hsps can be divided in accordance to their molecular weight into several families including Hsp100, Hsp90, Hsp70, Hsp60, and Hsp28, most of which are located in the cytoplasm, but also in mitochondria or the endoplasmic reticulum (Macario and Conway de Macario 2007). They are involved in thermotolerance, stabilization of inactive forms of hormone receptors, interaction with PKs to support their transit to plasma membranes, interactions with immunophilins and protein folding, stabilization, translocation, and degradation (Jindal 1996). In particular, the latter functions have led to their designation as *molecular chaperones*. Moreover, the Hsps' central role in cytoprotection and cell survival in stressful environments rendered them part of the so-called vitagenes, that is, genes coding proteins essential for maintaining *longevity assurance processes* (Rattan 1998).

On the transcriptional level, Hsps are regulated by heat shock protein transcription factors (HSFs) of which four types, HSFs 1–4, have been identified in vertebrates (Wu 1995; Nover et al. 2001). HSF1 has been suggested as the master regulator in mammalian cells in which it is expressed in most tissues and cell types (Akerfelt et al. 2010). In the absence of stressors it is kept as an inactive monomer in cells. External stimuli cause its trimerization and binding to heat shock elements in the DNA located upstream of Hsp genes. Next to heat, electrophiles and ROS are also able to activate HSF1. In the presence of those oxidative stimuli, the formation of a covalent intermolecular disulfide bond between two cysteine residues (C36 and C103) is triggered, which initiates trimerization and DNA binding of HSF1 (Lu et al. 2008). In contrast to that, an intramolecular disulfide bond in which three other cysteine residues participate (C153, C373, and C378) was discovered to prevent HSF1 trimerization and DNA binding (Lu et al. 2008). In addition, further posttranslational modifications (PTMs) such as phosphorylation involving various kinases as well as sumoylation and deacetylation are required to initiate HSF1-mediated gene transcription (Akerfelt et al. 2010). Overexpression of Hsps has detrimental consequences for cells. Therefore, HSF activity is tightly regulated by a negative feedback mechanism via Hsps and acetylation via p300-cyclic AMP response element-binding protein (CREB)-binding protein (CBP) of critical lysine residues within the DNA-binding domain corrupting its DNA affinity (Shi and Mello 1998; Westerheide et al. 2009). On the other hand, deacetylation by the $NAD^+$-dependent histone deacetylase SIRT1 is a critical step in the activation of HSF1 and the increased expression of Hsp genes in the presence of cellular stressors. This direct regulatory function of SIRT1 on HSF1 activity obviously links the effects of many nutritional components capable of enhancing SIRT1 activity (*vide infra*) to an elevated expression of Hsps (Calabrese et al. 2011).

### 4.2.3 Nuclear Factor-κB

The activation of the transcription factor NF-κB is elicited by a countless number of exogenous stimuli ranging from viruses, bacteria, and parasites to UV irradiation and reactive oxygen, nitrogen, and other chemical species (Baldwin 1996). One of the first physiologically essential functions of NF-κB has been unraveled in the development of the immune system. By means of suitable knockout and transgenic mice models, it was shown that defects in the NF-κB gene family are associated with impairments in the immune response related to disruption in B and T cell maturation, function, and proliferation (Doi et al. 1997; Gerondakis et al. 1998; Bendall et al. 1999; Gugasyan et al. 2000; Caamano and Hunter 2002). Especially the NF-κB-mediated expression of antiapoptotic signals is important for prematured lymphocytes to survive the presence of tumor necrosis factor-α (TNF-α)-induced death signals and ripen to fully competent immune cells. In this context, the inhibition of NF-κB by exogenous agents may be seen as less preferable because a compromised host defense can be expected (Lavon et al. 2000). However, constitutive activation of NF-κB can ignite a self-amplifying inflammatory response (Figure 4.4b). The increased expression of proinflammatory mediators, such as cytokines, chemokines, adhesion molecules, matrix metalloproteinases, cyclooxygenases, and inducible nitric oxide (NO) synthase recruits and activates more inflammatory cells further fueling the production and release of proinflammatory mediators (Li and Verma 2002). If at a certain time NF-κB activation cannot appropriately be counteracted by endogenous terminating signals (*vide infra*), cells will inevitably die. Consequently, the inflamed site becomes necrotic, disrupting organ and body function and ultimately leading to a diseased state.

In the manifestation and progression of many chronic disorders, such as cardiovascular, lung, liver, and kidney diseases, a key role for NF-κB overactivation could be defined (Kumar et al. 2004; Wong and Tergaonkar 2009). Therefore, inhibition of a persistent activation of NF-κB may be an important target of exogenous compounds, both, drugs and dietary compounds, to prevent and or ameliorate those diseases. Indeed, a large body of epidemiological studies indicates that, for example, diets rich in fruits and vegetables are associated with a lower risk of all kinds of chronic inflammatory diseases of the different organ systems, including cancer (Renaud and de Lorgeril 1992; Hertog et al. 1993; Steinmetz and Potter 1996; Ness and Powles 1997). To establish cause–effect relationships between the intake of specific (groups of) plant-derived compounds and the prevention or improvement of diseases, a considerable number of clinical trials have been conducted during the last decades (Hooper et al. 2008). Clinical research efforts have been accompanied by the accumulation of data from in vitro and in vivo experiments unraveling the molecular interaction of individual phytochemicals with the NF-κB regulatory network (Gupta et al. 2010; Chen 2011; Salminen et al. 2012; Golan-Goldhirsh and Gopas 2013).

The activity of NF-κB is orchestrated by a complex network with negative feedback mechanisms enabling auto-controlled regulation. NF-κB consists of five subunits that all possess a Rel homology domain. Via the transactivation domains of RelA (p65), RelB, and Rel (c-Rel) interaction with further proteins such as TATA-binding protein, CBP, and TFIIB, E1A-binding protein 300 KD (p300) is facilitated; NF-κB1 (p105) and NF-κB2 (p100) are longer precursors of the active DNA-binding forms p50 and p52, respectively (DiDonato et al. 2012; Ghosh and Hayden 2012; Perkins 2012). In the inactive state, NF-κB resides as dimers such as p50/RelA in the cytosol bound to a family of NF-κB inhibitor proteins, that is, IκBs. NF-κB activation through classical (i.e., canonical) pathways is initiated by phosphorylation of IκBs by IκB kinase (IKK). Typical inducers of the canonical pathway include cytokines such as TNF-α and interleukin-1, bacterial products such as lipopolysaccharide (LPS) binding to and activating Toll-like receptors (TLRs), and cellular stressors such as hypoxia, ROS, and DNA damage. IKK-mediated phosphorylation of IκB prompts its ubiquitination and proteasomal degradation. As a consequence, the NF-κB dimer (p50/RelA) becomes free and translocates into the nucleus where it binds to NF-κB-binding sites in the promoter regions of its target genes, inducing their transcription. Once in the nucleus, NF-κB activity can be modulated by posttranslational phosphorylation, acetylation, and methylation of RelA (p65), which determines the strength and duration of NF-κB activity. Phosphorylation of RelA is achieved by cyclic adenosine monophosphate (cAMP)-PKA, casein kinase II (CKII), and IKK. Acetylation by p300/CBP possessing histone acetylase activity maintains RelA's nuclear localization and suppresses interaction with IκBα. NF-κB activation can be terminated by IκBα, which translocates to the nucleus, removes the NF-κB complex from its DNA-binding site, and facilitates its transfer back to the cytoplasm. The transcription of IκBs and proteins that negatively regulate IKK activation is under the control of NF-κB and becomes increasingly expressed on prolonged NF-κB activation. In this way, NF-κB can downregulate itself.

The noncanonical pathway is activated during lymphoid organ development by specific receptor signals (e.g., lymphotoxin B, B-cell-activating factor, and CD40). Receptor activation stimulates NF-κB inducing kinase, which phosphorylates IKKα, which in turn phosphorylates p100, leading to its partial proteolysis and liberation of the active p52/RelB dimer (Gilmore 2006). It is obvious that phytochemicals can intervene in this network at various stages (Figure 4.5), for example, by modulating the following:

- Upstream of IKK, for example, cAMP–PKA, extracellular signal-regulated kinase–p38–MAPK–CKII, oxidative stress–induced C-Jun N-terminal kinase 1 (JNK1), and various other kinase signaling pathways (Gupta BBA 2010)
- IKK activity through direct binding to a specific cysteine residue (Cys179)
- IκB through phosphorylation and degradation, but also nuclear translocation
- NF-κB at specific cysteine residue (Cys38) as well as phosphorylation and acetylation
- NF-κB translocation to the nucleus
- NF-κB DNA-binding activity
- NF-κB transcriptional activity

A vast number of plant-derived substances have been identified under experimental conditions to interfere with one or several of these processes (extensively reviewed by Gupta et al. [2010], Salminen et al. [2012], and Golan-Goldhirsh and Gopas [2013]). Most of them belong to the group of phenolic compounds, followed by isoprenoids and a few alkaloids (Golan-Goldhirsh and Gopas 2013).

Beyond the aforementioned ways, we recently discovered in our laboratory a further mechanism by which natural compounds inhibit NF-κB, that is, through inhibition of the nuclear enzyme poly(ADP-ribose)-polymerase (PARP)-1. PARP enzymes transfer ADP-ribose units from $NAD^+$ to nuclear proteins such as histones, topoisomerase I and II, and to itself, thereby altering their structure and activities. In 1999, it was shown that PARP serves as an important cofactor of NF-κB-mediated gene

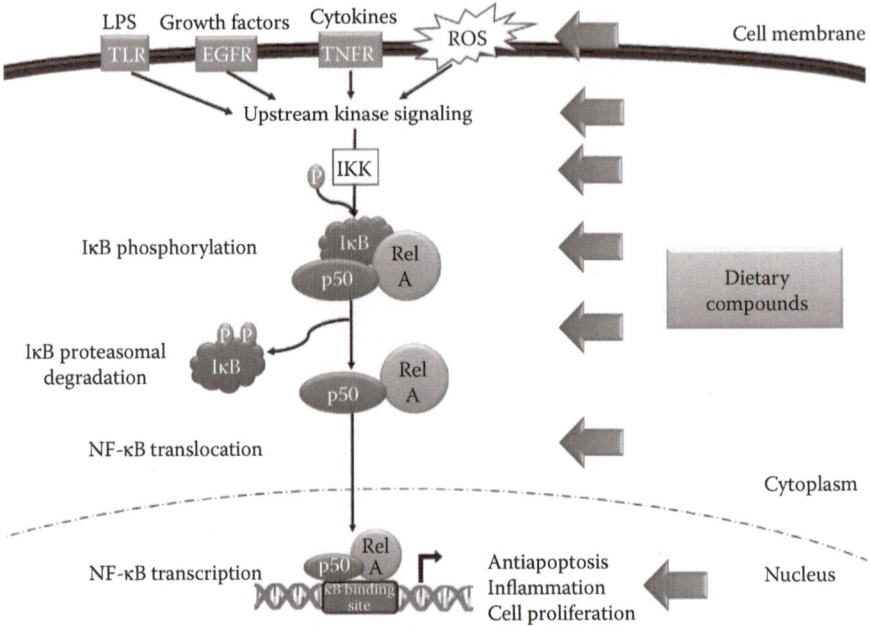

**FIGURE 4.5** In an inactive state, the transcription factor NF-κB resides as dimer (e.g., p50/RelA) in the cytosol bound to a family of NF-κB inhibitor proteins (IκBs). NF-κB activation is initiated by phosphorylation of IκBs by IKK. Typical IKK inducers are cytokines such as tumor necrosis factor-α (TNF-α) (via their receptors in the cell membrane, tumor necrosis factor receptor [TNFR]), bacterial products such as lipopolysaccharide (LPS) (via Toll-like receptors, TLRs), growth factors (via their receptors, epidermal growth factor receptor [EGFR]), and cellular stressors such as hypoxia and reactive oxygen species (ROS). IKK-mediated phosphorylation of IκB prompts its ubiquitination and proteasomal degradation. As a consequence, the NF-κB dimer (p50/RelA) becomes free and translocates into the nucleus, where it binds to NF-κB-binding sites in the promoter regions of its target genes, inducing their transcription. Dietary compounds have been shown to inhibit NF-κB activation at different levels by reduction of ROS levels, interference with kinase signaling upstream of IKK, decrease of IKK activity, interference with IκB and NF-κB at specific cysteine residues and with their phosphorylation and acetylation, inhibition of NF-κB translocation to the nucleus, decrease of NF-κB DNA-binding activity, and NF-κB transcriptional activity.

transcription (Hassa and Hottiger 1999). Accordingly, pharmaceutical inhibition of PARP-1 was revealed to significantly ameliorate the inflammatory state in animal models of endotoxic shock and pulmonary inflammation (Jagtap et al. 2002; Liaudet et al. 2002). In a screening of a broad range of dietary flavonoids, we identified among others myricetin, tricetin, quercetin, and fisetin, as well as theophylline and caffeine metabolites, as significant PARP-1 inhibitors in vitro (Moonen et al. 2005; Geraets et al. 2006, 2007a,b). In addition, we could demonstrate that the PARP-1 inhibiting flavonoids could significantly reduce pulmonary inflammation in mice (Geraets et al. 2009) and ex vivo LPS-induced cytokine production in the blood of healthy subjects as well as patients with chronic obstructive pulmonary disease and type 2 diabetes (Weseler et al. 2009). In contrast to synthetic PARP inhibitors being increasingly explored as drug candidates (Ekblad et al. 2013), the inhibitory action of plant-derived components has the advantage that PARP-1 activity is not fully blocked, thereby further granting vital DNA damage repair (Figure 4.4c).

Another remarkable effect of flavonoid-mediated NF-κB inhibition could be observed in human liposarcoma cells treated with the chemotherapeutic drug doxorubicin (Jacobs et al. 2011). A major drawback of doxorubicin is the resistance development of cancer cells toward this agent over time. One of the mechanisms responsible for this drug resistance is a doxorubicin-induced activation of NF-κB, leading to an increased expression of antiapoptotic genes. The inhibition of apoptosis let tumor cells survive even in the presence of such cytotoxic substances. Pretreatment of liposarcoma cells with semisynthetic flavonoid 7-mono-$O$-(β-hydroxyethyl)-rutoside significantly reduced doxorubicin-induced NF-κB activation, sensitizing tumor cells to apoptosis (Jacobs et al. 2011).

The regulatory cross talk between NF-κB, the deacetylase SIRT1, and FOXO transcription factors is also briefly addressed in the following paragraphs. In general, whereas SIRT 1 activates FOXO signaling, it inhibits NF-κB-mediated processes by deacetylation of the RelA subunit. The antagonistic effect of SIRT1 on both transcription factors appears to tune cellular systems toward an enhanced oxidative stress response (via activated FOXO signaling) and a suppressed inflammatory and atrophic response (via inhibited NF-κB signaling). This ability of SIRT1 has been related to longevity and protection against inflamm-aging (Salminen et al. 2008). It is easily conceivable that nutritional compounds that are able to activate SIRT1 will indirectly also attenuate NF-κB-mediated inflammatory processes.

Next to deacetylase, the modulation of the histone acetyltransferase (HAT) activity in the p300/CBP complex of NF-κB at the DNA-binding sites has been reported to be a target of several natural products including anacardic acid, curcumin, plumbagin, garcinol, gallic acid, and EGCG (Ghizzoni et al. 2011). Although most of these compounds have been proven to interfere also on other levels with NF-κB signaling, the inhibition of HATp300 was for all of them a crucial mechanism, resulting in downregulation of NF-κB-dependent gene expression.

Whereas an overwhelming amount of data from cell culture studies is available, indicating the interaction of many different dietary compounds at the various levels of the signaling processes of NF-κB activation eventually leading to its attenuation (Figure 4.5), clinical data is rather limited and less clearly proves similar actions in humans (Serafini et al. 2010). However, reduction in circulatory cytokines and chemokines supplemented with, for example, anthocyanidins (Karlsen et al. 2007), and monomeric and oligomeric proanthocyanidins (Weseler et al. 2011); flavanols such as quercetin (Boots et al. 2011); or $n$-3 polyunsaturated fatty acids (reviewed by Calder [2013]) suggests that the NF-κB inhibitory effects of nutritional components may indeed be of relevance under physiological conditions in humans.

### 4.2.4 SIRTUIN 1

Although the NAD$^+$-dependent histone deacetylases consist of seven SIRTs, currently most research has focused on elucidating the physiological and pathophysiological functions of SIRT1. The reason for this may be founded in its wide distribution in the body and the large spectrum of signaling pathways that are under its regulation. SIRT1 is involved in the control of the energy metabolism, stress response, and genomic stability and has been suggested as a key player in situations of energy restriction, aging, and cancer (Revollo and Li 2013). SIRT1 is located in the cytosol and the nucleus of brain, skeletal, muscle, heart, kidney, and uterus tissue (Villalba and Alcain 2012). As a deacetylase, SIRT1 removes acetyl groups from both histone and nonhistone proteins and transfers it to $NAD^+$, resulting in the formation of 2'-$O$-actetyl-ADP-ribose and nicotinamide (NAM). NAM is recycled back to $NAD^+$ through a salvage pathway in which NAM phosphoribosyltransferase first catalyzes the formation of NAM mononucleotide (NMN). In a second step, NMN adenyltransferase converts NMN back to $NAD^+$.

The activity of SIRT1 depends on the availability of $NAD^+$ and can be altered by the enzymes and intermediates of the $NAD^+$ salvage pathway. Therefore, it was suggested that the application of $NAD^+$ precursors such as NMN or NAM riboside as nutraceuticals may be an effective way to enhance systemic SIRT1 activity to prevent age-related loss of function of essential cell types such as pancreatic β-cells and neurons (Imai 2010). Indeed, preliminary data from animal models of impaired $NAD^+$ biosynthesis–induced pancreatic β-cell dysfunction indicate that NMN application is able to restore glucose-stimulated insulin secretion successfully (Ramsey et al. 2008; Imai 2010). Usually, the activity of SIRT1 is regulated on various subcellular levels by the availability of its substrates $NAD^+$ and acetylated proteins (*vide supra*), by chemical activators or inhibitors, by PTMs, by changes in its transcription factor–mediated gene expression, and by posttranscriptional events (Revollo and Li 2013) (Figure 4.6).

$NAD^+$ levels are not only dependent on de novo synthesis from tryptophan or recovery in the salvage pathway, but also on catalytic processes in which $NAD^+$ is used. PARPs are enzymes involved, for example, in DNA repair. Once activated, they can consume high amounts of $NAD^+$ while leading to poly(ADP-ribosyl)ation of target proteins, including themselves (Gibson and Kraus 2012; Canto et al. 2013). As a consequence, $NAD^+$ levels can drop so rapidly that cells may even fall into a deadly energy crisis (Pieper et al. 1999). Accordingly, inhibition of PARP-1, an isoform identified as a major cellular $NAD^+$ consumer, by gene knockout or pharmacological inhibition has been shown to not only restore $NAD^+$ levels but also SIRT1 activity in mice (Bai et al. 2011). In this context, the particularly interesting finding is that numerous methylxanthines and metabolites as well as flavonoids present in coffee and commonly ingested fruits and vegetables are efficient PARP inhibitors able to protect different cell types from PARP-induced $NAD^+$ depletion (Moonen et al. 2005; Geraets et al. 2006, 2007a,b). By this mechanism, it is likely that these nutritional components will also preserve SIRT1 activity.

At the protein level, SIRT1 deacetylase activity can be also modified by PTMs such as phosphorylation, methylation, sumoylation, nitrosylation, and protein–protein interactions (Revollo and Li 2013). Currently, most insights in the functional consequences of the multiple PTMs are available for phosphorylation events.

Various kinases have been revealed to phosphorylate SIRT1, such as JNK1. Under conditions of oxidative stress JNK1-mediated transfer of phosphate groups activated SIRT1-catalyzed deacetylation of histone H3, probably enhancing DNA integrity and making genes less accessible for damage by free radicals (Nasrin et al. 2009).

# Nutritional Components

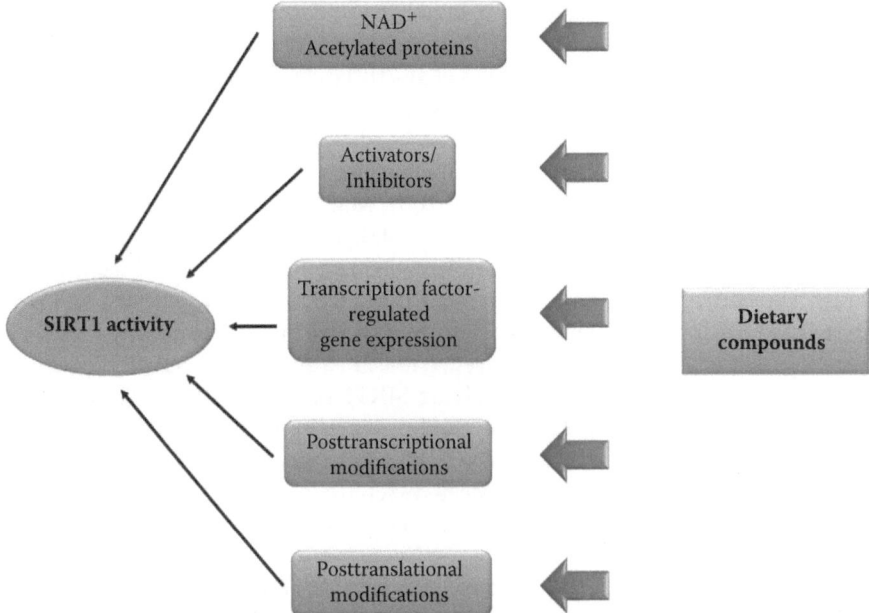

**FIGURE 4.6** The activity of the NAD$^+$-dependent histone deacetylase sirtuin 1 (SIRT1) is regulated at different levels, that is, by the availability of their substrates (acetylated proteins and NAD$^+$), by small molecules that activate or inhibit its activity, on gene expression level by the activity of transcription factors, for example, of the forkhead box (FOX) O family, by posttranscriptional via microRNAs and posttranslational modifications such as phosphorylation, methylation, nitrosylation, and sumoylation. Dietary compounds have been reported to interact directly or indirectly with the various regulatory levels, leading to alterations in SIRT1 activity.

In line with that finding, it was recently reported that activation of JNK1–SIRT1 signaling pathways is associated with protection of the heart against oxidative stress (Vinciguerra et al. 2012). Although short-term phosphorylation by JNK1 seems to increase SIRT1 activity, persistent JNK1 activation contributed to reduced SIRT1 activity by making it prone to ubiquitinylation and subsequent degradation (Gao et al. 2011). In view of the latter, an increasing number of in vitro and in vivo studies indicating that certain polyphenols can inhibit JNK1 signaling provide interesting possibilities to normalize situations of overactivated JNK activity with food-derived compounds (Dong 2000; Williams et al. 2004; Malemud 2007; Vauzour et al. 2010). Under genotoxic conditions, it was shown that phosphorylations by dual specificity tyrosine phosphorylation-regulated kinases and CKII increased the activity of SIRT1 toward acetylated p53 and the NF-κB subunit p65, respectively, and resulted in less apoptosis and increased cell survival (Kang et al. 2009; Dixit et al. 2012; Guo et al. 2012). Also, the cAMP–PKA pathway has been shown to stimulate SIRT1 phosphorylation in response to various stressors, leading to an increased oxidation of fatty acids and maintenance of energy homeostasis (Gerhart-Hines et al. 2011). Interestingly, food compounds such as caffeine, theophylline, and theobromine, as well as resveratrol and several flavonoids, are known to inhibit phosphodiesterases (PDEs), thereby blocking cAMP degradation to 5'-AMP and leading to elevated intracellular cAMP concentrations Park et al. 2012b, Rahimi et al. 2010 (refs). Hence, it may be conceivable that their ability to block PDE may also contribute via cAMP–PKA pathway to an activation of SIRT1 in a substrate- and gene-expression-independent manner.

In the last decennia, a couple of small chemicals have been reported as SIRT1 activators and inhibitors (Grozinger et al. 2001; Mai et al. 2005; Trapp et al. 2007; Alcain and Villalba 2009a,b; Freitag et al. 2011). The most prominent compound among the SIRT1 activators is resveratrol, a polyphenolic compound of the stilbene family. Typically, food sources of resveratrol are red wine, tea, peanuts, and soy (Burns et al. 2002). In a number of in vitro and in vivo studies it was observed that resveratrol extends the life span of yeast and mice in a similar way as caloric restriction does, that is, mechanistically involving SIRT1 activation (Howitz et al. 2003; Baur and Sinclair 2006; Barger et al. 2008). The efforts to unravel whether resveratrol directly or indirectly activates SIRT1 have generated some inconsistent findings and revealed that the link between resveratrol and SIRT1-mediated biological effects appears to be more complex and less clear than anticipated (Baur 2010; Hu et al. 2011).

The expression of the *SIRT1* gene is controlled by numerous transcription factors depending on the energy state. In response to fasting, CREB on activation via PKA and members of the FOXO transcription factor family induces the transcription of SIRT1. Conversely, in a high-energy state after feeding, the carbohydrate response/element-binding protein as well as the peroxisome proliferator-activated receptor γ (PPARγ) suppresses the transcription of the *SIRT1* (Nemoto et al. 2004; Han et al. 2010; Noriega et al. 2011). This unique consortium of transcription factors whose activity is at least partly influenced by the availability of carbohydrates and fats evince that macronutrients are also essentially involved in fine-tuning SIRT1 activity.

Posttranscriptionally, SIRT1 is affected by microRNAs (miRNAs), that is, short-strand, noncoding RNAs (ncRNAs) consisting of not more than approximately 22 nucleotides binding to mRNA and suppressing their translation. In various cell types, a distinct set of miRNAs was identified to regulate SIRT1 expression under different environmental conditions (Yamakuchi 2012). MiRNA-34a appears to be a key suppressor of SIRT1, though this finding is tissue-specific and becomes additionally complex through the observation that the expression of miRNAs is under transcriptional control of SIRT1 (Gao et al. 2010). In this context it is, however, interesting to note that dietary polyphenols are increasingly reported as natural modifiers of miRNA expression, although the physiological relevance cannot be estimated at the moment because of the current lack of in vivo and clinical data (Blade et al. 2013; Milenkovic et al. 2013).

Deacetylation catalyzed by SIRT1 is a key event in the regulation of the activity of several transcription factors, nuclear receptors, and enzymes such as Nrf2, NF-κB, FOXO, PPARγ, tumor suppressor p53, and endothelial NO synthase (Michan and Sinclair 2007; Yao and Rahman 2012). For example, SIRT1-mediated deacetylation of lysine residue 310 in the RelA–p65 protein complex of NF-κB inhibits its activation, which becomes potentiated in the presence of SIRT1-activating resveratrol

(Yeung et al. 2004). Likewise, the pathology of numerous chronic inflammatory and metabolic diseases as well as aging could be linked to decreased SIRT1 levels/ and activity (Yamamoto et al. 2007; Merksamer et al. 2013) and has initiated the search for effective SIRT activators (*vide supra*). On the other hand, in cancer SIRT1 overexpression is frequently found, increasing interest in the development of SIRT1 inhibitors (Alcain and Villalba 2009a,b; Villalba and Alcain 2012).

Although for most nutritional compounds it remains unclear whether they modulate SIRT1 activity directly or indirectly and which effects and mechanisms prevail under (patho)physiological conditions and in the presence of xenobiotic metabolism (Chung et al. 2010; Hu et al. 2011; Yao and Rahman 2012), it is likely that alterations in SIRT1 also contribute to the beneficial health outcomes for many of these compounds observed in humans. The role of SIRT1 in the defense and survival response to a multitude of environmental stressors turns the spotlight on this enzyme as a central molecular mediator in the ability to adapt.

### 4.2.5 Forkhead Box O Transcription Factors

Transcription factors of the FOXO family exert diverse effects in multiple signaling pathways involved in stress resistance, longevity, apoptosis, and atrophy (Greer and Brunet 2005). Their activity is stimuli dependently controlled via activation of different kinase signaling pathways. For example, in response to growth factors and insulin, the PI3K–Akt pathway becomes activated, leading to phosphorylation of specific FOXO sites that induce its degradation via the ubiquitin–proteasome pathway (Plas and Thompson 2003). Particular stressors such as oxidative stress, heat shock, and UV radiation are able to overcome Akt-mediated FOXO sequestration in the cytoplasm. By the activation of JNK1, these stimuli induce phosphorylation at different sites, which results in the translocation of FOXO into the nucleus, where it induces the expression of target genes. It is suggested that depending on the stressor, a specific battery of genes is induced which defines the fate of the cell, that is, cell-cycle arrest, stress resistance of apoptosis. (Greer and Brunet 2005).

Moreover, FOXO activity is also conferred by its acetylation state that is under control of SIRT1 (*vide supra*). SIRT1-mediated deacetylation of FOXO transcription factors differs between various FOXO types and does not necessarily comprise a feed-forward mechanism toward increased SIRT1 expression. Again, environmental stimuli impact decides whether SIRT1 deacetylation inhibits or activates, for example, FOXO3 (Brunet et al. 2004; Motta et al. 2004). In general, however, it appears that SIRT1 shifts FOXO's functional balance toward stress resistance rather than apoptosis, thereby extending the life span (Greer and Brunet 2005).

Next to the control of the expression of the *SIRT1*, FOXO3 also guards the expression of numerous important cellular defense enzymes. For example, it has been shown that SIRT1-catalyzed deacetylation of FOXO3 leads to an increased transcription of the DNA repair enzyme GADD45, the ROS detoxification enzymes catalase and MnSOD, and proteins involved in cell-cycle arrest (p27KIP1) and cell death (FasLigand and BIM) (Brunet et al. 1999; Medema et al. 2000; Dijkers et al. 2002; Kobayashi et al. 2005).

The translation of exogenous stressors into the expression of specific target genes coding for proteins that enable cells to cope to a certain extent with their changing environment also makes FOXO transcription factors important molecules that facilitate the cell's ability to adapt. Excitingly, a few recently published preliminary reports show that some polyphenols, such as EGCG, the major polyphenol in green tea; baicalein, a trihydroxyflavone; and resveratrol, can modulate FOXO activity in mammalian cells. First, it was suggested that the blood glucose–lowering effects of EGCG derive from its ability to mimic insulin-induced activation of the PI3K–Akt pathway by phosphorylation of FOXO (Anton et al. 2007). Also, in human skin fibroblasts, it was shown that EGCG modulates concentration-dependent FOXO transcription (Bartholome et al. 2010). Although at high concentrations (100 µM) EGCG induces insulin-like signaling through PI3K–Akt, leading to cytosolic degradation because of the production of hydrogen peroxide in the cell culture medium, at low concentrations (1 µM) it stimulates FOXO accumulation and DNA binding in the nucleus.

In human vascular endothelial cells, an intermediate concentration of 10 µM EGCG stimulated time-dependent phosphorylation of Akt, FOXO1, and AMP-activated protein kinase-$\alpha$. As a result, FOXO1 was kept in the cytosol and did not activate the expression of endothelin-1, which was found to possess a FOXO-binding domain in its promoter (Reiter et al. 2010). These findings were suggested to contribute to the cardiovascular benefits of EGCG observed in humans. The antitumor effects of EGCG and resveratrol on pancreatic and prostate cancer, respectively, could be linked in appropriate mice models to the ability of the compounds to arrest cell cycle, induce apoptosis, and reduce tumor growth and angiogenesis through regulation of FOXO transcription factors (Ganapathy et al. 2010; Shankar et al. 2013). Finally, the flavone baicalein was also reported to modulate FOXO transcription through phosphorylation via PI3K–Akt and acetylation via CBP and SIRT1 interactions (Kim et al. 2012). Moreover, baicalein could protect mice against irradiation-induced damage by attenuating NF-κB response and upregulating FOXO as well as catalase and SOD activities (Lee et al. 2011).

Although the exploration of the effects of nutritional components on Akt-FOXO signaling is still in its infancy, these examples show that modulation of FOXO under experimental conditions can be achieved by dietary polyphenols.

### 4.2.6 Metabolic Processes

Organisms are continuously exposed to potentially harmful exogenous chemical stimuli. These include many xenobiotics, such as drugs and dietary components. Drug-metabolizing enzymes are crucial in the biotransformation, elimination, and/or detoxification of xenobiotics. The biotransformation mechanisms are widely present in various organs and include the so-called phase I and phase II drug-metabolizing enzymes. These enzymes render xenobiotics more water-soluble, thus promoting their renal excretion. Uptake and efflux transporters are indicated as phase 0 and phase III mechanisms, respectively (Xu et al. 2005) (Figure 4.1).

The prime example of a phase I drug-metabolizing enzyme is cytochrome P450. In fact, it is a superfamily of heme-containing enzymes suggested to enable aerobic organisms to survive the ultimate toxic response early in evolution, that is, during the emergence of oxygen in the atmosphere of the earth. Initial life forms were not adapted to the presence of oxygen, which slowly appeared in the atmosphere. Cytochrome P450 has been suggested to be an ancient enzyme that could offer protection against reactive oxygen (Bast 1986), thereby conferring a very early form of adaptation. During evolution, these heme proteins began to use oxygen in several modes of functioning, namely as monooxygenase, oxidase, or peroxidase. The common presence of cytochrome P450 on the phylogenetic scale (Nebert et al. 1991) and in various cellular compartments (Neve and Ingelman-Sundberg 2010) seems indicative of an early role in evolution.

Phase II conjugation reactions are facilitated by many superfamilies of enzymes, including UGT, sulfotransferases, GSTs, N-acetyltransferases, and epoxide hydrolases. In many cases (but certainly not always), the functional group incorporated into the molecule (by phase I) is presented as an anchor for the conjugation (phase II) and excretion of the compound increases. Interplay between phase I and II reactions thus occurs (Bast and Haenen 1984).

Phase III transporters such as the ATP-binding cassette transporters P-glycoprotein (Pgp) and multidrug resistance–associated protein as well as ATP-independent organic anion–transporting polypeptides (OATPs) such as OATP2 can be found in various tissues. These transporters form a barrier against xenobiotic penetration in these tissues. They are, for example, expressed on the brush-border membrane of the intestine and are involved in excretion of xenobiotics in the lumen, preventing the absorption of xenobiotics (Dean et al. 2001). Nutritional compounds influence phase I and phase II enzymes as well as efflux transporters in many ways.

### 4.2.6.1 Phase I Metabolism

The transcription factor aryl hydrocarbon receptor (AhR) binds, among others, nutritional compounds, which increases the affinity for DNA (Long et al. 2012) and leads to an induction of CYP1 family genes. On the other hand, inhibition of the metabolism can also occur by dietary compounds (Harris et al. 2003). Other nuclear receptors regulated by dietary factors are the pregnane X receptor (PXR), PPAR, retinoid X receptor (RXR), liver X receptors (LXR), and farnesoid X receptor (Murray 2007), which all affect different isoforms of the cytochrome P450 enzyme family. Examples are manifold: polyphenolics modulating PXR (Kluth et al. 2007), vitamin E activating PXR (Cho et al. 2009), retinoic acid (D'Ambrosio et al. 2011) modulating RXR, and also via heterodimer formation PPARs and LXR, thus influencing a wide variety of response elements on DNA (Combs 2008).

### 4.2.6.2 Phase II Metabolism

Glutathione-S-transferases can be found in cytosolic, mitochondrial, and microsomal compartments. They are crucial in detoxification processes and are implicated in preventing cancer development (Hayes et al. 2005). The promoters of cytosolic and microsomal glutathione transferase genes contain AREs through which they are transcriptionally activated during exposure to Michael reaction acceptors and oxidative stress (Saeidnia and Abdollahi 2013). Interestingly, direct interaction between nutrients such as vitamin E and curcumin and glutathione transferase has also been described (Van Haaften et al. 2001; Hayeshi et al. 2007), but this inhibition has surprisingly not received much attention (van Haaften et al. 2003).

The phytochemical regulation of UGTs has been reviewed in depth (Saracino and Lampe 2007). Earlier described target genes containing xenobiotic response elements (XREs), AREs, and PXR response elements (PXREs) in their promoters are induced via AhR, Nrf2, and PXR, respectively. The target genes differentially induce various isoforms of UGT. The expression and function of UGT is modulated not only by gene regulation but also by PTMs (Ishii et al. 2010).

#### 4.2.6.3 Phase III Transporters and Cross Talk

Pgp is usually coexpressed and coinduced with the CYP3A isoform of cytochrome P450 and is regulated by PXR (Xu et al. 2005). PXR is also involved in inducing OATP2. Our understanding of the cross talk between the signaling pathways is developing. A link has been proposed between the AhR and *Nrf2* gene. *Nrf2* could serve as an AhR target gene and might be activated by (the AhR-regulated) CYP1A1-generated electrophiles (Kohle and Bock 2007). We increasingly recognize that the nuclear receptor control of gene transactivation is also under the influence of dietary factors. Hormones are also part of the picture, for example, corticosteroids control AhR target genes. The corticosteroid dexamethasone enhances the induction of specific isoenzymes of cytochrome P450 and of GST (Xiao et al. 1995). In recent studies on the effect of corticosteroids on the phase II enzyme GST, it was found that PXR interacts with factors binding to the ARE, which elicits the PXR-mediated response of the GST isoform GSTA2 (Falkner et al. 2001). A suppression of the cellular antioxidant defense capacity by glucocorticoids has also been observed (Kratschmar et al. 2012).

On the other hand, antioxidant dietary compounds have been reported to protect the oxidatively damaged corticosteroid response by preventing disruption of histone deacetylase 2 (HDAC2), an enzyme important in mediating corticosteroid response (Barnes 2009). Thus, maintaining the corticosteroid response preserves the adaptive response of the organism to stress.

### 4.2.7 Epigenetic Mechanisms

Already, the impressive data and follow-up studies from people of the Dutch famine in the last year of World War II highlight the essential role of nutrients for the health of pregnant mothers and their siblings even two generations later (Lumey and Van Poppel 1994; Roseboom et al. 2006; Painter et al. 2008; Veenendaal et al. 2013). At the latest, experiments by Waterland and Jirtle (2003) with transgenic Agouti mice elegantly illustrate that dietary components are capable to modulate health and disease via mechanisms acting *epi*, that is, on top, of the genome. Epigenetic processes lead to inheritable, though reversible, changes in gene expression without any alterations in the DNA sequence. Three fundamental mechanisms have been identified by which this phenomenon is achieved: (1) DNA methylation, (2) histone modifications, and (3) ncRNAs (Hamilton 2011).

DNA methylation takes place at cytosine-guanine (CpG) dinucleotides. The transfer of a methyl group to the C5 position of the pyrimidine ring of cytosine is enzymatically catalyzed by DNA methyltransferases (DNMTs) by using *S*-adenosyl methionine (SAM) as a methyl group donor. Two types of DNMTs are functionally well-described. DNMT1 is responsible for the maintenance of CpG methylation during cell division whereas DNMT3A/3B transfers novel methyl residues during cell development and differentiation (Jaenisch and Bird 2003). The majority (i.e., 60%–90%) of single CpGs spread throughout the genome are methylated, whereas CpG-rich regions, called CpG islands, at the 5'-end of many genes are often found unmethylated. CpG methylation is generally associated with a decreased transcriptional activity. This is achieved by the binding of methyl-CpG-binding domain proteins, which in turn attract chromatin-modifying proteins such as HDACs, leading to the formation of a compact, transcriptionally inactive chromatin, that is, heterochromatin. This is, for example, the case in CpG-rich promoter regions of oncogenes, where hypermethylation leads to a stable silencing of the expression of these genes. Loss of methylation, however, has been identified with its transcriptional activation and reported as an early event in the development of cancer (Robertson 2005; Ptak and Petronis 2008). Demethylation of CpGs can occur either passively during replication in the absence of a functional DNA methylation machinery or actively under involvement of various enzymes and the formation of 5-hydroxymethylcytosine as an intermediate (Tammen et al. 2013).

Histones are basic globular proteins around which DNA is wrapped and packed as chromatin in the nucleus. When the structure of the chromatin is tight and condensed it is termed heterochromatin, while when it is loose and accessible for the DNA transcription machinery, it is called euchromatin. The chromatin's density is determined by PTMs of lysine, arginine, serine, and threonine residues in the histone tails including acetylation, phosphorylation, sumoylation, ubiquitinylation, and glycosylation (Kouzarides 2007). The currently most intensively studied histone modifications are acetylation and deacetylation catalyzed by HATs and HDACs, respectively.

Finally, ncRNAs have been identified as regulators of chromatin architecture and retain the expression of epigenetic enzymes such as DNMTs, HDACs, SIRT, and polycomb proteins at the mRNA level (Chuang and Jones 2007). As their name suggests, ncRNAs do not code for proteins. They form a heterogenous group of RNAs of different lengths, such as miRNAs, tinyRNAs, and long-ncRNAs (lnc-RNAs), all exerting a regulatory effect in gene expression, RNA splicing, translation and turnover, and in other yet-to-be-discovered processes (Chuang and Jones 2007). A critical factor for the activity of DNA and histone-modifying enzymes is the availability of their cosubstrates that are intermediates of the energy metabolism, such as acetyl-CoA, NAD$^+$, and SAM (Vanden Berghe 2012).

To maintain the methyl group pool and eventually sufficient SAM levels, folic acid, choline, betaine, methionine, and vitamin B12 are required for the recycling and novel synthesis of SAM. Modifications in the intake of these nutrients was revealed to change genomic DNA methylation states. For example, folate and vitamin B deficiencies were associated with gene-specific DNA hypomethylation in rodents (Kim et al. 1997; Liu et al. 2007). On the other hand, Waterland and Jirtle (2003) showed in the viable yellow agouti ($A^{vy}$) mice that a mothers' supplementation with extra folic acid, vitamin B12, choline, and betaine increased CpG methylation in the offspring, resulting in inheritably phenotypic changes.

Other nutritional components have also been reported to modify epigenetic processes. Isoflavonoids such as genistein as well as EGCG appeared to inhibit the activity of DNMTs that was associated with hypomethylation in the promoter regions of specific genes and reactivation of their transcription in various cancer cell lines (Fang et al. 2005; Lee et al. 2005; Jha et al. 2010; Wang and Chen 2010; Nandakumar et al. 2011). Moreover, it was suggested that dietary polyphenols may influence indirectly DNA methylation by mildly reducing SAM levels, because they are extensively metabolized by catechol-$O$-methyltransferase into methylated metabolites (Fang et al. 2007). In humans, however, changes in DNA methylation on supplementation of nutrients provided less clear effects until now. Although, for example, an 8-week supplementation with 5 mg folic acid led to an increase in folate and SAM and a reduction in $S$-adenosylhomocysteine concentrations in blood, DNA methylation in peripheral mononuclear cells remained unaffected (Pizzolo et al. 2011).

Apigenin and various other flavonoids, curcumin, lycopene, and sulforaphane were also reported to inhibit DNMT, yet somewhat more weakly than genistein and EGCG (reviewed, e.g., by Park et al. [2012], and Vanden Berghe [2012] with references therein). Dietary polyphenols such as curcumin, resveratrol, and several flavonoids like quercetin and catechins are also able to modulate the activity of HATs and HDACs including SIRT (reviewed, e.g., by Park et al. [2012] and Vanden Berghe [2012]). Deacetylation of nonhistone proteins by SIRT1 as a crucial event in the regulation of metabolic processes, stress resistance, cell survival, and attenuation of inflamm-aging has been described earlier (*vide supra*). SIRT-mediated deacetylation of specific acetyl groups in lysine residues of histone proteins was recently reported (Vaquero et al. 2007) and a fundamental role in the maintenance of intact telomeric chromatin being essential for healthy ageing was suggested (Dang et al. 2009). It is thus conceivable that SIRT1 modulating nutritional components will also affect this epigenetic mechanism. Finally, miRNAs directed to specific epigenetic modifier enzymes have also been shown to be modulated by dietary polyphenols, thereby indirectly altering epigenetic processes (Milenkovic et al. 2012; Parasramka et al. 2012).

It becomes obvious that diet has a critical impact on epigenetic mechanisms. Remarkably, already in the earliest stages of our lives, that is, in utero, exposure to nutrients via our mother may have tremendous consequences for our health state and disease susceptibility, respectively, later in life. Almost all of the chronic diseases (including cancer) emerging in the course of lifetime are multifactorial and are substantially influenced by the external environment. Epigenetic mechanisms can remember these environmental stimuli (Feinberg 2008) by translating them into stable chemical modifications that result in gene expression changes and eventually altered body functions. Hence, disrupted epigenetic regulations are increasingly acknowledged as important pathophysiological mechanisms underlying these diseases and attract attention as novel molecular therapeutic targets. Next to the development of epigenetic drugs, research in the field of nutritional epigenetics is of greatest value. Currently still in its infancy, it challenges scientists with many unsolved questions, in particular, concerning practical applications and public health implications (Park et al. 2012). However, a targeted modulation of epigenetic mechanisms by nutritional components is an easily implementable measure to modulate the activity of genes for an optimized adaptation of cells to their ever-changing environment and thereby eventually maintaining an individuals' health.

## 4.3 CONCLUSIONS

During evolution we have developed strategies to survive. Fluctuating conditions necessitated our physiology to find ways to cope with changes. An intricate physiological network of regulatory pathways allows us to maintain health and prevent diseases. Apparently, with meticulously tuned stress response and survival pathways we can keep the physiological parameters within a certain range. We adapt our biochemical grid and thus preclude derangement.

The diet is in itself one of the main environmental factors we are exposed to. We need a continuous adaptation to this potentially toxic dietary exposure, that is, to prevent diet-induced inflammation, accumulation, and toxic insults of dietary agents. This is achieved by adaptive biochemical physiological responses to nutritional components. Increased understanding of how these mechanisms function on a molecular level will enable us to control these processes more specifically and to optimize the use of these subtle multitarget changes to our benefits. With adaptive responses induced by our diet, we are becoming able to improve human health.

## REFERENCES

Akerfelt, M., R. I. Morimoto, and L. Sistonen. 2010. "Heat shock factors: Integrators of cell stress, development and lifespan." *Nat Rev Mol Cell Biol* 11(8): 545–55. doi: 10.1038/nrm2938.

Albert, A. 1985. *Selective Toxicity: The Physico-chemical Basis of Therapy*, 7th ed. Chapman & Hall.

Alcain, F. J., and J. M. Villalba. 2009a. "Sirtuin activators." *Expert Opin Ther Pat* 19(4): 403–14. doi: 10.1517/13543770902762893.

Alcain, F. J., and J. M. Villalba. 2009b. "Sirtuin inhibitors." *Expert Opin Ther Pat* 19(3): 283–94. doi: 10.1517/13543770902755111.

Anonymous. 2009. "What is health? The ability to adapt." *Lancet* 373: 781.

Anton, S., L. Melville, and G. Rena. 2007. "Epigallocatechin gallate (EGCG) mimics insulin action on the transcription factor FOXO1a and elicits cellular responses in the presence and absence of insulin." *Cell Signal* 19(2): 378–83. doi: 10.1016/j.cellsig.2006.07.008.

Bai, P., C. Canto, H. Oudart, A. Brunyanszki, Y. Cen, C. Thomas, H. Yamamoto et al. 2011. "PARP-1 inhibition increases mitochondrial metabolism through SIRT1 activation." *Cell Metab* 13(4): 461–8. doi: 10.1016/j.cmet.2011.03.004.

Baird, L., and A. T. Dinkova-Kostova. 2011. "The cytoprotective role of the Keap1–Nrf2 pathway." *Arch Toxicol* 85(4): 241–72. doi: 10.1007/s00204-011-0674-5.

Baldwin, A. S. Jr. 1996. "The NF-kappa B and I kappa B proteins: New discoveries and insights." *Annu Rev Immunol* 14: 649–83. doi: 10.1146/annurev.immunol.14.1.649.

Barger, J. L., T. Kayo, J. M. Vann, E. B. Arias, J. Wang, T. A. Hacker, Y. Wang et al. 2008. "A low dose of dietary resveratrol partially mimics caloric restriction and retards aging parameters in mice." *PLoS One* 3(6): e2264. doi: 10.1371/journal.pone.0002264.

Barnes, P. J. 2009. "Histone deacetylase-2 and airway disease." *Ther Adv Respir Dis* 3(5): 235–43. doi: 10.1177/1753465809348648.

Bartholome, A., A. Kampkotter, S. Tanner, H. Sies, and L. O. Klotz. 2010. "Epigallocatechin gallate-induced modulation of FoxO signaling in mammalian cells and *C. elegans*: FoxO stimulation is masked via PI3K/Akt activation by hydrogen peroxide formed in cell culture." *Arch Biochem Biophys* 501(1): 58–64. doi: 10.1016/j.abb.2010.05.024.

Bast, A. 1986. "Is formation of reactive oxygen by cytochrome P-450 perilous and predictable?" *Trends Pharmacol Sci* 7: 266–70.

Bast, A., and G. R. Haenen. 2013. "Ten misconceptions about antioxidants." *Trends Pharmacol Sci* 34(8):430–6. doi: 10.1016/j.tips.2013.05.010.

Bast, A., and G. R. M. M. Haenen. 1984. "Cytochrome P-450 and glutathione: What is the significance of their interrelationship in lipid peroxidation?" *Trends Biochem Sci* 9(12): 510–3.

Baur, J. A. 2010. "Resveratrol, sirtuins, and the promise of a DR mimetic." *Mech Ageing Dev* 131(4): 261–9. doi: 10.1016/j.mad.2010.02.007.

Baur, J. A., and D. A. Sinclair. 2006. "Therapeutic potential of resveratrol: The in vivo evidence." *Nat Rev Drug Discov* 5(6): 493–506. doi: 10.1038/nrd2060.

Bendall, H. H., M. L. Sikes, D. W. Ballard, and E. M. Oltz. 1999. "An intact NF-kappa B signaling pathway is required for maintenance of mature B cell subsets." *Mol Immunol* 36(3): 187–95.

Biesalski, H. K., T. Grune, J. Tinz, I. Zollner, and J. B. Blumberg. 2010. "Reexamination of a meta-analysis of the effect of antioxidant supplementation on mortality and health in randomized trials." *Nutrients* 2(9): 929–49. doi: 10.3390/nu2090929.

Bindoli, A., and M. P. Rigobello. 2013. "Principles in redox signaling: From chemistry to functional significance." *Antioxid Redox Signal* 18(13): 1557–93. doi: 10.1089/ars.2012.4655.

Birringer, M. 2011. "Hormetics: Dietary triggers of an adaptive stress response." *Pharm Res* 28(11): 2680–94. doi: 10.1007/s11095-011-0551-1.

Blade, C., L. Baselga-Escudero, M. J. Salvado, and A. Arola-Arnal. 2013. "miRNAs, polyphenols, and chronic disease." *Mol Nutr Food Res* 57(1): 58–70. doi: 10.1002/mnfr.201200454.

Boots, A. W., M. Drent, V. C. de Boer, A. Bast, and G. R. Haenen. 2011. "Quercetin reduces markers of oxidative stress and inflammation in sarcoidosis." *Clin Nutr* 30(4): 506–12. doi: 10.1016/j.clnu.2011.01.010.

Boots, A. W., G. R. Haenen, and A. Bast. 2008. "Health effects of quercetin: From antioxidant to nutraceutical." *Eur J Pharmacol* 585(2–3): 325–37. doi: 10.1016/j.ejphar.2008.03.008.

Brigelius-Flohe, R., and A. Banning. 2006. "Part of the series: From dietary antioxidants to regulators in cellular signaling and gene regulation. Sulforaphane and selenium, partners in adaptive response and prevention of cancer." *Free Radic Res* 40(8): 775–87. doi: 10.1080/10715760600722643.

Brunet, A., A. Bonni, M. J. Zigmond, M. Z. Lin, P. Juo, L. S. Hu, M. J. Anderson, K. C. Arden, J. Blenis, and M. E. Greenberg. 1999. "Akt promotes cell survival by phosphorylating and inhibiting a Forkhead transcription factor." *Cell* 96(6): 857–68.

Brunet, A., L. B. Sweeney, J. F. Sturgill, K. F. Chua, P. L. Greer, Y. Lin, H. Tran et al. 2004. "Stress-dependent regulation of FOXO transcription factors by the SIRT1 deacetylase." *Science* 303(5666): 2011–5. doi: 10.1126/science.1094637.

Caamano, J., and C. A. Hunter. 2002. "NF-kappaB family of transcription factors: Central regulators of innate and adaptive immune functions." *Clin Microbiol Rev* 15(3): 414–29.

Calabrese, V., C. Cornelius, S. Cuzzocrea, I. Iavicoli, E. Rizzarelli, and E. J. Calabrese. 2011. "Hormesis, cellular stress response and vitagenes as critical determinants in aging and longevity." *Mol Aspects Med* 32(4–6): 279–304. doi: 10.1016/j.mam.2011.10.007.

Calder, P. C. 2013. "Omega-3 polyunsaturated fatty acids and inflammatory processes: Nutrition or pharmacology?" *Br J Clin Pharmacol* 75(3): 645–62. doi: 10.1111/j.1365-2125.2012.04374.x.

Canto, C., A. A. Sauve, and P. Bai. 2013. "Crosstalk between poly(ADP-ribose) polymerase and sirtuin enzymes." *Mol Aspects Med.* 34(6):1168–201. doi: 10.1016/j.mam.2013.01.004.

Carnelio, S., S. A. Khan, and G. Rodrigues. 2008. "Definite, probable or dubious: Antioxidants trilogy in clinical dentistry." *Br Dent J* 204(1): 29–32. doi: 10.1038/bdj.2007.1186.

Carnelio, S., S. A. Khan, and G. S. Rodrigues. 2007. "Free radicals and antioxidant therapy in clinical practice: To be or not to be?" *J Coll Physicians Surg Pak* 17(3): 173–4. doi: 03.2007/JCPSP.173174.

Chen, S. 2011. "Natural products triggering biological targets—a review of the anti-inflammatory phytochemicals targeting the arachidonic acid pathway in allergy asthma and rheumatoid arthritis." *Curr Drug Targets* 12(3): 288–301.

Cho, J. Y., D. W. Kang, X. Ma, S. H. Ahn, K. W. Krausz, H. Luecke, J. R. Idle, and F. J. Gonzalez. 2009. "Metabolomics reveals a novel vitamin E metabolite and attenuated vitamin E metabolism upon PXR activation." *J Lipid Res* 50(5): 924–37. doi: 10.1194/jlr.M800647-JLR200.

Chuang, J. C., and P. A. Jones. 2007. "Epigenetics and microRNAs." *Pediatr Res* 61(5 Pt 2): 24R–29R. doi: 10.1203/pdr.0b013e3180457684.

Chung, S., H. Yao, S. Caito, J. W. Hwang, G. Arunachalam, and I. Rahman. 2010. "Regulation of SIRT1 in cellular functions: Role of polyphenols." *Arch Biochem Biophys* 501(1): 79–90. doi: 10.1016/j.abb.2010.05.003.

Combs, G. F. Jr. 2008. *The Vitamins: Fundamental Aspects in Nutrition and Health*, 3rd ed. Elsevier Academic Press.

D'Ambrosio, D. N., R. D. Clugston, and W. S. Blaner. 2011. "Vitamin A metabolism: An update." *Nutrients* 3(1): 63–103. doi: 10.3390/nu3010063.

Dang, W., K. K. Steffen, R. Perry, J. A. Dorsey, F. B. Johnson, A. Shilatifard, M. Kaeberlein, B. K. Kennedy, and S. L. Berger. 2009. "Histone H4 lysine 16 acetylation regulates cellular lifespan." *Nature* 459(7248): 802–7. doi: 10.1038/nature08085.

Dean, M., Y. Hamon, and G. Chimini. 2001. "The human ATP-binding cassette (ABC) transporter superfamily." *J Lipid Res* 42(7): 1007–17.

DiDonato, J. A., F. Mercurio, and M. Karin. 2012. "NF-kappaB and the link between inflammation and cancer." *Immunol Rev* 246(1): 379–400. doi: 10.1111/j.1600-065X.2012.01099.x.

Dijkers, P. F., K. U. Birkenkamp, E. W. Lam, N. S. Thomas, J. W. Lammers, L. Koenderman, and P. J. Coffer. 2002. "FKHR-L1 can act as a critical effector of cell death induced by cytokine withdrawal: Protein kinase B-enhanced cell survival through maintenance of mitochondrial integrity." *J Cell Biol* 156(3): 531–42. doi: 10.1083/jcb.200108084.

Dinkova-Kostova, A. T., W. D. Holtzclaw, R. N. Cole, K. Itoh, N. Wakabayashi, Y. Katoh, M. Yamamoto, and P. Talalay. 2002. "Direct evidence that sulfhydryl groups of Keap1 are the sensors regulating induction of phase 2 enzymes that protect against carcinogens and oxidants." *Proc Natl Acad Sci USA* 99(18): 11908–13. doi: 10.1073/pnas.172398899.

Dixit, D., V. Sharma, S. Ghosh, V. S. Mehta, and E. Sen. 2012. "Inhibition of Casein kinase-2 induces p53-dependent cell cycle arrest and sensitizes glioblastoma cells to tumor necrosis factor (TNFalpha)-induced apoptosis through SIRT1 inhibition." *Cell Death Dis* 3: e271. doi: 10.1038/cddis.2012.10.

Doi, T. S., T. Takahashi, O. Taguchi, T. Azuma, and Y. Obata. 1997. "NF-kappa B RelA-deficient lymphocytes: Normal development of T cells and B cells, impaired production of IgA and IgG1 and reduced proliferative responses." *J Exp Med* 185(5): 953–61.

Dong, Z. 2000. "Effects of food factors on signal transduction pathways." *Biofactors* 12(1–4): 17–28.

Ehrlich, P. 1909. "Über den jetzigen Stand der Chemotherapie." *Ber Dtsch Chem Ges* 42: 17–47.

Ekblad, T., E. Camaioni, H. Schuler, and A. Macchiarulo. 2013. "PARP inhibitors: Polypharmacology versus selective inhibition." 280(15):3563–75. *FEBS J*. doi: 10.1111/febs.12298.

Falkner, K. C., J. A. Pinaire, G. H. Xiao, T. E. Geoghegan, and R. A. Prough. 2001. "Regulation of the rat glutathione S-transferase A2 gene by glucocorticoids: Involvement of both the glucocorticoid and pregnane X receptors." *Mol Pharmacol* 60(3): 611–9.

Fang, M., D. Chen, and C. S. Yang. 2007. "Dietary polyphenols may affect DNA methylation." *J Nutr* 137(Suppl 1): 223S–228S.

Fang, M. Z., D. Chen, Y. Sun, Z. Jin, J. K. Christman, and C. S. Yang. 2005. "Reversal of hypermethylation and reactivation of *p16INK4a*, *RARbeta*, and *MGMT* genes by genistein and other isoflavones from soy." *Clin Cancer Res* 11(19 Pt 1): 7033–41. doi: 10.1158/1078-0432.CCR-05-0406.

Feinberg, A. P. 2008. "Epigenetics at the epicenter of modern medicine." *JAMA* 299(11): 1345–50. doi: 10.1001/jama.299.11.1345.

Forman, H. J., K. J. Davies, and F. Ursini. 2013. "How do nutritional antioxidants really work: Nucleophilic tone and para-hormesis versus free radical scavenging in vivo." *Free Radic Biol Med.* 66:24–35 doi: 10.1016/j.freeradbiomed.2013.05.045.

Freitag, M., J. Schemies, T. Larsen, K. El Gaghlab, F. Schulz, T. Rumpf, M. Jung, and A. Link. 2011. "Synthesis and biological activity of splitomicin analogs targeted at human NAD(+)-dependent histone deacetylases (sirtuins)." *Bioorg Med Chem* 19(12): 3669–77. doi: 10.1016/j.bmc.2011.01.026.

Ganapathy, S., Q. Chen, K. P. Singh, S. Shankar, and R. K. Srivastava. 2010. "Resveratrol enhances antitumor activity of TRAIL in prostate cancer xenografts through activation of FOXO transcription factor." *PLoS One* 5(12): e15627. doi: 10.1371/journal.pone.0015627.

Gao, J., W. Y. Wang, Y. W. Mao, J. Graff, J. S. Guan, L. Pan, G. Mak, D. Kim, S. C. Su, and L. H. Tsai. 2010. "A novel pathway regulates memory and plasticity via SIRT1 and miR-134." *Nature* 466(7310): 1105–9. doi: 10.1038/nature09271.

Gao, Z., J. Zhang, I. Kheterpal, N. Kennedy, R. J. Davis, and J. Ye. 2011. "Sirtuin 1 (SIRT1) protein degradation in response to persistent c-Jun N-terminal kinase 1 (JNK1) activation contributes to hepatic steatosis in obesity." *J Biol Chem* 286(25): 22227–34. doi: 10.1074/jbc.M111.228874.

Geraets, L., A. Haegens, K. Brauers, J. A. Haydock, J. H. Vernooy, E. F. Wouters, A. Bast, and G. J. Hageman. 2009. "Inhibition of LPS-induced pulmonary inflammation by specific flavonoids." *Biochem Biophys Res Commun* 382(3): 598–603. doi: 10.1016/j.bbrc.2009.03.071.

Geraets, L., H. J. Moonen, K. Brauers, R. W. Gottschalk, E. F. Wouters, A. Bast, and G. J. Hageman. 2007a. "Flavone as PARP-1 inhibitor: Its effect on lipopolysaccharide induced gene-expression." *Eur J Pharmacol* 573(1–3): 241–8. doi: 10.1016/j.ejphar.2007.07.013.

Geraets, L., H. J. Moonen, K. Brauers, E. F. Wouters, A. Bast, and G. J. Hageman. 2007b. "Dietary flavones and flavonoles are inhibitors of poly(ADP-ribose)polymerase-1 in pulmonary epithelial cells." *J Nutr* 137(10): 2190–5.

Geraets, L., H. J. Moonen, E. F. Wouters, A. Bast, and G. J. Hageman. 2006. "Caffeine metabolites are inhibitors of the nuclear enzyme poly(ADP-ribose)polymerase-1 at physiological concentrations." *Biochem Pharmacol* 72(7): 902–10. doi: 10.1016/j.bcp.2006.06.023.

Gerhart-Hines, Z., J. E. Dominy Jr., S. M. Blattler, M. P. Jedrychowski, A. S. Banks, J. H. Lim, H. Chim, S. P. Gygi, and P. Puigserver. 2011. "The cAMP/PKA pathway rapidly activates SIRT1 to promote fatty acid oxidation independently of changes in NAD(+)." *Mol Cell* 44(6): 851–63. doi: 10.1016/j.molcel.2011.12.005.

Gerondakis, S., R. Grumont, I. Rourke, and M. Grossmann. 1998. "The regulation and roles of Rel/NF-kappa B transcription factors during lymphocyte activation." *Curr Opin Immunol* 10(3): 353–9.

Ghizzoni, M., H. J. Haisma, H. Maarsingh, and F. J. Dekker. 2011. "Histone acetyltransferases are crucial regulators in NF-kappaB mediated inflammation." *Drug Discov Today* 16(11–12): 504–11. doi: 10.1016/j.drudis.2011.03.009.

Ghosh, S., and M. S. Hayden. 2012. "Celebrating 25 years of NF-kappaB research." *Immunol Rev* 246(1): 5–13. doi: 10.1111/j.1600-065X.2012.01111.x.

Gibson, B. A., and W. L. Kraus. 2012. "New insights into the molecular and cellular functions of poly(ADP-ribose) and PARPs." *Nat Rev Mol Cell Biol* 13(7): 411–24. doi: 10.1038/nrm3376.

Gilmore, T. D. 2006. "Introduction to NF-kappaB: Players, pathways, perspectives." *Oncogene* 25(51): 6680–4. doi: 10.1038/sj.onc.1209954.

Golan-Goldhirsh, A., and J. Gopas. 2013. "Plant derived inhibitors of NF-κB." *Phytochem Rev* 1–15. doi: 10.1007/s11101-013-9293-5.

Greer, E. L., and A. Brunet. 2005. "FOXO transcription factors at the interface between longevity and tumor suppression." *Oncogene* 24(50): 7410–25. doi: 10.1038/sj.onc.1209086.

Grozinger, C. M., E. D. Chao, H. E. Blackwell, D. Moazed, and S. L. Schreiber. 2001. "Identification of a class of small molecule inhibitors of the sirtuin family of NAD-dependent deacetylases by phenotypic screening." *J Biol Chem* 276(42): 38837–43. doi: 10.1074/jbc.M106779200.

Gugasyan, R., R. Grumont, M. Grossmann, Y. Nakamura, T. Pohl, D. Nesic, and S. Gerondakis. 2000. "Rel/NF-kappaB transcription factors: Key mediators of B-cell activation." *Immunol Rev* 176: 134–40.

Guner, O. F., ed. 2000. *Pharmacophore Perception, Development, and Use in Drug Design*, Vol. 2, IUL Biotechnology Series. International University Line.

Guo, X., M. Kesimer, G. Tolun, X. Zheng, Q. Xu, J. Lu, J. K. Sheehan, J. D. Griffith, and X. Li. 2012. "The NAD(+)-dependent protein deacetylase activity of SIRT1 is regulated by its oligomeric status." *Sci Rep* 2: 640. doi: 10.1038/srep00640.

Gupta, S. C., C. Sundaram, S. Reuter, and B. B. Aggarwal. 2010. "Inhibiting NF-kappaB activation by small molecules as a therapeutic strategy." *Biochim Biophys Acta* 1799(10–12): 775–87. doi: 10.1016/j.bbagrm.2010.05.004.

Halliwell, B. 2013. "The antioxidant paradox: Less paradoxical now?" *Br J Clin Pharmacol* 75(3): 637–44. doi: 10.1111/j.1365-2125.2012.04272.x.

Hamilton, J. P. 2011. "Epigenetics: Principles and practice." *Dig Dis* 29(2): 130–5. doi: 10.1159/000323874.

Han, L., R. Zhou, J. Niu, M. A. McNutt, P. Wang, and T. Tong. 2010. "SIRT1 is regulated by a PPAR{gamma}-SIRT1 negative feedback loop associated with senescence." *Nucleic Acids Res* 38(21): 7458–71. doi: 10.1093/nar/gkq609.

Hanekamp, J. C., and A. Bast. 2007. "Hormesis in precautionary regulatory culture: Models preferences and the advancement of science." *Hum Exp Toxicol* 26(11): 855–73. doi: 10.1177/0960327107083414.

Hanekamp, J. C., A. Bast, and J. H. Kwakman. 2012. "Of reductionism and the pendulum swing: Connecting toxicology and human health." *Dose Response* 10(2): 155–76. doi: 10.2203/dose-response.11-018.Hanekamp.

Harris, R. Z., G. R. Jang, and S. Tsunoda. 2003. "Dietary effects on drug metabolism and transport." *Clin Pharmacokinet* 42(13): 1071–88.

Hassa, P. O., and M. O. Hottiger. 1999. "A role of poly (ADP-ribose) polymerase in NF-kappaB transcriptional activation." *Biol Chem* 380(7–8): 953–9. doi: 10.1515/BC.1999.118.

Hayes, J. D., J. U. Flanagan, and I. R. Jowsey. 2005. "Glutathione transferases." *Annu Rev Pharmacol Toxicol* 45: 51–88. doi: 10.1146/annurev.pharmtox.45.120403.095857.

Hayeshi, R., I. Mutingwende, W. Mavengere, V. Masiyanise, and S. Mukanganyama. 2007. "The inhibition of human glutathione *S*-transferases activity by plant polyphenolic compounds ellagic acid and curcumin." *Food Chem Toxicol* 45(2): 286–95. doi: 10.1016/j.fct.2006.07.027.

Hertog, M. G., E. J. Feskens, P. C. Hollman, M. B. Katan, and D. Kromhout. 1993. "Dietary antioxidant flavonoids and risk of coronary heart disease: The Zutphen Elderly Study." *Lancet* 342(8878): 1007–11.

Hollman, P. C., A. Cassidy, B. Comte, M. Heinonen, M. Richelle, E. Richling, M. Serafini, A. Scalbert, H. Sies, and S. Vidry. 2011. "The biological relevance of direct antioxidant effects of polyphenols for cardiovascular health in humans is not established." *J Nutr* 141(5): 989S–1009S. doi: 10.3945/jn.110.131490.

Hooper, L., P. A. Kroon, E. B. Rimm, J. S. Cohn, I. Harvey, K. A. Le Cornu, J. J. Ryder, W. L. Hall, and A. Cassidy. 2008. "Flavonoids, flavonoid-rich foods, and cardiovascular risk: A meta-analysis of randomized controlled trials." *Am J Clin Nutr* 88(1): 38–50.

Howitz, K. T., K. J. Bitterman, H. Y. Cohen, D. W. Lamming, S. Lavu, J. G. Wood, R. E. Zipkin et al. 2003. "Small molecule activators of sirtuins extend *Saccharomyces cerevisiae* lifespan." *Nature* 425(6954): 191–6. doi: 10.1038/nature01960.

Hu, Y., J. Liu, J. Wang, and Q. Liu. 2011. "The controversial links among calorie restriction, SIRT1, and resveratrol." *Free Radic Biol Med* 51(2): 250–6. doi: 10.1016/j.freeradbiomed.2011.04.034.

Huang, H. C., T. Nguyen, and C. B. Pickett. 2002. "Phosphorylation of Nrf2 at Ser-40 by protein kinase C regulates antioxidant response element-mediated transcription." *J Biol Chem* 277(45): 42769–74. doi: 10.1074/jbc.M206911200.

Hybertson, B. M., B. Gao, S. K. Bose, and J. M. McCord. 2011. "Oxidative stress in health and disease: The therapeutic potential of Nrf2 activation." *Mol Aspects Med* 32(4–6): 234–46. doi: 10.1016/j.mam.2011.10.006.

Imai, S. 2010. "A possibility of nutriceuticals as an anti-aging intervention: Activation of sirtuins by promoting mammalian NAD biosynthesis." *Pharmacol Res* 62(1): 42–7. doi: 10.1016/j.phrs.2010.01.006.

Ishii, Y., A. Nurrochmad, and H. Yamada. 2010. "Modulation of UDP-glucuronosyltransferase activity by endogenous compounds." *Drug Metab Pharmacokinet* 25(2): 134–48.

Jacobs, H., A. Bast, G. J. Peters, W. J. van der Vijgh, and G. R. Haenen. 2011. "The semisynthetic flavonoid monoHER sensitises human soft tissue sarcoma cells to doxorubicin-induced apoptosis via inhibition of nuclear factor-kappaB." *Br J Cancer* 104(3): 437–40. doi: 10.1038/sj.bjc.6606065.

Jacobs, H., M. Moalin, A. Bast, W. J. van der Vijgh, and G. R. Haenen. 2010. "An essential difference between the flavonoids monoHER and quercetin in their interplay with the endogenous antioxidant network." *PLoS One* 5(11): e13880. doi: 10.1371/journal.pone.0013880.

Jaenisch, R., and A. Bird. 2003. "Epigenetic regulation of gene expression: How the genome integrates intrinsic and environmental signals." *Nat Genet* 33(Suppl): 245–54. doi: 10.1038/ng1089.

Jagtap, P., F. G. Soriano, L. Virag, L. Liaudet, J. Mabley, E. Szabo, G. Hasko et al. 2002. "Novel phenanthridinone inhibitors of poly (adenosine 5'-diphosphate-ribose) synthetase: Potent cytoprotective and antishock agents." *Crit Care Med* 30(5): 1071–82.

Jha, A. K., M. Nikbakht, G. Parashar, A. Shrivastava, N. Capalash, and J. Kaur. 2010. "Reversal of hypermethylation and reactivation of the *RARbeta2* gene by natural compounds in cervical cancer cell lines." *Folia Biol* (Praha) 56(5): 195–200.

Jindal, S. 1996. "Heat shock proteins: Applications in health and disease." *Trends Biotechnol* 14(1): 17–20. doi: 10.1016/0167-7799(96)80909-7.

Kang, H., J. W. Jung, M. K. Kim, and J. H. Chung. 2009. "CK2 is the regulator of SIRT1 substrate-binding affinity, deacetylase activity and cellular response to DNA-damage." *PLoS One* 4(8): e6611. doi: 10.1371/journal.pone.0006611.

Kang, K. W., S. J. Lee, J. W. Park, and S. G. Kim. 2002. "Phosphatidylinositol 3-kinase regulates nuclear translocation of NF-E2-related factor 2 through actin rearrangement in response to oxidative stress." *Mol Pharmacol* 62(5): 1001–10.

Karlsen, A., L. Retterstol, P. Laake, I. Paur, S. K. Bohn, L. Sandvik, and R. Blomhoff. 2007. "Anthocyanins inhibit nuclear factor-kappaB activation in monocytes and reduce plasma concentrations of pro-inflammatory mediators in healthy adults." *J Nutr* 137(8): 1951–4.

Kim, D. H., J. M. Kim, E. K. Lee, Y. J. Choi, C. H. Kim, J. S. Choi, N. D. Kim, B. P. Yu, and H. Y. Chung. 2012. "Modulation of FoxO1 phosphorylation/acetylation by baicalin during aging." *J Nutr Biochem* 23(10): 1277–84. doi: 10.1016/j.jnutbio.2011.07.008.

Kim, Y. I., I. P. Pogribny, A. G. Basnakian, J. W. Miller, J. Selhub, S. J. James, and J. B. Mason. 1997. "Folate deficiency in rats induces DNA strand breaks and hypomethylation within the p53 tumor suppressor gene." *Am J Clin Nutr* 65(1): 46–52.

Kluth, D., A. Banning, I. Paur, R. Blomhoff, and R. Brigelius-Flohe. 2007. "Modulation of pregnane X receptor- and electrophile responsive element-mediated gene expression by dietary polyphenolic compounds." *Free Radic Biol Med* 42(3): 315–25. doi: 10.1016/j.freeradbiomed.2006.09.028.

Kobayashi, Y., Y. Furukawa-Hibi, C. Chen, Y. Horio, K. Isobe, K. Ikeda, and N. Motoyama. 2005. "SIRT1 is critical regulator of FOXO-mediated transcription in response to oxidative stress." *Int J Mol Med* 16(2): 237–43.

Kohle, C., and K. W. Bock. 2007. "Coordinate regulation of Phase I and II xenobiotic metabolisms by the Ah receptor and Nrf2." *Biochem Pharmacol* 73(12): 1853–62. doi: 10.1016/j.bcp.2007.01.009.

Kouzarides, T. 2007. "Chromatin modifications and their function." *Cell* 128(4): 693–705. doi: http://dx.doi.org/10.1016/j.cell.2007.02.005.

Kratschmar, D. V., D. Calabrese, J. Walsh, A. Lister, J. Birk, C. Appenzeller-Herzog, P. Moulin, C. E. Goldring, and A. Odermatt. 2012. "Suppression of the Nrf2-dependent antioxidant response by glucocorticoids and 11beta-HSD1-mediated glucocorticoid activation in hepatic cells." *PLoS One* 7(5): e36774. doi: 10.1371/journal.pone.0036774.

Kuehn, B. M. 2006. "Polypill could slash diabetes risks." *JAMA* 296(4): 377–80. doi: 10.1001/jama.296.4.377.

Kumar, A., Y. Takada, A. M. Boriek, and B. B. Aggarwal. 2004. "Nuclear factor-kappaB: Its role in health and disease." *J Mol Med* (Berl) 82(7): 434–48. doi: 10.1007/s00109-004-0555-y.

Lavon, I., I. Goldberg, S. Amit, L. Landsman, S. Jung, B. Z. Tsuberi, I. Barshack et al. 2000. "High susceptibility to bacterial infection, but no liver dysfunction, in mice compromised for hepatocyte NF-kappaB activation." *Nat Med* 6(5): 573–7. doi: 10.1038/75057.

Lee, E. K., J. M. Kim, J. Choi, K. J. Jung, D. H. Kim, S. W. Chung, Y. M. Ha, B. P. Yu, and H. Y. Chung. 2011. "Modulation of NF-kappaB and FOXOs by baicalein attenuates the radiation-induced inflammatory process in mouse kidney." *Free Radic Res* 45(5): 507–17. doi: 10.3109/10715762.2011.555479.

Lee, W. J., J. Y. Shim, and B. T. Zhu. 2005. "Mechanisms for the inhibition of DNA methyltransferases by tea catechins and bioflavonoids." *Mol Pharmacol* 68(4): 1018–30. doi: 10.1124/mol.104.008367.

Levonen, A. L., A. Landar, A. Ramachandran, E. K. Ceaser, D. A. Dickinson, G. Zanoni, J. D. Morrow, and V. M. Darley-Usmar. 2004. "Cellular mechanisms of redox cell signalling: Role of cysteine modification in controlling antioxidant defences in response to electrophilic lipid oxidation products." *Biochem J* 378(Pt 2): 373–82. doi: 10.1042/BJ20031049.

Li, Q., and I. M. Verma. 2002. "NF-kappaB regulation in the immune system." *Nat Rev Immunol* 2(10): 725–34. doi: 10.1038/nri910.

Liaudet, L., P. Pacher, J. G. Mabley, L. Virag, F. G. Soriano, G. Hasko, and C. Szabo. 2002. "Activation of poly(ADP-Ribose) polymerase-1 is a central mechanism of lipopolysaccharide-induced acute lung inflammation." *Am J Respir Crit Care Med* 165(3): 372–7. doi: 10.1164/ajrccm.165.3.2106050.

Linnewiel, K., H. Ernst, C. Caris-Veyrat, A. Ben-Dor, A. Kampf, H. Salman, M. Danilenko, J. Levy, and Y. Sharoni. 2009. "Structure activity relationship of carotenoid derivatives in activation of the electrophile/antioxidant response element transcription system." *Free Radic Biol Med* 47(5): 659–67. doi: 10.1016/j.freeradbiomed.2009.06.008.

Liu, Z., S. W. Choi, J. W. Crott, M. K. Keyes, H. Jang, D. E. Smith, M. Kim, P. W. Laird, R. Bronson, and J. B. Mason. 2007. "Mild depletion of dietary folate combined with other B vitamins alters multiple components of the Wnt pathway in mouse colon." *J Nutr* 137(12): 2701–8.

Lombardo, E., C. Sabellico, J. Hajek, V. Stankova, T. Filipsky, V. Balducci, P. De Vito et al. 2013. "Protection of cells against oxidative stress by nanomolar levels of hydroxyflavones indicates a new type of intracellular antioxidant mechanism." *PLoS One* 8(4): e60796. doi: 10.1371/journal.pone.0060796.

Long, M., T. Kruger, M. Ghisari, and E. C. Bonefeld-Jorgensen. 2012. "Effects of selected phytoestrogens and their mixtures on the function of the thyroid hormone and the aryl hydrocarbon receptor." *Nutr Cancer* 64(7): 1008–19. doi: 10.1080/01635581.2012.711419.

Lonn, E., J. Bosch, K. K. Teo, P. Pais, D. Xavier, and S. Yusuf. 2010. "The polypill in the prevention of cardiovascular diseases: Key concepts, current status, challenges, and future directions." *Circulation* 122(20): 2078–88. doi: 10.1161/CIRCULATIONAHA.109.873232.

Lu, M., H. E. Kim, C. R. Li, S. Kim, I. J. Kwak, Y. J. Lee, S. S. Kim et al. 2008. "Two distinct disulfide bonds formed in human heat shock transcription factor 1 act in opposition to regulate its DNA binding activity." *Biochemistry* 47(22): 6007–15. doi: 10.1021/bi702185u.

Lumey, L. H., and F. W. Van Poppel. 1994. "The Dutch famine of 1944–45: Mortality and morbidity in past and present generations." *Soc Hist Med* 7(2): 229–46.

Macario, A. J., and E. Conway de Macario. 2007. "Molecular chaperones: Multiple functions, pathologies, and potential applications." *Front Biosci* 12: 2588–600.

Mai, A., S. Massa, S. Lavu, R. Pezzi, S. Simeoni, R. Ragno, F. R. Mariotti, F. Chiani, G. Camilloni, and D. A. Sinclair. 2005. "Design, synthesis, and biological evaluation of sirtinol analogues as class III histone/protein deacetylase (Sirtuin) inhibitors." *J Med Chem* 48(24): 7789–95. doi: 10.1021/jm050100l.

Malemud, C. J. 2007. "Inhibitors of stress-activated protein/mitogen-activated protein kinase pathways." *Curr Opin Pharmacol* 7(3): 339–43. doi: 10.1016/j.coph.2006.11.012.

Medema, R. H., G. J. Kops, J. L. Bos, and B. M. Burgering. 2000. "AFX-like Forkhead transcription factors mediate cell-cycle regulation by Ras and PKB through p27kip1." *Nature* 404(6779): 782–7. doi: 10.1038/35008115.

Medina-Franco, J. L., M. A. Giulianotti, G. S. Welmaker, and R. A. Houghten. 2013. "Shifting from the single to the multitarget paradigm in drug discovery." *Drug Discov Today* 18(9–10): 495–501. doi: 10.1016/j.drudis.2013.01.008.

Merksamer, P. I., Y. Liu, W. He, M. D. Hirschey, D. Chen, and E. Verdin. 2013. "The sirtuins, oxidative stress and aging: An emerging link." *Aging* (Albany, NY) 5(3): 144–50.

Michan, S., and D. Sinclair. 2007. "Sirtuins in mammals: Insights into their biological function." *Biochem J* 404(1): 1–13. doi: 10.1042/BJ20070140.

Milenkovic, D., C. Deval, E. Gouranton, J. F. Landrier, A. Scalbert, C. Morand, and A. Mazur. 2012. "Modulation of miRNA expression by dietary polyphenols in apoE deficient mice: A new mechanism of the action of polyphenols." *PLoS One* 7(1): e29837. doi: 10.1371/journal.pone.0029837.

Milenkovic, D., B. Jude, and C. Morand. 2013. "miRNA as molecular target of polyphenols underlying their biological effects." *Free Radic Biol Med*. 64:40–51. doi: 10.1016/j.freeradbiomed.2013.05.046.

Moonen, H. J., L. Geraets, A. Vaarhorst, A. Bast, E. F. Wouters, and G. J. Hageman. 2005. "Theophylline prevents NAD+ depletion via PARP-1 inhibition in human pulmonary epithelial cells." *Biochem Biophys Res Commun* 338(4): 1805–10. doi: 10.1016/j.bbrc.2005.10.159.

Motta, M. C., N. Divecha, M. Lemieux, C. Kamel, D. Chen, W. Gu, Y. Bultsma, M. McBurney, and L. Guarente. 2004. "Mammalian SIRT1 represses forkhead transcription factors." *Cell* 116(4): 551–63.
Murray, M. 2007. "Role of signalling systems in the effects of dietary factors on the expression of mammalian CYPs." *Expert Opin Drug Metab Toxicol* 3(2): 185–96. doi: 10.1517/17425255.3.2.185.
Na, H. K., and Y. J. Surh. 2008. "Modulation of Nrf2-mediated antioxidant and detoxifying enzyme induction by the green tea polyphenol EGCG." *Food Chem Toxicol* 46(4): 1271–8. doi: 10.1016/j.fct.2007.10.006.
Nandakumar, V., M. Vaid, and S. K. Katiyar. 2011. "(−)-Epigallocatechin-3-gallate reactivates silenced tumor suppressor genes, *Cip1/p21* and *p16INK4a*, by reducing DNA methylation and increasing histones acetylation in human skin cancer cells." *Carcinogenesis* 32(4): 537–44. doi: 10.1093/carcin/bgq285.
Nasrin, N., V. K. Kaushik, E. Fortier, D. Wall, K. J. Pearson, R. de Cabo, and L. Bordone. 2009. "JNK1 phosphorylates SIRT1 and promotes its enzymatic activity." *PLoS One* 4(12): e8414. doi: 10.1371/journal.pone.0008414.
Nebert, D. W., D. R. Nelson, M. J. Coon, R. W. Estabrook, R. Feyereisen, Y. Fujii-Kuriyama, F. J. Gonzalez et al. 1991. "The P450 superfamily: Update on new sequences, gene mapping, and recommended nomenclature." *DNA Cell Biol* 10(1): 1–14.
Nemoto, S., M. M. Fergusson, and T. Finkel. 2004. "Nutrient availability regulates SIRT1 through a forkhead-dependent pathway." *Science* 306(5704): 2105–8. doi: 10.1126/science.1101731.
Ness, A. R., and J. W. Powles. 1997. "Fruit and vegetables, and cardiovascular disease: A review." *Int J Epidemiol* 26(1): 1–13.
Neve, E. P., and M. Ingelman-Sundberg. 2010. "Cytochrome P450 proteins: Retention and distribution from the endoplasmic reticulum." *Curr Opin Drug Discov Devel* 13(1): 78–85.
Niture, S. K., R. Khatri, and A. K. Jaiswal. 2013. "Regulation of Nrf2—an update." *Free Radic Biol Med.* 66:36–44. doi: 10.1016/j.freeradbiomed.2013.02.008.
Noriega, L. G., J. N. Feige, C. Canto, H. Yamamoto, J. Yu, M. A. Herman, C. Mataki, B. B. Kahn, and J. Auwerx. 2011. "CREB and ChREBP oppositely regulate SIRT1 expression in response to energy availability." *EMBO Rep* 12(10): 1069–76. doi: 10.1038/embor.2011.151.
Nover, L., K. Bharti, P. Doring, S. K. Mishra, A. Ganguli, and K. D. Scharf. 2001. "Arabidopsis and the heat stress transcription factor world: How many heat stress transcription factors do we need?" *Cell Stress Chaperones* 6(3): 177–89.
Ohta, T., K. Iijima, M. Miyamoto, I. Nakahara, H. Tanaka, M. Ohtsuji, T. Suzuki et al. 2008. "Loss of Keap1 function activates Nrf2 and provides advantages for lung cancer cell growth." *Cancer Res* 68(5): 1303–9. doi: 10.1158/0008-5472.CAN-07-5003.
Painter, R. C., C. Osmond, P. Gluckman, M. Hanson, D. I. Phillips, and T. J. Roseboom. 2008. "Transgenerational effects of prenatal exposure to the Dutch famine on neonatal adiposity and health in later life." *BJOG* 115(10): 1243–9. doi: 10.1111/j.1471-0528.2008.01822.x.
Parasramka, M. A., E. Ho, D. E. Williams, and R. H. Dashwood. 2012. "MicroRNAs, diet, and cancer: New mechanistic insights on the epigenetic actions of phytochemicals." *Mol Carcinog* 51(3): 213–30. doi: 10.1002/mc.20822.
Park, L. K., S. Friso, and S. W. Choi. 2012a. "Nutritional influences on epigenetics and age-related disease." *Proc Nutr Soc* 71(1): 75–83. doi: 10.1017/S0029665111003302.
Park, S. J., F. Ahmad, A. Philp, K. Baar, T. Williams, H. Luo, H. Ke, H. Rehmann, R. Taussig, A. L. Brown, M. K. Kim, M. A. Beaven, A. B. Burgin, V. Manganiello, and J. H. Chung. 2012b. "Resveratrol ameliorates aging-related metabolic phenotypes by inhibiting cAMP phosphodiesterases." *Cell* 148(3):421-33. doi: 10.1016/j.cell.2012.01.017.
Perkins, N. D. 2012. "The diverse and complex roles of NF-kappaB subunits in cancer." *Nat Rev Cancer* 12(2): 121–32. doi: 10.1038/nrc3204.

Pieper, A. A., A. Verma, J. Zhang, and S. H. Snyder. 1999. "Poly (ADP-ribose) polymerase, nitric oxide and cell death." *Trends Pharmacol Sci* 20(4): 171-81.

Pizzolo, F., H. J. Blom, S. W. Choi, D. Girelli, P. Guarini, N. Martinelli, A. M. Stanzial, R. Corrocher, O. Olivieri, and S. Friso. 2011. "Folic acid effects on *s*-adenosylmethionine, *s*-adenosylhomocysteine, and DNA methylation in patients with intermediate hyperhomocysteinemia." *J Am Coll Nutr* 30(1): 11–8.

Plas, D. R., and C. B. Thompson. 2003. "Akt activation promotes degradation of tuberin and FOXO3a via the proteasome." *J Biol Chem* 278(14): 12361–6. doi: 10.1074/jbc.M213069200.

Ptak, C., and A. Petronis. 2008. "Epigenetics and complex disease: From etiology to new therapeutics." *Annu Rev Pharmacol Toxicol* 48: 257–76. doi: 10.1146/annurev.pharmtox.48.113006.094731.

Ramsey, K. M., K. F. Mills, A. Satoh, and S. Imai. 2008. "Age-associated loss of Sirt1-mediated enhancement of glucose-stimulated insulin secretion in beta cell-specific Sirt1-overexpressing (BESTO) mice." *Aging Cell* 7(1): 78–88. doi: 10.1111/j.1474-9726.2007.00355.x.

Rattan, S. I. 1998. "The nature of gerontogenes and vitagenes. Antiaging effects of repeated heat shock on human fibroblasts." *Ann NY Acad Sci* 854: 54–60.

Reiter, C. E., J. A. Kim, and M. J. Quon. 2010. "Green tea polyphenol epigallocatechin gallate reduces endothelin-1 expression and secretion in vascular endothelial cells: Roles for AMP-activated protein kinase, Akt, and FOXO1." *Endocrinology* 151(1): 103–14. doi: 10.1210/en.2009-0997.

Renaud, S., and M. de Lorgeril. 1992. "Wine, alcohol, platelets, and the French paradox for coronary heart disease." *Lancet* 339(8808): 1523–6.

Revollo, J. R., and X. Li. 2013. "The ways and means that fine tune Sirt1 activity." *Trends Biochem Sci* 38(3): 160–7. doi: 10.1016/j.tibs.2012.12.004.

Robertson, K. D. 2005. "DNA methylation and human disease." *Nat Rev Genet* 6(8): 597–610. doi: 10.1038/nrg1655.

Roseboom, T., S. de Rooij, and R. Painter. 2006. "The Dutch famine and its long-term consequences for adult health." *Early Hum Dev* 82(8): 485–91. doi: 10.1016/j.earlhumdev.2006.07.001.

Ruijters, E. J., A. R. Weseler, C. Kicken, G. R. Haenen, and A. Bast. 2013. "The flavanol (–)-epicatechin and its metabolites protect against oxidative stress in primary endothelial cells via a direct antioxidant effect." *Eur J Pharmacol*. 715(1–3):147–53. doi: 10.1016/j.ejphar.2013.05.029.

Saeidnia, S., and M. Abdollahi. 2013. "Antioxidants: Friends or foe in prevention or treatment of cancer: The debate of the century." *Toxicol Appl Pharmacol* 271(1): 49–63. doi: 10.1016/j.taap.2013.05.004.

Salminen, A., A. Kauppinen, and K. Kaarniranta. 2012. "Phytochemicals suppress nuclear factor-kappaB signaling: Impact on health span and the aging process." *Curr Opin Clin Nutr Metab Care* 15(1): 23–8. doi: 10.1097/MCO.0b013e32834d3ae7.

Salminen, A., J. Ojala, J. Huuskonen, A. Kauppinen, T. Suuronen, and K. Kaarniranta. 2008. "Interaction of aging-associated signaling cascades: Inhibition of NF-kappaB signaling by longevity factors FoxOs and SIRT1." *Cell Mol Life Sci* 65(7–8): 1049–58. doi: 10.1007/s00018-008-7461-3.

Sang, S., J. D. Lambert, J. Hong, S. Tian, M. J. Lee, R. E. Stark, C. T. Ho, and C. S. Yang. 2005. "Synthesis and structure identification of thiol conjugates of (–)-epigallocatechin gallate and their urinary levels in mice." *Chem Res Toxicol* 18(11): 1762–9. doi: 10.1021/tx0501511.

Saracino, M. R., and J. W. Lampe. 2007. "Phytochemical regulation of UDP-glucuronosyltransferases: Implications for cancer prevention." *Nutr Cancer* 59(2): 121–41. doi: 10.1080/01635580701458178.

Serafini, M., I. Peluso, and A. Raguzzini. 2010. "Flavonoids as anti-inflammatory agents." *Proc Nutr Soc* 69(3): 273–8. doi: 10.1017/S002966511000162X.
Shankar, S., L. Marsh, and R. K. Srivastava. 2013. "EGCG inhibits growth of human pancreatic tumors orthotopically implanted in Balb C nude mice through modulation of FKHRL1/FOXO3a and neuropilin." *Mol Cell Biochem* 372(1–2): 83–94. doi: 10.1007/s11010-012-1448-y.
Shi, Y., and C. Mello. 1998. "A CBP/p300 homolog specifies multiple differentiation pathways in *Caenorhabditis elegans*." *Genes Dev* 12(7): 943–55.
Sies, H. 2010. "Polyphenols and health: Update and perspectives." *Arch Biochem Biophys* 501(1): 2–5. doi: 10.1016/j.abb.2010.04.006.
Steinmetz, K. A., and J. D. Potter. 1996. "Vegetables, fruit, and cancer prevention: A review." *J Am Diet Assoc* 96(10): 1027–39. doi: 10.1016/S0002-8223(96)00273-8.
Surh, Y. J. 2003. "Cancer chemoprevention with dietary phytochemicals." *Nat Rev Cancer* 3(10): 768–80. doi: 10.1038/nrc1189.
Suzuki, T., H. Motohashi, and M. Yamamoto. 2013. "Toward clinical application of the Keap1–Nrf2 pathway." *Trends Pharmacol Sci* 34(6): 340–6. doi: 10.1016/j.tips.2013.04.005.
Taguchi, K., H. Motohashi, and M. Yamamoto. 2011. "Molecular mechanisms of the Keap1–Nrf2 pathway in stress response and cancer evolution." *Genes Cells* 16(2): 123–40. doi: 10.1111/j.1365-2443.2010.01473.x.
Tammen, S. A., S. Friso, and S. W. Choi. 2013. "Epigenetics: The link between nature and nurture." *Mol Aspects Med* 34(4): 753–64. doi: 10.1016/j.mam.2012.07.018.
Trapp, J., R. Meier, D. Hongwiset, M. U. Kassack, W. Sippl, and M. Jung. 2007. "Structure–activity studies on suramin analogues as inhibitors of NAD+-dependent histone deacetylases (sirtuins)." *ChemMedChem* 2(10): 1419–31. doi: 10.1002/cmdc.200700003.
Uruno, A., and H. Motohashi. 2011. "The Keap1–Nrf2 system as an in vivo sensor for electrophiles." *Nitric Oxide* 25(2): 153–60. doi: 10.1016/j.niox.2011.02.007.
Van Haaften, R. I., C. T. Evelo, J. Penders, M. P. Eijnwachter, G. R. Haenen, and A. Bast. 2001. "Inhibition of human glutathione *S*-transferase P1-1 by tocopherols and alpha-tocopherol derivatives." *Biochim Biophys Acta* 1548(1): 23–8.
Van Haaften, R. I., G. R. Haenen, C. T. Evelo, and A. Bast. 2003. "Effect of vitamin E on glutathione-dependent enzymes." *Drug Metab Rev* 35(2–3): 215–53. doi: 10.1081/DMR-120024086.
Vanden Berghe, W. 2012. "Epigenetic impact of dietary polyphenols in cancer chemoprevention: Lifelong remodeling of our epigenomes." *Pharmacol Res* 65(6): 565–76. doi: 10.1016/j.phrs.2012.03.007.
Vaquero, A., R. Sternglanz, and D. Reinberg. 2007. "NAD+-dependent deacetylation of H4 lysine 16 by class III HDACs." *Oncogene* 26(37): 5505–20. doi: 10.1038/sj.onc.1210617.
Vauzour, D., A. Rodriguez-Mateos, G. Corona, M. J. Oruna-Concha, and J. P. Spencer. 2010. "Polyphenols and human health: Prevention of disease and mechanisms of action." *Nutrients* 2(11): 1106–31. doi: 10.3390/nu2111106.
Veenendaal, M. V., R. C. Painter, S. R. de Rooij, P. M. Bossuyt, J. A. van der Post, P. D. Gluckman, M. A. Hanson, and T. J. Roseboom. 2013. "Transgenerational effects of prenatal exposure to the 1944-45 Dutch famine." *BJOG* 120(5): 548–53. doi: 10.1111/1471-0528.12136.
Villalba, J. M., and F. J. Alcain. 2012. "Sirtuin activators and inhibitors." *Biofactors* 38(5): 349–59. doi: 10.1002/biof.1032.
Vinciguerra, M., M. P. Santini, C. Martinez, V. Pazienza, W. C. Claycomb, A. Giuliani, and N. Rosenthal. 2012. "mIGF-1/JNK1/SirT1 signaling confers protection against oxidative stress in the heart." *Aging Cell* 11(1): 139–49. doi: 10.1111/j.1474-9726.2011.00766.x.
Wald, N. J., and M. R. Law. 2003. "A strategy to reduce cardiovascular disease by more than 80%." *BMJ* 326(7404): 1419. doi: 10.1136/bmj.326.7404.1419.

Wang, Z., and H. Chen. 2010. "Genistein increases gene expression by demethylation of WNT5a promoter in colon cancer cell line SW1116." *Anticancer Res* 30(11): 4537–45.

Waterland, R. A., and R. L. Jirtle. 2003. "Transposable elements: Targets for early nutritional effects on epigenetic gene regulation." *Mol Cell Biol* 23(15): 5293–300.

Weseler, A. R., and A. Bast. 2012. "Pleiotropic-acting nutrients require integrative investigational approaches: The example of flavonoids." *J Agric Food Chem* 60(36): 8941–6. doi: 10.1021/jf3000373.

Weseler, A. R., L. Geraets, H. J. Moonen, R. J. Manders, L. J. van Loon, H. J. Pennings, E. F. Wouters, A. Bast, and G. J. Hageman. 2009. "Poly(ADP-ribose) polymerase-1-inhibiting flavonoids attenuate cytokine release in blood from male patients with chronic obstructive pulmonary disease or type 2 diabetes." *J Nutr* 139(5): 952–7. doi: 10.3945/jn.108.102756.

Weseler, A. R., E. J. Ruijters, M. J. Drittij-Reijnders, K. D. Reesink, G. R. Haenen, and A. Bast. 2011. "Pleiotropic benefit of monomeric and oligomeric flavanols on vascular health—a randomized controlled clinical pilot study." *PLoS One* 6(12): e28460. doi: 10.1371/journal.pone.0028460.

Westerheide, S. D., J. Anckar, S. M. Stevens Jr., L. Sistonen, and R. I. Morimoto. 2009. "Stress-inducible regulation of heat shock factor 1 by the deacetylase SIRT1." *Science* 323(5917): 1063–6. doi: 10.1126/science.1165946.

World Health Organization. 2006. "Constitution of the World Health Organization, Basic documents, Supplement, October 2006." WHO, Geneva. Available at: http://www.who.int/governance/eb/who_constitution_en.pdf

Williams, R. J., J. P. Spencer, and C. Rice-Evans. 2004. "Flavonoids: Antioxidants or signalling molecules?" *Free Radic Biol Med* 36(7): 838–49. doi: 10.1016/j.freeradbiomed.2004.01.001.

Wong, E. T., and V. Tergaonkar. 2009. "Roles of NF-kappaB in health and disease: Mechanisms and therapeutic potential." *Clin Sci* (London) 116(6): 451–65. doi: 10.1042/CS20080502.

Wu, C. 1995. "Heat shock transcription factors: Structure and regulation." *Annu Rev Cell Dev Biol* 11: 441–69. doi: 10.1146/annurev.cb.11.110195.002301.

Xiao, G. H., J. A. Pinaire, A. D. Rodrigues, and R. A. Prough. 1995. "Regulation of the Ah gene battery via Ah receptor-dependent and independent processes in cultured adult rat hepatocytes." *Drug Metab Dispos* 23(6): 642–50.

Xu, C., C. Y. Li, and A. N. Kong. 2005. "Induction of phase I, II and III drug metabolism/transport by xenobiotics." *Arch Pharm Res* 28(3): 249–68.

Yamakuchi, M. 2012. "MicroRNA regulation of SIRT1." *Front Physiol* 3: 68. doi: 10.3389/fphys.2012.00068.

Yamamoto, H., K. Schoonjans, and J. Auwerx. 2007. "Sirtuin functions in health and disease." *Mol Endocrinol* 21(8): 1745–55. doi: 10.1210/me.2007-0079.

Yao, H., and I. Rahman. 2012. "Perspectives on translational and therapeutic aspects of SIRT1 in inflammaging and senescence." *Biochem Pharmacol* 84(10): 1332–9. doi: 10.1016/j.bcp.2012.06.031.

Yeung, F., J. E. Hoberg, C. S. Ramsey, M. D. Keller, D. R. Jones, R. A. Frye, and M. W. Mayo. 2004. "Modulation of NF-kappaB-dependent transcription and cell survival by the SIRT1 deacetylase." *EMBO J* 23(12): 2369–80. doi: 10.1038/sj.emboj.7600244.

Yu, R., W. Lei, S. Mandlekar, M. J. Weber, C. J. Der, J. Wu, and A. N. Kong. 1999. "Role of a mitogen-activated protein kinase pathway in the induction of phase II detoxifying enzymes by chemicals." *J Biol Chem* 274(39): 27545–52.

Zhao, C. R., Z. H. Gao, and X. J. Qu. 2010. "Nrf2–ARE signaling pathway and natural products for cancer chemoprevention." *Cancer Epidemiol* 34(5): 523–33. doi: 10.1016/j.canep.2010.06.012.

# 5 Periodic Fasting and Hormesis

*Yan Y. Lam and Eric Ravussin*

## CONTENTS

5.1 Introduction ........................................................................................................ 79
5.2 Overview of Periodic Fasting ............................................................................. 80
5.3 Effects of Periodic Fasting ................................................................................. 81
    5.3.1 Energy Balance and Cardiometabolic Health ......................................... 81
    5.3.2 Cognitive Health, Aging, and Lifespan .................................................. 83
5.4 Hormesis in Periodic Fasting ............................................................................. 84
5.5 Periodic Fasting as a Health Improvement Strategy .......................................... 86
References .................................................................................................................. 87

## 5.1 INTRODUCTION

Dietary interventions, which typically involve changes in energy intake and/or macronutrient composition, are common strategies to improve health. For example, a hypocaloric low-fat diet is an integral part of lifestyle modifications to achieve weight loss and to prevent or delay diabetes in high-risk populations (Knowler et al. 2002, 2009). The dietary fatty acid profile is known to modify cardiovascular disease risks, with saturated fat being deleterious whereas long-chain polyunsaturated fatty acids appear to be neutral or even beneficial (Baum et al. 2012). By reducing total energy consumption (20%–50%) without compromising nutritional adequacy, calorie restriction is perhaps the dietary intervention with the strongest evidence to confer overall health benefits. Calorie restriction has been shown to improve cardiovascular risk (Sung and Dyck 2012). The literature also largely supports its effect on delaying aging and extending average and maximal lifespan in many species, including nonhuman primates (Weindruch and Walford 1982; Anderson et al. 2003; Colman et al. 2009), although it has recently been reported otherwise (Mattison et al. 2012).

There is evidence for calorie restriction to modify energy metabolism in humans by reducing energy expenditure (Heilbronn et al. 2006) and increasing insulin sensitivity (Murphy et al. 2012). Such a regimen, however, requires constant dietary restriction, and adherence is often poor in free-living conditions. It is also difficult to keep people motivated to continue following a hypocaloric diet after achieving some health benefits and this becomes one of the main reasons for relapses. These limitations make calorie restriction unlikely to be feasible in the long term for both weight management and health maintenance.

Periodic fasting refers to a reduction in meal frequency and is often used to achieve energy deficit. Interestingly, recent studies suggest that periodic fasting, which may or may not be associated with changes in overall energy intake, closely resembles the effects of calorie restriction on chronic disease prevention and improving longevity (Mattson and Wan 2005; Varady and Hellerstein 2007). This suggests that in addition to energy intake per se, eating pattern is also an important determinant of how the diet affects health. Specifically, there is evidence for the alternations between fast and feast to activate hormetic mechanisms that increase stress resistance and therefore overall cellular function. This chapter summarizes the latest literature to elucidate the role of hormesis in mediating the metabolic effects of periodic fasting. We also explore the feasibility of wide adoption of such a dietary regimen to improve health at the population level.

## 5.2 OVERVIEW OF PERIODIC FASTING

Periodic fasting, also known as intermittent fasting, includes calorie restriction ranging from continuous absolute fast for days to skipping meals on a regular basis (Stipp 2013). Over the centuries, periodic fasting has been observed for religious and therapeutic reasons (Johnstone 2007; Trepanowski and Bloomer 2010): the Islamic Ramadan fasting requires all healthy practicing adult Muslims to refrain from consuming any food or liquid from sunrise to sunset during the holy month of Ramadan; Buddhist monks who follow the Vinaya rules do not eat or drink every day from noon to dawn; Christians observe fasting at their own behest and the practice has evolved over time and varies among different churches. Advent and Lent fast in the Catholic Church, for example, used to limit food intake to one meal daily (evening meal with no meat, fish, eggs, and butter), which has evolved to the modern practice of eating less food and abstinence from meat and meat products. Therapeutic use of periodic fasting in the inpatient setting to achieve rapid weight loss has been reported in the 1960s (Duncan 1962; Gilliland 1968). It typically involved continuous fasting for 4–14 days, during which only water and energy-free beverages were allowed, and followed by a hypocaloric diet. Thereafter, some patients were instructed to fast 1 day per week to facilitate progressive weight reduction. The long-term success of such a drastic weight loss regimen, however, was disappointing with up to 50% of participants returning to their initial weight.

Today periodic fasting remains one of the most popular weight loss strategies in the population. According to the National Health Interview Survey, about 10% of American adults skipped meals when trying to lose weight (Kruger et al. 2004); data from the 2003 Youth Risk Behavior Survey revealed that 11.6% of normal weight adolescents have fasted (no food intake for ≥24 hours) in the past 30 days to lose weight or to prevent from gaining weight (Talamayan et al. 2006). In addition to weight management, a modified form of periodic fasting as part of integrative medicine in inpatient setting has been used with favorable clinical outcomes and improvements in disease-related complaints (Michalsen et al. 2005, 2013). Fasting prior to cancer treatment has been recently trialed and there is evidence for a single episode of fasting (48–140 hours) to reduce side effects of chemotherapy (Safdie et al. 2009).

Despite the long history of periodic fasting, its effects on health and the associated molecular mechanisms have not been adequately characterized. The wide range of regimens, for example, the duration and frequency of fasting, makes it difficult to generalize the effects of fasting. There is a sizable body of literature on Ramadan fasting, which is arguably the most standardized form of fasting in free-living conditions. However, the findings are often inconclusive as the study outcomes, including energy intake and macronutrient consumption, are highly variable across different Muslim populations who have specific food choices (el Ati et al. 1995; Trepanowski and Bloomer 2010). Alternate-day fasting (ADF) is one of the most studied forms of periodic fasting in experimental settings. By definition it consists of alternating 24-hour periods of absolute fasting ("fast" day) and ad libitum food intake ("feast" day) (Heilbronn et al. 2005a,b), although a modified version of ADF that allows limited energy intake on the fast days has also been used and is often called ADMF for alternate-day modified fast (Bhutani et al. 2010).

## 5.3 EFFECTS OF PERIODIC FASTING

### 5.3.1 Energy Balance and Cardiometabolic Health

During periodic fasting, the average energy consumption should, in theory, be compensated by a self-regulated increase in calorie intake during the feast days to achieve overall energy homeostasis. This has indeed been shown in some animal studies, in which on feast days ADF mice ate twice the amount of food consumed over a 24-hour period by their control counterparts, thus maintaining similar average daily food intake and body weight (Anson et al. 2003; Descamps et al. 2005; Froy et al. 2009). In other cases, however, a reduction of daily calorie consumption ($\approx$30%) and body weight ($\approx$20%) in rodents following ADF over 3–6 months has been reported (Wan et al. 2003; Singh et al. 2012).

Compensation of energy intake during periodic fasting appears to be generally not achievable in humans. Three-week ADF induced significant weight loss (2.5 ± 0.5% of initial body weight) in healthy nonobese individuals who were not motivated to lose weight (Heilbronn et al. 2005b). In an 8-week ADMF study (25% of daily energy requirement was provided on the fast days) in obese humans, ad libitum intake on the feast days did not exceed the energy need for the day, which subsequently resulted in an overall 37% caloric restriction and an average weight loss of 5.6 kg over the study duration (Klempel et al. 2010). A similar modified intermittent fasting study (a 6-month regimen consisting of weekly cycles of consuming 25% daily energy requirement for 2 days per week followed by ad libitum intake for 5 days per week) also induced a comparable reduction in energy intake ($\approx$30%) in overweight women (Harvie et al. 2011). Taken together, periodic fasting often leads to unintentional reduction in overall energy intake and body weight, which may then elicit effects overlapping with calorie restriction.

There is evidence for periodic fasting to improve glucose metabolism. ADF improved insulin sensitivity parameters in healthy mice (Anson et al. 2003; Wan et al. 2010; Lu et al. 2011) and was protective in rodent models that were prone to developing diabetes (Pedersen et al. 1999; Beigy et al. 2013). Similarly, dietary regimens that involved intermittent modified fasting (cycles of a very-low-calorie diet and ad libitum feeding) induced weight loss and improved insulin sensitivity in overweight/obese women (Harvie et al. 2011) and in patients with type 2 diabetes (Williams et al. 1998). The effect of periodic fasting on metabolic glucose processing in healthy humans, however, varied considerably across studies. A 22-day ADF regimen appeared to improve insulin sensitivity in men but negatively influenced glucose tolerance in women in response to a meal challenge (Heilbronn et al. 2005a); 14 days of ADF increased insulin sensitivity in men as determined by an euglycemic hyperinsulinemic clamp (Halberg et al. 2005). This effect, however, could not be replicated in another study with an identical dietary regimen as well as comparable subject population and outcome assessment (Soeters et al. 2009).

Periodic fasting has also been shown to confer cardioprotective benefits. Three days of water-only fasting reduced infarct size and trended to reduce the incidence and duration of arrhythmias in rats with acute myocardial infarction (Snorek et al. 2012). After being subjected to chronic myocardial ischemia, rats that then commenced an ADF regimen exhibited less severe damage as evidenced by a decrease in myocardium hypertrophy and fibrosis. The enhanced recovery and survival of these animals were at least partly due to an upregulation of angiogenesis (Katare et al. 2009). The protective effect of ADF is further supported by the reduced severity in damages (e.g., a reduction in the size of ischemic infarct, cardiomyocyte apoptosis, and functional decline) in rats subjected to experimental myocardial infarction (Ahmet et al. 2005; Wan et al. 2010). Furthermore, an improvement in cardiovascular risk factors (e.g., blood pressure, heart rate, and cholesterol esters in aorta and liver) was observed in healthy (Wan et al. 2003; Mulas et al. 2005) and diabetic (Tikoo et al. 2007) animals subjected to ADF.

The effect of periodic fasting on cardiovascular health is less studied in humans, but its relationship with a favorable blood lipid profile and blood pressure reduction (Aksungar et al. 2005; Johnson et al. 2007; Harvie et al. 2011; Zare et al. 2011) parallels the data from animal studies. In a population study, coronary angiography patients who observed a form of religious fasting (fasting 1 day per month) had a lower risk of coronary artery disease after adjustment for traditional risk factors including smoking, diabetes, and hyperlipidemia, although behavioral discipline associated with the religion is a possible confounder of the benefits of routine periodic fasting (Horne et al. 2008).

Given the overlapping effects of periodic fasting and calorie restriction, the laboratory of Mattson conducted a crossover study to investigate the impact of reduced meal frequency per se on metabolic parameters (Carlson et al. 2007; Stote et al. 2007). Healthy individuals (body mass index 23.4 ± 0.5 kg/m$^2$) were required to consume a weight maintenance diet in either three meals per day or one meal per day with all food provided and intake closely monitored in an attempt to maintain an isocaloric condition. Body weight and fat mass were significantly lower after having one meal per day as compared to having three meals per day but neither was statistically different from the baseline (Stote et al. 2007). Restricting food intake to one meal daily appeared to negatively impact glucose metabolism (Carlson et al. 2007) and its effect on cardiovascular risk factors was mixed (Stote et al. 2007). Overall, these results suggest that reducing meal frequency may have no major benefits in healthy humans, but its effect on metabolically unhealthy individuals is unknown.

## 5.3.2 Cognitive Health, Aging, and Lifespan

Often used as a means to restrict calorie intake, periodic fasting is most studied for its effects on cognitive function, aging, and longevity. Periodic fasting has consistently been shown to ameliorate deleterious sequelae in experimental models of brain injury/dysfunction. Compared to ad libitum controls, ADF rats (weighed ≈30% less) had reduced cortical/striatal infarct volume and reduced behavioral deficits when subjected to induced focal ischemic brain injury (stroke) (Yu and Mattson 1999). Similarly, ADF mice (weighed ≈15% less) recovered from motor function deficits that were chemically induced to mimic the pathology and behavioral symptoms of Parkinson's disease, and the improved outcome was associated with preservation of dopaminergic neurons in the substantia nigra (Duan and Mattson 1999). It should be noted, however, that ADF in the aforementioned studies resulted in spontaneous reduction in overall food intake and therefore also elicited the effects of calorie restriction. Indeed, Arumugam et al. (2010) showed that the protective effect of ADF on stroke outcomes was evident only in groups of mice that lost weight in response to ADF, although age may have been a possible confounder in this study. On the contrary, dissociated effects of periodic fasting and calorie restriction have also been demonstrated. In a transgenic mouse model of Alzheimer's disease, both 40% calorie restriction and ADF were protective against learning and memory deficits, although only the former resulted in significant weight loss (Halagappa et al. 2007). Even when food intake and body weight remained similar to ad libitum controls, Anson et al. (2003) showed that ADF was more effective than caloric restriction in reducing excitotoxin-induced hippocampal neuronal damage in mice. These data suggest that altered meal pattern per se can modulate neuronal function independent of energy consumption.

Periodic fasting has also been shown to ameliorate age-related impairment in brain function and the associated mortality. ADF reversed dendritic spine loss in the aging rat neocortex (Moroi-Fetters et al. 1989) and increased the protein expression of brain-derived neurotrophic factor (BDNF; a marker of neuronal survival) in the cortex of senescence-accelerated mice (Tajes et al. 2010). These data are consistent with the effect of ADF on reducing the functional decline associated with aging, as evidenced by the improvement in motor skills, learning, and memory in 24-month-old male albino Wistar rats (Singh et al. 2012).

The literature largely supports a positive impact of periodic fasting on both average and maximal lifespan. ADF increased both median and maximal lifespan in male wild-type mice (no effect on the female counterparts) (Arum et al. 2009) and prevented fatal lymphoblastic neoplasia in a mouse model of accelerated senescence (Descamps et al. 2005). ADF did not reduce the food intake of any of these mice (Descamps et al. 2005; Arum et al. 2009), which again provides evidence that the effects of periodic fasting are not necessarily mediated by energy restriction. This accords with a study in *Caenorhabditis elegans*, which showed that periodic fasting extended lifespan to a greater extent than calorie restriction, and a combination of the two did not further improve longevity (Honjoh et al. 2009). The authors concluded that the two dietary regimens acted via different mechanisms that may converge on common pathways (e.g., insulin/insulin growth factor–like signaling) and lead to at least some of the overlapping effects of the two dietary regimens (Honjoh et al. 2009).

## 5.4 HORMESIS IN PERIODIC FASTING

Hormesis can be defined as the mounting of a physiological defense at the cellular and tissue/organ level against a mild stress (Masoro 2000). The effects of periodic fasting closely resemble those of a hormetic response. For example, ADF reduces oxidative molecular damage to proteins and lipids in aging rats (Castello et al. 2010; Singh et al. 2012) as well as in rodent models of experimentally induced diabetes (Tikoo et al. 2007) and oxidative stress (Hfaiedh et al. 2008). Circulating levels of markers of oxidative stress were also reduced in humans who undertook a modified periodic fasting regimen (Johnson et al. 2007; Harvie et al. 2011). It has been proposed that the alternating periods of fast and feast create a unique metabolic state that ameliorates deleterious insult and at the same time enhances defense mechanisms against cellular damage. The longer total fasting duration, as opposed to the typical overnight fast, may decrease reactive oxygen species (ROS) production during energy substrate use. Fasting promotes the use of fat as the primary energy substrate and the subsequent increase in mitochondrial β-oxidation (even during the feast days), in theory, should generate fewer ROS as electron transfer bypass complex I (Guarente 2008). Consistent with this notion, ADF mice had significant fluctuations in respiratory quotient on the feast days (Duffy et al. 1997); ADF also led to an overall increase in fat oxidation by ≈20% in healthy individuals (Heilbronn et al. 2005b). A direct effect of periodic fasting on ROS production, however, is yet to be demonstrated.

It is also apparent that periodic fasting confers benefits by increasing cellular resistance to adverse conditions. Mice underwent 8 days of ADF before surgically induced sepsis had less severe organ injury, reduced proinflammatory cytokines (interleukin-6 and tumor necrosis factor-α), and better survival outcomes (Hasegawa et al. 2012). ADF also improved the functional outcomes of a transgenic mouse model of Alzheimer's disease despite a lack of any reductions in extracellular amyloid β-peptide deposition and tau phosphorylation in the brain (Halagappa et al. 2007). Periodic fasting increases resistance to oxidative damage by increasing the activity and/or the expression of proteins that are responsible for handling stress response. ADF restored superoxide dismutase activity in diabetic rats to the level of controls (Tikoo et al. 2007). Similarly, ADF increased antioxidant activity in the spleen of senescence-accelerated OF-1 mice, an effect associated with an attenuated response to succinate-induced ROS production in the spleen mitochondria ex vivo (Descamps et al. 2005).

Induction of heat-shock protein (HSP) is another avenue by which periodic fasting protects cells against metabolic insult. Fasting during Ramadan significantly increased the circulating level of HSP70 in healthy humans as soon as 3 days into the dietary regimen and it continued to increase throughout the 25-day study (Zare et al. 2011). Similarly, the neuroprotective effect of ADF in rodents is closely associated with an increased expression of HSP in various regions of the brain and the preservation of mitochondrial function of neurons (Duan and Mattson 1999; Yu and Mattson

1999; Arumugam et al. 2010; Tajes et al. 2010). Finally, periodic fasting induces the expression of BDNF, which protects neurons from oxidative and metabolic stress (Duan et al. 2001; Tajes et al. 2010).

While it is generally accepted that the effect of periodic fasting is a consequence of a hormetic response, little is known about how the alternations between fast and feast actually induce a stress that subsequently elicits the cytoprotective effect of hormesis. One thing for certain is that this requires regular intervals of energy deprivation as increasing rather than reducing meal frequency is without effect on energy substrate utilization (Munsters and Saris 2012; Ohkawara et al. 2013) and oxidative stress (Heden et al. 2013). The hypothalamic–pituitary–adrenal axis is believed to be the first responder to periodic fasting. Fluctuations in blood glucose level trigger short intervals of glucocorticoid secretion to adapt to changes in fuel availability (Leakey et al. 1994). Whether glucocorticoids directly mediate the effects of periodic fasting is not clear, but they are known to adjust the circulating insulin concentration, which leads to intermittent storage and use of glycogen and fat. It has been proposed that the alternating periods of anabolism and catabolism may improve the metabolic efficiency of the liver and peripheral tissues (Fabry and Tepperman 1970; Leakey et al. 1994). The occasional flux of energy substrate metabolites may also enhance protective functions. For example, many of the glycolytic intermediates (aldehydes) are cytotoxic. They can overwhelm the cellular defense mechanisms and make continuous glycolysis unsustainable. Intermittent glycolytic flux, however, would have these metabolites (at a lower concentration and shorter duration) serve as hormetic agents to provoke transient HSP activity (Hipkiss 2007).

Another avenue by which periodic fasting may induce a mild stress is ROS production. Both extremes of the feeding spectrum (energy overload [Bournat and Brown 2010] and glucose undersupply [Ristow and Zarse 2010]) generate ROS in the mitochondria. In the case of periodic fasting, the cycles between fast and feast would result in short-term spikes of ROS concentrations. This, together with the fluctuations in the activity of the hypothalamic–pituitary–adrenal axis and glycolytic flux, presents multiple forms of transient mild stressors, which may result in an adaptive response (or preconditioning) and enhance overall cellular stress resistance (Anson et al. 2003).

At the molecular level, sirtuin 1 (SIRT1) has been suggested to mediate at least some of the hormetic effect of periodic fasting. SIRT1 is an $NAD^+$-dependent deacetylase that regulates energy substrate use in metabolically active tissues. Its activation has been implicated to positively influence neuronal function and lifespan (Wang et al. 2010; Zhang et al. 2011; Houtkooper et al. 2012). In rodent models, ADF ameliorates cardiovascular risk factors (Tikoo et al. 2007) and the aging process in the brain (Tajes et al. 2010), and in both cases the effects are associated with an increase in SIRT1 expression. Heilbronn and others (2005a) were the first to provide evidence for ADF to increase SIRT1 in skeletal muscle at the transcriptional level in humans. A follow-up study further showed that HepG2 cells increased SIRT1 protein expression and heat-shock resistance when treated with serum from ADF individuals, suggesting that periodic fasting may lead to the release of some circulating factors that enhance the stress responsiveness of the peripheral tissues (Allard et al. 2008). ADF-induced SIRT1 expression may also improve energy metabolism by promoting mitochondrial biogenesis. SIRT1 is known to deacetylate and activate peroxisome proliferator-activated

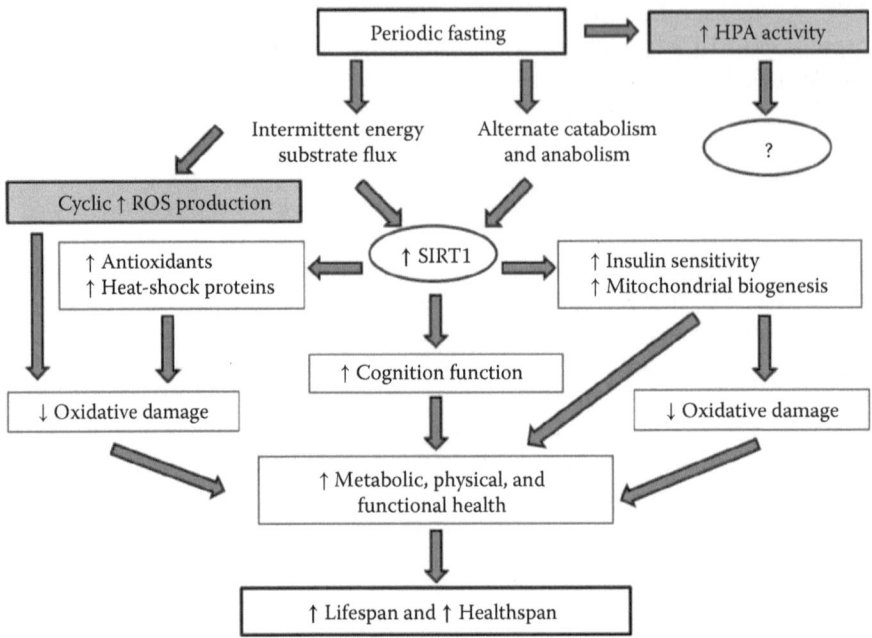

**FIGURE 5.1** Possible mechanisms by which periodic fasting confers health benefits. Both extremes of the feeding spectrum (fasting and overfeeding) increase mitochondrial metabolism and lead to cyclic increased production of reactive oxygen species (ROS). Periodic fasting stimulates the activity of the hypothalamic–pituitary–adrenal axis (HPA) and triggers short intervals of glucocorticoid secretion, although how it mediates the effects of periodic fasting is not clear. Intermittent flux of energy substrates and the alternating catabolism and anabolism increase the expression and probably the activity of sirtuin 1 (SIRT1), which improves cellular defense, energy metabolism, and cognition function. Together with its effect on reducing oxidative molecular damage, SIRT1 positively influences metabolic, physical and functional health, and therefore increases lifespan and healthspan.

receptor gamma coactivator-1α, which subsequently increases mitochondrial gene transcription, including those required for electron transport and respiration (recently reviewed by Brenmoehl and Hoeflich [2013]). It is therefore logical to hypothesize an association between periodic fasting, SIRT1 activation, and improvement in mitochondrial function in peripheral tissues. Indeed, given the role of SIRT1 as one of the major molecular links between calorie restriction and its associated benefits (Bordone and Guarente 2005; Nogueiras et al. 2012), it is tempting to assume similar causality between SIRT1 activation and the effects of periodic fasting. The role of SIRT1-regulated pathways in increasing healthspan is summarized in Figure 5.1.

## 5.5 PERIODIC FASTING AS A HEALTH IMPROVEMENT STRATEGY

Periodic fasting was initially proposed as an easier way to achieve calorie restriction as most people fail to fully compensate the energy deficit during fasting. Perhaps somewhat surprisingly, recent data suggest that reducing meal frequency per se not only meets, (Mager et al. 2006; Harvie et al. 2011) but in some cases exceeds (Anson et al. 2003; Allard et al. 2008) the benefits of calorie restriction. Periodic fasting may well be a more preferable long-term strategy than chronic calorie restriction to increase healthspan. There is no doubt that adherence to periodic fasting is much better than that to chronic caloric restriction. The important question, however, is how the population would adapt to this eating pattern with complete fast every other day. ADF is known to confer positive health outcomes in rodents (Anson et al. 2003; Wan et al. 2003) but a 22-day feasibility study revealed that this eating pattern is unlikely to be sustainable in humans in free-living conditions (Heilbronn et al. 2005b). Other forms of periodic fasting, for example consuming a small portion ($\approx 25\%$) of daily energy need on alternate days (Bhutani et al. 2010) or on several days of the week (Harvie et al. 2011) and allowing ad libitum intake for the remaining days, appear to improve adherence.

Future studies that determine how the span of fasting and postprandial periods affects hormetic response and how it attunes with the circadian patterns (known to be affected by eating time [Iraki et al. 1997; Froy et al. 2009; BaHammam et al. 2010; Bahijri et al. 2013]) would also be important to inform on meal patterns to promote better health. Finally, the largely overlapping effects of periodic fasting and calorie restriction do not imply that they are mutually exclusive but in fact provide alternatives to allow greater flexibility to tailor dietary interventions that best suit individual needs. Also, these regimens would need to be combined with overall lifestyle modifications, for example, a diet of high nutritional value (Klempel et al. 2013) and increased physical activity (Bhutani et al. 2013), to achieve maximal health benefits.

## REFERENCES

Ahmet, I., Wan, R., Mattson, M.P., Lakatta, E.G., Talan, M. 2005. Cardioprotection by intermittent fasting in rats. *Circulation* 112, 3115–21.

Aksungar, F.B., Eren, A., Ure, S., Teskin, O., Ates, G. 2005. Effects of intermittent fasting on serum lipid levels, coagulation status and plasma homocysteine levels. *Ann Nutr Metab* 49, 77–82.

Allard, J.S., Heilbronn, L.K., Smith, C., Hunt, N.D., Ingram, D.K., Ravussin, E., de Cabo, R. 2008. In vitro cellular adaptations of indicators of longevity in response to treatment with serum collected from humans on calorie restricted diets. *PLoS One* 3, e3211.

Anderson, R.M., Bitterman, K.J., Wood, J.G., Medvedik, O., Sinclair, D.A. 2003. Nicotinamide and PNC1 govern lifespan extension by calorie restriction in *Saccharomyces cerevisiae*. *Nature* 423, 181–5.

Anson, R.M., Guo, Z., de Cabo, R., Iyun, T., Rios, M., Hagepanos, A., Ingram, D.K., Lane, M.A., Mattson, M.P. 2003. Intermittent fasting dissociates beneficial effects of dietary restriction on glucose metabolism and neuronal resistance to injury from calorie intake. *Proc Natl Acad Sci USA* 100, 6216–20.

Arum, O., Bonkowski, M.S., Rocha, J.S., Bartke, A. 2009. The growth hormone receptor gene-disrupted mouse fails to respond to an intermittent fasting diet. *Aging Cell* 8, 756–60.

Arumugam, T.V., Phillips, T.M., Cheng, A., Morrell, C.H., Mattson, M.P., Wan, R. 2010. Age and energy intake interact to modify cell stress pathways and stroke outcome. *Ann Neurol* 67, 41–52.

BaHammam, A., Alrajeh, M., Albabtain, M., Bahammam, S., Sharif, M. 2010. Circadian pattern of sleep, energy expenditure, and body temperature of young healthy men during the intermittent fasting of Ramadan. *Appetite* 54, 426–9.

Bahijri, S., Borai, A., Ajabnoor, G., Abdul Khaliq, A., Alqassas, I., Al-Shehri, D., Chrousos, G. 2013. Relative metabolic stability, but disrupted circadian cortisol secretion during the fasting month of Ramadan. *PLoS One* 8, e60917.

Baum, S.J., Kris-Etherton, P.M., Willett, W.C., Lichtenstein, A.H., Rudel, L.L., Maki, K.C., Whelan, J., Ramsden, C.E., Block, R.C. 2012. Fatty acids in cardiovascular health and disease: A comprehensive update. *J Clin Lipidol* 6, 216–34.

Beigy, M., Vakili, S., Berijani, S., Aminizade, M., Ahmadi-Dastgerdi, M., Meshkani, R. 2013. Alternate-day fasting diet improves fructose-induced insulin resistance in mice. *J Anim Physiol Anim Nutr* (Berl) 97, 1125–31.

Bhutani, S., Klempel, M.C., Berger, R.A., Varady, K.A. 2010. Improvements in coronary heart disease risk indicators by alternate-day fasting involve adipose tissue modulations. *Obesity* (Silver Spring) 18, 2152–9.

Bhutani, S., Klempel, M.C., Kroeger, C.M., Trepanowski, J.F., Varady, K.A. 2013. Alternate day fasting and endurance exercise combine to reduce body weight and favorably alter plasma lipids in obese humans. *Obesity* (Silver Spring) 21, 1370–9.

Bordone, L., Guarente, L. 2005. Calorie restriction, SIRT1 and metabolism: Understanding longevity. *Nat Rev Mol Cell Biol* 6, 298–305.

Bournat, J.C., Brown, C.W. 2010. Mitochondrial dysfunction in obesity. *Curr Opin Endocrinol Diabetes Obes* 17, 446–52.

Brenmoehl, J., Hoeflich, A. 2013. Dual control of mitochondrial biogenesis by sirtuin 1 and sirtuin 3. *Mitochondrion* 13, 755–61.

Carlson, O., Martin, B., Stote, K.S., Golden, E., Maudsley, S., Najjar, S.S., Ferrucci, L. et al. 2007. Impact of reduced meal frequency without caloric restriction on glucose regulation in healthy, normal-weight middle-aged men and women. *Metabolism* 56, 1729–34.

Castello, L., Froio, T., Maina, M., Cavallini, G., Biasi, F., Leonarduzzi, G., Donati, A., Bergamini, E., Poli, G., Chiarpotto, E. 2010. Alternate-day fasting protects the rat heart against age-induced inflammation and fibrosis by inhibiting oxidative damage and NF-κB activation. *Free Radic Biol Med* 48, 47–54.

Colman, R.J., Anderson, R.M., Johnson, S.C., Kastman, E.K., Kosmatka, K.J., Beasley, T.M., Allison, D.B. et al. 2009. Caloric restriction delays disease onset and mortality in rhesus monkeys. *Science* 325, 201–4.

Descamps, O., Riondel, J., Ducros, V., Roussel, A.M. 2005. Mitochondrial production of reactive oxygen species and incidence of age-associated lymphoma in OF1 mice: Effect of alternate-day fasting. *Mech Ageing Dev* 126, 1185–91.

Duan, W., Guo, Z., Mattson, M.P. 2001. Brain-derived neurotrophic factor mediates an excitoprotective effect of dietary restriction in mice. *J Neurochem* 76, 619–26.

Duan, W., Mattson, M.P. 1999. Dietary restriction and 2-deoxyglucose administration improve behavioral outcome and reduce degeneration of dopaminergic neurons in models of Parkinson's disease. *J Neurosci Res* 57, 195–206.

Duffy, P.H., Leakey, J.E., Pipkin, J.L., Turturro, A., Hart, R.W. 1997. The physiologic, neurologic, and behavioral effects of caloric restriction related to aging, disease, and environmental factors. *Environ Res* 73, 242–8.

Duncan, G.G. 1962. Intermittent fasts in the correction and control of intractable obesity. *Trans Am Clin Climatol Assoc* 74, 121–9.

el Ati, J., Beji, C., Danguir, J. 1995. Increased fat oxidation during Ramadan fasting in healthy women: An adaptive mechanism for body-weight maintenance. *Am J Clin Nutr* 62, 302–7.

Fabry, P., Tepperman, J. 1970. Meal frequency—A possible factor in human pathology. *Am J Clin Nutr* 23, 1059–68.

Froy, O., Chapnik, N., Miskin, R. 2009. Effect of intermittent fasting on circadian rhythms in mice depends on feeding time. *Mech Ageing Dev* 130, 154–60.

Gilliland, I.C. 1968. Total fasting in the treatment of obesity. *Postgrad Med J* 44, 58–61.

Guarente, L. 2008. Mitochondria—A nexus for aging, calorie restriction, and sirtuins? *Cell* 132, 171–6.

Halagappa, V.K., Guo, Z., Pearson, M., Matsuoka, Y., Cutler, R.G., Laferla, F.M., Mattson, M.P. 2007. Intermittent fasting and caloric restriction ameliorate age-related behavioral deficits in the triple-transgenic mouse model of Alzheimer's disease. *Neurobiol Dis* 26, 212–20.

Halberg, N., Henriksen, M., Soderhamn, N., Stallknecht, B., Ploug, T., Schjerling, P., Dela, F. 2005. Effect of intermittent fasting and refeeding on insulin action in healthy men. *J Appl Physiol* 99, 2128–36.

Harvie, M.N., Pegington, M., Mattson, M.P., Frystyk, J., Dillon, B., Evans, G., Cuzick, J. et al. 2011. The effects of intermittent or continuous energy restriction on weight loss and metabolic disease risk markers: A randomized trial in young overweight women. *Int J Obes* (London) 35, 714–27.

Hasegawa, A., Iwasaka, H., Hagiwara, S., Asai, N., Nishida, T., Noguchi, T. 2012. Alternate day calorie restriction improves systemic inflammation in a mouse model of sepsis induced by cecal ligation and puncture. *J Surg Res* 174, 136–41.

Heden, T.D., Liu, Y., Sims, L.J., Whaley-Connell, A.T., Chockalingam, A., Dellsperger, K.C., Kanaley, J.A. 2013. Meal frequency differentially alters postprandial triacylglycerol and insulin concentrations in obese women. *Obesity* (Silver Spring) 21, 123–9.

Heilbronn, L.K., Civitarese, A.E., Bogacka, I., Smith, S.R., Hulver, M., Ravussin, E. 2005a. Glucose tolerance and skeletal muscle gene expression in response to alternate day fasting. *Obes Res* 13, 574–81.

Heilbronn, L.K., de Jonge, L., Frisard, M.I., DeLany, J.P., Larson-Meyer, D.E., Rood, J., Nguyen, T. et al. 2006. Effect of 6-month calorie restriction on biomarkers of longevity, metabolic adaptation, and oxidative stress in overweight individuals: A randomized controlled trial. *JAMA* 295, 1539–48.

Heilbronn, L.K., Smith, S.R., Martin, C.K., Anton, S.D., Ravussin, E. 2005b. Alternate-day fasting in nonobese subjects: Effects on body weight, body composition, and energy metabolism. *Am J Clin Nutr* 81, 69–73.

Hfaiedh, N., Allagui, M.S., Carreau, S., Zourgui, L., Feki, A., Croute, F. 2008. Impact of dietary restriction on peroxidative effects of nickel chloride in Wistar rats. *Toxicol Mech Methods* 18, 597–603.

Hipkiss, A.R. 2007. Dietary restriction, glycolysis, hormesis and ageing. *Biogerontology* 8, 221–4.

Honjoh, S., Yamamoto, T., Uno, M., Nishida, E. 2009. Signalling through RHEB-1 mediates intermittent fasting-induced longevity in *C. elegans*. *Nature* 457, 726–30.

Horne, B.D., May, H.T., Anderson, J.L., Kfoury, A.G., Bailey, B.M., McClure, B.S., Renlund, D.G. et al. 2008. Usefulness of routine periodic fasting to lower risk of coronary artery disease in patients undergoing coronary angiography. *Am J Cardiol* 102, 814–9.

Houtkooper, R.H., Pirinen, E., Auwerx, J. 2012. Sirtuins as regulators of metabolism and healthspan. *Nat Rev Mol Cell Biol* 13, 225–38.

Iraki, L., Bogdan, A., Hakkou, F., Amrani, N., Abkari, A., Touitou, Y. 1997. Ramadan diet restrictions modify the circadian time structure in humans. A study on plasma gastrin, insulin, glucose, and calcium and on gastric pH. *J Clin Endocrinol Metab* 82, 1261–73.

Johnson, J.B., Summer, W., Cutler, R.G., Martin, B., Hyun, D.H., Dixit, V.D., Pearson, M. et al. 2007. Alternate day calorie restriction improves clinical findings and reduces markers of oxidative stress and inflammation in overweight adults with moderate asthma. *Free Radic Biol Med* 42, 665–74.

Johnstone, A.M. 2007. Fasting—The ultimate diet? *Obes Rev* 8, 211–22.

Katare, R.G., Kakinuma, Y., Arikawa, M., Yamasaki, F., Sato, T. 2009. Chronic intermittent fasting improves the survival following large myocardial ischemia by activation of BDNF/VEGF/PI3K signaling pathway. *J Mol Cell Cardiol* 46, 405–12.

Klempel, M.C., Bhutani, S., Fitzgibbon, M., Freels, S., Varady, K.A. 2010. Dietary and physical activity adaptations to alternate day modified fasting: Implications for optimal weight loss. *Nutr J* 9, 35.

Klempel, M.C., Kroeger, C.M., Norkeviciute, E., Goslawski, M., Phillips, S.A., Varady, K.A. 2013. Benefit of a low-fat over high-fat diet on vascular health during alternate day fasting. *Nutr Diabetes* 3, e71.

Knowler, W.C., Barrett-Connor, E., Fowler, S.E., Hamman, R.F., Lachin, J.M., Walker, E.A., Nathan, D.M. 2002. Reduction in the incidence of type 2 diabetes with lifestyle intervention or metformin. *N Engl J Med* 346, 393–403.

Knowler, W.C., Fowler, S.E., Hamman, R.F., Christophi, C.A., Hoffman, H.J., Brenneman, A.T., Brown-Friday, J.O., Goldberg, R., Venditti, E., Nathan, D.M. 2009. 10-year follow-up of diabetes incidence and weight loss in the Diabetes Prevention Program Outcomes Study. *Lancet* 374, 1677–86.

Kruger, J., Galuska, D.A., Serdula, M.K., Jones, D.A. 2004. Attempting to lose weight: Specific practices among U.S. adults. *Am J Prev Med* 26, 402–6.

Leakey, J.E., Chen, S., Manjgaladze, M., Turturro, A., Duffy, P.H., Pipkin, J.L., Hart, R.W. 1994. Role of glucocorticoids and "caloric stress" in modulating the effects of caloric restriction in rodents. *Ann NY Acad Sci* 719, 171–94.

Lu, J., E, L., Wang, W., Frontera, J., Zhu, H., Wang, W.T., Lee, P. et al. 2011. Alternate day fasting impacts the brain insulin-signaling pathway of young adult male C57BL/6 mice. *J Neurochem* 117, 154–63.

Mager, D.E., Wan, R., Brown, M., Cheng, A., Wareski, P., Abernethy, D.R., Mattson, M.P. 2006. Caloric restriction and intermittent fasting alter spectral measures of heart rate and blood pressure variability in rats. *FASEB J* 20, 631–7.

Masoro, E.J. 2000. Hormesis is the beneficial action resulting from the response of an organism to a low-intensity stressor. *Hum Exp Toxicol* 19, 340–1.

Mattison, J.A., Roth, G.S., Beasley, T.M., Tilmont, E.M., Handy, A.M., Herbert, R.L., Longo, D.L. et al. 2012. Impact of caloric restriction on health and survival in rhesus monkeys from the NIA study. *Nature* 489, 318–21.

Mattson, M.P., Wan, R. 2005. Beneficial effects of intermittent fasting and caloric restriction on the cardiovascular and cerebrovascular systems. *J Nutr Biochem* 16, 129–37.

Michalsen, A., Hoffmann, B., Moebus, S., Backer, M., Langhorst, J., Dobos, G.J. 2005. Incorporation of fasting therapy in an integrative medicine ward: Evaluation of outcome, safety, and effects on lifestyle adherence in a large prospective cohort study. *J Altern Complement Med* 11, 601–7.

Michalsen, A., Li, C., Kaiser, K., Ludtke, R., Meier, L., Stange, R., Kessler, C. 2013. In-patient treatment of fibromyalgia: A controlled nonrandomized comparison of conventional medicine versus integrative medicine including fasting therapy. *Evid Based Complement Alternat Med* 2013, 908610.

Moroi-Fetters, S.E., Mervis, R.F., London, E.D., Ingram, D.K. 1989. Dietary restriction suppresses age-related changes in dendritic spines. *Neurobiol Aging* 10, 317–22.

Mulas, M.F., Demuro, G., Mulas, C., Putzolu, M., Cavallini, G., Donati, A., Bergamini, E., Dessi, S. 2005. Dietary restriction counteracts age-related changes in cholesterol metabolism in the rat. *Mech Ageing Dev* 126, 648–54.

Munsters, M.J., Saris, W.H. 2012. Effects of meal frequency on metabolic profiles and substrate partitioning in lean healthy males. *PLoS One* 7, e38632.

Murphy, J.C., McDaniel, J.L., Mora, K., Villareal, D.T., Fontana, L., Weiss, E.P. 2012. Preferential reductions in intermuscular and visceral adipose tissue with exercise-induced weight loss compared with calorie restriction. *J Appl Physiol* 112, 79–85.

Nogueiras, R., Habegger, K.M., Chaudhary, N., Finan, B., Banks, A.S., Dietrich, M.O., Horvath, T.L., Sinclair, D.A., Pfluger, P.T., Tschop, M.H. 2012. Sirtuin 1 and sirtuin 3: Physiological modulators of metabolism. *Physiol Rev* 92, 1479–514.

Ohkawara, K., Cornier, M.A., Kohrt, W.M., Melanson, E.L. 2013. Effects of increased meal frequency on fat oxidation and perceived hunger. *Obesity* (Silver Spring) 21, 336–43.

Pedersen, C.R., Hagemann, I., Bock, T., Buschard, K. 1999. Intermittent feeding and fasting reduces diabetes incidence in BB rats. *Autoimmunity* 30, 243–50.

Ristow, M., Zarse, K. 2010. How increased oxidative stress promotes longevity and metabolic health: The concept of mitochondrial hormesis (mitohormesis). *Exp Gerontol* 45, 410–8.

Safdie, F.M., Dorff, T., Quinn, D., Fontana, L., Wei, M., Lee, C., Cohen, P., Longo, V.D. 2009. Fasting and cancer treatment in humans: A case series report. *Aging* (Albany, NY) 1, 988–1007.

Singh, R., Lakhanpal, D., Kumar, S., Sharma, S., Kataria, H., Kaur, M., Kaur, G. 2012. Late-onset intermittent fasting dietary restriction as a potential intervention to retard age-associated brain function impairments in male rats. *Age* (Dordr) 34, 917–33.

Snorek, M., Hodyc, D., Sedivy, V., Durisova, J., Skoumalova, A., Wilhelm, J., Neckar, J., Kolar, F., Herget, J. 2012. Short-term fasting reduces the extent of myocardial infarction and incidence of reperfusion arrhythmias in rats. *Physiol Res* 61, 567–74.

Soeters, M.R., Lammers, N.M., Dubbelhuis, P.F., Ackermans, M., Jonkers-Schuitema, C.F., Fliers, E., Sauerwein, H.P., Aerts, J.M., Serlie, M.J. 2009. Intermittent fasting does not affect whole-body glucose, lipid, or protein metabolism. *Am J Clin Nutr* 90, 1244–51.

Stipp, D. 2013. Is fasting good for you? *Sci Am* 308, 23–4.

Stote, K.S., Baer, D.J., Spears, K., Paul, D.R., Harris, G.K., Rumpler, W.V., Strycula, P. et al. 2007. A controlled trial of reduced meal frequency without caloric restriction in healthy, normal-weight, middle-aged adults. *Am J Clin Nutr* 85, 981–8.

Sung, M.M., Dyck, J.R. 2012. Age-related cardiovascular disease and the beneficial effects of calorie restriction. *Heart Fail Rev* 17, 707–19.

Tajes, M., Gutierrez-Cuesta, J., Folch, J., Ortuno-Sahagun, D., Verdaguer, E., Jimenez, A., Junyent, F., Lau, A., Camins, A., Pallas, M. 2010. Neuroprotective role of intermittent fasting in senescence-accelerated mice P8 (SAMP8). *Exp Gerontol* 45, 702–10.

Talamayan, K.S., Springer, A.E., Kelder, S.H., Gorospe, E.C., Joye, K.A. 2006. Prevalence of overweight misperception and weight control behaviors among normal weight adolescents in the United States. *Sci World J* 6, 365–73.

Tikoo, K., Tripathi, D.N., Kabra, D.G., Sharma, V., Gaikwad, A.B. 2007. Intermittent fasting prevents the progression of type I diabetic nephropathy in rats and changes the expression of Sir2 and p53. *FEBS Lett* 581, 1071–8.

Trepanowski, J.F., Bloomer, R.J. 2010. The impact of religious fasting on human health. *Nutr J* 9, 57.

Varady, K.A., Hellerstein, M.K. 2007. Alternate-day fasting and chronic disease prevention: A review of human and animal trials. *Am J Clin Nutr* 86, 7–13.

Wan, R., Ahmet, I., Brown, M., Cheng, A., Kamimura, N., Talan, M., Mattson, M.P. 2010. Cardioprotective effect of intermittent fasting is associated with an elevation of adiponectin levels in rats. *J Nutr Biochem* 21, 413–7.

Wan, R., Camandola, S., Mattson, M.P. 2003. Intermittent fasting and dietary supplementation with 2-deoxy-D-glucose improve functional and metabolic cardiovascular risk factors in rats. *FASEB J* 17, 1133–4.

Wang, J., Fivecoat, H., Ho, L., Pan, Y., Ling, E., Pasinetti, G.M. 2010. The role of Sirt1: At the crossroad between promotion of longevity and protection against Alzheimer's disease neuropathology. *Biochim Biophys Acta* 1804, 1690–4.

Weindruch, R., Walford, R.L. 1982. Dietary restriction in mice beginning at 1 year of age: Effect on life-span and spontaneous cancer incidence. *Science* 215, 1415–8.

Williams, K.V., Mullen, M.L., Kelley, D.E., Wing, R.R. 1998. The effect of short periods of caloric restriction on weight loss and glycemic control in type 2 diabetes. *Diabetes Care* 21, 2–8.

Yu, Z.F., Mattson, M.P. 1999. Dietary restriction and 2-deoxyglucose administration reduce focal ischemic brain damage and improve behavioral outcome: Evidence for a preconditioning mechanism. *J Neurosci Res* 57, 830–9.

Zare, A., Hajhashemi, M., Hassan, Z.M., Zarrin, S., Pourpak, Z., Moin, M., Salarilak, S., Masudi, S., Shahabi, S. 2011. Effect of Ramadan fasting on serum heat shock protein 70 and serum lipid profile. *Singapore Med J* 52, 491–5.

Zhang, F., Wang, S., Gan, L., Vosler, P.S., Gao, Y., Zigmond, M.J., Chen, J. 2011. Protective effects and mechanisms of sirtuins in the nervous system. *Prog Neurobiol* 95, 373–95.

# 6 Iron, Metabolic Syndrome, and Hormesis

*Kupper A. Wintergerst and Lu Cai*

## CONTENTS

6.1 Introduction ................................................................................................. 93
6.2 Iron Homeostasis, Biological Function, and Toxicity ................................... 94
6.3 Iron Deficiency ............................................................................................. 96
    6.3.1 Iron Deficiency and Insulin Resistance .............................................. 96
    6.3.2 Iron Deficiency and Glycated Hemoglobin Levels ........................... 97
    6.3.3 Iron Deficiency and Cardiovascular Diseases ................................... 97
6.4 Iron Overload and Insulin Resistance or Type 2 Diabetes .......................... 99
6.5 Summary ..................................................................................................... 101
Acknowledgments .............................................................................................. 102
References .......................................................................................................... 102

## 6.1 INTRODUCTION

Iron is one of the essential metals required for a variety of molecules to maintain their normal structures and functions and for cells to live, grow, and proliferate. However, maintaining a safe dose level of iron is important for normal functioning of our body because iron overdose often causes toxicity in several organs and iron deficiency causes anemia and other diseases (Anderson and Wang 2012). Therefore, abnormal homeostasis of iron is associated with several pathogenic conditions, such as metabolic syndrome (Beutler et al. 2003; Liu et al. 2009).

The term *metabolic syndrome* dates back to at least the late 1950s, but came into common usage in the late 1970s to describe various associations of risk factors with diabetes that had been noted as early as the 1920s. In 1988, in his Banting lecture, Gerald M. Reaven proposed insulin resistance as the underlying factor and named the constellation of abnormalities *syndrome X*. The terms *metabolic syndrome*, *insulin resistance syndrome*, and *syndrome X* are now used specifically to define a constellation of abnormalities associated with increased risk for the development of type 2 diabetes and atherosclerotic vascular disease, for example, heart disease and stroke. For diagnosis of metabolic syndrome, the 1999 World Health Organization criteria require the presence of any one of diabetes mellitus, impaired glucose tolerance, impaired fasting glucose, or insulin resistance, with two of the following: (1) blood pressure: ≥140/90 mmHg; (2) dyslipidemia: triglycerides: 1.695 mmol/L and high-density lipoprotein cholesterol ≤0.9 mmol/L (male) or ≤1.0 mmol/L (female);

(3) central obesity: the ratio of waist to hip >0.90 (male) or >0.85 (female) or body mass index >30 kg/m$^2$; or (4) microalbuminuria: urinary albumin excretion ratio ≥20 μg/min or albumin/creatinine ratio ≥30 mg/g.

Recently, several reviews have discussed the role of iron in the development of diabetes and diabetic complications (Swaminathan et al. 2007; Liu et al. 2009; Dongiovanni et al. 2011; Grunblatt et al. 2011; Bao et al. 2012) with different focuses; therefore, this chapter briefly outlines iron homeostasis, biological function, and toxicity, and then mainly updates the experimental and epidemiological studies showing the hormetic effect of iron on metabolic syndrome.

## 6.2 IRON HOMEOSTASIS, BIOLOGICAL FUNCTION, AND TOXICITY

Iron plays a crucial role in oxygen sensing and transport, electron transfer, and catalysis (Bruick 2003; Rouault 2006). A significant fraction of cellular iron is associated with proteins in the form of heme, a common prosthetic group composed of protoporphyrin IX and a ferrous ($Fe^{2+}$) ion. The insertion of $Fe^{2+}$ into protoporphyrin IX, catalyzed by ferrochelatase in the mitochondria, defines the terminal step of the heme biosynthetic pathway. Heme is then exported to the cytosol for incorporation into hemoproteins. The most abundant mammalian hemoproteins, hemoglobin and myoglobin, serve as oxygen carriers in the erythroid tissue and in the muscle, respectively. Other hemoproteins include various cytochromes and enzymes, such as oxygenase, peroxidase, nitric oxide (NO) synthase, or guanylate cyclase (Bruick 2003; Kikuchi et al. 2005; Rouault 2006).

Iron is chemically active in forming a variety of coordination complexes with organic ligands in a dynamic and flexible mode, and has potential to switch between $Fe^{2+}$ and ferric ion ($Fe^{3+}$). The efficiency of $Fe^{2+}$ as an electron donor and $Fe^{3+}$ as an electron acceptor is a fundamental feature for many biochemical reactions and renders iron to be an essential mineral and nutrient. However, the ability of iron to cycle between its two stable oxidation states potentially generates reactive oxygen species or reactive nitrogen species (ROS or RNS) such as hydroxyl radical (OH$^•$) through Fenton and Haber–Weiss reactions (Cai et al. 1998; Eaton and Qian 2002; Papanikolaou and Pantopoulos 2005). Heme iron may catalyze the formation of radicals, mainly via formation of oxoferryl intermediates. Finally, $Fe^{2+}$ can also contribute as a reactant, rather than as a catalyst, to free radical generation by a direct interaction with oxygen, through ferryl ($Fe^{2+}$–O) or perferryl ($Fe^{2+}$–$O_2$) iron intermediates (see reviews by Eaton and Qian [2002]; Papanikolaou and Pantopoulos [2005]).

ROS and RNS, including superoxide, NO, peroxynitrite, hydrogen peroxide, and hydroxyl radical, are highly reactive and potentially damaging to cells and tissues (Eaton and Qian 2002; Papanikolaou and Pantopoulos 2005), although at low concentrations these species may also act as second messengers, gene regulators, and/or mediators of cellular activation (Cai 2006). To control and balance the production of ROS and RNS, the cell builds up a set of antioxidants and detoxifying enzymes, such as superoxide dismutase, catalase, and glutathione peroxidase, that can scavenge excessive ROS or RNS. An increase in the steady-state levels of

ROS and/or RNS beyond the antioxidant capacity of the organism, called oxidative (or nitrosative) stress, is encountered in many pathological conditions, such as diabetes and diabetes-related cardiovascular diseases (Cai 2006; Chyun and Young 2006; Rouault 2006).

These features may determine the hormetic effect of iron on the biological endpoints. For instance, bone metabolism has a close relationship with iron homeostasis. The hormetic effect of iron on osteoblast cells in vitro in terms of cell proliferation and apoptotic cell death was reported recently (Zhao et al. 2012). They used human osteoblast cells (hFOB1.19) for the incubation with a medium supplemented with 0–200 µmol/L ferric ammonium citrate as donor iron and 0–20 µmol/L deferoxamine (DFO) as iron chelator to remove free iron from cells. As shown in Figure 6.1,

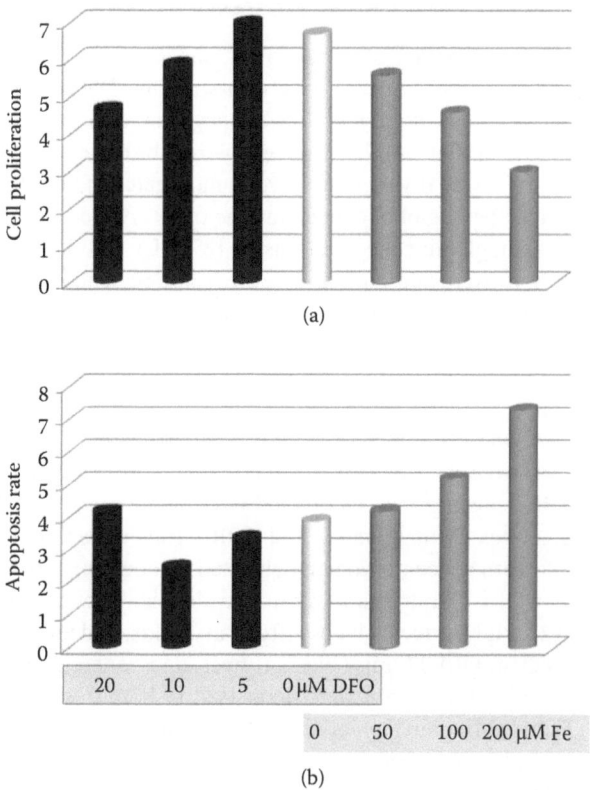

**FIGURE 6.1** The proliferation (a) and apoptotic cell death (b) of osteoblasts after exposure to ferric ammonium citrate at various concentrations. The proliferation was inhibited with the increase in iron (Fe) from ferric ammonium citrate in a concentration-dependent manner; however, it was promoted at 5 µmol/L but inhibited at 10 and 20 µmol/L of iron chelator (deferoxamine [DFO]). The apoptosis rate was increased with the increase in Fe from ferric ammonium citrate in a concentration-dependent manner, and it was decreased at 5 and 10 µmol/L of DFO, but increased at 20 µmol/L of DFO. The graph was made by the authors based on published data (Zhao et al. 2012). Black bars indicate iron deficiency, the white bar indicates the control, and gray bars indicate iron overload.

excessive iron inhibited osteoblast proliferation and induced apoptotic cell death in a concentration-dependent manner; in contrast, low iron concentrations produced a biphasic manner on osteoblasts: mild low iron promoted osteoblast proliferation and decreased apoptotic cell death, but serious low iron inhibited osteoblast proliferation and increased apoptotic cell death. Osteogenesis was optimal (hormetic) in certain iron concentrations (Figure 6.1). The mechanism underlying biological activity invoked by excessive iron may be attributed to increased intracellular ROS levels. Therefore, whether there is a hormetic effect of iron in metabolic syndrome is the next interesting issue to be discussed in Sections 6.3 and 6.4.

## 6.3 IRON DEFICIENCY

### 6.3.1 Iron Deficiency and Insulin Resistance

Iron deficiency is the reduction of storage iron, shown by reduced plasma ferritin concentration and decreased bone marrow iron staining. Deficiency of iron in the system can be derived from blood loss, low dietary iron, or a disease condition that inhibits iron uptake (Liu et al. 2009). For investigating the effect of iron deficiency on insulin resistance, rats were divided into severe, moderate, and mild iron-deficiency anemia groups (mean hemoglobin concentrations of 59, 79, and 107 g/L) and a control group (mean hemoglobin concentrations of 137 g/L). Basal plasma glucose and insulin concentrations were similar between the iron-deficiency anemia and control rats, but an increase in insulin sensitivity was observed in all three anemic groups (Borel et al. 1993).

Enhancing insulin sensitivity by reducing iron level was also noticed in diabetic models. For instance, in a type 2 diabetes rat model, Otsuka Long-Evans Tokushima Fatty (OLETF) rats and age-matched Long-Evans Tokushima Otsuka (LETO) rats, systemic iron levels were reduced either by feeding with low-iron diet from 15 weeks of age or by phlebotomy at 29 weeks of age by removing 4 and 3 ml of blood from the tail vein every week (Minamiyama et al. 2010). Rats were killed at 43 weeks of age. The plasma ferritin concentration and glycated hemoglobin (HbA1c) levels were markedly higher in OLETF rats, but lower in iron-deficient diet and phlebotomy rats. Decreased levels of triglycerides, glucose, free fatty acids, and total cholesterol were found in iron-deficient OLETF rats. Plasma, liver, and pancreas lipid peroxidation and hepatic superoxide production were decreased in both iron-deficient and phlebotomy OLETF rats (Minamiyama et al. 2010). A similar improvement of diabetic complications was also observed in an ob/ob type 2 diabetes mouse model when these mice were fed a low-iron diet (Cooksey et al. 2010).

As already mentioned, iron has a critical role in erythropoiesis. When erythroblasts proliferate and differentiate, they require large amounts of iron for hemoglobin synthesis. Therefore, maternal iron deficiency may affect fetal development and many physiological functions may be altered in the offspring (Furuta et al. 2012). Mouse offspring exposed to maternal iron deficiency during the gestational period were found to develop a susceptibility to high-fat diet (HFD)-induced obesity, and also developed glucose intolerance and increases in blood pressure (Bourque et al. 2012). These symptoms are indicative of metabolic syndrome. These results provided insight into the human situation where the offspring of mothers with iron deficiency at pregnancy have a high risk of developing metabolic syndrome in adulthood under Western diet conditions.

In summary, iron deficiency may increase insulin sensitivity at certain conditions, but increases susceptibility to HFD-induced metabolic syndrome under other conditions. Since there are only few studies focusing on the influence of iron deficiency on insulin sensitivity, further studies are required to explore the effect of severity and period of iron deficiency on insulin sensitivity and resistance.

### 6.3.2 Iron Deficiency and Glycated Hemoglobin Levels

The major form of glycated hemoglobin in diabetic patients is HbA1c. In nondiabetic subjects, HbA1c reaches a steady state with 3.0%–6.5% of the hemoglobin, whereas the HbA1c fraction is abnormally elevated in patients with chronic hyperglycemia and correlates positively with glycemic control. It has been reported that iron deficiency is able to enhance HbA1c in nondiabetic individuals (Tarim et al. 1999; El-Agouza et al. 2002; Coban et al. 2004). For instance, a survey was performed on 730 university students; 81 students were anemic, of which 47 were iron deficient with an average HbA1c of $6.2 \pm 0.6\%$. When they were treated with oral iron and followed up for 20 weeks, HbA1c fell significantly to $5.3 \pm 0.5\%$ (El-Agouza et al. 2002). Recently, a National Health and Nutrition Examination Survey (1999–2006) also revealed the association between iron deficiency and HbA1c levels among adults without diabetes (Kim et al. 2010). In this study, it was found that among women ($n = 6666$), 13.7% had iron deficiency and 4.0% had iron-deficiency anemia. Among women with iron deficiency, 316 had HbA1c $\geq 5.5\%$ and 32 had HbA1c $\geq 6.5\%$.

An early study showed that iron deficiency in a 68-year-old female patient with diabetes was also associated with a marked rise in HbA1c (from 10.7% to 15.4%), and if the iron-deficiency anemia was corrected, the HbA1c level fell to 11.0% (Davis et al. 1983). A prospective study including 37 patients with type 1 diabetes (11 patients were iron deficient and the remaining 26 were iron sufficient) showed that patients with iron deficiency had higher levels of HbA1c than those without iron deficiency. After iron supplementation for 3 months, these patients showed a significant decrease in HbA1c levels (Tarim et al. 1999).

In summary, these studies suggest that among nondiabetic or diabetic individuals with a similar level of glycemia, iron-deficiency anemia is associated with higher concentrations of HbA1c. Iron replacement therapy decreases HbA1c levels in both diabetic and nondiabetic individuals (Davis et al. 1983; Tarim et al. 1999; El-Agouza et al. 2002; Coban et al. 2004).

### 6.3.3 Iron Deficiency and Cardiovascular Diseases

As introduced at the beginning of this chapter, the risk of cardiovascular diseases is also one symptom of metabolic syndrome, and the potential role of iron deficiency on cardiac pathogenesis is being discussed (Dec 2009). This author stated that anemia ranges in prevalence from below 10% among patients with mild heart failure symptoms to over 40% for those with advanced disease. Iron deficiency is a potential cause of anemia and the prevalence of iron deficiency has been reported to range from 5% to 21%. Iron is essential not only for erythropoiesis but also for several bioenergetic processes in skeletal muscle. Chronic iron deficiency may, by itself, reduce exercise capacity and cause ultrastructural alterations in cardiomyocytes (Brownlie et al. 2004). Therefore, whether iron deficiency is associated with cardiovascular diseases has been explored recently (Dec 2009).

In the first study, 40 patients with anemia received treatment with intravenous iron or saline infusion for 5 weeks. In the intravenous iron group as compared with the saline infusion group, the mean hemoglobin level increased by 1.4 g/dL, and there was a significant improvement in the Minnesota Living with Heart Failure score, decrease in levels of C-reactive protein and N-terminal pro–brain natriuretic peptide, and an increase in left ventricular ejection fraction and distance on the six-minute walk test (Toblli et al. 2007).

In the Ferric Iron Sucrose in Heart Failure (FERRIC-HF) trial, 35 patients with iron deficiency received intravenous iron, aimed at improving the ferritin level, or saline infusion. Intravenous iron loading decreased the New York Heart Association (NYHA) functional class and improved the patients' global assessment score. The peak oxygen uptake rose significantly in patients who had anemia but not in those who did not have anemia (Okonko et al. 2008).

The Ferinject Assessment in Patients with Iron Deficiency and Chronic Heart Failure (FAIR-HF) was a multicenter trial that evaluated the efficacy of intravenous iron infusion on symptoms and submaximal exercise capacity in a cohort of patients with mild or moderate heart failure because of left ventricular systolic dysfunction (Anker et al. 2009). The trial enrolled 459 patients with NYHA functional class II or III symptoms, a depressed left ventricular ejection fraction, and documented iron deficiency. According to a randomized, placebo-controlled design, patients were assigned to receive 200 mg of intravenous iron or infused saline weekly until their iron stores were replete. Then, intravenous iron or placebo infusions were continued every 4 weeks up to week 24. The primary endpoints were the self-reported Patient Global Assessment and the NYHA functional class at week 24. Secondary endpoints included the distance on the six-minute walk test and health-related quality-of-life validated surveys at weeks 4, 12, and 24.

Ferric carboxymaltose therapy rapidly increased ferritin levels to be within the normal range. The administration of intravenous iron, as compared with placebo, convincingly improved the self-reported Patient Global Assessment and the NYHA functional class. For the self-reported Patient Global Assessment, 50% of the treated patients reported that they were much or moderately improved, as compared with only 28% of the control patients. The degree of improvement in both endpoints was similar in patients with anemia and those without anemia. Furthermore, significant improvement in the secondary endpoints, including an increase of more than 30 m in the six-minute walk distance, was also observed (Anker et al. 2009). This improvement of cardiac function by intravenous iron treatment was further proved to be independent of anemia status (Comin-Colet et al. 2013). Therefore, the approach taken in this study may have merit in patients with moderately symptomatic heart failure and documented iron deficiency (Dec 2009).

In a later prospective observational study, 546 patients with stable systolic chronic heart failure were enrolled with the finding that iron deficiency is common and constitutes a strong, independent predictor of unfavorable outcome. In a multiple logistic model, iron deficiency was more prevalent in women, those in the advanced NYHA class, with higher plasma N-terminal pro–brain natriuretic peptide and higher serum high-sensitivity C-reactive protein. At the end of follow-up (mean duration: 731 ± 350 days), there were 153 (28%) deaths and 30 (6%) heart transplantations.

In multivariable models, iron deficiency (but not anemia) was related to an increased risk of death or heart transplantations (Jankowska et al. 2010). As a single case report, it was found that iron deficiency is also a risk factor for cerebral venous thrombosis (Huang et al. 2010). In a study that investigated the role of iron deficiency in cyanotic congenital heart disease, 25 iron-deficient cyanotic congenital heart disease patients were prospectively studied. Oral ferrous fumarate was titrated to a maximum dose of 200 mg thrice daily. Three months of iron replacement therapy in iron-deficient cyanotic congenital heart disease patients was safe and resulted in significant improvement in exercise tolerance and quality of life (Tay et al. 2011). However, in community-dwelling U.S. adults with self-reported heart failure in the National Health and Nutrition Examination Survey III, although iron deficiency was common and was associated with markers of anemia and inflammation, their data did not support a direct relationship between iron deficiency and mortality (Parikh et al. 2011).

In summary, although there were certain discrepancies, most of the earlier mentioned studies seem to indicate the importantly etiological role of iron deficiency in the pathogenesis of various cardiac dysfunctions and damage as well as the risk of cardiovascular diseases (Attanasio et al. 2012).

## 6.4 IRON OVERLOAD AND INSULIN RESISTANCE OR TYPE 2 DIABETES

Iron overload means an increase of storage iron, which involves primarily hepatocytes in the case of digestive hyperabsorption (hemochromatosis and dyserythropoiesis) and macrophages in the case of transfusional excess. An early study in 36 healthy subjects indicated a significant correlation of serum ferritin with insulin resistance (Fernandez-Real et al. 1998). This correlation was confirmed by subsequently epidemiological studies; that is, increased heme iron (derived from animal products) was significantly associated with an increased risk of insulin resistance and type 2 diabetes among a variety of populations, including Asians, Americans, and Europeans (Kim et al. 2000, 2011; Hua et al. 2001; Jehn et al. 2004, 2007; D'Souza et al. 2005; Rajpathak et al. 2006; Wrede et al. 2006; Forouhi et al. 2007; Vari et al. 2007; Ashourpour et al. 2010; Yu et al. 2012). This association between iron overload and insulin resistance or type 2 diabetes as well as controversial reports were discussed in detail before (Rajpathak et al. 2009). Therefore, we will just outline several features regarding this association in this section.

Serum ferritin was found to be an important and independent predictor of the development of insulin resistance (Brudevold et al. 2008) and diabetes (Forouhi et al. 2007; Freixenet et al. 2011). For instance, in 40 patients with metabolic syndrome, the ferritin level, but not increased iron load, was found to be positively associated with insulin resistance (Brudevold et al. 2008). Another report also stated that the baseline serum ferritin in 360 patients with type 2 diabetes was significantly higher than that in 758 control participants after adjustment for other known risk factors (age, body mass index, sex, family history, physical activity, smoking habit) and dietary factors (Forouhi et al. 2007).

Increased serum ferritin is frequently observed in nonalcoholic fatty liver disease that is often accompanied by hepatic insulin resistance. As reported earlier, iron depletion by phlebotomy in 64 patients with nonalcoholic fatty liver disease significantly decreased insulin resistance as compared with 64 nonphlebotomy patients with nonalcoholic fatty liver disease who were matched for age, sex, ferritin, obesity, and alanine aminotransferase levels and underwent lifestyle modifications (Valenti et al. 2007). On the basis of this study, iron depletion seemed to improve insulin resistance more effectively than lifestyle changes alone.

Dietary consumption of red meat is considered to increase heme iron intake; therefore, whether risks of insulin resistance and type 2 diabetes are associated with dietary consumption of various types of meat has been extensively investigated (see the reviews by Micha et al. [2012]; Rajpathak et al. [2009]; The InterAct [2013]). Diets high in processed meat were related to increased risk of type 2 diabetes (Schulze et al. 2003). In this study, however, serum iron level was not available. In contrast, vegetarians, who had lower body iron stores than meat eaters, were more insulin sensitive than meat eaters (Hua et al. 2001). In this study, six male meat eaters were phlebotomized to reduce body iron stores to a level similar to that of vegetarians, which resulted in a significant reduction of serum ferritin concentration along with a significant improvement of insulin resistance (Hua et al. 2001).

These studies seem to indicate the importance of serum iron derived from dietary meat in the development of insulin resistance. However, there was another study showing a contrast. These authors further subdivided heme iron intake in two groups based on red meat and sources other than red meat, and then examined each source to the risk of diabetes (Jiang et al. 2004). Diabetes risk was increased with increasing heme iron intakes from red meat, but not with those from sources other than red meat, such as chicken and fish. This finding suggests that heme iron may not be a causative factor for metabolic syndrome or type 2 diabetes, but rather may be one of several metabolic syndrome–related influencing factors. This concept was supported by the positive association of hyperferritinemia with insulin resistance and fatty liver in patients without iron overload (Brudevold et al. 2008).

In addition, there was a significant difference between men and women and between different ethnic populations for the association of heme iron levels with insulin sensitivity or risk of type 2 diabetes (Sheu et al. 2003; Acton et al. 2006; Shi et al. 2006). A study showed that the mean serum ferritin concentration was significantly greater in women with diabetes in athletic groups and in Native American men with diabetes than in those without diabetes (Acton et al. 2006). One previous study (Sheu et al. 2003) observed that a relationship between serum ferritin levels and insulin resistance exists in women but not in men, which is supported by studies from China (Shi et al. 2006; Yu et al. 2012). In contrast to the above studies, a study from Korea with 12,090 subjects (6,378 men and 5,712 women; age, 20–89 years) indicated that iron overload is associated with insulin resistance in men, but not in women (Kim et al. 2011). A study of the population of North China indicated that associations among higher serum ferritin level, higher heme iron intake, and elevated risk of diabetes were found without gender difference (Luande et al. 2008).

In Chinese women, body mass index also affected the risk of meat consumption for type 2 diabetes (Villegas et al. 2006). High consumption of total unprocessed meat was related to a modest reduction in the risk of type 2 diabetes among normal-weight women, but was associated with a modest increase in the risk of type 2 diabetes among obese women. However, poultry consumption was not associated with a high risk of type 2 diabetes among obese participants. Therefore, processed meat consumption was associated with an increased risk of developing type 2 diabetes, particularly for obese participants, suggesting the important contribution of components of processing to the risk of diabetes (Villegas et al. 2006).

In summary, these studies imply that heme iron levels derived from different diets, including red meat, were positively associated with the increased risk of insulin resistance and diabetes under most conditions; however, whether it is a direct cause remains a concern since the correlation of increased heme iron with insulin resistance or diabetes did not exist under certain other conditions. For instance, iron overload significantly increases the risk of insulin resistance or type 2 diabetes in Asian women (but not in Asian men), and also in certain ethnic men such as Native American men with diabetes, probably because certain unrecognized confounding factors exist in these populations. Therefore, a diet rich in iron is similar to a diet rich in fat and energy, which is known to be associated with metabolic syndrome. Both of them are associated with insulin resistance, modified hepatic lipid and iron metabolism, and increased mitochondrial dysfunction and oxidative stress (Choi et al. 2013).

## 6.5 SUMMARY

Iron is a very essential mineral for the body. Either deficiency or overload will contribute certain pathogenic effects on metabolic syndrome. As summarized in Figure 6.2, temporal mild iron deficiency seems to prevent the development of insulin resistance

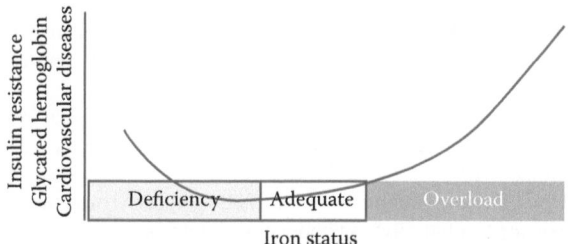

**FIGURE 6.2** A scheme showing the hormetic effect of iron on metabolic syndrome. On the basis of the available experimental and epidemiological data, temporal mild iron deficiency seems to prevent the development of insulin resistance in the normal population and to sensitive insulin response in type 2 diabetes. On the other hand, chronic iron deficiency enhances glycated hemoglobin levels and the risk of cardiovascular diseases. Iron overload increases the prevalence of insulin resistance, type 2 diabetes, and cardiovascular diseases.

in normal population and to sensitive insulin response in type 2 diabetes, but on the other hand, chronic iron deficiency enhances the risk of cardiovascular diseases. Iron overload increases the prevalence of insulin resistance and even type 2 diabetes. All these experimental and clinical observations strongly suggest the hormetic effect of iron on the metabolic syndrome.

## ACKNOWLEDGMENTS

The work was supported in part by research grants from the American Diabetes Association (02-07-JF-10, 05-07-CD-02, and 01-11-BS-17) and a Starting-Up Fund for Chinese-American Research Institute for Diabetic Complications from Wenzhou Medical University, Wenzhou, China.

## REFERENCES

Acton, R. T., Barton, J. C., Passmore, L. V., Adams, P. C., Speechley, M. R., Dawkins, F. W., Sholinsky, P. et al. 2006. Relationships of serum ferritin, transferrin saturation, and HFE mutations and self-reported diabetes in the Hemochromatosis and Iron Overload Screening (HEIRS) study. *Diabetes Care* 29, 2084–9.

Anderson, G. J., and Wang, F. 2012. Essential but toxic: Controlling the flux of iron in the body. *Clin Exp Pharmacol Physiol* 39, 719–24.

Anker, S. D., Comin Colet, J., Filippatos, G., Willenheimer, R., Dickstein, K., Drexler, H., Luscher, T. F. et al. 2009. Ferric carboxymaltose in patients with heart failure and iron deficiency. *N Engl J Med* 361, 2436–48.

Ashourpour, M., Djalali, M., Djazayery, A., Eshraghian, M. R., Taghdir, M., and Saedisomeolia, A. 2010. Relationship between serum ferritin and inflammatory biomarkers with insulin resistance in a Persian population with type 2 diabetes and healthy people. *Int J Food Sci Nutr* 61, 316–23.

Attanasio, P., Ronco, C., Anker, S. D., Cicoira, M., and von Haehling, S. 2012. Role of iron deficiency and anemia in cardio-renal syndromes. *Semin Nephrol* 32, 57–62.

Bao, W., Rong, Y., Rong, S., and Liu, L. 2012. Dietary iron intake, body iron stores, and the risk of type 2 diabetes: A systematic review and meta-analysis. *BMC Med* 10, 119.

Beutler, E., Hoffbrand, A. V., and Cook, J. D. 2003. Iron deficiency and overload. *Hematology Am Soc Hematol Educ Program*, 40–61.

Borel, M. J., Beard, J. L., and Farrell, P. A. 1993. Hepatic glucose production and insulin sensitivity and responsiveness in iron-deficient anemic rats. *Am J Physiol* 264, E380–90.

Bourque, S. L., Komolova, M., McCabe, K., Adams, M. A., and Nakatsu, K. 2012. Perinatal iron deficiency combined with a high-fat diet causes obesity and cardiovascular dysregulation. *Endocrinology* 153, 1174–82.

Brownlie, T., Utermohlen, V., Hinton, P. S., and Haas, J. D. 2004. Tissue iron deficiency without anemia impairs adaptation in endurance capacity after aerobic training in previously untrained women. *Am J Clin Nutr* 79, 437–43.

Brudevold, R., Hole, T., and Hammerstrom, J. 2008. Hyperferritinemia is associated with insulin resistance and fatty liver in patients without iron overload. *PLoS One* 3, e3547.

Bruick, R. K. 2003. Oxygen sensing in the hypoxic response pathway: Regulation of the hypoxia-inducible transcription factor. *Genes Dev* 17, 2614–23.

Cai, L. 2006. Suppression of nitrative damage by metallothionein in diabetic heart contributes to the prevention of cardiomyopathy. *Free Radic Biol Med* 41, 851–61.

Cai, L., Tsiapalis, G., and Cherian, M. G. 1998. Protective role of zinc-metallothionein on DNA damage in vitro by ferric nitrilotriacetate (Fe-NTA) and ferric salts. *Chem Biol Interact* 115, 141–51.

Choi, J. S., Koh, I. U., Lee, H. J., Kim, W. H., and Song, J. 2013. Effects of excess dietary iron fat on glucose lipid metabolism. *J Nutr Biochem.* 24, 1634–44. doi: 10.1016/j.jnutbio.2013.02.004.

Chyun, D. A., and Young, L. H. 2006. Diabetes mellitus and cardiovascular disease. *Nurs Clin North Am* 41, 681–95.

Coban, E., Ozdogan, M., and Timuragaoglu, A. 2004. Effect of iron deficiency anemia on the levels of hemoglobin A1c in nondiabetic patients. *Acta Haematol* 112, 126–8.

Comin-Colet, J., Lainscak, M., Dickstein, K., Filippatos, G. S., Johnson, P., Luscher, T. F., Mori, C., Willenheimer, R., Ponikowski, P., and Anker, S. D. 2013. The effect of intravenous ferric carboxymaltose on health-related quality of life in patients with chronic heart failure and iron deficiency: A subanalysis of the FAIR-HF study. *Eur Heart J* 34, 30–8.

Cooksey, R. C., Jones, D., Gabrielsen, S., Huang, J., Simcox, J. A., Luo, B., Soesanto, Y., Rienhoff, H., Abel, E. D., and McClain, D. A. 2010. Dietary iron restriction or iron chelation protects from diabetes and loss of beta-cell function in the obese (ob/ob lep−/−) mouse. *Am J Physiol Endocrinol Metab* 298, E1236–43.

D'Souza, R. F., Feakins, R., Mears, L., Sabin, C. A., and Foster, G. R. 2005. Relationship between serum ferritin, hepatic iron staining, diabetes mellitus, and fibrosis progression in patients with chronic hepatitis C. *Aliment Pharmacol Ther* 21, 519–24.

Davis, R. E., McCann, V. J., and Nicol, D. J. 1983. Influence of iron-deficiency anaemia on the glycosylated haemoglobin level in a patient with diabetes mellitus. *Med J Aust* 1, 40–1.

Dec, G. W. 2009. Anemia and iron deficiency—New therapeutic targets in heart failure? *N Engl J Med* 361, 2475–7.

Dongiovanni, P., Fracanzani, A. L., Fargion, S., and Valenti, L. 2011. Iron in fatty liver and in the metabolic syndrome: A promising therapeutic target. *J Hepatol* 55, 920–32.

Eaton, J. W., and Qian, M. 2002. Molecular bases of cellular iron toxicity. *Free Radic Biol Med* 32, 833–40.

El-Agouza, I., Abu Shahla, A., and Sirdah, M. 2002. The effect of iron deficiency anaemia on the levels of haemoglobin subtypes: Possible consequences for clinical diagnosis. *Clin Lab Haematol* 24, 285–9.

Fernandez-Real, J. M., Ricart-Engel, W., Arroyo, E., Balanca, R., Casamitjana-Abella, R., Cabrero, D., Fernandez-Castaner, M., and Soler, J. 1998. Serum ferritin as a component of the insulin resistance syndrome. *Diabetes Care* 21, 62–8.

Forouhi, N. G., Harding, A. H., Allison, M., Sandhu, M. S., Welch, A., Luben, R., Bingham, S., Khaw, K. T., and Wareham, N. J. 2007. Elevated serum ferritin levels predict new-onset type 2 diabetes: Results from the EPIC-Norfolk prospective study. *Diabetologia* 50, 949–56.

Freixenet, N., Vilardell, C., Llaurado, G., Gimenez-Palop, O., Berlanga, E., Gutierrez, C., Caixas, A., Vendrell, J., and Gonzalez-Clemente, J. M. 2011. Men with hyperferritinemia and diabetes in the Mediterranean area do not have a higher iron overload than those without diabetes. *Diabetes Res Clin Pract* 91, e33–6.

Furuta, M., Funabashi, T., and Akema, T. 2012. Maternal iron deficiency heightens fetal susceptibility to metabolic syndrome in adulthood. *Endocrinology* 153, 1003–4.

Grunblatt, E., Bartl, J., and Riederer, P. 2011. The link between iron, metabolic syndrome, and Alzheimer's disease. *J Neural Transm* 118, 371–9.

Hua, N. W., Stoohs, R. A., and Facchini, F. S. 2001. Low iron status and enhanced insulin sensitivity in lacto-ovo vegetarians. *Br J Nutr* 86, 515–9.

Huang, P. H., Su, J. J., and Lin, P. H. 2010. Iron deficiency anemia—A rare etiology of sinus thrombosis in adults. *Acta Neurol Taiwan* 19, 125–30.

The InterAct, C. 2013. Association between dietary meat consumption and incident type 2 diabetes: The EPIC-InterAct study. *Diabetologia* 56, 47–59.

Jankowska, E. A., Rozentryt, P., Witkowska, A., Nowak, J., Hartmann, O., Ponikowska, B., Borodulin-Nadzieja, L. et al. 2010. Iron deficiency: An ominous sign in patients with systolic chronic heart failure. *Eur Heart J* 31, 1872–80.

Jehn, M., Clark, J. M., and Guallar, E. 2004. Serum ferritin and risk of the metabolic syndrome in U.S. adults. *Diabetes Care* 27, 2422–8.

Jehn, M. L., Guallar, E., Clark, J. M., Couper, D., Duncan, B. B., Ballantyne, C. M., Hoogeveen, R. C., Harris, Z. L., and Pankow, J. S. 2007. A prospective study of plasma ferritin level and incident diabetes: The Atherosclerosis Risk in Communities (ARIC) Study. *Am J Epidemiol* 165, 1047–54.

Jiang, R., Ma, J., Ascherio, A., Stampfer, M. J., Willett, W. C., and Hu, F. B. 2004. Dietary iron intake and blood donations in relation to risk of type 2 diabetes in men: A prospective cohort study. *Am J Clin Nutr* 79, 70–5.

Kikuchi, G., Yoshida, T., and Noguchi, M. 2005. Heme oxygenase and heme degradation. *Biochem Biophys Res Commun* 338, 558–67.

Kim, C., Bullard, K. M., Herman, W. H., and Beckles, G. L. 2010. Association between iron deficiency and A1C Levels among adults without diabetes in the National Health and Nutrition Examination Survey, 1999–2006. *Diabetes Care* 33, 780–5.

Kim, C. H., Kim, H. K., Bae, S. J., Park, J. Y., and Lee, K. U. 2011. Association of elevated serum ferritin concentration with insulin resistance and impaired glucose metabolism in Korean men and women. *Metabolism* 60, 414–20.

Kim, N. H., Oh, J. H., Choi, K. M., Kim, Y. H., Baik, S. H., Choi, D. S., and Kim, S. J. 2000. Serum ferritin in healthy subjects and type 2 diabetic patients. *Yonsei Med J* 41, 387–92.

Liu, Q., Sun, L., Tan, Y., Wang, G., Lin, X., and Cai, L. 2009. Role of iron deficiency and overload in the pathogenesis of diabetes and diabetic complications. *Curr Med Chem* 16, 113–29.

Luande, C., Li, H., Li, S. J., Zhao, Z., Li, X., and Liu, Z. M. 2008. Body iron stores and dietary iron intake in relation to diabetes in adults in North China. *Diabetes Care* 31, 285–6.

Micha, R., Michas, G., and Mozaffarian, D. 2012. Unprocessed red and processed meats and risk of coronary artery disease and type 2 diabetes—An updated review of the evidence. *Curr Atheroscler Rep* 14, 515–24.

Minamiyama, Y., Takemura, S., Kodai, S., Shinkawa, H., Tsukioka, T., Ichikawa, H., Naito, Y., Yoshikawa, T., and Okada, S. 2010. Iron restriction improves type 2 diabetes mellitus in Otsuka Long-Evans Tokushima fatty rats. *Am J Physiol Endocrinol Metab* 298, E1140–9.

Okonko, D. O., Grzeslo, A., Witkowski, T., Mandal, A. K., Slater, R. M., Roughton, M., Foldes, G. et al. 2008. Effect of intravenous iron sucrose on exercise tolerance in anemic and nonanemic patients with symptomatic chronic heart failure and iron deficiency FERRIC–HF: A randomized, controlled, observer-blinded trial. *J Am Coll Cardiol* 51, 103–12.

Papanikolaou, G., and Pantopoulos, K. 2005. Iron metabolism and toxicity. *Toxicol Appl Pharmacol* 202, 199–211.

Parikh, A., Natarajan, S., Lipsitz, S. R., and Katz, S. D. 2011. Iron deficiency in community-dwelling US adults with self-reported heart failure in the National Health and Nutrition Examination Survey III: Prevalence and associations with anemia and inflammation. *Circ Heart Fail* 4, 599–606.

Rajpathak, S., Ma, J., Manson, J., Willett, W. C., and Hu, F. B. 2006. Iron intake and the risk of type 2 diabetes in women: A prospective cohort study. *Diabetes Care* 29, 1370–6.

Rajpathak, S. N., Crandall, J. P., Wylie-Rosett, J., Kabat, G. C., Rohan, T. E., and Hu, F. B. 2009. The role of iron in type 2 diabetes in humans. *Biochim Biophys Acta* 1790, 671–81.

Rouault, T. A. 2006. The role of iron regulatory proteins in mammalian iron homeostasis and disease. *Nat Chem Biol* 2, 406–14.

Schulze, M. B., Manson, J. E., Willett, W. C., and Hu, F. B. 2003. Processed meat intake and incidence of type 2 diabetes in younger and middle-aged women. *Diabetologia* 46, 1465–73.

Sheu, W. H., Chen, Y. T., Lee, W. J., Wang, C. W., and Lin, L. Y. 2003. A relationship between serum ferritin and the insulin resistance syndrome is present in non-diabetic women but not in non-diabetic men. *Clin Endocrinol (Oxf)* 58, 380–5.

Shi, Z., Hu, X., Yuan, B., Pan, X., Meyer, H. E., and Holmboe-Ottesen, G. 2006. Association between serum ferritin, hemoglobin, iron intake, and diabetes in adults in Jiangsu, China. *Diabetes Care* 29, 1878–83.

Swaminathan, S., Fonseca, V. A., Alam, M. G., and Shah, S. V. 2007. The role of iron in diabetes and its complications. *Diabetes Care* 30, 1926–33.

Tarim, O., Kucukerdogan, A., Gunay, U., Eralp, O., and Ercan, I. 1999. Effects of iron deficiency anemia on hemoglobin A1c in type 1 diabetes mellitus. *Pediatr Int* 41, 357–62.

Tay, E. L., Peset, A., Papaphylactou, M., Inuzuka, R., Alonso-Gonzalez, R., Giannakoulas, G., Tzifa, A. et al. 2011. Replacement therapy for iron deficiency improves exercise capacity and quality of life in patients with cyanotic congenital heart disease and/or the Eisenmenger syndrome. *Int J Cardiol* 151, 307–12.

Toblli, J. E., Lombrana, A., Duarte, P., and Di Gennaro, F. 2007. Intravenous iron reduces NT-pro-brain natriuretic peptide in anemic patients with chronic heart failure and renal insufficiency. *J Am Coll Cardiol* 50, 1657–65.

Valenti, L., Fracanzani, A. L., Dongiovanni, P., Bugianesi, E., Marchesini, G., Manzini, P., Vanni, E., and Fargion, S. 2007. Iron depletion by phlebotomy improves insulin resistance in patients with nonalcoholic fatty liver disease and hyperferritinemia: Evidence from a case-control study. *Am J Gastroenterol* 102, 1251–8.

Vari, I. S., Balkau, B., Kettaneh, A., Andre, P., Tichet, J., Fumeron, F., Caces, E., Marre, M., Grandchamp, B., and Ducimetiere, P. 2007. Ferritin and transferrin are associated with metabolic syndrome abnormalities and their change over time in a general population: Data from an Epidemiological Study on the Insulin Resistance Syndrome (DESIR). *Diabetes Care* 30, 1795–801.

Villegas, R., Shu, X. O., Gao, Y. T., Yang, G., Cai, H., Li, H., and Zheng, W. 2006. The association of meat intake and the risk of type 2 diabetes may be modified by body weight. *Int J Med Sci* 3, 152–9.

Wrede, C. E., Buettner, R., Bollheimer, L. C., Scholmerich, J., Palitzsch, K. D., and Hellerbrand, C. 2006. Association between serum ferritin and the insulin resistance syndrome in a representative population. *Eur J Endocrinol* 154, 333–40.

Yu, F. J., Huang, M. C., Chang, W. T., Chung, H. F., Wu, C. Y., Shin, S. J., and Hsu, C. C. 2012. Increased ferritin concentrations correlate with insulin resistance in female type 2 diabetic patients. *Ann Nutr Metab* 61, 32–40.

Zhao, G. Y., Zhao, L. P., He, Y. F., Li, G. F., Gao, C., Li, K., and Xu, Y. J. 2012. A comparison of the biological activities of human osteoblast hFOB1.19 between iron excess and iron deficiency. *Biol Trace Elem Res* 150, 487–95.

# 7 Radiation Exposure

*Alexander Vaiserman*

## CONTENTS

7.1 Introduction .................................................................................................. 107
7.2 Units of Radiation Dose: Basic Terms and Definitions ............................... 109
7.3 Low-Dose Radiation Effects in Human Studies .......................................... 109
    7.3.1 Health Effects in Professionally Exposed Groups ........................... 110
        7.3.1.1 Health Status in Medical Professional Groups Exposed to Radiation in Their Work Environments ...... 111
        7.3.1.2 Nuclear Worker Cohort Studies ......................................... 113
        7.3.1.3 Long-Term Follow-Ups in Nuclear Test Participants ......... 116
    7.3.2 Diagnostic and Therapeutic Medical Exposures ............................. 116
        7.3.2.1 Diagnostic Radiation Procedures ....................................... 117
        7.3.2.2 Radiotherapy ...................................................................... 118
    7.3.3 Environmental Background Radiation ............................................. 120
    7.3.4 Residential Radon and Lung Cancer Risk ....................................... 121
    7.3.5 Long-Term Health Consequences of Accidents in Nuclear Power Plants ...................................................................................... 123
    7.3.6 Health Status in Atomic Bomb Survivors ........................................ 128
7.4 Mechanisms of Radiation Hormesis ............................................................ 130
    7.4.1 Immune System Response ................................................................. 131
    7.4.2 Deoxyribonucleic Acid Repair .......................................................... 133
    7.4.3 Free Radical Scavenging .................................................................. 134
    7.4.4 Heat-Shock Response ........................................................................ 135
    7.4.5 Autophagy and Apoptosis ................................................................. 135
    7.4.6 Epigenetic Alterations ....................................................................... 136
7.5 Concluding Remarks .................................................................................... 139
Acknowledgment .................................................................................................. 140
References ............................................................................................................. 140

## 7.1 INTRODUCTION

For nearly 100 years since the discovery of radioactivity, a number of experimental and epidemiologic studies have provided increasing information on the health risks of ionizing radiation. At the end of the nineteenth century and the beginning of the twentieth century, radiation exposure was considered to be a promising therapeutic approach for the treatment of cancer and other diseases. *Mild radium therapy*, involving the oral or parenteral administration of microgram quantities of radium and its daughter isotopes, was commonly used (Macklis 1990). The death in 1932 of the prominent millionaire Eben M. Byers from radium poisoning after consuming a large quantity of Radithor (a popular and expensive mixture of radium-226 and radium-228 in distilled water) helped to bring about an end to the era of mild radium therapy. The event alerted the public and much of the medical profession about the potential harmful effects of internal radium.

During the first decades of the twentieth century, consensus emerged that there is no harm below a certain *threshold* dose (Calabrese 2009). Since the second half of the last century, the *linear no-threshold* (LNT) risk model, according to which the risk of cancer proceeds in a linear fashion at lower doses without a threshold and the smallest dose has the potential to cause a small increase in risk to humans, has come to dominate (see BEIR 2006). The LNT model of radiation carcinogenesis was initially proposed to minimize cancer risk from low-level radiation. This model is based on the following assumptions: every dose, no matter how low, carries with it some cancer risk; risk per unit dose is constant and independent of dose rate (the radiation dose absorbed per unit of time); and risk is additive and increases as dose increases. The LNT model reflects the simplified assumption that each radioactive particle or photon can cause DNA mutations, which, in turn, can lead to cancer.

During the past decades, however, the LNT hypothesis has come under attack within the scientific community, because there is insufficient scientific evidence to support it (Jaworowski 2008; Tubiana et al. 2009; Jargin 2012; Mossman 2012; Ricci et al. 2012). A large body of data has been accumulated showing that health hazards caused by ordinary levels of radiation (natural background exposures, medical x-rays, etc.) are much lower than LNT-based projections. Moreover, the LNT hypothesis ignores the fact that life evolved on Earth under conditions of a natural background radiation level, which was significantly higher than today (Zagórski and Kornacka 2012). According to the calculations by Karam and Leslie (1999), this level, 3.5 billion years ago, was about three times higher than its current level. Therefore, all living beings survived via evolutionarily derived adaptive responses to the harsher radiation environment (Jaworowski 2008). Such adaptive response phenotypes would likely be invoked by ionizing radiation doses somewhat higher than current natural levels. Furthermore, a large number of studies have demonstrated that the mechanisms of action in the low-dose (LD) region are different from those seen in the high-dose region for many biological impacts. When radiation is delivered at an LD rate (i.e., over a longer time period), it is much less effective in producing biological effects than when the same dose is delivered in a short time period. Therefore, the health risks due to LD rate effects may be overestimated.

A growing body of experimental and epidemiological evidence does not support the use of the LNT model for estimating cancer risk at low levels of exposure to ionizing radiation (Tubiana et al. 2009). On the contrary, there is substantial evidence supporting the hormetic dose-response relationship with low doses of ionizing radiation being protective and high doses causing harm (Luckey 2008; Scott 2008; Jaworowski 2010a; Calabrese et al. 2013). LD ionizing radiation is among the conditions that bring about biologically beneficial effects by initially causing low-level molecular damage, which then leads to the activation of one or more stress response pathways and thereby strengthens homeodynamics (Rattan 2012). Within the past decade, there has been an increasing interest in the problem of radiation hormesis, largely due to the fact that in recent years there has been a shift in the field of radiobiology from a DNA-centric view of how radiation damage occurs to a more biological view that incorporates concepts of hormesis, nonlinear systems, bioenergy field theory, uncertainty, and homeodynamics (Mothersill and Seymour 2012a). In this chapter, recent clinical and epidemiological data concerning the effects of low-level ionizing radiation on the risk of developing cancer and noncancer diseases are presented and discussed.

## 7.2 UNITS OF RADIATION DOSE: BASIC TERMS AND DEFINITIONS

Radioactivity or the strength of a radioactive source is measured in units of becquerel (Bq), 1 Bq = 1 event of radiation emission per second. X-rays ionize atoms and molecules in tissues through the deposition of energy. This ionization is the first step in a series of events that may have a biologic effect. Absorbed dose is a measure of the energy deposited per unit mass and provides a means of predicting the potential for biologic effects. Absorbed dose is measured in grays (Gy) or milligrays (mGy), 1 Gy = 1000 mGy. 1 Gy is equivalent to an energy deposition of 1 J/kg of tissue. To take into account the fact that not all types of radiation produce the same effects in humans, the concept of dose equivalent has been introduced. Dose equivalent is the product of the absorbed dose and a radiation weighting factor and is expressed in sieverts (Sv) or millisieverts (mSv), 1Sv = 1000 mSv. For x- or γ-rays, the radiation weighting factor is 1.0. Thus, an absorbed dose of 1 Gy is equivalent to 1 Sv (Verdun et al. 2008). Equivalent dose is often simply referred to as dose in everyday use of radiation terminology. The old unit of dose equivalent or dose was rem, 1 rem = 1000 millirem = 10 mSv.

## 7.3 LOW-DOSE RADIATION EFFECTS IN HUMAN STUDIES

Over the past few decades, the effects of exposure to low levels of ionizing radiation and the shape of the response curve at such low doses have repeatedly been investigated and debated in a number of epidemiological and clinical studies. Although the effects of high doses were well-defined in these studies, those attributed to low doses present a much bigger investigative challenge. Most such studies lack statistical power, involve various complicated confounding factors, and use incorrect methodological approaches (Scott 2008; Scott et al. 2008; Ogata 2011). One faulty approach is throwing away some of the radiation doses, which can abolish threshold doses. Another such approach is averaging doses over wide dose intervals and including irradiated persons in the unirradiated group, which can abolish hormetic responses.

A number of epidemiological studies provide direct evidence of the effects of radiation exposure on human populations. However, there are problems in interpreting the evidence that they provide in the context of radiation protection in general. One of the most important of these problems arises because many of the studies involved populations that received relatively high radiation exposures in a relatively short time. This can substantially misrepresent results because it is known that the effects of radiation exposure are proportionately greater at high doses and dose rates than at the low doses and dose rates that are usually more relevant to occupational, medical, and public exposures (Wall et al. 2006). As noted by Little and others (2008), "There are good radiobiological reasons for considering the moderate and LD studies separately, since the mechanisms that operate for doses in this range are likely to be very different to those that are relevant at higher (e.g., radiotherapeutic) doses."

Another important methodological problem is using the so-called ecological or geographical methodology, which implies comparison of the health indicators averaged over areas with similar environmental, social, and economic conditions but with different levels of ionizing radiation or radioactive contamination. This approach, according to the opinion of international experts, can lead to biased conclusions due to the so-called ecological fallacy occurring when data about a group is used to conclude information about an individual (Goodman 1959; Le Bourg 2013). The meta-analyses of different diseases from exposure to LD ionizing radiation and estimates of potential population mortality risks are generally limited by heterogeneity among studies, the possibility of uncontrolled confounding in some occupational groups by lifestyle factors, and higher-dose (>0.5 Sv) groups generally driving the observed trends (Little et al. 2010, 2012).

The occupational studies that include both unexposed and exposed subjects are considered to be especially valuable since a dose-response evaluation can be made of the relation between radiation exposure and health outcomes. Typically, study populations in retrospective cohort studies include persons who have worked with radiation in medical facilities or in the nuclear industry or patients with cancer or other diseases who have been treated with radiation (BEIR 2006). To date, the most extensive and methodologically valid surveys of health effects of radiation have been conducted in Japanese atomic bomb survivors, persons exposed to fallout radiation from nuclear tests or accidents, radiation workers receiving occupational exposure, populations living in areas characterized by above-average natural background radiation, and patients exposed for diagnostic purposes or treated with radiation for malignant and nonmalignant disease (Boice 2012).

### 7.3.1 Health Effects in Professionally Exposed Groups

The risk of cancer and other chronic diseases among physicians and other professional groups occupationally exposed to ionizing radiation has been a subject of extensive study since the 1940s. The main research activities have been focused on cancer incidence and mortality among radiologists and radiotherapists, military personnel exposed to nuclear weapons tests, and nuclear industry workers in different countries around the world. The types of radiation exposure were distinguished in various cohorts exposed to ionizing radiation in the workplace, with different contributions from photons, neutrons, and α- and β-particles. Studies in professionally exposed groups are of particular importance for radiation protection because most workers are exposed to protracted LD radiation exposure, which is a type of exposure that has considerable importance for public radiation protection.

As noted in the Biological Effects of Ionizing Radiation (BEIR) (2006) report by the National Research Council, "studies of some occupationally exposed groups, particularly in the nuclear industry, are well suited for direct estimation of the effects of low doses and low dose rates of ionizing radiation for the following reason: large numbers of workers have been employed in this industry since its beginning in the early to mid-1940s (more than 1 million workers worldwide); these populations are relatively stable; and by law, individual real-time monitoring of potentially exposed personnel has been carried out in most countries with the use of personal dosimeters (at least for external higher-energy exposures) and the measurements have been kept." The epidemiological research of occupational exposure to ionizing radiation, however, faces a number of obstacles with respect to assessing the dose-response relationship in the LD region. The statistical power necessary to detect adverse health effects from the low doses commonly encountered in occupational settings requires a large number of exposed workers and sufficiently long follow-up to account for the latency periods. Therefore, most of the follow-ups of individual cohorts of workers have insufficient statistical power (BEIR 2006). The lack of individual dose estimates and a suitable comparison group are major limitations in many such studies. The usefulness of many analyses involving external comparison groups is also limited due to the *healthy worker effect*, because of the fact that only the healthiest workers would voluntarily work in radiation areas (Carpenter et al. 1990).

### 7.3.1.1 Health Status in Medical Professional Groups Exposed to Radiation in Their Work Environments

The risk of cancer and noncancer diseases among radiologists exposed to ionizing radiation in the workplace has been a subject of study since the 1940s, when increased mortality from leukemia was reported among radiologists compared to mortality among other medical specialists (Dublin and Spiegelman 1948). Such radiation-associated health problems, including an excess of skin cancer and leukemia, together with an increased risk of cancer deaths as well as deaths from all causes, were detected among radiologists and radiologic technologists in the first half of the twentieth century. These problems disappeared after the 1950s, when the improvement of radiation safety standards for radiation workers in the United States and other Western countries and the concomitant reduction in radiation exposure occurred (Yoshinaga et al. 2004). Specifically, radiological protection standards were changed from 30,000 mSv/year in 1902 to 50 mSv/year in 1957.

Yoshinaga and others (2004), reviewing the epidemiologic data on cancer risks from eight cohorts of over 270,000 radiologists and radiologic technologists in various countries, revealed an increased mortality due to leukemia among early workers employed before 1950, when radiation exposures were high. However, after the implementation of the recommendations of the International Commission on Radiological Protection in 1930 (dose limit of 500 mSv/year), this excess mortality decreased and disappeared entirely among the physicians who started working after 1950 (Matanoski et al. 1987; Doll et al. 2005; Linet et al. 2010). The pattern of cancer mortality has changed for those radiologists who joined specialty societies after 1940. Whereas among early entrants young radiologists had higher mortality rates than those of other specialists, among later entrants young radiologists have lower mortality rates. However, as these later-entrant radiologists age, their rates exceed those of other specialists (Matanoski et al. 1987). The authors concluded that radiation exposure could produce a protective effect among radiologists. Berrington and others (2001) also found that the mortality of British radiologists who had registered since 1954 was remarkably low in comparison with that of medical practitioners as a whole; this was true for both cancer and all other causes of death combined. In the study by Cameron (2002), post-1955 radiologists had a 32% lower standardized mortality ratio (SMR) for deaths from all causes and a 36% lower SMR for noncancer deaths than those of other physicians. The SMR for cancer was 29% lower (not significant) than that for all male physicians. The author stressed that "the British radiology data show that moderate doses of radiation are beneficial rather than a risk to health."

In the study by Mohan and others (2003), mortality risks in relation to work characteristics were evaluated in a nationwide cohort of U.S. radiologic technologists ($n = 146,022$). In this professional cohort, SMRs were low for all causes (SMR = 0.76) and for all cancers (SMR = 0.82) compared to those in the general U.S. population. The relative risks (RRs) were higher for all cancers and for breast cancer among those radiologic technologists who were first employed prior to 1940, compared to those who were first employed in 1960 or later, and risks declined with more recent calendar year of first employment (Mohan et al. 2002, 2003). Risk for the combined category of acute lymphocytic, acute myeloid, and chronic myeloid leukemias was also increased among those first employed prior to 1950 compared to those first employed in 1950 or later. In the study by Doody and others (2006), the adjusted breast cancer risks for female radiologic technologists who first worked in the 1960s, 1950s, 1940s, from 1935 to 1939, and before 1935 were 1.0, 1.2, 1.0, 1.8, and 2.9, respectively, compared to those who began working in 1970 or later. The RR rose with the number of years worked before 1940 and was elevated significantly among those who began working before the age of 17 years (RR = 2.6) but was not related to the total years worked in the 1940s or later. In a study by Hauptmann and others (2003), an increased mortality from diseases of the circulatory system with occupational radiation exposure was shown in U.S. radiologic technologists, but this was only for those who were first employed before 1950 when radiation doses were likely high. In addition, by examining the risk of childhood cancer (<20 years) among 105,950 offspring born during 1921–1984 to U.S. radiologic technologist cohort members, Johnson and others (2008) found no convincing evidence of any increased risk of childhood cancer in the offspring in association with parental occupational radiation exposure.

Interventional cardiologists are another medical professional group routinely exposed to radiation in the workplace. The effects of LD x-ray radiation exposure on chromosomal damage and on selected indices of cellular and humoral immunity in interventional cardiologists were studied by Zakeri and others (2010). It has been found that, while cytogenetic results showed higher chromosomal damage, some immune responses were stimulated or modulated immunologically in this occupational group. A recent study by Russo and others (2012) found that chronic exposure to LD radiation is associated in interventional cardiologists with an altered redox balance mirrored by an increase in hydrogen peroxide and with two possible adaptive cellular responses, namely, enhanced antioxidant defense and increased susceptibility to apoptotic induction, which might efficiently remove genetically damaged cells.

# Radiation Exposure

By summarizing the epidemiological data regarding the risk for cancer mortality in radiologists and other professional groups exposed to radiation in their workplace, Tubiana (2008) concluded that the lowest potentially cancer-inducing dose is about 500 mSv.

### 7.3.1.2 Nuclear Worker Cohort Studies

The large number of nuclear industry workers in different countries provides a possibility for large-scale epidemiological studies around the world. The most comprehensive study of occupational radiation exposure in the United States was the Nuclear Shipyard Worker Study conducted in 1980–1988. The radiation workers in this study were exposed to external cobalt-60. Three large cohorts were compared in the research: a high-dose cohort of 27,872 nuclear workers (cumulative doses more than 5 mGy), an LD cohort of 10,348 workers (cumulative doses less than 5 mGy), and an unexposed (control) cohort of 32,510 shipyard workers with the same ages and jobs. Although this study was designed to search for adverse effects of occupational LD-rate γ radiation, no risks have been reported. Surprisingly, the high-dose-exposed workers, on the contrary, demonstrated health benefits compared to the controls. Specifically, the high-dose-exposed cohort had a death rate from all causes 24% lower than the controls. In addition, their death rates from cancer and cardiovascular disease were also significantly lower than those of the controls (Sponsler and Cameron 2005). Similarly, the combined analyses by Gilbert and others (1989, 1993) of mortality in nuclear facility employees who were continuously exposed to low doses of ionizing radiation with an average dose of <50 mSv (primarily neutrons) at the Hanford Site, Washington D.C.; Oak Ridge National Laboratory, Tennessee; and Rocky Flats Nuclear Weapons Plant, Colorado, provided no evidence of a correlation between radiation exposure and mortality from all cancers or from leukemia.

Of 11 other specific types of cancer analyzed, multiple myeloma was the only cancer found to exhibit a statistically significant correlation with radiation exposure. Estimates of the excess risk of all cancer types and of leukemia, based on the combined data, were negative. The authors concluded that "estimates obtained through extrapolation from high-dose data do not seriously underestimate risks of LD exposure, but leave open the possibility that extrapolation may overestimate risks." Similarly, in an analysis of the mortality data of 46,970 workers employed during 1948–1999 at Rocketdyne/Atomics International, California, a low standardized mortality ratio for all causes of death (SMR: 0.82; 95% confidence interval [CI]: 0.78–0.85) was detected, indicating that Rocketdyne radiation workers were healthier than the general population and were less likely to die (Boice et al. 2011).

An epidemiological study in three corporations covering the Canadian nuclear fuel cycle examining the effect of occupational radiation exposure among nuclear industry workers, found that cancer mortality in this population was significantly lower than national average (Abbatt et al. 1983). In another Canadian cohort study of mortality among 954 military personnel exposed to LD ionizing radiation during nuclear reactor cleanup operations in Chalk River, Ontario, no excess of death from leukemia or thyroid cancer was found compared to unexposed controls (Raman et al. 1987). In a large cohort study of Canadian nuclear power industry workers ($n = 45,468$), Zablotska and others (2004) found a much lower cancer mortality rate than in the general population. For all solid cancers combined, a significant reduction in risk in the 1–49 mSv category was indicated compared to the lowest dose group (<1 mSv) with an RR of 0.7. Above 100 mSv, the risk appeared to increase. A recent reanalysis of cancer mortality in Canadian nuclear energy workers conducted by the Canadian Nuclear Safety Commission (CNSC 2011) confirmed that there is no increase in risk of solid cancer mortality among Canadian nuclear power plant workers for any time period or among the Atomic Energy of Canada Limited nuclear energy workers first employed after 1965. Although the data suggest an increased solid cancer mortality risk for workers first employed before 1965 (1956–1964), a comparison using the Canadian Mortality Database showed lower rates of all causes of death and cancer mortality for this group than for the general Canadian population.

Furthermore, in a recent study in Korea there was no epidemiological evidence for increased risk of cancer due to radiation in adults residing near nuclear power plants (Ahn et al. 2012). In the United Kingdom, a negative association between radiation exposure and mortality from cancers, in particular leukemia (excluding chronic lymphatic leukemia) and multiple myeloma, was found for radiation workers (Kendall et al. 1992). Tokarskaya and others (1997) have determined the dose-response relationship for α-radiation-induced lung cancers (adenocarcinoma, squamous carcinoma, and small cell carcinoma) using data for 500 workers employed at the Mayak nuclear enterprise, Chelyabinsk region, Russia, and chronically exposed by inhalation to $^{239}$Pu. A trend for decreasing risk was observed for the lowest levels of plutonium incorporation. Specifically, lung cancer incidence, corrected for smoking, at body burdens of 343, 1180, and 4200 Bq was significantly reduced to 0.56, 0.59, and 0.83, respectively, compared to internal controls (Figure 7.1). No clear-cut dose-response relationship for lung cancer induction by chronic external γ-irradiation was obtained. It is noteworthy that in this cohort of

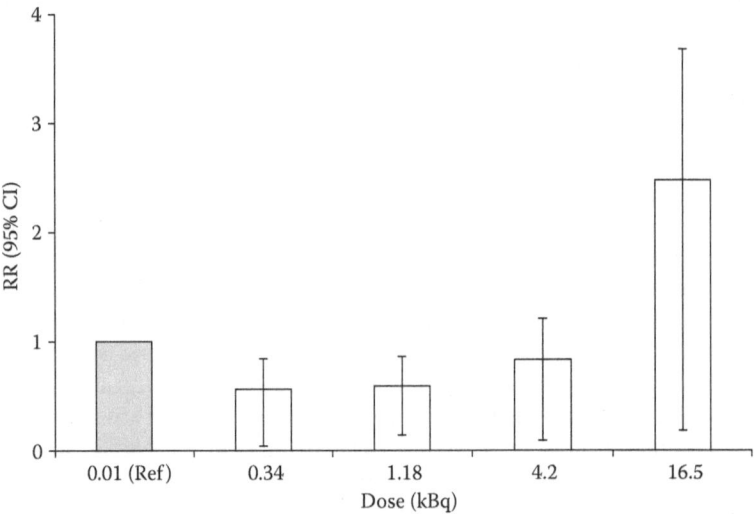

**FIGURE 7.1** Lung cancer incidence rates, corrected for smoking, in Mayak nuclear enterprise workers. (Adapted from Tokarskaya et al., *Health Phys*, 73, 899–905, 1997.)

nuclear workers lung cancer induction by cigarette smoking had a linear dependence: smoking of one pack of cigarettes per day for 5 years increased the risk of lung cancer twofold. In a more recent research by Azizova and others (2010), incidence of and mortality from cardiovascular diseases were studied in a cohort of 12,210 workers first employed at one of the main plants of the Mayak nuclear facility during 1948–1958 and followed up to 2000. In this study, no statistically significant trend with external γ-ray dose or for internal liver dose after adjustment for external dose was obtained. A second analysis by Azizova and others (2012) has been conducted based on an extended cohort and 5 additional years of follow-up. This cohort includes 18,763 workers first employed at the Mayak nuclear facility in 1948–1972 and followed up to the end of 2005. New findings in ischemic heart disease incidence revealed a statistically significant decrease among workers exposed to external γ-ray doses of 0.2–0.5 Gy in relation to external doses below 0.2 Gy.

Several cross-country surveys have been conducted in this field around the world. The largest international study of the nuclear worker cohort to date has found little evidence for an association between low doses of external ionizing radiation and chronic lymphocytic leukemia mortality (Vrijheid et al. 2008). Initially, a study of cancer mortality among 85,000 nuclear industry workers in three countries did not detect any increase in solid tumor incidence and only found a small increase in leukemia for those exposed to high doses above 400 mSv (Cardis et al. 1995). Later, in a more comprehensive study of 407,391 nuclear plant workers in 15 countries (Australia, Belgium, Canada, Finland, France, Hungary, Japan, Korea, Lithuania, Slovakia, Spain, Sweden, Switzerland, the United Kingdom, and the United States), Cardis and others (2005) concluded that 1%–2% of cancer deaths among the cohort can be due to radiation. However, in a subsequent study by the same authors of a larger cohort (410,000 workers), no cancer excess for cumulative doses below 150 mSv was observed (Cardis et al. 2007; Vrijheid et al. 2007).

To study the long-term cancer risks following prenatal radiation exposure, Schonfeld and others (2012) evaluated the association between in utero radiation exposure and the risk of solid cancer and leukemia mortality among 8000 offspring born during 1948–1988 of female workers at the Mayak Nuclear Facility. The mothers' cumulative γ-radiation uterine dose during pregnancy served as a surrogate for the fetal dose. The results of the study provide no evidence that in utero LD γ-radiation exposure can increase solid cancer or leukemia mortality risk. Similarly, the results by Draper and others (1997) suggested that paternal preconception irradiation did not cause childhood leukemia and non-Hodgkin lymphoma in the offspring of radiation workers in the United Kingdom.

On the basis of generalizing the research done in different countries of the world, a technical report from the Electric Power Research Institute (EPRI) (2009) concluded that "from an epidemiological perspective, individual radiation doses of less than 100 mSv in a single exposure are too small to allow detection of any statistically significant excess cancers in the presence of naturally occurring cancers. The doses received by nuclear power plant workers fall into this category because exposure is accumulated over many years with an average annual dose about 100 times less than 100 mSv." Since 1983, the U.S. nuclear industry has monitored more than 100,000 radiation workers each year, and no workers have been exposed to more than 50 mSv in a year since 1989 (EPRI 2009).

### 7.3.1.3 Long-Term Follow-Ups in Nuclear Test Participants

In the United Kingdom, a long-term follow-up was conducted to study mortality and cancer incidence among men who took part in atmospheric nuclear weapon tests and experimental programs during the 1950s–1960s. The results of this follow-up study showed that participation in the United Kingdom's atmospheric nuclear weapons tests had no detectable effect on expectation of life or on subsequent risk of developing cancer or other fatal diseases (Darby et al. 1993). Furthermore, participation in the test program did not seem, in itself, to have caused any detectable effect on the participants' expectation of life (Darby et al. 1988). In a more recent study by Muirhead and others (2003), it was reported that overall levels of mortality and cancer incidence in U.K. nuclear weapons test participants were similar to those in a matched control group and overall mortality remained lower than expected from national rates. Mortality rates among test participants from causes other than cancer were generally similar to those among the controls.

A recent paper by Grosche and others (2011) reported an analysis of the Semipalatinsk historical cohort ($n = 19{,}545$) exposed to radioactive fallout from nuclear testing in the vicinity of the Semipalatinsk Nuclear Test Site, Kazakhstan. Radiation dose estimates in this cohort range from 0 to 630 mGy (whole-body external). In this study, the exposed population showed a high mortality from cardiovascular disease. Rates of mortality from cardiovascular disease in the exposed group substantially exceeded those of the comparison group. However, further analysis revealed that a dose-response relationship that was found when analyzing the entire cohort could be completely explained by differences between the baseline rates in exposed and unexposed groups. When taking this difference into account, no statistically significant dose-response relationship for all cardiovascular disease, for heart disease, or for stroke was found. The authors concluded that within this population and at the level of doses estimated, there is no detectable risk of radiation-related mortality from cardiovascular disease.

### 7.3.2 Diagnostic and Therapeutic Medical Exposures

The growing use of high-dose diagnostic x-ray examinations and fluoroscopy-guided procedures represents a tremendous advantage for diagnosis and treatment in modern medicine. Along with the increasing use of these procedures comes a concern about the long-term consequences of radiation exposure during fluoroscopy. Although the radiation dose for a single procedure is low, many patients may be subjected to repeated examinations and treatments over time, which may result in relatively high cumulative doses. Many of the epidemiologic studies of health effects following medical radiation exposures are characterized by cohort design, large population size, long-term follow-up of the cohort, well-characterized dose estimates for individuals, and a wide range of doses in order to estimate a dose-response relationship, which provide a good opportunity for studying the effects of LD radiation exposures (Kleinerman 2006).

To date, research in patients who were irradiated for diagnostic or therapeutic reasons has provided substantial information for the assessment of radiation-induced cancer risk. Generally, patients received high doses of radiation on the order of 40–60 Gy to the targeted region, aimed at producing cell killing (BEIR 2006).

These high doses of therapeutic radiation are, however, delivered directly to the target tissues, and other nonneighboring tissues usually receive low doses of radiation (100 mGy or less). The use of such studies to estimate the effect of LD exposures raises a number of questions because partial-body exposures possibly result in a different risk than an equivalent whole-body uniform exposure.

Diagnostic radiation procedures, in contrast, generally result in small doses to target organs. A number of such procedures, however, have resulted in sizable doses to specific tissues, and studies of patients who have undergone these examinations provide valuable information on radiation risks. Computed tomography (CT) can deliver sizable doses, typically on the order of tens of millisieverts per examination; cumulative doses can reach on the order of 100 mSv (UNSCEAR 2000).

### 7.3.2.1 Diagnostic Radiation Procedures

Several articles have recently appeared in both scientific and public literature that predict thousands of cancers around the world due to the radiation exposure resulting from medical imaging. As a consequence, imaging examinations may be delayed in some cases, leading to a much greater risk to patients than that associated with diagnostic radiation exposures (Hendee and O'Connor 2012; Cohen 2013). In fact, only low doses are really used in such procedures. So, the effective dose values used in CT generally range from 1 to 20 mSv, while an excess of cancers has never been detected for doses below 100 mSv. However, some authors stress the risk of cancer after diagnostic radiation exposure on the basis of the LNT model. For example, Brenner and Hall (2007) suggested on the basis of the LNT hypothesis that in a few decades about 1.5%–2% of all cancers in the U.S. population might be caused by CT scan usage.

Some data confirm that weekly delivered low doses of radiation can have cumulative effects. In several surveys, a breast cancer excess was detected in girls and young women after repeated chest fluoroscopic procedures for chronic tuberculosis or scoliosis (Miller et al. 1989; Boice et al. 1991; Moory Doody et al. 2000; Ron 2003). However, as stressed by Tubiana and others (2008), after repeated x-ray examinations a cancer excess was found only for cumulative doses greater than about 500 mSv. A number of scientific observations contradict the assumption that diagnostic x-ray examinations are dangerous. Several studies have not detected any increase in both leukemia and solid tumors in x-ray-examined patients (Boice et al. 1991; Kleinerman 2006). Furthermore, in the Canadian Breast Fluoroscopy Study (Miller et al. 1989), in which breast cancer mortality was estimated in 32,710 women examined by multiple fluoroscopy between 1930 and 1952, the breast cancer RR in the groups exposed to radiation ranging from 100–200 mGy and 200–300 mGy was reduced compared to controls by 34% and 16%, respectively" (Figure 7.2).

Scott and others (2008) stated in their commentary that CT scans may reduce rather than increase the risk of cancer. It has been suggested that the radiation exposure associated with diagnostic x-rays, CT scans, and mammograms can reduce cancer risk by stimulating the removal of precancerous neoplastically transformed cells and other genomically unstable cells and via preventing metastasis of existing cancer (Cheda et al. 2004; Nowosielska et al. 2006a,b; Liu 2003, 2006; Bauer 2007).

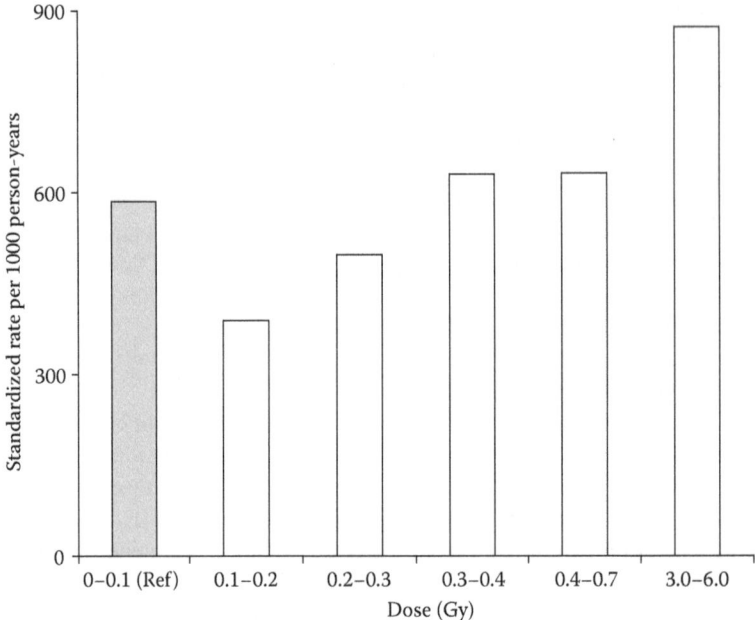

**FIGURE 7.2** Breast cancer mortality in the Canadian Breast Fluoroscopy Study, by dose category. (Adapted from Miller et al., *N Engl J Med*, 321, 1285–9, 1989. With permission.)

Data from many studies suggest that for high-energy γ-ray photons the hormetic zone includes doses in the range 1–100 mGy (Scott and Di Palma 2007). Doses currently associated with routine diagnostic x-ray procedures fall in this zone and therefore may likely be protective against cancer and some other diseases.

Nevertheless, some uncertainty about the effects of diagnostic radiation exposures still remains, so they must be applied with caution. This especially applies to pregnant women because the fetus is particularly vulnerable to the effects of ionizing radiation, and severe defects can result from fetal radiation exposures >100 mGy (Groen et al. 2012). It is obvious, however, that such doses are achieved only very rarely (for reference, 100 mGy is equivalent to ~1000 chest x-rays). Given these points, diagnostic radiation procedures should be performed during pregnancy only with an understanding of the maternal and fetal risks and benefits, comparative advantages of different modalities, and unique anatomic and physiologic issues associated with pregnancy (Wang et al. 2012).

### 7.3.2.2 Radiotherapy

Nowadays, more than half of all cancer patients receive radiotherapy as a part of their treatment (Kumar 2012). However, the efficacy of conventional radiation therapy is substantially limited by the resistance of tumors and tissue toxicity. In the last decade, several studies have shown that protocols using LD radiation are more effective in providing local tumor control with negligible normal tissue toxicity (Farooque et al. 2011). It has been found that LD radiation can stimulate antioxidant capacity, repair of DNA damage, apoptosis, and induction of immune responses, which might be collectively responsible for providing effective local tumor control.

LD radioimmunotherapy is currently considered a promising strategy for the prevention and treatment of cancer (Pollycove and Feinendegen 2008, 2011). Data from many studies indicate that LD radiation can stimulate the immune system in both animals and men (Farooque et al. 2011). It has been repeatedly demonstrated that radiation-induced changes in immune activity show an inverse J-shaped (hormetic) curve, indicating LD stimulation and high-dose suppression. Such a stimulation of immunity by LD radiation concerns most anticancer parameters, including antibody formation, natural killer activity, secretion of interferons and other cytokines, as well as other cellular changes (Liu 2006). Extensive preclinical studies in animal models have found that LD radiation can retard tumor growth, decrease cancer metastasis, and inhibit carcinogenesis induced by high-dose radiation. These effects of LD radiation on cancer development and progression have been shown to be related to the stimulation of immunity.

Experimental and clinical studies in the United States and Japan have shown that LD radioimmunotherapy is likely to be much more effective than chemotherapy in the treatment of non-Hodgkin's lymphoma (Pollycove and Feinendegen 2008). In two clinical trials by Harvard University, Cambridge, Massachusetts, stimulation of the immune system by low-level total-body and half-body irradiation was used on an experimental basis for the medical treatment of lymphosarcoma (non-Hodgkin's lymphoma). Four years after the start of the 1976 clinical trial, there was 70% survival of LD-irradiated patients, whereas there was only 40% survival of matched control patients treated with chemotherapy (Chaffey et al. 1976). In the 1979 trial, 4-year survival rates were similar: 74% and 52% for LD-irradiated patients and patients treated with chemotherapy, respectively (Choi et al. 1979). These data were later confirmed in experimental and clinical studies on cancer control with total-body or upper-half-body irradiation conducted in Tohoku University, Sendai, Japan (Sakamoto et al. 1997). Subsequently, it has also been shown that LD single-fraction palliative radiotherapy is an effective treatment for other cancers including indolent non-Hodgkin lymphoma (Chan et al. 2011), cutaneous T-cell lymphoma (Thomas et al. 2013), and primary orbital marginal zone lymphoma (Tran et al. 2013).

Formenti and Demaria (2013) concluded in their recent review that

> the therapeutic application of ionizing radiation has been largely based on its cytocidal power combined with the ability to selectively target tumors. Radiotherapy effects on survival of cancer patients are generally interpreted as the consequence of improved local control of the tumor, directly decreasing systemic spread. Experimental data from multiple cancer models have provided sufficient evidence to propose a paradigm shift, whereby some of the effects of ionizing radiation are recognized as contributing to systemic antitumor immunity. Recent examples of objective responses achieved by adding radiotherapy to immunotherapy in metastatic cancer patients support this view. Therefore, the traditional palliative role of radiotherapy in metastatic disease is evolving into that of a powerful adjuvant for immunotherapy. This combination strategy adds to the current anticancer arsenal and offers opportunities to harness the immune system to extend survival, even among metastatic and heavily pretreated cancer patients.

LD pretreatment could be a useful tool in radiation therapy. Such a pretreatment can induce an adaptive response, which will yield increased protection when a large therapeutic dose is applied. Blankenbecler (2010) stated in his analytical review that the resultant immediate damage will thereby be reduced as well as the probability that the high-dose therapy itself will induce a subsequent secondary cancer.

LD radiation may likely also have clinical implications for noncancer diseases, such as neurodegenerative diseases as well as diabetes and diabetic cardiovascular complications. Wang and others (2008) reviewed data indicating that preexposure of mice to LD radiation reduces the incidence of alloxan-induced diabetes and also delays the onset of hyperglycemia in diabetes-prone nonobese diabetic mice. The authors suggested that mechanisms by which LD radiation prevents diabetes can include the induction of pancreatic antioxidants to prevent oxidative damage of β-cells and immunomodulation to preserve pancreatic function. It has also been shown that LD radiation may have a neuroprotective effect. In a recent study by Otani and others (2012), it has been found that LD and LD-rate irradiation can delay neurodegeneration in a mouse model of retinitis pigmentosa (a hereditary, progressive neurodegenerative disease that leads to blindness). Multiple rounds of irradiation strengthened this neuroprotective effect. In this study, LD γ-radiation in a LD-rate condition prevented photoreceptor cell apoptosis and caused upregulation (563%) of the antioxidative gene peroxiredoxin-2 compared to control.

The results of some animal studies also suggest that most doses of human exposure to ionizing radiation, including radiotherapy regimens, may be unlikely to result in transgenerational instability in the offspring of irradiated fathers (Mughal et al. 2012).

### 7.3.3 Environmental Background Radiation

The existing published literature on the health effects of environmental background radiation primarily consists of reports that are descriptive in nature and ecologic in design. The preponderance of this type of study is due to the fact that they are relatively easy to carry out and are usually based on existing data. However, the BEIR (2006) report by the National Research Council stated that

> weaknesses associated with studies of this type make them of limited value in assessing risk. The primary limitation is that the unit of analysis is not the individual; thus, generally little or no information is available that is specific to the individual circumstances of the people under study. Of most concern in this regard is the definition of radiation exposure. Ecologic studies generally do not include estimates of individual exposure or radiation dose. Either aggregate population estimates are used to define population dose for groups of people, or surrogate indicators such as distance or geographic location are used to define the likelihood or potential for exposure or, in some cases, an approximate magnitude or level of exposure. This approach has serious limitations. It implies, for example, that residents who live within a fixed distance from a facility are assumed to have received higher radiation doses than those who live at greater distances or than individuals in the larger population as a whole who do not live in the vicinity of the facility. Further, it assumes that everyone within the boundary that defines exposure (or a given level of exposure) is equally exposed or has the same opportunity for exposure. In most situations, such assumptions are unlikely to be accurate, and variability in exposure of individuals within the population may be

substantially greater than the exposure attributed on a population basis. The resulting almost certain misclassification of exposure can lead to a substantial overestimation or underestimation of the association of the exposure with the disease under study. Similarly, there is usually no information available in ecologic studies regarding other factors that might influence the risk of developing the disease(s) under study (i.e., other risk factors). Thus, there is no way to evaluate the impact of such factors in relation to the potential effect of radiation exposure. This inability to evaluate or account for the potential confounding effect of other important factors, or the modifying effect of such factors on risk, makes the ecologic approach of limited use in deriving quantitative estimates of radiation risk. Nevertheless, although such studies may be limited by an ecologic design, they can be informative in assessing risk (Davis 2012).

Large-scale epidemiological studies analyzing the frequency of health effects in high natural background radiation areas mainly considered the risk of cancer and noncancer diseases on the basis of incidence or mortality data. Overall, such studies show no adverse health effects in populations residing in high background radiation areas compared to LD populations (Hendry et al. 2009). Furthermore, several lines of evidence were obtained that the level of natural background radiation is inversely related to cancer mortality (Hart and Hyun 2012). In a large-scale study in the United States, the mortality rate due to all cancers has been shown to be lower in states with a higher annual radiation dose (Frigerio and Stowe 1976). In a Chinese study, which compared areas with an average radiation exposure of 2.31 mSv/y and areas with 0.96 mSv/y average exposure, the cancer mortality rate was lower in the high background radiation areas (High Background Radiation Research Group 1980). In another Chinese study, the cancer mortality rate was also lower in the areas with a relatively high background radiation compared to that in areas with lower levels of background radiation (Wei et al. 1990). Similarly, in India the cancer incidence and mortality rates were significantly lower in areas with a high background radiation level than in areas with a low level (Nambi and Soman 1987).

More recently, no increase in cancer incidence or mortality associated with a high radiation background has been found in Yangjiang, China (Tao et al. 2000), and in Kerala, India (Nair et al. 2009). Furthermore, the high level of natural chronic background radiation in the Kerala coast did not show any significant effect on telomere length in newborns (Das et al. 2012). Thus, most of the epidemiological studies have failed to show any adverse health effects in populations living in high background radiation areas. By generalizing existing data, Cameron (2005) provocatively stated that "we need increased background radiation to improve our health."

### 7.3.4 Residential Radon and Lung Cancer Risk

Radon gas is the leading source of natural background radiation exposure and the second leading cause of lung cancer, which is the leading cause of cancer-related deaths worldwide (Sethi et al. 2012). A causal relationship between lung cancer and radon exposure has been repeatedly demonstrated, primarily in underground miners (BEIR 1999). However, as exposure levels and dose rates for miners are typically much higher than those for residential exposures, the extrapolation of risk to that for lower residential exposures may involve considerable uncertainty, and the application of data from miners is generally not capable of properly describing the situation of people in houses or other aboveground buildings (Scott 2011a; Rockwell 2013). Moreover, the extrapolation of risk from miner data is problematic, because miners are known to be cigarette smokers much more often than the average population. In most of the miner populations studied, most men were smokers. A good example is the study by Becker (2003). He reported that in the early uranium-mining period (1946–1954) the miners' smoking rate in the Schneeberg area, Germany, was estimated to have been above 90%. It is a very important point because the effect of smoking on lung cancer risk is much greater than that of radon exposure alone (Darby and Hill 2003). In the BEIR (1999) report, it is stated that smoking increases lung cancer risk by a factor of 10–20 and radon by 0.2–0.3, which implies that the smoking risk is about 50 times higher than the radon risk. Lung cancer risk has been shown to be significantly higher for smokers than for nonsmokers: more than 85% of radon-induced lung cancer deaths are among smokers (Lantz et al. 2013). Summarizing the findings of these types of studies, Neuberger and Gesell (2002) concluded that radon may contribute to lung cancer risk in current smokers in high residential radon environments.

The results of several studies suggest that residential radon exposure would likely even prevent lung cancer. To date, the most extensive studies investigating the link between residential radon levels and risk of lung cancer are those by Dr. Bernard Cohen (1993, 1995, 1997) at the University of Pittsburgh, Pennsylvania. After correcting for smoking, a significant negative correlation between average residential radon levels and mortality rates from lung cancer in nearly 2000 counties housing more than 90% of the U.S. population has been shown in his investigation. A U-shaped dose-response relationship between 1950–1954 county rates of lung cancer mortality and U.S. residential radon in white women who predominantly never smoked was obtained in an analysis conducted by Bogen and Cullen (2002). In another U.S. study, the comparison of three Rocky Mountain states (Idaho, Colorado, and New Mexico) with three Gulf states (Louisiana, Mississippi, and Alabama) revealed a strong negative correlation between natural radon levels and mortality from lung cancer (Jagger 1998).

A hormetic dose-response relationship between residential radon exposure and lung cancer risk was indicated in the study by Thompson and others (2008). Specifically, for the average radon exposures of 25, 50, 75, 150, and 250 Bq/m$^3$, the odds ratios for lung cancer adjusted by years of residency, smoking, education, income, and years of job exposure were 0.53, 0.31, 0.47, 0.22, and 2.50, respectively (Figure 7.3). In a more recent study by Thompson and others (2010), a statistically significant decrease in cancer risk with increased exposure was found for values ≤157 Bq/m$^3$ normalized to the reference exposure of 4.4 Bq/m$^3$, the lowest radon concentration measured. A lower rate of cancer mortality has also been reported among residents of the Misasa spa area, Japan, where there is a high background radon level, compared to that in the whole Japanese population (Mifune et al. 1992). Due to the presence of many radon spas, the urban residents in Misasa are exposed to much greater levels of radiation than the rural ones. By comparing cancer mortality rates in urban and rural areas in Misasa, it has been shown that urban populations have significantly lower cancer mortality rates than those living in the suburbs (Kondo 1993).

# Radiation Exposure

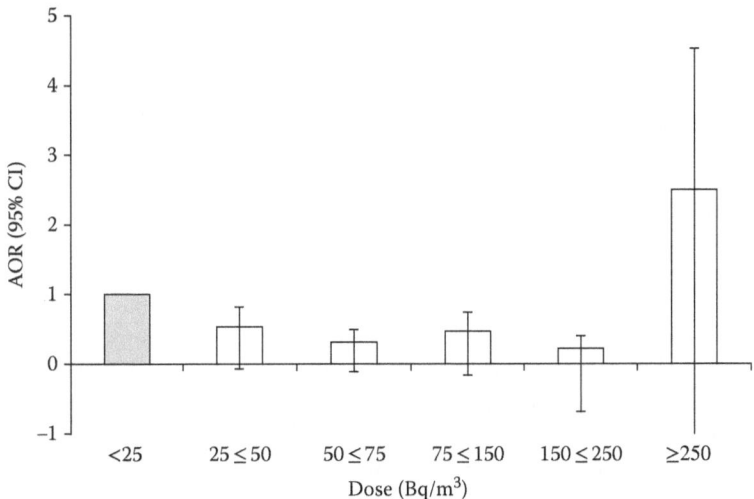

**FIGURE 7.3** Adjusted odds ratios (AORs) for lung cancer versus residential radon exposure compared to reference group. (Adapted from Thompson et al., *Health Phys*, 94, 228–41, 2008. With permission.)

Recently, Scott (2011a), referring to a review of 15 ecologic studies by Stidley and Samet (1993), where a positive association between radon levels and lung cancer was found for 7 studies, no association was found for 6 studies, and a hormetic response was found for 2 studies, pointed out that we can get a positive association, no association, or a negative association depending on how the reference exposure group is selected relative to the hormetic zone (zone of maximal protection). Specifically, if the radon level for the reference group is significantly below the zone of maximal protection, then a negative association (hormetic effect) would be expected for a range of radon exposure levels for the test groups. If the reference group is within the zone of maximal protection, then no association may be observed for the test groups if they also fall within the indicated zone. If the reference group is near the upper end of the zone of maximal protection, then a positive association between radon level and lung cancer occurrence would be expected irrespective of the choice of test dose groups.

## 7.3.5 Long-Term Health Consequences of Accidents in Nuclear Power Plants

Since the beginning of the industrial use of nuclear energy, the expansion of these technologies has been accompanied by numerous crises and persistent public health concerns (Sovacool 2008). The impact of potential nuclear power plant breakdowns on public health has been the key concern for governmental agencies and civilians alike. In the past 40 years, the world has seen three disastrous nuclear accidents, namely, in the Three Mile Island nuclear power station, Dauphin, Pennsylvania, in 1979; Chernobyl nuclear power plant in the former Ukrainian Republic of the Soviet Union in 1986; and Fukushima Daiichi nuclear power plant in Japan in 2011.

The short- and long-term health risks associated with these disasters became the subject of a comprehensive study.

The Three Mile Island accident has received the least attention among them. Although this accident was serious, the average dose of radiation exposure to about 2 million people was low, only about 1.7 millirem/year for this area. Sell and Gilles (2012) noted that this dose is similar to the additional background radiation dose that a resident of Denver, Colorado, receives every week compared to the residents of the area around Three Mile Island. This nuclear power plant accident prompted the Pennsylvania Department of Health to initiate a large-cohort mortality study ($n = 32,135$) in the accident area. In the long-term follow-up of the residents in the Three Mile Island accident area, no consistent evidence that radioactivity released during this nuclear accident had a significant impact on the overall mortality in the cohort studied was obtained (Talbott et al. 2003). Overall cancer mortality in this cohort was similar to that in the local population. There was also a decreasing trend in heart disease mortality with dose for males and females in the study.

The health effects of the Chernobyl accident have been most extensively studied compared to the consequences of other incidents in nuclear power plants. Due to this accident, large areas of Belarus, Ukraine, and Russia were contaminated in varying degrees. The main radionuclides released were iodine-131 and cesium-137. In the spring and summer of 1986, 116,000 people were evacuated from the area surrounding the Chernobyl reactor to noncontaminated areas, and another 220,000 people were relocated in later years.

To date, the consequences of the Chernobyl accident have been the subject of unprecedented response by the scientific community and speculation and exaggeration by some media organizations (Balonov 2012). The most frequently reported negative health consequence arising from radiation exposure due to this accident is early death from acute radiation sickness among the emergency workers ("liquidators") receiving high doses of irradiation (Wakeford 2011). The other evident consequence of Chernobyl radiation exposure is post-Chernobyl radiogenic thyroid cancer in children. As stated in the United Nations Scientific Committee on the Effects of Atomic Radiation (UNSCEAR 2008) report, it may be because the doses from radioiodine were so high (up to tens of grays), the number of children receiving high thyroid doses so large, and spontaneous incidence rates in children so low (a few cases per million population per year) that the effect of radiation was detected in both analytical and ecological studies. As for cancers other than thyroid cancer that might be in excess as a result of the Chernobyl accident, there is little strong scientific evidence of an effect of radiation exposure. Childhood leukemia is an obvious candidate for an excess of cases and some evidence does exist for an effect of Chernobyl in the heavily contaminated regions of the former USSR (Davis et al. 2006), but control selection problems could be responsible for this association (Noshchenko et al. 2010).

An International Atomic Energy/World Health Organization/United Nations Development Program press release (IAEA/WHO/UNDP 2005) stressed that "most emergency workers and people living in contaminated areas received relatively low whole body radiation doses, comparable to natural background levels. As a consequence, no evidence or likelihood of decreased fertility among the affected population has been found, nor has there been any evidence of increases in congenital malformations that can be attributed to radiation exposure." Cardis and others (2006) noted in their report that no carcinogenic effects were obtained in individuals exposed to doses below 100 mSv due to radiation from the Chernobyl accident.

However, this accident was thought to have caused social unrest and mental damage that had far more impact than the damage caused by radiation exposure (Takamura and Yamashita 2011). The relocation proved a "deeply traumatic experience" for some 350,000 people who were moved out of the affected areas. Although 116,000 were moved from the most heavily impacted area immediately after the accident, later relocations did little to reduce radiation exposure. Persistent myths and misperceptions about the threat of radiation have resulted in "paralyzing fatalism" among the residents of affected areas.

However, some current scientific publications still suggest that there are catastrophic consequences of Chernobyl radiation exposure on human health, including increasing death rates. As an example, in the volume № 1181 of the *Annals of the New York Academy of Sciences*, which is a translation of a book originally published in Russian (Yablokov et al. 2009), the authors, based on the LNT assumption, concluded that 985,000 additional deaths that occurred globally in the years 1986–2004 may be a consequence of the Chernobyl accident. Such obvious exaggeration of radiation-induced health effects is clearly due to the biased methodology used by the authors. Specifically, a key postulate of modern radiation epidemiology, requiring proof of a radiation dose-effect relationship, is summarily rejected; the selection of articles is unbalanced; and articles where the effect of radiation is not found are ignored in the paper by Yablokov and others (Jargin 2010; Balonov 2012).

The common methodological weakness that characterizes the post-Chernobyl studies is the lack of direct measurements, particularly for individual dosimetry, that makes the reconstruction of doses received by epidemiological research subjects quite complicated. In most of the works referenced by Yablokov and others (2009), ecological or geographical methodology is used instead of the currently preferred cohort or case-control approaches, because the latter require the reconstruction of individual doses. It must also be taken into account that after the Chernobyl accident incidence and mortality were more intensively studied in affected populations by most of the medical research institutions in trying to find any changes that could be attributed to Chernobyl. Therefore, health screening and diagnostic techniques are not similar in "contaminated" and "clean" areas, and distinctions in incidence and mortality in different regions may not reflect actual differences in the population health status, but rather they are the result of a systematic bias caused by differences in diagnostic capabilities between the surveys.

There are also some obvious problems in attempting to determine trends in health indicators over time, since the economic depression that followed the breakup of the Soviet Union led to the collapse of the healthcare system and to increased morbidity and mortality in post-USSR countries (Meslé 2004). It is noteworthy that the increase in mortality was more pronounced in the far east of Russia, which was hardly affected by the Chernobyl accident (Men et al. 2003). This makes it difficult to identify Chernobyl-related health effects (Balonov 2012). Furthermore, in studies similar to those reported in the survey by Yablokov and others (2009), it is very difficult to distinguish the effects of high and low doses of radiation. As an example,

Ivanov and coworkers (Ivanov 2007; Ivanov et al. 2004, 2009) stated in their articles that an increased solid cancer risk is evident among the Chernobyl emergency workers who stayed in the 30 km zone around the Chernobyl nuclear power plant during 1986–1987. However, it is not possible to differentiate between the effects of low and high doses from these studies, because individuals exposed to a very wide range of cumulative doses (0.001–0.3 Gy) were all included in the research (Ivanov et al. 2004). A closer examination of the data from this study rather suggests a hormetic dose-response relationship with a solid cancer risk rate below the natural level in the LD (79 mGy) group (Figure 7.4).

The report by the Chernobyl Forum (2003–2005) concluded that psychosomatic disorders and screening effects were the only detectable health consequences among the general population. By summarizing the authoritative and expert assessments, the World Nuclear Association (2009) states that "fighting the panic and mass hysteria could be regarded as the most important countermeasure to protect the public against the effects of a similar accident should it occur again." A retrospective analysis of the results obtained from the study of the long-term consequences of the Chernobyl accident for people's health demonstrated that using the LNT assumption as a basis for protection measures and radiation dose limitations was counterproductive and lead to the suffering and pauperization of millions of inhabitants of contaminated areas.

Several experiments have been devoted to the study of the effects of Chernobyl radiation in animal models. For example, the research by Rodgers and Holmes (2008) aimed to assess genotoxicity after exposure to Chernobyl radiation as a function of dose and dose rate and induction of a radioadaptive response (a reduced effect of a high challenging dose of irradiation after a smaller, inducing dose) following a priming dose administered at varying dose rates in a mouse model. In this study, male BALB/c mice were exposed to environmental radiation contamination in the Red Forest near the Chernobyl Nuclear Power Plant. The results of the study indicated that subacute environmental exposures of 10 cGy γ-radiation resulted in indistinguishable

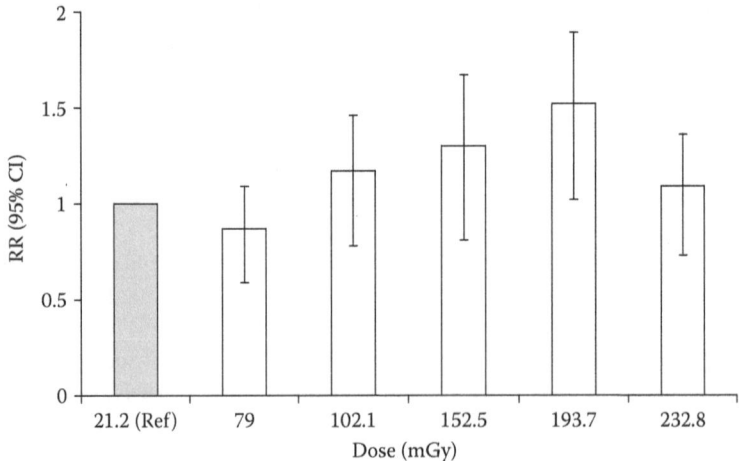

**FIGURE 7.4** Solid cancer risk in Chernobyl emergency workers. (Adapted from Ivanov et al., *Radiat Environ Biophys*, 43, 35–42, 2004. With permission.)

levels of chromosomal damage compared to controls. A radioadaptive response was observed in all experimental groups exposed to a subsequent acute challenge dose of 1.5 Gy, demonstrating that LD rates of low linear energy transfer (LET) radiation are effective in reducing genetic damage from a subsequent acute low-LET radiation exposure. Furthermore, these data demonstrate that the subchronic exposure to low levels of low-LET radiation in an environmental setting can cause beneficial effects and does not support the LNT hypothesis. To gain insight into the mechanism of this radioadaptive response, the liver, bone marrow, and skeletal muscle tissues were excised from mice in order to examine transcriptional responses in genes suspected to play a role in adaptation. Preliminary experiments utilizing archived tissue samples have revealed that expression of the radical scavenging gene, *SOD1*, was significantly elevated in mice receiving a subacute 10 cGy dose; expression of this gene, however, was not elevated in mice receiving the same dose administered acutely.

Evidence for the potentially hormetic effects of LD radiation exposure from accidents at nuclear power plants were also obtained in several epidemiological studies. In the study by Kostyuchenko and Krestina (1994), changes in sex and age structure, mortality, and reproductive function of 7854 persons exposed to radiation by thermal explosion in a radioactive waste storage facility in the former Soviet Union (Siberia) in 1957 were investigated. It was shown that the exposure groups of 496, 120, and 40 mGy corresponded to 28%, 39%, and 27% lower cancer mortality rates, respectively, compared to those living in nearby villages that were not exposed (Figure 7.5).

Another observation indicating the possibility of beneficial effects following the Chernobyl disaster is that a 15%–30% deficit of solid cancer mortality was found among the Russian emergency workers and a 5% deficit of solid cancer incidence was found in populations residing in the most contaminated areas compared to the general population of Russia (Jaworowski 2010b).

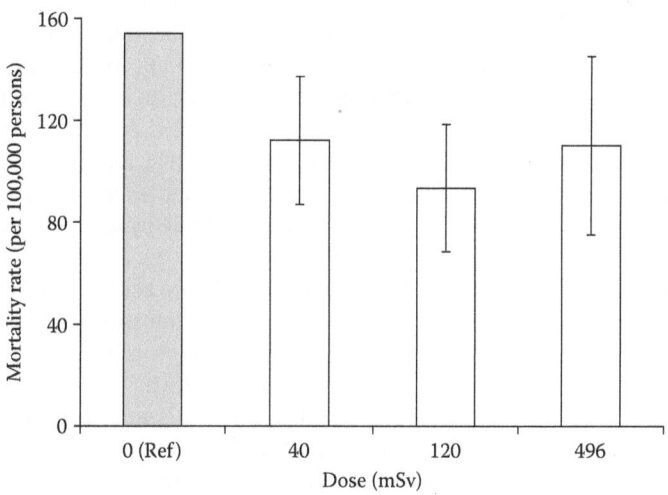

**FIGURE 7.5** Tumor-related mortality rates in 1958–1987 in the population evacuated from the east Ural radioactive trace area. (Adapted from Kostyuchenko, V. A., and L. Krestina, *Sci Total Environ*, 142, 119–25, 1994. With permission.)

The nuclear incident in the Fukushima Daiichi (Fukushima I) nuclear power plant in Japan, which was caused by the magnitude 9.0 earthquake and the subsequent tsunami in March 2011, was comparable in scale to the 1986 Chernobyl accident. In contrast to the Chernobyl reactor accident, the estimates of radiation doses received by the Japanese population as a result of the Fukushima accident suggest very low uptakes of radioactive iodine, which was a major determinant of the increase in the incidence of childhood thyroid cancer following Chernobyl accident contamination. Boice (2012) stressed that "apart from as regards the extreme psychological stress caused by the horrific loss of life following the tsunami and the large-scale evacuation from homes and villages, such studies have limited to no chance of providing information on possible health risks following LD exposures received gradually over time—the estimated doses (to date) are just too small." When assessing the potential radiological harm to Fukushima recovery workers, Scott (2011b) stated that for effective doses up to and somewhat exceeding 250 mSv hormetic effects are orders of magnitude more likely than cancer induction.

### 7.3.6 Health Status in Atomic Bomb Survivors

To date, the cohort of atomic bomb survivors of Hiroshima and Nagasaki, Japan, is the largest and most intensively studied radiation-exposed population group in the world (Doss 2012a). Most of the current population-based cancer risk estimates following radiation exposure are based primarily on these cohort data, because of the wide and well-characterized range of doses received. These individual doses were estimated on the basis of position and shielding for each person at the time of the A-bombing. Among 86,543 persons exposed to radiation in A-bombed Japanese cities, 45,148 received up to 10 mSv and served as "in city controls" (Luckey 2008).

This cohort has been studied systematically during the past several decades to report the health effects of radiation (see, e.g., Pierce et al. 1996; Pierce and Preston 2000; Preston et al. 2003; Ozasa 2012). Leukemia was the first cancer to be associated with A-bomb radiation exposure, with preliminary indications of excess among the survivors within the first 5 years after the bombings; an excess of solid cancers became apparent approximately 10 years after radiation exposure (Little 2009). In most of the reports published, it has been concluded that the data are consistent with a linear, dose-dependent increase in excess cancers without a threshold dose. Generally, studies of A-bomb survivors indicate that acute irradiation with absorbed doses of more than 0.1–0.2 Gy per person increased cancer risk in the exposed groups proportionally with dose (Pierce et al. 1996; Pierce and Preston 2000). Doses above 0.5 Gy were associated with an elevated risk of both stroke and heart disease in the large Life Span Study cohort of atomic bomb survivors, but the degree of risk at lower doses was found to be unclear (Shimizu et al. 2010).

However, on closer examination of the data and observing reduced cancers for the survivors exposed to low doses, an alternative interpretation of the data has been given by some researchers (Doss 2012a). For example, in the study by Preston and others (2007) it was shown that the excess RR of solid cancer increased linearly with dose up to 2 Gy, however, this increase was not statistically significant at doses below approximately 150 mGy. No evidence for increase in risk after absorbed doses below 0.1 Gy has been obtained (Heidenreich et al. 1997; Preston et al. 2004). No increase in the number of deaths due to cancers has been found for those individuals who received doses lower than 200 mSv (UNSCEAR 1994). A number of studies done on atomic bomb survivors have shown no demonstrable genetic or hereditary effects among individuals exposed to radiation (Neel et al. 1990; Otake et al. 1990; Yoshimoto et al. 1990; UNSCEAR 2000), even though low to moderate radiation exposure has been repeatedly shown to be useful in inducing genomic instability in laboratory animals (Dubrova et al. 2000; Barber et al. 2002, 2009; Abouzeid et al. 2012). On the basis of these data, Neel and others (1990) suggested that human beings are probably less sensitive to the genetic effects of radiation than has been assumed based on extrapolations from experiments with mice.

Furthermore, a trend toward reduction in leukemia has been reported for the survivors exposed to LD radiation (Land 1980; Pierce et al. 1996) (Figure 7.6). Mortality caused by leukemia, lung cancer, and colon cancer as well as the overall cancer mortality level were lower in A-bomb survivors at doses below 100 mSv than in age-matched control cohorts (Luckey 1991; Kondo 1993; Huilgol 2012). Moreover, there is some other evidence for increased health in the LD-exposed population, including reduced mutation level and solid tissue cancer mortality rate, as well as increased average life span (for a review, see Luckey 2008).

Inconsistent data have been obtained regarding the effect of A-bomb radiation on the longevity of atomic bomb survivors. In the study by Mine and others (1990), 290 male subjects exposed to 50–149 cGy of A-bomb radiation showed significantly lower mortality from noncancerous diseases than age-matched unexposed males. The results by Cologne and Preston (2000), however, do not support claims that survivors exposed to low to intermediate doses of radiation live longer than comparable unexposed individuals. While reexamining the effect of radiation on life expectancy in the cohort of atomic-bomb survivors ($n = 120{,}321$), they have indicated that

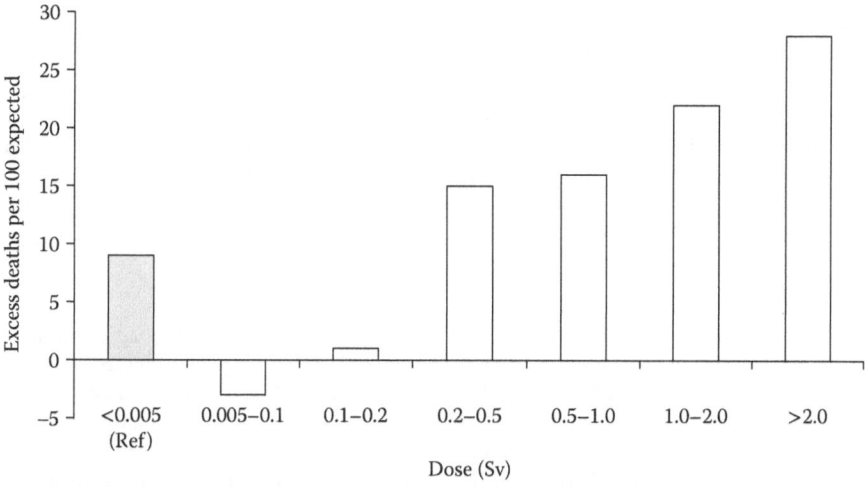

**FIGURE 7.6** Leukemia mortality among atomic bomb survivors in 1950–1990, by dose category. (Adapted from Pierce et al., *Radiat Res*, 146, 1–27, 1996. With permission.)

median life expectancy decreased with increasing radiation dose at a rate of about 1.3 years/Gy but declined more rapidly at high doses.

In his recent paper published in the *Dose-Response* journal, Dr. Mohan Doss (2012a) at the Fox Chase Cancer Center, Philadelphia, Pennsylvania, noted that

> a recent update on the atomic bomb survivor cancer mortality data has concluded that excess relative risk (ERR) for solid cancers increases linearly with dose and that zero dose is the best estimate for the threshold, apparently validating the present use of the LNT model for estimating the cancer risk from LD radiation. A major flaw in the standard ERR formalism for estimating cancer risk from radiation (and other carcinogens) is that it ignores the potential for a large systematic bias in the measured baseline cancer mortality rate, which can have a major effect on the ERR values. Cancer rates are highly variable from year to year and between adjacent regions and so the likelihood of such a bias is high. Calculations show that a correction for such a bias can lower the ERRs in the atomic bomb survivor data to negative values for intermediate doses. This is consistent with the phenomenon of radiation hormesis, providing a rational explanation for the decreased risk of cancer observed at intermediate doses for which there is no explanation based on the LNT model.

## 7.4 MECHANISMS OF RADIATION HORMESIS

The LNT model implies a constant carcinogenic effect per unit dose irrespective of dose and dose rate, assuming constant ability of the body to compensate for damages. However, there is ample evidence that this assumption is not correct, as DNA repair can be inhibited after high-dose irradiation and, on the contrary, stimulated after LD irradiation (Cohen 1994). The radiation hormesis model, unlike the LNT model, assumes that adaptive mechanisms may be induced by LD radiation, and these mechanisms may prevent both spontaneous and toxicant-related cancers as well as other adverse health effects (Calabrese et al. 2007). The overall effect may result from the confrontation between initial DNA damage, which increases linearly with the dose, and cell defense mechanisms, which are more effective at low doses and less effective when the dose increases. The system's response to low-level radiation exposure can evolve from damage to the basic molecular level and also from adaptive responses that may occur on the whole-body level. The balance between damage and protection depends on tissue dose; at single doses below 0.1 Gy, the benefits outweigh the detriments (Feinendegen et al. 2007). Thus, if the protection operates properly the dose–carcinogenic effect relationship is not linear and a threshold or hormetic effect can be observed (Calabrese 2013; Calabrese et al. 2013). Morgan and Sowa (2009) noted in their review that

> while the nature of radiation-induced DNA damage as well as imperfect repair forms a strong basis for the LNT dose-response for cancer induction, things may change when higher-order protective functions are taken into account. Misrepaired, mutated cells could be eliminated by apoptosis, cell competition, and immunological surveillance. If such protective functions work more efficiently at lower doses, the dose-response would be non-linear, and there might be a threshold. Furthermore, some phenomena mainly found in cultured cells could possibly complicate the situation, including adaptive responses in cells preexposed to a low dose of radiation, bystander effects in cells

that are not directly irradiated, and genomic instability that manifests in the progeny of irradiated cells many generations after exposure. Currently it is unclear to what extent these three phenomena are active in vivo, and how they are inter-related.

The cell defense strategy depends on the dose, dose rate, and amount of damage in neighboring cells (Tubiana 2008). There are plenty of experimental evidences that LD and LD-rate radiation may activate a system of cooperative protective processes in the body. Low-level radiation exposure has repeatedly been found to be able to stimulate many intracellular and intercellular signaling pathways, which leads to the activation of natural protection against cancer and other genomic instability–associated diseases. The key components of radiation-induced hormetic response are induction of DNA repair pathways; elimination of abnormal cells, such as preneoplastic cells, by apoptosis; activation of immune functions; synthesis of stress proteins; scavenging of free radicals; activation of membrane receptors; secretion of cytokines and growth factors; and compensatory cell proliferation (for detailed reviews, see Feinendegen 2005; Scott and Di Palma 2007; Scott et al. 2007).

### 7.4.1 Immune System Response

The basic premise of the LNT model is that even a single track of ionizing radiation can cause mutations leading to cancer (Hall and Giaccia 2006). However, the emergence of mutations seems to be an essential but not a determinative factor in the occurrence of clinical cancer. Autopsy studies have shown that the presence of cancer cells is not a decisive factor in the occurrence of clinical cancer (Doss 2012b). For example, in the Japanese study by Imaida and others (1997), the percentage of patients in a geriatric hospital who had mutations indicative of cancer, as determined from full-body autopsies, was approximately 40%. This rate was relatively unchanged for the age ranges covering 48–94 years. The incidences did not significantly differ among all age groups. Thus, there was no age dependency in the incidence of cancer. The cancer mortality rate in Japan was, however, increased by a factor of approximately 10 for a similar age range, indicating that the presence of cancer cells is not a decisive factor in clinical cancer development.

The suppression of the immune system more than doubles the cancer risk in organ transplant patients, indicating its key role in keeping occult cancers in check (Doss 2012b). Data from several studies suggest that the immune system, in addition to destroying tumor cells and sculpting tumor immunogenicity, can also restrain cancer growth for extended time periods, maintaining cancers in an equilibrium state and preventing undetected (occult) cancers from becoming clinical cancers (Koebel et al. 2007; Schneider and Tacke 2012). Jones (2011) suggested that the age-related increase in cancer incidence could be caused by a decline in immune function with age rather than the occurrence of mutations.

Since suppression of the immune system is known to be associated with an increased risk of cancer, it is logical to infer that improving immune system response may reduce cancer incidence (Doss 2013). The effect of radiation on immune system response is not linear, since high-dose radiation is well-known to suppress the immune system, whereas LD radiation can stimulate it. Stimulation of immune functions by LD radiation was evident in many animal models (for a review, see Liu 2006). A significant reduction in lymphoma incidence after chronic, fractionated LD total-body irradiation has been reported in some animal studies (Covelli et al. 1989; Ishii et al. 1996). For example, in the study by Ishii and others (1996), spontaneous lymphoma incidence was decreased to 48.6% by 150 mGy x-irradiation delivered twice a week for 40 weeks. This antitumor effect was accompanied by a significant increase in mean survival time from 283 days for the control animals to 316 days for the irradiated animals. The augmentation of the immune system was proposed by the authors as one possible mechanism for explaining the observed effects.

The possibility of LD radiation to prevent and inhibit noncancer diseases, such as autoimmune diseases, has been shown as well. In the study by Ina and Sakai (2004), chronic LD-rate γ-irradiation at 0.35 or 1.2 mGy/h resulted in significant life extension of MRL-*lpr/lpr* mice carrying a deletion in the apoptosis-regulating *Fas* gene that markedly shortens life span due to severe autoimmune disease. The prolongation of life span was associated with immunological modifications including a significant increase in CD8+ T-cells and a significant decrease in CD+ CD45R/B220+ as well as CD45R/B220+ CD40+ cells, along with the amelioration of total-body lymphadenopathy, splenomegaly, proteinuria, and kidney and brain syndromes.

A good example that demonstrates the obvious importance of adaptive response in carcinogenesis is the repeatedly reported occasional inhibition of distant untreated tumors through immune stimulation following radiation therapy to a tumor, known as an abscopal effect (Demaria et al. 2004; Okuma et al. 2011). The activation of antitumor immune response has been proposed as a key mechanism for this effect (Stamell et al. 2013). Since LD radiation may improve the immune system, the observed effects are most likely to be from adaptive response in the parts of the body incidentally exposed to LD radiation during the radiation therapy. Doss (2013) stresses in his recent paper that the immune origin of the abscopal effect suggests that the elevation of immune system response from total-body or half-body LD radiation may be useful in arresting the carcinogenic process in patients who have been diagnosed with early-stage cancer in screening programs.

Several studies have demonstrated that LD radiation has significant immune-modifying effects that lead to a coordinated and integrated anti-inflammatory response, revealing an anti-inflammatory phenotype (Calabrese and Calabrese 2013a). Whereas x-irradiation with high doses is established to exert proinflammatory effects, LD radiotherapy with single fractions below 1.0 Gy and a total dose below 12 Gy is clinically known to exert anti-inflammatory and analgesic effects on several inflammatory diseases and painful degenerative disorders (Rödel et al. 2007). For example, in a recent critical review of the biomedical literature by Calabrese and Calabrese (2013b) it was noted that exposure to low doses of ionizing radiation led to the occurrence of a generalized anti-inflammatory phenotype that prevented and reversed arthritic changes in multiple animal models in a manner quantitatively similar to that of well-known pharmaceutical agents. In vitro studies revealed a variety of mechanisms related to the anti-inflammatory effect of LD radiotherapy, in particular the modulation of cytokine and adhesion molecule expression on activated endothelial cells and leukocytes and of nitric oxide production and oxidative burst in activated macrophages and native granulocytes (Rödel et al. 2007). On the basis of these findings, LD radiation has been proposed by several authors as a promising treatment to stimulate immunity against cancer and other diseases (Cheda et al. 2004, 2009; Nowosielska et al. 2006a,b; Doss 2012a,b, 2013).

### 7.4.2 Deoxyribonucleic Acid Repair

DNA double-strand breaks (DSBs) are well-known as a serious threat to genome stability and cell viability. In a series of papers, Feinendegen and Pollycove suggest that the hazard of radiation exposure is negligible in comparison with the DNA damage that results from oxidative processes during normal metabolism (Feinendegen and Pollycove 2001; Pollycove and Feinendegen 2003; Feinendegen et al. 2004; Feinendegen 2005). Their hypothesis is based on data suggesting that various forms of nonradiation DNA damage in tissues far outweigh corresponding DNA damage from LD radiation exposure at the level of, and well above, background radiation (Pollycove and Feinendegen 2003). The authors suggested that low-level radiation causes a dual effect on cellular DNA.

One effect (called the "direct effect") is the absorption of radiation energy to DNA directly, inducing structural alterations in DNA. In addition, the interaction of radiation with water molecules in cells produces water-derived free radicals that lead to DNA damage indirectly. The probability of DNA damage per energy deposition event is relatively low and increases proportionally to the dose absorbed. At background radiation exposure, this damage to DNA is orders of magnitude lower than that from endogenous sources, such as reactive oxygen species (ROSs). In proliferating human cells, the DNA DSBs induced by radiation doses of up to 200 mGy are shown to be completely repaired after 24 hours. Moreover, some evidence is presented that the level of DSBs in cultures irradiated in low doses decreases to that of nonirradiated cell cultures if the cells are allowed to proliferate after irradiation, probably due to the elimination of the cells carrying unrepaired DSBs (Rothkamm and Lobrich 2003). This is in contrast to the effect of high-dose radiation, which generally results in residual DSBs. Thus, while the initial induction of DSBs shows a linear dose-response relationship, DNA damage caused by LD radiation has little chance to persist in the cells (Suzuki and Yamashita 2012).

The other effect is adaptive protection against DNA damage, which occurs after LD exposure and during LD-rate exposure to ionizing radiation (Feinendegen et al. 2004). This adaptive protection causes DNA damage prevention and increases DNA repair capacity. It decreases steadily at doses above about 100–200 mGy and is not observed any more after acute exposures of more than about 500 mGy (Feinendegen 2005). The authors hypothesized that LD radiation exposure may likely stimulate a specific DNA repair mechanism to reduce both spontaneous and radiation-induced damage to below the spontaneous level, thus causing a hormetic effect (Feinendegen and Pollycove 2001; Pollycove and Feinendegen 2003; Feinendegen et al. 2004). Such a persisting protective response can reduce the steady-state level of nonradiation DNA damage, thereby reducing deleterious outcomes such as cancer and aging.

According to recent data by Neumaier and others (2012), the formation of so-called ionizing radiation-induced foci (RIFs) may be one of the plausible mechanisms playing an important role in the cellular adaptive response to LD irradiation. RIFs are characterized by the local recruitment of DNA damage–sensing proteins such as p53-binding proteins. It has been shown in this study that RIF induction rate increases with increasing radiation dose, whereas the rate at which RIFs disappear decreases. The authors found that multiple DSBs, 1 to 2 μm apart, can rapidly cluster into repair centers. They also observed that the absolute RIF yield was surprisingly much smaller at higher doses: 15 RIF/Gy after 2 Gy exposure compared to approximately 64 RIF/Gy after 0.1 Gy exposure.

### 7.4.3 Free Radical Scavenging

An increase in the capacity of free radical scavenging systems is another potentially important mechanism of radiation hormesis. It is well-known that oxidative stress resulting from the excess generation of ROSs can cause cell damage and can be a major contributing factor in many types of diseases (Taverne et al. 2013). It has been repeatedly demonstrated in animal studies that LD radiation exposure may activate the endogenous antioxidant defense systems and thus may play a role in reducing the oxidative damage produced by ROSs, as well as in reducing the risk of several diseases potentially driven by ROSs (Pathak et al. 2007; Durović et al. 2008). LD radiation has been shown to be able to significantly increase the level of endogenous antioxidants, for example, SOD, glutathione (GSH), Gpx, GSH reductase, and catalase (Wang et al. 2008) in different mice tissues, including spleen, liver, pancreas, and brain (Yamaoka et al. 1998, 1999, 2004; Kojima et al. 1998, 1999; Takahashi et al. 2000; Kataoka 2013).

The important molecular pathway responsible for radiation hormesis is probably the activation of the nuclear factor erythroid 2-related factor (Nrf2) by radiation-caused oxidative stress. Nrf2 is a transcription factor that is involved in the transcriptional regulation of many antioxidant genes and in the control of the redox homeostatic gene regulatory network (Jaiswal 2004; Deramaudt et al. 2013). Under oxidative stress, the Nrf2 signaling pathway is activated to enhance the expression of a multitude of antioxidant genes, such as heme oxygenase, superoxide dismutase, GSH, and catalase (Calabrese et al. 2008; Rattan 2012). Lewis and others (2010) hypothesized that this pathway is the master regulator of the aging process and can play a critical role in the determination of species longevity.

Nrf2 was shown to bind the antioxidant DNA response element (ARE) to activate important cellular cytoprotective defense systems. McDonald and others (2010) reported that single doses of ionizing radiation from 2 to 8 Gy may activate ARE-dependent transcription in breast cancer cells in a dose-dependent manner but only after a 5-day delay. In vitro clonogenic survival assays and in vivo sublethal whole-body irradiation tests showed that Nrf2 deletion increased radiation sensitivity. The authors concluded that these results suggest that the Nrf2–ARE pathway is important to maintain resistance to irradiation, but it operates as a second-tier antioxidant adaptive response system activated by radiation only under specific circumstances, including those that may be highly relevant to tumor response during standard clinical dose-fractionated radiation therapy.

Doss (2012b) recently pointed out that LD radiation can lead to a slightly increased production of free radicals stimulating the body's defensive mechanisms such as increased antioxidant capacity and the immune system. In this respect, LD irradiation may be a good alternative to the administration of exogenous antioxidants, which has failed to reduce diseases in clinical trials possibly because there may not have been sufficient bioavailability of antioxidants in the relevant organs to reduce oxidative damage. LD radiation, in contrast, may elevate the endogenous production of antioxidants in the relevant organs. Therefore, it seems a promising application in the treatment of different diseases caused by oxidative damage.

### 7.4.4 Heat-Shock Response

The heat-shock response can be another stress response pathway triggering hormesis (Verbeke et al. 2001; Rattan 2012; Lagisz et al. 2013). The heat-shock response is the coordinated induction of a set of proteins (the so-called heat shock proteins [HSPs]) in response to a variety of cellular stresses. The chaperone protein network controls both initial protein folding and subsequent maintenance of proteins in the cell (Liberek et al. 2008).

Induction of HSPs is the obvious common response to most stresses, including high-dose radiation. Several studies have shown that the expression of HSPs can be induced by LD irradiation as well. In a study by Ibuki and others (1998), LD irradiation induced the synthesis of some proteins that were attributed to the enhancement of the colony-forming ability of myeloid leukemia cells. Specifically, irradiation led to the expression of hsp70 mRNA. Furthermore, the irradiated cells showed thermoresistance 1 hour after LD preirradiation and also showed radioresistance 4 hours after irradiation. The role of inducible Hsp70 and Hsp25 in adaptive response by irradiation in a low priming dose (1 cGy) in radiation-induced fibrosarcoma cells has been shown in the study by Lee and others (2002).

### 7.4.5 Autophagy and Apoptosis

It is known that LD and LD-rate radiation are able to induce apoptosis (Krueger et al. 2007). The removal of damaged or senescent cells by apoptosis and compensatory cell proliferation are considered by some authors to be the key components of radiation-induced hormetic response. The significant role of apoptotic cell death in radioinduced aging and life span modulations in *Drosophila melanogaster* has been shown in a series of investigations by Moskalev and Zainullin in the Komi Scientific Center, Russia (Zainullin and Moskalev 2001; Moskalev and Zainullin 2001, 2002, 2003, 2004; Moskalev et al. 2006). Chronic irradiation (accumulated doses between 0.6 and 0.8 Gy) has been shown to change the life span in male *D. melanogaster*: mean life span was increased in wild-type strains and decreased in mutant strains defective in DNA repair and displaying a higher sensitivity to apoptosis induction (Zainullin and Moskalev 2001). In a *Drosophila* strain with dysfunction of the pro-apoptosis gene reaper (*rpr*), life extension was observed after LD irradiation and/or treatment with the apoptosis inductor etoposide (Moskalev and Zainullin 2001).

In a study by Moskalev and others (2006), a relationship between radioinduced apoptosis in larval ganglion cells and aging in *D. melanogaster* was observed. In this study, the authors suggested that the elimination of damaged or unwanted cells by the induction of apoptosis can slow down aging and extend life span. In a study by Mirzaie-Joniani and others (2002), where HeLa Hep2 cells were exposed to γ-radiation (i.e., 0.5, 1, 2, 5, 10, and 15 Gy) from a cobalt-60 radiation source at a dose rate of 0.80 Gy/min, radiation doses below 2 Gy did not cause any significant apoptosis, but between 5 and 15 Gy, significant apoptosis was observed, with peak values at 5 Gy. When the HeLa Hep2 cells were exposed to doses of 2, 5, and 10 Gy at a 10-fold lower dose rate (0.072 Gy/min), the rate of apoptosis was unchanged up to the dose of 2 Gy and was significantly increased at doses above 2 Gy. In a recent study, low doses of radiation were administered to the eyes in a retinal degeneration mouse model, which prevented photoreceptor cell apoptosis both morphologically and functionally (Otani et al. 2012). These changes were accompanied by a significant neuroprotective effect.

Radiation-induced ROS generation is also known to be a prominent cause of autophagy ("self-eating"), a catabolic process involving the degradation of a cell's own components through the lysosomal machinery, which has been shown to be related to both aging retardation and cancer prevention (Szumiel 2012). The functional relationship between autophagy and apoptosis is complex, since under certain circumstances autophagy constitutes a stress adaptation that avoids cell death and suppresses apoptosis, whereas in other cellular settings it constitutes an alternative cell-death pathway (Maiuri et al. 2007; Cheng and Yang 2013). Autophagy and apoptosis may be triggered by common upstream signals, and sometimes this can result in combined autophagy and apoptosis; in other instances, the cell switches between the two responses in a mutually exclusive manner. Under the conditions of radiation-induced oxidative stress, a balance between signaling functions and damaging effects of ROSs is likely the most important factor that decides the fate of the mammalian cell (Szumiel 2012).

Hormesis induced by LD ionizing radiation is often mirrored by its stimulation of cell proliferation. The LD ionizing radiation was shown to be able to significantly enhance the proliferation of cultured mesenchymal stem cells via the activation of the mitogen-activated protein kinases/extracellular-signal-regulated kinases pathway, which is known to play an important role in cell growth (Liang et al. 2011). Fliedner and others (2012) showed that hemopoietic stem and early progenitor cells have the capacity to tolerate and adapt to being repetitively hit by energy deposition events. According to the authors, their data are compatible with the *injured stem cell hypothesis*, stating that radiation-injured stem cells, depending on dose rate, may continue to deliver clones of functional cells that maintain homeostasis of hemopoiesis throughout their lives.

### 7.4.6 Epigenetic Alterations

The system response to acute or chronic LD radiation exposure can evolve from damage to the basic molecular level and from adaptive responses that may occur at the whole-organism level (Feinendegen et al. 2007). The damage may propagate to successive higher levels of organization, meeting protective barriers, which may be upregulated by adaptive responses. The balance between health risks and benefits of LD exposure for a given individual may become predictable by gene expression profiles in irradiated and nonirradiated cells of the individual. Whole-organism epigenetic transcriptional reorganization is likely a common mechanism underlying different hormesis-like responses, including radiation hormesis.

Epigenetic alterations are heritable changes that affect gene function without modifying the DNA sequence (Bird 2007). The major epigenetic mechanisms include DNA methylation, histone modifications, and alterations in the expression of noncoding RNAs.

In several studies, wide-scale changes in gene expression were found after LD radiation exposures. For example, in a study by Feinendegen (2005), the exposure of human skin fibroblasts culture to a single dose of 20 mGy γ-radiation changed the expression of more than 100 genes. These genes included stress response genes and were different from the group of genes in parallel cultures exposed to 500 mGy. A similar pattern of expression among a total of 1574 genes developed in a γ-irradiated mouse brain. In a study by Seong and others (2011), the life span extension in *D. melanogaster* induced by LD radiation was accompanied by genome-wide changes in expression profile. In response to radiation, approximately 13% of the genes exhibited changes in expression levels; among them, many aging-related genes were significantly regulated. Radiation-induced epigenetic alterations accompanied by adaptive phenotypic changes were revealed in a recent research by Bernal and others (2013). In this study, conducted in the viable yellow agouti ($A^{vy}$) mouse model, LD ionizing radiation resulted in significantly increased DNA methylation at the agouti ($A^{vy}$) locus in male offspring exposed to doses between 0.7 and 7.6 cGy. The offspring's coat color was concomitantly shifted toward pseudoagouti. Maternal dietary antioxidant supplementation mitigated both DNA methylation changes and coat color shift in the irradiated offspring. The authors concluded that LD ionizing radiation exposure during gestation can elicit epigenetic alterations that lead to positive adaptive phenotypic changes (radiation hormesis).

The important role of epigenetic effectors is evident in a variety of hormesis-like responses, including bystander effects, or nontargeted effects, of radiation (Mothersill and Seymour 2012b; Merrifield and Kovalchuk 2013). Radiation-induced bystander effects occur in cells that are not directly hit by radiation tracks but receive signals from hit cells (Mancuso et al. 2012; Kadhim et al. 2013). Such signals may be secreted and/or shed by irradiated cells or communicated via molecules transported between cells by cell to cell gap junctions (Dauer et al. 2010). Many of the bystander effects are detrimental (e.g., induction of mutations, chromosome aberrations, chromosomal instability, and oncogenic transformation). However, protective effects have also been reported, for example, radiation-induced adaptive responses and induction of terminal differentiation and apoptosis of potentially damaged cells (Coates et al. 2004).

Initially, the bystander effects were considered to be a more in vitro phenomenon; however, they have now been established for in vivo models as well (Brooks 2004). The aforementioned abscopal phenomenon implies that bystander effects in vivo can coordinate a complex interplay of different regulation pathways for the induction of a whole organism's protective response. Scott and others (2009) have suggested that via bystander signaling pathways large numbers of cells participate in adaptive responses through mild stress-related epigenetic reprogramming (epireprogramming). Such epireprogramming can prime the cells for augmented defense against future, more threatening insults. Evidence was also found for an interorganismic bystander interaction (Mothersill and Seymour 2012b). It has been shown that irradiated fishes can release signals in the water that enhance defense against cancer in other fishes swimming in the same water (Smith et al. 2007). Mothersill and others (2010) found a clear legacy effect of early-life bystander irradiation in adult fish. Their data indicate that bystander signals from irradiated to nonirradiated fish can be transmitted in vivo and once induced are persistent during the animals' life span.

A growing body of evidence suggests that epigenetic transcriptional regulation may be involved in radiation-induced bystander effects (Kovalchuk and Baulch 2008; Ilnytskyy and Kovalchuk 2011; Kovalchuk et al. 2012; Mothersill and Seymour 2012b). In a study in which mice were unilaterally exposed to x-ray irradiation, Koturbash and others (2006) found a significant reduction in the levels of the de novo DNA methyltransferases DNMT3a and DNMT3b and a concurrent increase in the levels of the maintenance DNA methyltransferase DNMT1 in mouse bystander tissues. Furthermore, the levels of two methyl-binding proteins known to be involved in transcriptional silencing, methyl CpG binding protein 2 (MeCP2) and MBD2, were also found increased in bystander tissues. To investigate the possibility that localized x-ray irradiation induces persistent bystander effects in distant tissue, Koturbash and others (2007) in their subsequent research monitored the induction of epigenetic changes in rat spleen tissue 24 hours and 7 months after localized cranial exposure to 20 Gy of x-rays. Such a radiation exposure led to the induction of bystander effects in lead-shielded distant spleen tissue. This exposure caused a profound epigenetic dysregulation in the bystander spleen tissue that manifested as a significant loss of global DNA methylation, alterations in methylation of long interspersed nucleotide element-1 retrotransposable elements and downregulation of DNA methyltransferases and the methyl-binding protein MeCP2. In addition, irradiation significantly altered the expression of miR-194 putatively targeting both DNA methyltransferase-3a and MeCP2.

In a study by Ilnytskyy and others (2008), radiation exposure resulted in a significant deregulation of microRNA expression in murine spleen and thymus tissues. Among the regulated microRNAs, they found that changes in the expression of miR-34a and miR-7 may be involved in important protective mechanisms counteracting radiation cytotoxicity. A significant increase in the expression of tumor-suppressor miR-34a, accompanied by a decrease in the expression of its target oncogenes NOTCH1, MYC, E2F3, and cyclin D1, has been observed. Additionally, whereas miR-7 was significantly downregulated, lymphoid-specific helicase, a pivotal regulator of DNA methylation and genome stability, was significantly upregulated. The authors hypothesize that these cellular changes may constitute an attempt to counteract radiation-induced hypomethylation. Further, Ilnytskyy and others (2009) analyzed changes in global DNA methylation and microRNAome in skin and spleen of animals subjected to whole-body or cranial exposure to 0.5 Gy of x-rays. Both acute and fractionated radiation exposures resulted in a significant loss of global DNA methylation in the exposed and bystander spleens. These changes were paralleled by a reduction in the expression of the methyl-binding protein MeCP2. Irradiation also induced tissue-specific microRNAome alterations in skin and spleen.

## 7.5 CONCLUDING REMARKS

The current radiological standards for limiting exposure to ionizing radiation in humans are based on the LNT model, which is a linear extrapolation from the dose-response data of high doses of ionizing radiation and states that the health risks are directly proportional to the dose at all dose levels. The BEIR report (2006) on the health risks from exposure to low levels of ionizing radiation pointed out that the LNT model is the most practical model to estimate radiation risks, especially for radiation protection purposes. However, according to the results of many studies across the world, the assumptions based on the LNT model are not compatible with the actual observed health effects of low doses of radiation. Many scientists are questioning whether LNT-based regulations are the best way to protect and promote public health (Cuttler and Pollycove 2009; Mossman 2012). These regulations ignore the fact that biological mechanisms of natural radiation protection may reduce damages caused by LD radiation.

Data from many recent studies provide evidence to support the hormetic model, which, unlike the LNT model, assumes that adaptive mechanisms can be stimulated by LD radiation and the induction of these mechanisms may reduce the level of naturally occurring damages, such as those from free radical oxidation and cellular-level inflammation. This means that health hazards from ordinarily encountered radiation exposures, such as background radiation, medical x-rays, and so on, are much lower than those given by the LNT-based estimates. Among humans, there is no evidence of a carcinogenic effect for acute irradiation at doses less than 100 mSv and for protracted irradiation at doses less than 500 mSv (Tubiana et al. 2009). Moreover, there is abundant evidence that exposure to LD and LD-rate radiation can be beneficial rather than detrimental and can thereby lead to a hormetic effect. The EPRI (2009) report stresses that

> research into the health effects of LD radiation should continue and should use holistic, systems-based approaches to develop models that define the shape of the dose-response relationships in the LD regions. Risk models should fuse the latest radiobiology and epidemiology results to produce a comprehensive understanding of radiation risk that addresses both damage (likely with a linear effect) and response (possibly with nonlinear consequences). While recent scientific advances have provided much new information in the LD region, they have also raised additional research questions. New research in areas such as systems biology can provide mechanistic understandings of LD and low dose-rate effects needed to estimate human cancer risks. Therefore, it is essential that research into LD radiation biology, dose reconstruction, and epidemiology continue in order to provide opportunities for continuous improvement in the scientific support of future regulatory and policy actions.

Currently, many authors consider that radiation safety norms are exceedingly restrictive and should be revised to become more realistic and workable. In his recent paper, Jargin (2012) concluded that there are no evidence-based contraindications to fivefold elevation of the total equivalent effective doses to individual members of the public (up to 5 mSv/year) and doubling of the limits for professional exposures. As has been noted by Doss (2012b) in his recent paper, a paradigm shift may be advisable in radiation safety from the present one based on the LNT model to one that recognizes the presence of a beneficial adaptive response from LD radiation.

## ACKNOWLEDGMENT

The authors thank Oksana Zabuga for her assistance in the preparation of this chapter.

## REFERENCES

Abbatt, J. D., Hamilton T. R., and Weeks J. L. 1983. Epidemiological studies in three corporations covering the Canadian nuclear fuel cycle. In: *Biological Effects of Low-Level Radiation*. IAEA Symposium Proceedings, 351–61. Vienna, Austria.

Abouzeid Ali, H. E., Barber, R. C., and Dubrova, Y. E. 2012. The effects of maternal irradiation during adulthood on mutation induction and transgenerational instability in mice. *Mutat Res* 732(1–2): 21–5.

Ahn, Y. O., Li, Z. M., and KREEC Study Group. 2012. Cancer risk in adult residents near nuclear power plants in Korea—a cohort study of 1992–2010. *J Korean Med Sci* 27(9): 999–1008.

Azizova, T. V., Muirhead, C. R., Druzhinina, M. B., et al. 2010. Cardiovascular diseases in the cohort of workers first employed at Mayak PA in 1948–1958. *Radiat Res* 174(2): 155–68.

Azizova, T. V., Muirhead, C. R., Moseeva, M. B., et al. 2012. Ischemic heart disease in nuclear workers first employed at the Mayak PA in 1948–1972. *Health Phys* 103(1): 3–14.

Balonov, M. I. 2012. On protecting the inexperienced reader from Chernobyl myths. *J Radiol Prot* 32: 181.

Barber, R. C., Hardwick, R. J., Shanks, M. E., et al. 2009. The effects of in utero irradiation on mutation induction and transgenerational instability in mice. *Mutat Res* 664(1–2): 6–12.

Barber, R. C., Plumb, M. A., Boulton, E., Roux, I., and Dubrova, Y. E. 2002. Elevated mutation rates in the germ line of first- and second-generation offspring of irradiated male mice. *Proc Natl Acad Sci USA* 99: 6877–82.

Bauer, G. 2007. Low dose radiation and intercellular induction of apoptosis: potential implications for the control of oncogenesis. *Int J Radiat Biol* 83: 873–88.

Becker, K. 2003. Health effects of high radon environments in central Europe: another test for the LNT hypothesis? Nonlinearity. *Biol Toxicol Med* 1: 3–35.

BEIR (Biological Effects of Ionizing Radiation) VI Report. 1999. *Health Effects of Exposure to Radon*. National Research Council, National Academy Press: Washington, DC.

BEIR (Biological Effects of Ionizing Radiation) VII Phase 2 Report. 2006. *Health Risks from Exposure to Low Levels of Ionizing Radiation*. National Research Council, National Academy Press: Washington, DC.

Bernal, A. J., Dolinoy, D. C., Huang, D., et al. 2013. Adaptive radiation-induced epigenetic alterations mitigated by antioxidants. *FASEB J* 27: 665–71.

Berrington, A., Darby, S. C., Weiss, H. A., and Doll, R. 2001. 100 years of observation on British radiologists: mortality from cancer and other causes 1897–1997. *Br J Radiol* 74: 507–19.

Bird, A. 2007. Perceptions of epigenetics. *Nature* 427: 164–7.

Blankenbecler, R. 2010. Low-dose pretreatment for radiation therapy. *Dose-Response* 8(4): 534–42.

Bogen, K. T., and Cullen, J. 2002. Residential radon in U.S. counties vs. lung cancer in women who predominantly never smoked. *Environ Geochem Health* 24: 229–47.

Boice, J. D. 2012. Radiation epidemiology: a perspective on Fukushima. *J Radiol Prot* 32(1): N33–40.

Boice, J. D., Cohen, S. S., Mumma, M. T., et al. 2011. Updated mortality analysis of radiation workers at Rocketdyne (Atomics International), 1948–2008. *Radiat Res* 176(2): 244–58.

Boice, J. D., Morin, M. M., and Glass, A. G. 1991. Diagnostic x-ray procedures and risk of leukemia, lymphoma and multiple myeloma. *JAMA* 265: 1290–4.
Brenner, D. J., and Hall, E. J. 2007. Computed tomography—an increasing source of radiation exposure. *N Engl J Med* 357: 2277–84.
Brooks, A. L. 2004. Evidence for "bystander effects" in vivo. *Hum Exp Toxicol* 23: 67–70.
Calabrese, E. J. 2009. Getting the dose-response wrong: why hormesis became marginalized and the threshold model accepted. *Arch Toxicol* 83: 227–47.
Calabrese, E. J. 2013. Hormesis: toxicological foundations and role in aging research. *Exp Gerontol* 48(1): 99–102.
Calabrese, E. J., and Calabrese, V. 2013a. Reduction of arthritic symptoms by low dose radiation therapy (LD-RT) is associated with an anti-inflammatory phenotype. *Int J Radiat Biol* 89(4): 278–86.
Calabrese, E. J., and Calabrese, V. 2013b. Low dose radiation therapy (LD-RT) is effective in the treatment of arthritis: animal model findings. *Int J Radiat Biol* 89(4): 287–94.
Calabrese, E. J., Bachmann, K. A., Bailer, A. J., et al. 2007. Biological stress response terminology: integrating the concepts of adaptive response and preconditioning stress within a hormetic dose-response framework. *Toxicol Appl Pharmacol* 222: 122–8.
Calabrese, E. J., Iavicoli, I., and Calabrese, V. 2013. Hormesis: its impact on medicine and health. *Hum Exp Toxicol* 32(2): 120–52.
Calabrese, V., Cornelius, C., Mancuso, C., et al. 2008. Cellular stress response: a novel target for chemoprevention and nutritional neuroprotection in aging, neurodegenerative disorders and longevity. *Neurochem Res* 33(12): 2444–71.
Cameron, J. R. 2002. Radiation increased the longevity of British radiologists. *Br J Radiol* 75: 637–9.
Cameron, J. R. 2005. Moderate dose rate ionizing radiation increases longevity. *Br J Radiol* 78: 11–3.
Cardis, E., Gilbert, E. S., and Carpenter, L. 1995. Effects of low doses and low dose rates of external ionizing radiation: cancer mortality among nuclear industry workers in three countries. *Radiat Res* 142: 117–32.
Cardis, E., Howe, G., and Ron, E. 2006. Cancer consequences of the Chernobyl accident: 20 years after. *J Radiol Prot* 26: 127–40.
Cardis, E., Vrijheid, M., and Blettner, M. 2007. The 15-country collaborative study of cancer risk among radiation workers in the nuclear industry. *Radiat Res* 167: 396–416.
Cardis, E., Vrijheid, M., Blettner, M., Gilbert, E., Hakama, M., and Hill, C. 2005. Risk of cancer after low doses of ionising radiation: retrospective cohort study in 15 countries. *BMJ* 331: 77–80.
Carpenter, L., Beral, V., Fraser, P., and Booth, M. 1990. Health related selection and death rates in the United Kingdom Atomic Energy Authority workforce. *Brit J Ind Med* 47: 248–58.
Chaffey, J. T., Rosenthal, D. S., Moloney, W. D., and Hellman, S. 1976. Total body irradiation as treatment for lymphosarcoma. *Int J Radiat Oncol Biol Phys* 1: 399–405.
Chan, E. K., Fung, S., Gospodarowicz, M., et al. 2011. Palliation by low-dose local radiation therapy for indolent non-Hodgkin lymphoma. *Int J Radiat Oncol Biol Phys* 81(5): e781–6.
Cheda, A., Nowosielska, E. M., Wrembel-Wargocka, J., and Janiak, M. K. 2009. Single or fractionated irradiations of mice with low doses of x-rays stimulate innate immune mechanisms. *Int J Low Rad* 6: 325–42.
Cheda, A., Wrembel-Wargocka, J., Lisiak, E., Nowosielska, E. M., Marciniak, M., and Janiak, M. K. 2004. Single low doses of x rays inhibit the development of experimental tumor metastases and trigger the activities of NK cells in mice. *Radiat Res* 161(3): 335–40.
Cheng, Y., and Yang, J. M. 2013. Autophagy and apoptosis: rivals or mates? *Chin J Cancer* 32(3): 103–5.

Chernobyl Forum. 2003–2005. *Chernobyl's Legacy: Health, Environmental and Socio-Economic Impacts and Recommendations to the Governments of Belarus, the Russian Federation and Ukraine.* Second revised version. www.iaea.org/Publications/Booklets/Chernobyl/chernobyl.pdf (accessed April, 2006).

Choi, N. C., Timothy, A. R., Kaufman, S. D., Carey, R. W., and Aisenberg, A. C. 1979. Low dose fractionated whole body irradiation in the treatment of advanced non-Hodgkin's lymphoma. *Cancer* 43: 1636–42.

CNSC (Canadian Nuclear Safety Commission). 2011. Verifying Canadian Nuclear Energy Worker Radiation Risk: A Reanalysis of Cancer Mortality in Canadian Nuclear Energy Workers (1957–1994) Summary Report. 2011. INFO-0811.

Coates, P. J., Lorimore, S. A., and Wright, E. G. 2004. Damaging and protective cell signaling in the untargeted effects of ionizing radiation. *Mutat Res* 568: 5–20.

Cohen, B. L. 1993. Relationship between exposure to radon and various types of cancer. *Health Phy* 65: 529–31.

Cohen, B. L. 1994. Dose-response relationship for radiation carcinogenesis in the low-dose region. *Int Arch Occup Environ Health* 66: 71–5.

Cohen, B. L. 1995. Test of the linear-no threshold theory of radiation carcinogenesis for inhaled radon decay products. *Health Phys* 68: 157–74.

Cohen, B. L. 1997. Lung cancer rate vs. mean radon level in U.S. counties of various characteristics. *Health Phys* 72(1): 114–9.

Cohen, M. D. 2013. Radiation risks of medical imaging. *Radiology* 266(3): 995.

Cologne, J. B., and Preston, D. L. 2000. Longevity of atomic-bomb survivors. *Lancet* 2000(9226): 303–7.

Covelli, V., Di Majo, V., Coppola, M., and Rebessi, S. 1989. The dose-response relationships for myeloid leukemia and malignant lymphoma in BC3F1 mice. *Radiat Res* 119(3): 553–61.

Cuttler, J. M., and Pollycove, M. 2009. Nuclear energy and health: and the benefits of LD radiation hormesis. *Dose-Response* 7: 52–89.

Darby, S. C., and Hill, D. C. 2003. Health effects of residential radon: a European perspective at the end of 2002. *Radiat Prot Dosim* 104: 321–9.

Darby, S. C., Kendall, G. M., Fell, T. P., et al. 1988. A summary of mortality and incidence of cancer in men from the United Kingdom who participated in the United Kingdom's atmospheric nuclear weapon tests and experimental programmes. *BMJ* 296(6618): 332–8.

Darby, S. C., Kendall, G. M., Fell, T. P., et al. 1993. Further follow up of mortality and incidence of cancer in men from the United Kingdom who participated in the United Kingdom's atmospheric nuclear weapon tests and experimental programmes. *BMJ* 307(6918): 1530–5.

Das, B., Saini, D., and Seshadri, M. 2012. No evidence of telomere length attrition in newborns from high level natural background radiation areas in Kerala coast, south west India. *Int J Radiat Biol* 88(9): 642–7.

Dauer, L. T., Brooks, A. L., Hoel, D. G., Morgan, W. F., Stram, D., and Tran, P. 2010. Review and evaluation of updated research on the health effects associated with low-dose ionising radiation. *Radiat Prot Dosimetry* 140: 103–36.

Davis, S. 2012. Health risks associated with environmental radiation exposures. *J Radiol Prot* 32(1): N21–5.

Davis, S., Day, R. W., Kopecky, K. J., et al. 2006. Childhood leukaemia in Belarus, Russia, and Ukraine following the Chernobyl power station accident: results from an international collaborative population-based case-control study. *Int J Epidemiol* 35: 386–96.

Demaria, S., Ng, B., Devitt, M. L., et al. 2004. Ionizing radiation inhibition of distant untreated tumors (abscopal effect) is immune mediated. *Int J Radiat Oncol Biol Phys* 58(3): 862–70.

Deramaudt, T. B., Dill, C., and Bonay, M. 2013. Regulation of oxidative stress by Nrf2 in the pathophysiology of infectious diseases. *Med Mal Infect* 43(3): 100–7.

Doll, R., Berrington, A., and Darby, S. C. 2005. Low mortality of British radiologists. *Br J Radiol* 78: 1057–8.

Doody, M. M., Freedman, D. M., Alexander, B. H., et al. 2006. Breast cancer incidence in U.S. radiologic technologists. *Cancer* 106(12): 2707–15.

Doss, M. 2012a. Evidence supporting radiation hormesis in atomic bomb survivor cancer mortality data. *Dose-Response* 10(4): 584–92.

Doss, M. 2012b. Shifting the paradigm in radiation safety. *Dose-Response* 10(4): 562–83.

Doss, M. 2013. The importance of adaptive response in cancer prevention and therapy. *Med Phys* 40(3): 030401.

Draper, G. J., Little, M. P., Sorahan, T., et al. 1997. Cancer in the offspring of radiation workers: a record linkage study. *BMJ* 315(7117): 1181–8.

Dublin, L. I., and Spigelman, M. 1948. Mortality among medical specialists. *J Am Med Dir Assoc* 137: 1519–24.

Dubrova, Y. E., Plumb, M., Gutierrez, B., Boulton, E., and Jeffreys, A. J. 2000. Transgenerational mutation by radiation. *Nature* 405: 37.

Durović, B., Spasić-Jokić, V., and Durović, B. 2008. Influence of occupational exposure to low-dose ionizing radiation on the plasma activity of superoxide dismutase and glutathione level. *Vojnosanit Pregl* 65(8): 613–8.

EPRI (Electric Power Research Institute). 2009. Program on Technology Innovation: Evaluation of Updated Research on the Health Effects and Risks Associated with LD Ionizing Radiation. Electric Power Research Institute (EPRI), Palo Alto, California, 1019227.

Farooque, A., Mathur, R., Verma, A., et al. 2011. Low-dose radiation therapy of cancer: role of immune enhancement. *Expert Rev Anticancer Ther* 11(5): 791–802.

Feinendegen, L. E. 2005. Evidence for beneficial low level radiation effects and radiation hormesis. *Br J Radiol* 78: 3–7.

Feinendegen, L. E., and Pollycove, M. 2001. Biologic responses to low doses of ionizing radiation: detriment versus hormesis. Part 1. Dose response of cells and tissues. *J Nucl Med* 42: 17N–27N.

Feinendegen, L. E., Pollycove, M., and Neumann, R. D. 2007. Whole-body responses to low-level radiation exposure. New concepts in mammalian radiobiology. *Exp Hematol* 35: 37–46.

Feinendegen, L. E., Pollycove, M., and Sondhaus, C. A. 2004. Responses to low doses of ionizing radiation in biological systems. *Nonlinearity Biol Toxicol Med* 2(3): 143–71.

Fliedner, T. M., Graessle, D. H., Meineke, V., and Feinendegen, L. E. 2012. Hemopoietic response to low dose-rates of ionizing radiation shows stem cell tolerance and adaptation. *Dose-Response* 10(4): 644–63.

Formenti, S. C., and Demaria, S. 2013. Combining radiotherapy and cancer immunotherapy: a paradigm shift. *J Natl Cancer Inst* 105(4): 256–65.

Frigerio, N. A., and Stowe, R. S. 1976. Carcinogenic and genetic hazard from background radiation. In: *Biological and Environmental Effects of Low-Level Radiation*. IAEA Symposium Proceedings (v. 2), 385–93. Vienna, Austria.

Gilbert, E. S., Cragle, D. L., and Wiggs, L. D. 1993. Updated analyses of combined mortality data for workers at the Hanford Site, Oak Ridge National Laboratory, and Rocky Flats Weapons Plant. *Radiat Res* 136: 408–21.

Gilbert, E. S., Fry, S. A., Wiggs, L. D., Voelz, G. L., Cragle, D. L., and Petersen, G. R. 1989. Analyses of combined mortality data on workers at the Hanford Site, Oak Ridge National Laboratory, and Rocky Flats Nuclear Weapons Plant. *Radiat Res* 120: 19–35.

Goodman, L.A. 1959. Some alternatives to ecological correlation. *Am J Sociol* 64: 610–25.
Groen, R. S., Bae, J. Y., and Lim, K. J. 2012. Fear of the unknown: ionizing radiation exposure during pregnancy. *Am J Obstet Gynecol* 206(6): 456–62.
Grosche, B., Lackland, D. T., Land, C. E., et al. 2011. Mortality from cardiovascular diseases in the Semipalatinsk historical cohort, 1960–1999, and its relationship to radiation exposure. *Radiat Res* 176(5): 660–9.
Hall, E. J., and Giaccia, A. J. 2006. *Radiobiology for the Radiologist*. Philadelphia, PA: Lippincott Williams and Wilkins.
Hart, J., and Hyun, S. 2012. Cancer mortality, state mean elevations, and other selected predictors. *Dose-Response* 10(1): 58–65.
Hauptmann, M., Mohan, A. K., Doody, M. M., Linet, M. S., and Mabuchi, K. 2003. Mortality from diseases of the circulatory system in radiologic technologists in the United States. *Am J Epidemiol* 157(3): 239–48.
Heidenreich, W. F., Paretzke, H. G., and Jacob, P. 1997. No evidence for increased tumor rates below 200 mSv in the atomic bomb survivor data. *Radiat Environ Biophys* 36: 205–7.
Hendee, W. R., and O'Connor, M. K. 2012. Radiation risks of medical imaging: separating fact from fantasy. *Radiology* 264(2): 312–21.
Hendry, J. H., Simon, S. L., Wojcik, A., et al. 2009. Human exposure to high natural background radiation: what can it teach us about radiation risks? *J Radiol Prot* 29: A29–42.
High Background Radiation Research Group. 1980. Health survey in high background radiation areas in China. *Science* 209: 877–80.
Huilgol, N. G. 2012. Hormesis: a peep in to the human nature. *J Cancer Res Ther* 8: 175.
IAEA/WHO/UNDP. 2005. "Chernobyl: The True Scale of the Accident," IAEA/WHO/UNDP press release, September 5.
Ibuki, Y., Hayashi, A., Suzuki, A., and Goto, R. 1998. Low-dose irradiation induces expression of heat shock protein 70 mRNA and thermo- and radio-resistance in myeloid leukemia cell line. *Biol Pharm Bull* 21(5): 434–9.
Ilnytskyy, Y., and Kovalchuk, O. 2011. Non-targeted radiation effects—an epigenetic connection. *Mutat Res* 714(1–2): 113–25.
Ilnytskyy, Y., Koturbash, I., and Kovalchuk, O. 2009. Radiation-induced bystander effects in vivo are epigenetically regulated in a tissue-specific manner. *Environ Mol Mutagen* 50: 105–13.
Ilnytskyy, Y., Zemp, F. J., Koturbash, I., and Kovalchuk, O. 2008. Altered microRNA expression patterns in irradiated hematopoietic tissues suggest a sex-specific protective mechanism. *Biochem Biophys Res Commun* 377: 41–5.
Imaida, K., Hasegawa, R., Kato, T., et al. 1997. Clinicopathological analysis on cancers of autopsy cases in a geriatric hospital. *Pathol Int* 47: 293–300.
Ina, Y., and Sakai, K. 2004. Prolongation of life span associated with immunological modification by chronic low-dose-rate irradiation in MRL-lpr/lpr mice. *Radiat Res* 161: 168–73.
Ishii, K., Hosoi, Y., Yamada, S., Ono, T., and Sakamoto, K. 1996. Decreased incidence of thymic lymphoma in AKR mice as a result of chronic, fractionated low-dose total-body x irradiation. *Radiat Res* 146(5): 582–5.
Ivanov, V. K. 2007. Late cancer and noncancer risks among Chernobyl emergency workers of Russia. *Health Phys* 93: 470–9.
Ivanov, V. K., Gorsky, A. I., Kashcheev, V. V., Maksioutov, M. A., and Tumanov, K. A. 2009. Latent period in induction of radiogenic solid tumors in the cohort of emergency workers. *Radiat Environ Biophys* 8(3): 247–52.
Ivanov, V. K., Groski, A. I., Tsyb, A. F., Ivanov, S. I., Naumenko, R. N., and Ivanova, L. V. 2004. Solid cancer incidence among the Chernobyl emergency workers residing in Russia: estimation of radiation risks. *Radiat Environ Biophys* 43: 35–42.

Jagger, J. 1998. Natural background radiation and cancer death in Rocky Mountain states and Gulf Coast states. *Health Phys* 75: 428–30.

Jaiswal, A. K. 2004. Nrf2 signaling in coordinated activation of antioxidant gene expression. *Free Radic Biol Med* 36(10): 1199–207.

Jargin, S. V. 2010. Overestimation of Chernobyl consequences: poorly substantiated information. *Radiat Environ Biophys* 49: 743–5.

Jargin, S. V. 2012. Hormesis and radiation safety norms. *Hum Exp Toxicol* 31(7): 671–5.

Jaworowski, Z. 2008. The paradigm that failed. *Int J Low Radiation* 5: 151–5.

Jaworowski, Z. 2010a. Radiation hormesis—a remedy for fear. *Hum Exp Toxicol*. 29(4): 263–70.

Jaworowski, Z. 2010b. Observations on the Chernobyl disaster and LNT. *Dose-Response* 8(2): 148–71.

Johnson, K. J., Alexander, B. H., Doody, M. M., et al. 2008. Childhood cancer in the offspring born in 1921–1984 to US radiologic technologists. *Br J Cancer* 99(3): 545–50.

Jones, C. 2011. Failing to adapt—the ageing immune system's role in cancer pathogenesis. *Rev Clin Gerontol* 21: 209–18.

Kadhim, M., Salomaa, S., Wright, E., et al. 2013. Non-targeted effects of ionising radiation—implications for low dose risk. *Mutat Res* 752(2): 84–98.

Karam, P. A., and Leslie, S. A. 1999. Calculations of background beta-gamma radiation dose through geologic time. *Health Phys* 77: 662–7.

Kataoka, T. 2013. Study of antioxidative effects and anti-inflammatory effects in mice due to low-dose x-irradiation or radon inhalation. *J Radiat Res* [In press].

Kendall, G. M., Muirhead, C. R., MacGibbon, B. H., O'Hagan, J. A., Conquest, A. J., and Goodill, A. A. 1992. Mortality and occupational exposure to radiation: first analysis of the National Registry for Radiation Workers. *BMJ* 304: 220–5.

Kleinerman, R. A. 2006. Cancer risks following diagnostic and therapeutic radiation exposure in children. *Pediatr Radiol* 2: 121–5.

Koebel, C. M., Vermi, W., Swann, J. B., et al. 2007. Adaptive immunity maintains occult cancer in an equilibrium state. *Nature* 450: 903–7.

Kojima, S., Matsuki, O., Nomura, T., Kubodera, A., and Yamaoka, K. 1998. Elevation of mouse liver glutathione level by low-dose gamma-ray irradiation and its effect on CCl4-induced liver damage. *Anticancer Res* 18(4A): 2471–6.

Kojima, S., Matsuki, O., Nomura, T., Yamaoka, K., Takahashi, M., and Niki, E. 1999. Elevation of antioxidant potency in the brain of mice by low-dose gamma-ray irradiation and its effect on 1-methyl-4-phenyl-1,2,3,6-tetrahydropyridine (MPTP)-induced brain damage. *Free Radic Biol Med* 26: 388–95.

Kondo, S. 1993. *Health Effects of Low Level Radiation*. Madison, WI: Medical Physics Pub. Corp.

Kostyuchenko, V. A., and Krestina, L. 1994. Long-term irradiation effects in the population evacuated from the east-Urals radioactive trace area. *Sci Total Environ* 142: 119–25.

Koturbash, I., Boyko, A., Rodriguez-Juarez, R., et al. 2007. Role of epigenetic effectors in maintenance of the long-term persistent bystander effect in spleen in vivo. *Carcinogenesis* 28: 1831–8.

Koturbash, I., Rugo, R. E., Hendricks, C. A., et al. 2006. Irradiation induces DNA damage and modulates epigenetic effectors in distant bystander tissue in vivo. *Oncogene* 25: 4267–75.

Kovalchuk, A., Lowings, M., Rodriguez-Juarez, R., et al. 2012. Epigenetic bystander-like effects of stroke in somatic organs. *Aging (Albany NY)* 4(3): 224–34.

Kovalchuk, O., and Baulch, J. E. 2008. Epigenetic changes and nontargeted radiation effects—is there a link? *Environ Mol Mutagen* 49: 16–25.

Krueger, S. A., Joiner, M. C., Weinfeld, M., Piasentin, E., and Marples, B. 2007. Role of apoptosis in low-dose hyper-radiosensitivity. *Radiat Res* 167: 260–7.

Kumar, S. 2012. Second malignant neoplasms following radiotherapy. *Int J Environ Res Public Health* 9(12): 4744–59.

Lagisz, M., Hector, K. L., and Nakagawa, S. 2013. Life extension after heat shock exposure: assessing meta-analytic evidence for hormesis. *Ageing Res Rev* 12(2): 653–60.

Land, C. E. 1980. Estimating cancer risks from low doses of ionizing radiation. *Science* 209: 1197–203.

Lantz, P. M., Mendez, D., and Philbert, M. A. 2013. Radon, smoking, and lung cancer: the need to refocus radon control policy. *Am J Public Health* 103(3): 443–7.

Le Bourg, E. 2013. Obsolete ideas and logical confusions can be obstacles for biogerontology research. *Biogerontology* 14: 221–7.

Lee, Y. J., Park, G. H., Cho, H. N., et al. 2002. Induction of adaptive response by low-dose radiation in RIF cells transfected with Hspb1 (Hsp25) or inducible Hspa (Hsp70). *Radiat Res* 157(4): 371–7.

Lewis, K. N., Mele, J., Hayes, J. D., and Buffenstein, R. 2010. Nrf2, a guardian of healthspan and gatekeeper of species longevity. *Integr Comp Biol* 50(5): 829–43.

Liang, X., So, Y. H., Cui, J., et al. 2011. The low-dose ionizing radiation stimulates cell proliferation via activation of the MAPK/ERK pathway in rat cultured mesenchymal stem cells. *J Radiat Res* 52(3): 380–6.

Liberek, K., Lewandowska, A., and Zietkiewicz, S. 2008. Chaperones in control of protein disaggregation. *EMBO J* 27(2): 328–35.

Linet, M. S., Kim, K. P., Miller, D. L., Kleinerman, R. A., Simon, S. L., and Berrington de Gonzalez, A. 2010. Historical review of occupational exposures and cancer risks in medical radiation workers. *Radiat Res* 174(6): 793–808.

Little, M. P. 2009. Cancer and non-cancer effects in Japanese atomic bomb survivors. *J Radiol Prot* 29: 43–59.

Little, M. P., Azizova, T. V., Bazyka, D., et al. 2012. Systematic review and meta-analysis of circulatory disease from exposure to low-level ionizing radiation and estimates of potential population mortality risks. *Environ Health Perspect* 120(11): 1503–11.

Little, M. P., Tawn, E. J., Tzoulaki, I., et al. 2008. A systematic review of epidemiological associations between low and moderate doses of ionizing radiation and late cardiovascular effects, and their possible mechanisms. *Radiat Res* 169: 99–109.

Little, M. P., Tawn, E. J., Tzoulaki, I., et al. 2010. Review and meta-analysis of epidemiological associations between low/moderate doses of ionizing radiation and circulatory disease risks, and their possible mechanisms. *Radiat Environ Biophys* 49(2): 139–53.

Liu, S-Z. 2003. Nonlinear dose-response relationships in the immune system following exposure to ionizing radiation: mechanisms and implications. *Nonlinearity Biol Toxicol Med* 1: 71–92.

Liu, S-Z. 2006 Cancer control related to stimulation of immunity by low-dose radiation. *Dose-Response* 5(1): 39–47.

Luckey, T. D. 1991. *Radiation Hormesis*. Boca Raton, FL: CRC Press.

Luckey, T. D. 2008. Atomic bomb health benefits. *Dose-Response* 6: 369–82.

Macklis, R. M. 1990. Radithor and the era of mild radium therapy. *JAMA* 264: 614–8.

Maiuri, M. C., Zalckvar, E., Kimchi, A., and Kroemer, G. 2007. Self-eating and self-killing: crosstalk between autophagy and apoptosis. *Nat Rev Mol Cell Biol* 8(9): 741–52.

Mancuso, M., Pasquali, E., Giardullo, P., et al. 2012. The radiation bystander effect and its potential implications for human health. *Curr Mol Med* 12(5): 613–24.

Matanoski, G. M., Sternberg, A., and Elliott, E. A. 1987. Does radiation exposure produce a protective effect among radiologists? *Health Phys* 52: 637–43.

McDonald, J. T., Kim, K., Norris, A. J., et al. 2010. Ionizing radiation activates the Nrf2 antioxidant response. *Cancer Res* 70(21): 8886–95.

Men, T., Brennan, P., Boffetta, P., and Zaridze, D. 2003. Russian mortality trends for 1991–2001: analysis by cause and region. *BMJ* 327(7421): 964.

Merrifield, M., and Kovalchuk, O. 2013. Epigenetics in radiation biology: a new research frontier. *Front Genet* 4: 40.

Meslé, F. 2004. Mortality in central and eastern Europe: long-term trends and recent upturns. *Demograp Res* Special Collection 2(3): 45–70.

Mifune, M., Sobue, T., Arimoto, H., Komoto, Y., Kondo, S., and Tanooka, H. 1992. Cancer mortality survey in a spa area (Misasa, Japan) with a high radon background. *Jpn J Cancer Res* 83: 1–5.

Miller, A. B., Howe, G. R., and Sherman, G. J. 1989. Mortality from breast cancer after irradiation during fluoroscopic examinations in patients being treated for tuberculosis. *N Engl J Med* 321: 1285–9.

Mine, M., Okumura, Y., Ichimaru, M., Nakamura, T., Kondo, S. 1990. Apparently beneficial effect of low to intermediate doses of A-bomb radiation on human lifespan. *Int J Radiat Biol*. 58(6): 1035–43.

Mirzaie-Joniani, H., Eriksson, D., Sheikholvaezin, A., et al. 2002. Apoptosis induced by low-dose and low-dose-rate radiation. *Cancer* 94(4): 1210–4.

Mohan, A. K., Hauptmann, M., Freedman, D. M., et al. 2003. Cancer and other causes of mortality among radiologic technologists in the United States. *Int J Cancer* 103: 259–67.

Mohan, A. K., Hauptmann, M., Linet, M. S., et al. 2002. Breast cancer mortality among female radiologic technologists in the United States. *J Natl Cancer Inst* 94(12): 943–8.

Moory Doody, M., Lonstein, J. E., and Stovall, M. 2000. U.S. Scoliosis Cohort Study collaborators. Breast cancer mortality following diagnostic x-rays: findings from the U.S. Scoliosis Cohort Study. *Spine* 25: 2052–63.

Morgan, W. F., and Sowa, M. B. 2009. Non-targeted effects of ionizing radiation: implications for risk assessment and the radiation dose response profile. *Health Phys* 97: 426–32.

Moskalev, A. A., and Zainullin, V. G. 2001. Role of apoptotic cell death in radioinduced aging in *Drosophila melanogaster*. [Article in Russian]. *Radiats Biol Radioecol* 41(6): 650–2.

Moskalev, A. A., and Zainullin, V. G. 2002. Longevity of *Drosophila* after exposure to low doses of radiation and etoposide. [Article in Russian]. *Adv Gerontol* 10: 51–63.

Moskalev, A. A., and Zainullin, V. G. 2003. The role of reaper-dependent apoptosis in radiation-induced life-span alterations in *Drosophila melanogaster*. [Article in Russian]. *Radiats Biol Radioecol* 43(2): 242–4.

Moskalev, A. A., and Zainullin, V. G. 2004. Aging rate after continual low dose irradiation of drosophila strains with apoptosis deregulation. [Article in Russian]. *Radiats Biol Radioecol* 44(2): 156–61.

Moskalev, A. A., Iatskiv, A. S., and Zainullin, V. G. 2006. Effect of low-dose irradiation on the lifespan in various strains of *Drosophila melanogaster*. [Article in Russian]. *Genetika* 42: 773–82.

Mossman, K. L. 2012. The LNT debate in radiation protection: science vs. policy. *Dose-Response* 10(2): 190–202.

Mothersill, C., and Seymour, C. 2012a. Changing paradigms in radiobiology. *Mutat Res* 750(2): 85–95.

Mothersill, C., and Seymour, C. 2012b. Are epigenetic mechanisms involved in radiation-induced bystander effects? *Front Genet* 3: 74.

Mothersill, C., Smith, R. W., Saroya, R., et al. 2010. Irradiation of rainbow trout at early life stages results in legacy effects in adults. *Int J Radiat Biol* 86: 817–28.

Mughal, S. K., Myazin, A. E., Zhavoronkov, L. P., Rubanovich, A. V., and Dubrova, Y. E. 2012. The dose and dose-rate effects of paternal irradiation on transgenerational instability in mice: a radiotherapy connection. *PLoS One* 7(7): e41300.

Muirhead, C. R., Bingham, D., Haylock, R. G., et al. 2003. Follow up of mortality and incidence of cancer 1952–98 in men from the UK who participated in the UK's atmospheric nuclear weapon tests and experimental programmes. *Occup Environ Med* 60(3): 165–72.

Nair, R. R., Rajan, B., Akiba, S., et al. 2009. Background radiation and cancer incidence in Kerala, India–Karanagappally cohort study. *Health Phys* 96: 55–66.

Nambi, K. S. V., and Soman, S. D. 1987. Environmental radiation and cancer in India. *Health Phys* 52: 653–7.

Neel, J. V., Schull, W. J., Awa, A. A., et al. 1990. The children of parents exposed to atomic bombs: estimates of the genetic doubling dose of radiation for humans. *Am J Hum Genet* 46(6): 1053–72.

Neuberger, J. S., and Gesell, T. F. 2002. Residential radon exposure and lung cancer: risk in nonsmokers. *Health Phys* 83(1): 1–18.

Neumaier, T., Swenson, J., Pham, C., et al. 2012. Evidence for formation of DNA repair centers and dose-response nonlinearity in human cells. *Proc Natl Acad Sci U S A* 109(2): 443–8.

Noshchenko, A. G., Bondar, O. Y., and Drozdova, V. D. 2010. Radiation-induced leukemia among children aged 0–5 years at the time of the Chernobyl accident. *Int J Cancer* 127: 412–26.

Nowosielska, E. W., Wrembel-Wargocka, J., Cheda, A., and Janiak, M. K. 2006a. A single low-dose irradiation with x-rays stimulates NK cells and macrophages to release factors related to the cytotoxic functions of these cells. *Centr Eur J Immunol* 31: 51–7.

Nowosielska, E. M., Wrembel-Wargocka, J., Cheda, A., Lisiak, E., and Janiak, M. K. 2006b. Enhanced cytotoxic activity of macrophages and suppressed tumor metastases in mice irradiated with low doses of X-rays. *J Radiat Res* 47: 229–36.

Ogata, H. 2011. A review of some epidemiological studies on cancer risk from LD radiation or other carcinogenic agents. *Radiat Prot Dosimetry* 146(1–3): 268–71.

Okuma, K., Yamashita, H., Niibe, Y., Hayakawa, K., and Nakagawa, K. 2011. Abscopal effect of radiation on lung metastases of hepatocellular carcinoma: a case report. *J Med Case Rep* 5: 111.

Otake, M., Schull, W. J., and Neel, J. V. 1990. Congenital malformations, stillbirths, and early mortality among the children of atomic bomb survivors: a reanalysis. *Radiat Res* 122: 1–11.

Otani, A., Kojima, H., Guo, C., Oishi, A., and Yoshimura, N. 2012. Low-dose-rate, low-dose irradiation delays neurodegeneration in a model of retinitis pigmentosa. *Am J Pathol* 180(1): 328–36.

Ozasa, K., Shimizu, Y., Suyama, A., et al. 2012. Studies of the mortality of atomic bomb survivors, report 14, 1950–2003: an overview of cancer and noncancer diseases. *Radiat Res* 177: 229–43.

Pathak, C. M., Avti, P. K., Kumar, S., Khanduja, K. L., and Sharma, S. C. 2007. Whole body exposure to low-dose gamma radiation promotes kidney antioxidant status in Balb/c mice. *J Radiat Res* 48: 113–20.

Pierce, D. A., and Preston, D. L. 2000. Radiation-related cancer risks at low doses among atomic bomb survivors. *Radiat Res* 154: 178–86.

Pierce, D. A., Shimizu, Y., Preston, D. L., Vaeth, M., and Mabuchi, K. 1996. Studies of the mortality of atomic bomb survivors. Report 12, Part I. Cancer: 1950–1990. *Radiat Res* 146: 1–27.

Pollycove, M., and Feinendegen, L. E. 2003. Radiation-induced versus endogenous DNA damage: possible effect of inducible protective responses in mitigating endogenous damage. *Hum Exp Toxicol* 22(6): 290–306.

Pollycove, M., and Feinendegen, L. E. 2008. Low-dose radioimmuno-therapy of cancer. *Hum Exp Toxicol* 27(2): 169–75.

Pollycove, M., and Feinendegen, L. E. 2011. Low-dose radiotherapy of disease. *Health Phys* 100(3): 322–4.
Preston, D. L., Pierce, D. A., Shimizu, Y., et al. 2004. Effect of recent changes in atomic bomb survivor dosimetry on cancer mortality risk estimates. *Radiat Res* 162: 377–89.
Preston, D. L., Ron, E., Tokuoka, S., et al. 2007. Solid cancer incidence in atomic bomb survivors: 1958–1998. *Radiat Res* 168: 1–64.
Preston, D. L., Shimizu, Y., Pierce, D. A., Suyama, A., and Mabuchi, K. 2003. Studies of mortality of atomic bomb survivors. Report 13: Solid cancer and noncancer disease mortality: 1950–1997. *Radiat Res* 160: 381–407.
Raman, S., Dulberg, C. S., Spasoff, R. A., and Scott, T. 1987. Mortality among Canadian military personnel exposed to low-dose radiation. *Can Med Assoc J* 136: 1051–6.
Rattan, S. I. 2012. Rationale and methods of discovering hormetins as drugs for healthy ageing. *Expert Opin Drug Discov* 7(5): 439–48.
Ricci, P. F., Straja, S. R., and Cox, A. L. Jr. 2012. Changing the risk paradigms can be good for our health: j-shaped, linear and threshold dose-response models. *Dose-Response* 10(2): 177–89.
Rockwell, T. 2013. Human lung cancer risks from radon: influence from bystander and adaptive response non-linear dose response effects. *Radiat Prot Dosimetry* 154: 262–3.
Rödel, F., Keilholz, L., Herrmann, M., Sauer, R., and Hildebrandt, G. 2007. Radiobiological mechanisms in inflammatory diseases of low-dose radiation therapy. *Int J Radiat Biol* 83(6): 357–66.
Rodgers, B. E., and Holmes, K. M. 2008. Radio-adaptive response to environmental exposures at Chernobyl. *Dose-Response* 6: 209–21.
Ron, E. 2003. Cancer risks from medical radiation. *Health Phys* 85: 47–59.
Rothkamm, K., and Lobrich, M. 2003. Evidence for a lack of DNA double-strand break repair in human cells exposed to very low x-ray doses. *Proc Natl Acad Sci U S A* 100: 5057–62.
Russo, G. L., Tedesco, I., Russo, M., Cioppa, A., Andreassi, M. G., and Picano, E. 2012. Cellular adaptive response to chronic radiation exposure in interventional cardiologists. *Eur Heart J* 33(3): 408–14.
Sakamoto, K., Myogin, M., Hosoi, Y., et al. 1997. Fundamental and clinical studies on cancer control with total or upper half body irradiation. *J Jpn Soc Ther Radiol Oncol* 9: 161–75.
Schneider, C., and Tacke, F. 2012. Distinct anti-tumoral functions of adaptive immune cells in liver cancer. *Oncoimmunology* 1(6): 937–9.
Schonfeld, S. J., Tsareva, Y. V., Preston, D. L., et al. 2012. Cancer mortality following in utero exposure among offspring of female Mayak Worker Cohort members. *Radiat Res* 178(3): 160–5.
Scott, B. R. 2008. It's time for a new low-dose-radiation risk assessment paradigm—one that acknowledges hormesis. *Dose-Response* 6: 333–51.
Scott, B. R. 2011a. Residential radon appears to prevent lung cancer. *Dose-Response* 9(4): 444–64.
Scott, B. R. 2011b. Assessing potential radiological harm to Fukushima recovery workers. *Dose-Response* 9: 301–12.
Scott, B. R., and Di Palma, J. 2007. Sparsely ionizing diagnostic and natural background radiations are likely preventing cancer and other genomic-instability-associated diseases. *Dose-Response* 5: 230–55.
Scott, B. R., Belinsky, S. A., Leng, S., Lin, Y., Wilder, J. A., and Damiani, L. A. 2009. Radiation-stimulated epigenetic reprogramming of adaptive-response genes in the lung: an evolutionary gift for mounting adaptive protection against lung cancer. *Dose-Response* 7: 104–31.

Scott, B. R., Haque, M., and Di Palma, J. 2007. Biological basis for radiation hormesis in mammalian cellular communities. *Int J Low Radiation* 4: 1–16.

Scott, B. R., Sanders, C. L., Mitchel, R. E., and Boreham, D. R. 2008. CT scans may reduce rather than increase the risk of cancer. *J Am Phys Surg* 13: 9–11.

Sell, T. K., and Gilles, K. 2012. Radiological disasters: what's the difference? *Biosecur Bioterror* 10(4): 412–6.

Seong, K. M., Kim, C. S., Seo, S. W., et al. 2011. Genome-wide analysis of low-dose-irradiated male *Drosophila melanogaster* with extended longevity. *Biogerontology* 12(2): 93–107.

Sethi, T. K., El-Ghamry, M. N., and Kloecker, G. H. 2012. Radon and lung cancer. *Clin Adv Hematol Oncol* 10(3): 157–64.

Shimizu, Y., Kodama, K., Nishi, N., et al. 2010. Radiation exposure and circulatory disease risk: Hiroshima and Nagasaki atomic bomb survivor data, 1950–2003. *BMJ* 340: b5349.

Smith, R. W., Wang, J., Bucking, C. P., Mothersill, C. E., and Seymour, C. B. 2007. Evidence for a protective response by the gill proteome of rainbow trout exposed to x-ray induced bystander signals. *Proteomics* 7: 4171–80.

Sovacool, B. K. 2008. The costs of failure: a preliminary assessment of major energy accidents, 1907–2007. *Energy Policy* 36: 1802–20.

Sponsler, R., and Cameron, J. R. 2005. Nuclear Shipyard Worker Study (1980–1988): a large cohort exposed to low-dose-rate gamma radiation. *Int J Low Radiation* 1: 463–78.

Stamell, E. F., Wolchok, J. D., Gnjatic, S., Lee, N. Y., and Brownell, I. 2013. The abscopal effect associated with a systemic anti-melanoma immune response. *Int J Radiat Oncol Biol Phys* 85(2): 293–5.

Stidley, C. A., and Samet, J. M. 1993. A review of ecologic studies of lung cancer and indoor radon. *Health Phys* 65(3): 234–5.

Suzuki, K., and Yamashita, S. 2012. Low-dose radiation exposure and carcinogenesis. *Jpn J Clin Oncol* 42(7): 563–8.

Szumiel, I. 2012. Radiation hormesis: autophagy and other cellular mechanisms. *Int J Radiat Biol* 88(9): 619–28.

Takahashi, M., Kojima, S., Yamaoka, K., and Niki, E. 2000. Prevention of type I diabetes by low-dose gamma irradiation in NOD mice. *Radiat Res* 154: 680–5.

Takamura, N., and Yamashita, S. 2011. Lessons from Chernobyl. *Fukushima J Med Sci* 57(2): 81–5.

Talbott, E. O., Youk, A. O., McHugh-Pemu, K. P., and Zborowski, J. V. 2003. Long-term follow-up of the residents of the Three Mile Island accident area: 1979–1998. *Environ Health Perspect* 111: 341–8.

Tao, Z., Zha, Y., and Akiba, S. 2000. Cancer mortality in the high background radiation areas of Yangjiang, China during the period between 1979 and 1995. *J Radiat Res* 41: 31–41.

Taverne, Y. J., Bogers, A. J., Duncker, D. J., and Merkus, D. 2013. Reactive oxygen species and the cardiovascular system. *Oxid Med Cell Longev.* 2013: 862423.

Thomas, T. O., Agrawal, P., Guitart, J., et al. 2013. Outcome of patients treated with a single-fraction dose of palliative radiation for cutaneous T-cell lymphoma. *Int J Radiat Oncol Biol Phys* 85(3): 747–53.

Thompson, R. E. 2010. Epidemiological evidence for possible radiation hormesis from radon exposure: a case-control study conducted in Worcester, MA. *Dose-Response* 9(1): 59–75.

Thompson, R. E., Nelson, D. F., Popkin, J. H., and Popkin, A. 2008. Case-control study of lung cancer risk from residential radon exposure in Worcester County, Massachusetts. *Health Phys* 94: 228–41.

Tokarskaya, Z. B., Okladnikova, N. D., Belyaeva, Z. D., and Drozhko, E. G. 1997. Multifactorial analysis of lung cancer dose response relationship for workers at the Mayak Nuclear Enterprise. *Health Phys* 73: 899–905.
Tran, K. H., Campbell, B. A., Fua, T., et al. 2013. Efficacy of low dose radiotherapy for primary orbital marginal zone lymphoma. *Leuk Lymphoma* 54(3): 491–6.
Tubiana, M., Feinendegen, L. E., Yang, Ch., and Kaminski, J. M. 2009. The linear no-threshold relationship is inconsistent with radiation biologic and experimental data. *Radiology* 251: 13–22.
Tubiana, M. 2008. The 2007 Marie Curie prize: the linear no threshold relationship and advances in our understanding of carcinogenesis. *Int J Low Radiation* 5: 173–204.
Tubiana, M., Nagataki, S., and Feinendegen, L. E. 2008. Computed tomography and radiation exposure. *N Engl J Med* 358: 850–3.
UNSCEAR (United Nations Scientific Committee on the Effects of Atomic Radiation). 1994. Sources and Effects of Ionizing Radiation. Report to the General Assembly with Scientific Annexes. United Nations Scientific Committee on the Effects of Atomic Radiation, New York.
UNSCEAR (United Nations Scientific Committee on the Effects of Atomic Radiation). 2000. Sources and Effects of Ionising Radiation, Volume 2: Annex G. Biological effects at low radiation doses. United Nations Scientific Committee on the Effects of Atomic Radiation, New York.
UNSCEAR (United Nations Scientific Committee on the Effects of Atomic Radiation). 2008. Sources and Effects of Ionizing Radiation. Report to the General Assembly with Annexes. Annex D. Health effects due to radiation from the Chernobyl accident. United Nations Scientific Committee on the Effects of Atomic Radiation, New York.
Verbeke, P., Fonager, J., Clark, B. F. C., and Rattan, S. I. S. 2001. Heat shock response and ageing: mechanisms and applications. *Cell Biol Int* 25: 845–57.
Verdun, F. R., Bochud, F., Gudinchet, F., et al. 2008. Radiation risk: what you should know to tell your patient. *Radiographics* 28: 1807–16.
Vrijheid, M., Cardis, E., and Blettner, M. 2007. The 15-country collaborative study of cancer risk among radiation workers in the nuclear industry: design, epidemiological methods and descriptive results. *Radiat Res* 167: 361–79.
Vrijheid, M., Cardis, E., Ashmore, P., et al. 2008. Ionizing radiation and risk of chronic lymphocytic leukemia in the 15-country study of nuclear industry workers. *Radiat Res* 170(5): 661–5.
Wakeford, R. 2011. The silver anniversary of the Chernobyl accident. Where are we now? *J Radiol Prot* 31: 1.
Wall, B. F., Kendall, G. M., Edwards, A. A., Bouffler, S., Muirhead, C. R., and Meara, J. R. 2006. What are the risks from medical x-rays and other low dose radiation? *Br J Radiol* 79: 285–94.
Wang, G. J., Li, X. K., Sakai, K., and Lu, C. 2008. Low-dose radiation and its clinical implications: diabetes. *Hum Exp Toxicol* 27(2): 135–42.
Wang, P. I., Chong, S. T., Kielar, A. Z., et al. 2012. Imaging of pregnant and lactating patients: part 2, evidence-based review and recommendations. *AJR Am J Roentgenol* 198(4): 785–92.
Wei, L. X., Zha, Y. R., and Tao, Z. F. 1990. Epidemiological investigation of radiological effects in high background radiation areas of Yangjiang, China. *Radiat Res* 31: 119–36.
World Nuclear Association. 2009. Health impacts: Chernobyl accident appendix 2. http://www.world-nuclear.org/info/chernobyl/health_impacts.html (updated November, 2009).

Yablokov, A. V., Nesterenko, V. B., and Nesterenko, A. V. 2009. *Chernobyl: Consequences of the Catastrophe for People and the Environment. Annals of the New York Academy of Sciences* (Vol. 1181): New York.

Yamaoka, K., Kataoka, T., Nomura, T., et al. 2004. Inhibitory effects of prior low-dose x-ray irradiation on carbon tetrachloride-induced hepatopathy in acatalasemic mice. *J Radiat Res* 45: 89–95.

Yamaoka, K., Kojima, S., and Nomura, T. 1999. Changes of SOD-like substances in mouse organs after low-dose x-ray irradiation. *Physiol Chem Phys Med NMR* 31: 23–8.

Yamaoka, K., Kojima, S., Takahashi, M., Nomura, T., and Iriyama, K. 1998. Change of glutathione peroxidase synthesis along with that of superoxide dismutase synthesis in mice spleens after low-dose x-ray irradiation. *Biochim Biophys Acta* 1381: 265–70.

Yoshimoto, Y., Neel, J. V., Schull, W. J., et al. 1990. Malignant tumors during the first 2 decades of life in the offspring of atomic bomb survivors. *Am J Hum Genet* 46: 1041–52.

Yoshinaga, S., Mabuchi, K., and Sigurdson, A. J. 2004. Cancer risks among radiologists and radiologic technologists: review of epidemiologic studies. *Radiology* 233: 313–21.

Zablotska, L. B., Ashmore, J. P., and Jowe, G. R. 2004. Analysis of mortality among Canadian nuclear power industry workers after chronic low-dose exposure to ionizing radiation. *Radiat Res* 161: 633–41.

Zagórski, Z. P., and Kornacka, E. M. 2012. Ionizing radiation: friend or foe of the origins of life? *Orig Life Evol Biosph* 42(5): 503–5.

Zainullin, V. G., and Moskalev, A. A. 2001. Radiation-induced changes in the life span of laboratory *Drosophila melanogaster* strains. *Russ J Genet* 37: 1094–5 [in Russian].

Zakeri, F., Hirobe, T., and Akbari Noghabi, K. 2010. Biological effects of LD ionizing radiation exposure on interventional cardiologists. *Occup Med (Lond)* 60(6): 464–9.

# 8 Thermal Hydrotherapy as Adaptive Stress Response
## Hormetic Significance, Mechanisms, and Therapeutic Implications

G. Scapagnini, S. Davinelli,
N.A. Fortunati, D. Zella, and M. Vitale

## CONTENTS

8.1 Thermal Therapies: From Hyperthermia to Balneotherapy ......................... 153
8.2 Therapeutic Applications of Hydrothermal Hormesis ................................. 155
8.3 Hydrogen Sulfide as a Hormetic Cytoprotectant in Thermal Waters:
    Therapeutic and Molecular Targets ............................................................. 159
8.4 Health Effects of Radon-Thermal Water: An Example of Hormesis ........... 161
8.5 Conclusions ................................................................................................. 161
Acknowledgment .................................................................................................. 161
References ............................................................................................................. 162

## 8.1 THERMAL THERAPIES: FROM HYPERTHERMIA TO BALNEOTHERAPY

The use of thermal therapy for preventive and therapeutic purposes has its origin in antiquity, dating back thousands of years. The definition of *thermotherapy* has been used to refer to all treatments based on the transfer of thermal energy. In general, the transfer of heat from the environment to the body happens through different physical processes: conduction, convection, or radiation. Conductive heating is defined as heat transfer without noticeable movement in the conducting medium, thus direct contact takes place between the heat source and the target tissues (i.e., hot packs). Convective heating is the transfer of heat by the movement of the transferring heating medium, usually air or a fluid (i.e., hydrotherapy and traditional sauna). Radiant heat is a type of conversion heating; high-energy photons penetrate the tissues, and this energy is converted into heat (i.e., infrared lamps and far-infrared sauna). Wavelengths producing temperature rises in tissues range from the spectrum of far infrared to visible yellow; longer wavelengths of light, from green to ultraviolet, produce photochemical reactions that do not significantly raise tissue temperature (Ritter 1996).

The normal temperature of a human body is approximately 37°C, and a constant internal temperature has to be maintained to preserve physiologic functions. Several adaptive systems at the molecular and cellular levels are promptly activated in the body to conserve thermal homeostasis. The biologic effects of heating depend on the duration of exposure, frequency, tissue characteristics, and energy power density (Ritter 2006). Biophysiological responses to thermal treatments are directly proportional to the extent of the environmental change. If the energy absorbed is insufficient to stimulate the tissue, there is no effect. By contrast, tissue damage or a detrimental effect may occur if an excess of energy is absorbed (Prentice 1999). The modern discipline of thermal biology and in particular its clinical hyperthermia field is based on this last effect, and emerged in the 1970s as complementary treatment to radiation and certain chemotherapeutic agents (Dewey et al. 1977; Hahn 1982). Thermal dosimetric concepts were established in the mid-1980s (Sapareto and Dewey 1984), providing compelling evidence that elevated temperatures (>42°C) can damage and kill cancer cells in animal model systems (Dewey et al. 1980) and in humans (Lyng et al. 1991) with minimal injury to normal tissues (van der Zee 2002). Nowadays, thermal medicine includes several areas with emerging technologies and new therapeutic strategies.

In the past decade, significant progress has been made in this controversial scientific field, and the most valuable advances are related to different medical practices such as hyperthermia, hypothermia, cryotherapy, and thermal ablation. The next generation of clinical thermal therapy involves cutting-edge technologies to control the delivery of thermal energy to specific regions in the body. For instance, new strategies include laser, microwave, radio frequency, or nanoparticle-mediated thermal therapy. However, it should be noted that the approaches mentioned earlier are largely deep heat–based cancer treatments, but thermal therapy can be used in different forms and combinations. Therefore, even though hyperthermia has gained much attention as an alternative noninvasive technique to deactivate cancer cells, there are several therapeutic superficial thermal procedures used to reduce muscle pain due to arthritis, improve hemodynamic functions or endothelial dysfunction, and counteract proinflammatory processes (Steiner et al. 1979; Tei et al. 1995; Kihara et al. 2002; Bender et al. 2005). This healing aspect of mild heat relates to the capacity of living systems to respond to stress, and accumulating evidences show that transfer of thermal energy into or out of the body causes physiological changes because of molecular or cellular responses to adaptation (Sramek et al. 2000). It is clear that the relationship between dosage and effect is crucial, but since the hormetic effect has been documented across a broad range of biological models (Calabrese and Blain 2011), thermal treatment as a manifestation of hormetic response needs to be considered (Rattan and Demirovic 2010).

Hormetic responses to heat shock are particularly interesting because thermotolerance, as a form of adaptation to thermal stress factor, is known to be related to a precise molecular mechanism: the induction of heat shock proteins (HSPs) (Kuennen et al. 2011). HSPs are highly conserved and ubiquitous molecular chaperones with multifunctional roles to prevent proteotoxic stress–induced denaturation of other proteins through *holding and folding* pathways (Gething and Sambrook 1992; Netzer and Hartl 1998). There are strong evidences indicating that HSPs can protect various cell types from different challenges and slow down programmed cell death pathways, which would therefore postpone cell aging and death (Calderwood et al. 2009). Although mild heat stress has been widely reported to delay aging and prolong longevity in cells and animals, some doubts about the efficacy of this approach to extend life span at the whole-organism level have been recently raised by a comparative meta-analysis performed to quantify the effect of heat shock exposure on longevity (Lagisz et al. 2013). Nevertheless, the possibility that a short and mild elevation of body temperature, obtained with different procedures, might be healing and beneficial for several conditions is part of the story of humanity.

Considering that the stress adaptation concept has been an object of considerable investigation, an emerging new biomedical framework is to investigate whether exposure to mild stressors might apply to humans as a pro-healthy intervention. The phenomenon of hormesis has been extensively reviewed in the biomedical literature (Kendig et al. 2010), but its clinical application to thermal stress is still lacking. In this chapter, we focus more closely on a specific thermal modality: hydrothermal therapy. In the time of Hippocrates, heating the body by using hot water was considered beneficial and was used to prevent/treat a variety of diseases (Jackson 1990). Although balneotherapy, hydrotherapy, and climatology are not fully recognized as independent medical specialties on a global international level, they are used worldwide in the context of spa protocols by millions of people as a general procedure for well-being or for treatment of specific pathologies. The aim of this chapter is to discuss the more characteristic studies regarding hydrothermal therapy and its health outcomes while employing a hormetic perspective. In particular, we address this topic by providing an overview of experimental observations that thermal effects produced by water in all its various forms at temperatures above or below that of the body trigger healthy physiological changes and stimulate endogenous defense systems. Furthermore, we describe the dose-dependent therapeutic effect of hydrogen sulfide ($H_2S$) and radon waters used to counteract several painful diseases.

## 8.2 THERAPEUTIC APPLICATIONS OF HYDROTHERMAL HORMESIS

Thermal baths are considered an integral part of health traditions in many cultures, from Indian Ayurvedic medicine to the therapeutic baths in Eastern Europe. Throughout the ages and from nation to nation, thermal medicine has been applied in many different ways and for several medical applications (van Tubergen and van der Linden 2002). However, because of the great heterogeneity of the techniques and methods of treatment developed in different places and different environments, it is extremely difficult to consider thermal treatment on a scientific basis (Gutenbrunner et al. 2010). Spa treatments, for instance, consist of multiple techniques, but are based on the healing effects of water. Unfortunately, the lack of exact definitions profoundly inhibits preparation of systematic reviews and meta-analyses. Furthermore, the efficacy of spa treatments has been supported mostly by empirical data, and relatively few data generated by randomized clinical trials are available to date.

The term balneotherapy comes from the Latin *balneum* (bath). The term is classically used for bathing in medical mineral waters, medical peloids, and natural gases. The use of plain water (i.e., tap water) for therapeutic use is called hydrotherapy, and the use of climatic factors for therapy is called climatotherapy. Bathing is usually combined with other treatments, such as physical exercise, massages, diets, and behavioral techniques. The water that is used by each individual spa has a unique entity in terms of chemical composition, and several studies have shown that the minerals contained in thermal waters are essential parts of their healing properties. Nevertheless, it is clear that water temperature has a central role for the metabolic effects of balneotherapy. Although confusion persists regarding the adjective *thermal*, we assume that water below 20°C is defined as *cold water*, between 20°C and 30°C as *hypothermal*, between 30°C and 40°C as *homeothermal*, and above 40°C as *hyperthermal*.

To date, there have been very few scientific investigations aimed at identifying a hormetic response during hydrothermal treatments. Nevertheless, most of the current literature regarding the potential benefits of modern hydrothermal therapy might be in accordance with the theory of hormesis. Although there are technical difficulties in detecting molecular and cellular events in hormetic responses, it is possible to speculate that a mild hydrothermal shock may induce a hormetic effect that allows an increased activity of cytoprotective and restorative proteins. Although they are likely underestimated in the hydrothermal research, HSPs might have a critical role in balneotherapy beneficial effects. Studies on rats provided first evidences of an increase in HSPs expression after bathing for 15 minutes in hot water (42°C) (Yamashita et al. 1998). Balneotherapy has also shown significant beneficial effect on vasculogenesis through the attenuation of accumulation of inflammatory cells in adventitia and suppression of neointimal thickening in cuff-injured arteries. This effect appears to be mediated by enhanced expression of HSP-72, induced by hydrothermal treatment and reduction of oxidative stress (Okada et al. 2004).

The possibility that balneotherapy in humans has some beneficial effects on antioxidant status by the activation of an adaptive defensive response has been explored in several studies (Bender et al. 2013). A controlled pilot study has shown that balneotherapy, with different mineral water (alkaline or alkaline-chlorine) or hot tap water, has a positive influence on the antioxidant system of healthy subjects, but different effects on antioxidant enzymes and oxidative stress biomarker levels have been reported depending on the mineral water composition (Bender et al. 2007). Balneotherapy for 30 minutes in warmed water, both tap or mineral (<42°C) for 15 sessions, has shown to ameliorate oxidative status, protein C levels, and serum lipids, even though no changes for HSP-60 levels were seen (Oláh et al. 2010). A clinically controlled trial on type 2 diabetic patients showed that 4 weeks of twice or thrice daily balneotherapy at a spa in Hokkaido (Japan), with the water temperature between 39°C and 40°C, resulted in partial improvement of platelet glutathione metabolism (Ohtsuka et al. 1996). The authors concluded that balneotherapy is beneficial for patients whose platelet antioxidant defense system is damaged, such as those with diabetes mellitus and coronary heart disease.

The possibility that thermal baths' ability to improve some aspects of an organism's antioxidant capacity represents a hormetic phenomenon, which is suggested by the evidence that whole-body immersion in very hot water (42°C or higher), used in some Japanese traditional *onsen* (hot springs), induces opposite effects. Immersion in water at 42°C for 10 minutes causes oxidative stress in the human body by increasing the levels of lipid peroxides and decreasing the activities of glutathione peroxides in erythrocytes (Ohtsuka et al. 1994). Moreover, whole-body bathing in very hot water (42°C or higher for 10–15 minutes) is reported to induce a marked increase of blood viscosity and an enhancement of the blood coagulation system (Shirakura and Kubota 1992). In determining the involvement of activated platelets in frequent thrombosis after hot-spring bath, Take et al. (1996) found that plasma levels of β-thromboglobulin began to rise at 5 minutes and elevated significantly 10 minutes after the start of a 47°C hot-spring bath at Kusatsu (Japan). This effect was not observed with bathing at 42°C. Acute hyperthermal stress due to hot-spring water may also decrease fibrinolytic capacity, leading to the occurrence of thrombotic events (Tamura et al. 1996).

Another traditional spa protocol is mud therapy, known also as *fangotherapy*, which refers to thermal techniques using a mix of mineral/thermal water together with inorganic and/or organic material derived from geological processes in the form of packs or baths. Each of the used materials (mud, clay, and/or peat) has its own special properties but, in general, their principal characteristic is to hold heat and to release it slowly; they have been successfully used for thermal application against chronic conditions. The treatment generally begins with mud applied to the body at a temperature of around 40°C–46°C. Full-body applications are left in place for 20 minutes whereas spot treatments are left in place for up to 30 minutes. Mudpacks have gained a relevant place as a nonpharmacological tool in certain clinical settings, such as degenerative articular processes and selected skin disorders. In such conditions it has been demonstrated that thermal mudpack therapy induces a reduction in circulating levels of prostaglandin E2, leukotriene B4, interleukin-1β, tumor necrosis factor-α, matrix metalloproteinases-3, leptin, and adiponectin (Bellometti and Galzigna 1998; Bellometti et al. 2005; Ardiç et al. 2007; Fioravanti et al. 2011), which are important mediators of inflammation. Mudpacks, similar to thermal baths, have also been shown to have positive effects on the oxidant/antioxidant system, resulting in a reduced release of reactive oxygen species (Benedetti et al. 2010).

Recently, we have explored the ability of total-body thermal mud treatment, or thermal steam bath in a natural grotto, to induce changes in blood levels of HSP-70. Twenty healthy subjects aged 45 years and over were enrolled into this pilot study. Blood samples were collected after 2 hours from the first treatment session and after 1 week of treatments repeated once a day. A quantitative real-time PCR protocol was developed to determine HSP-70 levels. The analysis of gene expression data showed a direct correlation between HSP-70 levels and thermal treatment, supporting the relationships between hormetic pathways and thermal treatments (Scapagnini et al. 2012).

Sauna therapy is another heating technique that has been used for hundreds of years in the Scandinavian region. It has been now adopted worldwide as a standard health activity. Traditional saunas use either wood stoves or 220-V heaters to heat the air to approximately 85°C, which then heats the occupant, mainly via convection. In the last years far-infrared sauna (120-V infrared elements, similar to the infrared warmers on neonatal resuscitation beds), to radiate heat with a wavelength of around 10 μm, has been successfully introduced. As infrared heat penetrates more deeply than warmed air, users develop a more vigorous sweat at a lower temperature than they would in traditional saunas. Studies document the effectiveness of sauna therapy for people with hypertension, congestive heart failure, and for post-myocardial infarction care. One of the molecular mechanisms by which repeated sauna therapy improves endothelial function is the induction of endothelial nitric oxide synthase, as demonstrated in hamsters exposed to a far-infrared ray dry sauna system (39°C for 15 minutes) for 4 weeks (Ikeda et al. 2001). Repeated sauna therapy performed in patients at risk for lifestyle-related diseases has been found to significantly improve vascular endothelial function and to reduce both body weight and fasting plasma glucose (Tei and Tanaka 1996).

Siewert et al. (1994) investigated the effects of regular sauna bathing on blood pressure and found that a 3-month period of biweekly sauna bathing lowered mean blood pressure from 166/101 to 143/92 mmHg in 46 hypertensive patients. In another study, a 60°C far-infrared-ray dry sauna, for 15 minutes once a day for 2 weeks performed on patients with at least one coronary risk factor, was able to reduce systolic blood pressure and improve urinary 8-epi-prostaglandin F2α levels (a gold standard biomarker of oxidative stress) (Masuda et al. 2004). Michalsen et al. (2003) suggests a beneficial adaptive response in patients with chronic heart failure undergoing a hydrotherapeutic program, consisting of a well-structured combination of warm and cold thermal applications. Recently, an interesting study attempted to determine the effects of exposure to cold (bath in ice-cold water) and heat (sauna) on the activity of markers of cellular damage such as lysosomal enzymes. On the basis of the data obtained, it has been argued that regular winter swimming combined with sauna may lead to adaptation and thermotolerance, which protects cellular membranes against the deleterious effects of thermal and other stress factors (Mila-Kierzenkowska et al. 2012). The body reacts defensively against the reduced temperature by increasing many of its metabolic functions. Conversely, increased blood flow and muscle tone occur when the cold stress is over.

An adaptation to repeated hot and cold stress may be involved in increased tolerance to stress and diseases. It was shown that the neuroendocrine and immune systems are challenged after thermal stress. Individuals at rest, routinely and briefly subjected to hydrothermal treatment, have enhanced blood concentrations of several analytes involved in the immune system (Dugué and Leppänen 2000). However, in addition to hot/cold immersion therapy, hydrothermal treatments are applied in many rehabilitative programs to promote healing. Hormetic stress response was observed during physical activity (Ristow et al. 2009), and physical training in a pool with a water temperature of 32°C (15 minute sessions) has been shown to provide cardiorespiratory adaptation in patients with chronic obstructive pulmonary disease (COPD). In spite of the restriction of lung function in water due to the hydrostatic pressure, the COPD population performed the training in the pool without clinically relevant discomfort (Perk et al. 1996). Furthermore, a randomized clinical trial produced positive results assessing the effectiveness of low-intensity aquatic exercise in those with COPD (de Souto Araujo et al. 2012).

Although cold baths are used to reduce swelling and inflammation by constricting blood vessels, hyperthermia and the local application of heat (mud 40°C–42°C followed by immersion in 37°C–38°C thermal water) have been shown to prevent inflammation. These anti-inflammatory effects may result from the deregulation of hypothalamic–pituitary–adrenal (HPA) axis function induced by thermal treatment. Significantly and in accordance with the low/modest stress observed in hormetic dose-responses, no effect was observed on the HPA axis function in subjects who received immersion in homeothermal water (34.5°C–35°C) (Cozzi et al. 1995).

## 8.3 HYDROGEN SULFIDE AS A HORMETIC CYTOPROTECTANT IN THERMAL WATERS: THERAPEUTIC AND MOLECULAR TARGETS

Thermal waters have several chemical and physical properties and they are generally rich in sulfur, $H_2S$, and sulfates (Valitutti et al. 1990). $H_2S$ is a toxic gas and an environmental hazard, but it is endogenously produced as a signaling molecule in processes such as neuromodulation or smooth muscle relaxation. $H_2S$ is emitted from volcanoes or a polluted environment, but it is found dissolved in thermal sulfuric waters at variable concentrations ranging from 1 to 500 μM (Reigstad et al. 2011). $H_2S$ has long been studied to evaluate its toxic properties, but recently it has attracted new scientific interest due to numerous studies that have recognized $H_2S$ as an important endogenous gasotransmitter with diverse biological activities. It is well-established that thermal waters rich in $H_2S$ exert attractive therapeutic effects able to efficiently counteract pathological conditions related to inflammation, metabolic disorders, hypertension, and many other deleterious consequences involved in the progression of diseases (Wang 2010; Predmore et al. 2012). In addition, $H_2S$ possesses antibacterial and antifungal properties (Nasermoaddeli and Kagamimori 2005). It should be noted that the therapeutic outcomes of thermal water are correlated to the concentrations of chemical compounds, and to achieve an effective therapeutic response, different diseases may require varying concentrations (Matz et al. 2003). The hormetic dose-response appears to be useful to describe the therapeutic response elicited by $H_2S$. The medical constraints of $H_2S$ are broad-spectrum toxicity, and thus it has a very narrow therapeutic window. The hydrothermal therapeutic procedure with $H_2S$ must be considered carefully, paying particular attention to the duration of exposure. In this context, the low level of $H_2S$ concentrations found in most hydrothermal natural sources would appear to enhance the fitness of the endogenous defense system and its cytoprotective activity.

Mild stressors and hormetic components have diverse molecular targets affecting many pathways (Calabrese et al. 2010), and $H_2S$ at different concentrations may be involved in each of these pathways to trigger protective responses. The transcription factor Nrf2 (nuclear factor erythroid 2-related factor 2), a critical regulator of cellular defense mechanisms, may have a crucial role in hormetic processes induced by $H_2S$ in hydrothermal treatment. A global mapping study revealed the regulatory network governed by the Nrf2, proving its central role in cell survival response and its pivotal activity in regulating a diverse array of cytoprotective genes, including those encoding detoxifying enzymes (Malhotra et al. 2010). Interestingly, it was hypothesized that Nrf2 serves as a converging node for cellular signaling pathways of gasotransmitters, and this may attribute additional protective effects to $H_2S$ (Liu et al. 2012). Furthermore, it was also reported that the cardioprotective effects of $H_2S$

(Sodha et al. 2008) are mediated in large part by Nrf2, enhancing the resistance to the oxidative stress associated with myocardial ischemia/reperfusion injury (Calvert et al. 2009). $H_2S$ can also induce upregulation of heme oxygenase-1 (HO–1), an Nrf2 target gene that has been shown to have many protective properties (Calabrese et al. 2004; Szabó 2007).

Moreover, beneficial hormetic effects induced by $H_2S$ may be increased additively or even synergistically through the action of HSPs induced by hydrothermal treatment. The hydrothermal research field is still in its infancy in the investigation of potential additive or synergistic roles of $H_2S$, HSPs, and Nrf2. However, the advantages would be related to enhancing the modest hormetic effects of hydrothermal therapy and its therapeutic exploitation. A proposed hydrothermal stress response mechanism for increased HSP-70 (highly inducible in response to multiple stressors) expression and activation of Nrf2 is shown in Figure 8.1.

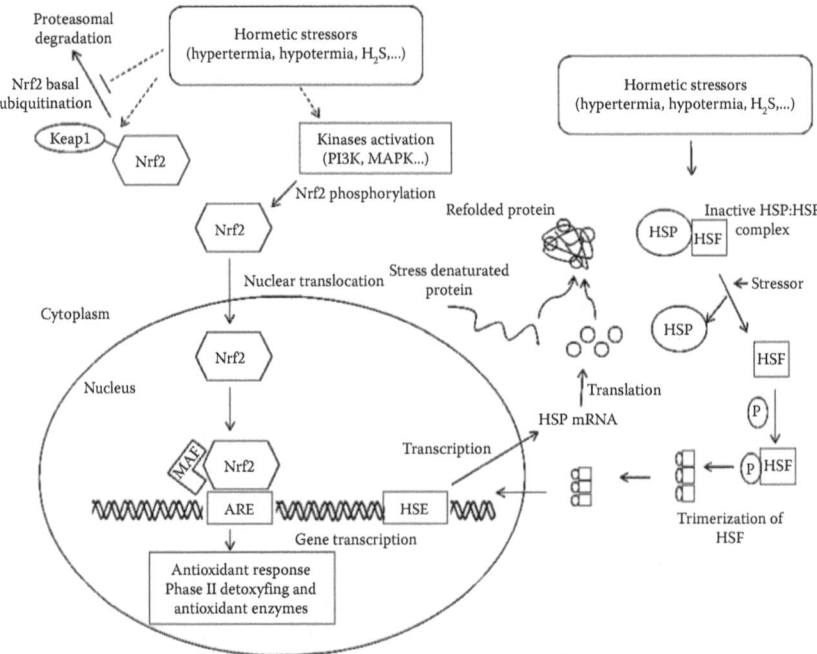

**FIGURE 8.1** Potential signaling pathways involved in the cellular adaptative response and modulated by hormetic thermal treatments. Under conditions of chemical/oxidative stress, nuclear factor erythroid 2-related factor 2 (Nrf2) dissociates from Keap1 and activates antioxidant responsive element (ARE)-regulated genes. The activation of the Nrf2–ARE pathway leads to the transcriptional activation of cytoprotective genes. In addition, thermal hormetic signals activate the inducible form of the 70-kDa heat shock protein (HSP-70) synthesis. Heat shock factors (HSFs), present in the cytosol, are maintained in an inactive state by HSPs. Hormetic stressors are thought to activate HSFs, causing them to separate from HSPs. HSFs are phosphorylated (P) by protein kinases and form trimers in the cytosol. These HSF complexes enter the nucleus and are able to bind the heat shock elements (HSEs) in the promoter region of the HSP-70 gene. HSP-70 mRNA is transcribed and leaves the nucleus for the cytosol, where new HSP-70 is synthesized.

## 8.4 HEALTH EFFECTS OF RADON-THERMAL WATER: AN EXAMPLE OF HORMESIS

Thermal water may also be radioactive. Radioactivity of thermal and mineral waters is mainly due to the radon content. During the last century, high concentrations of radon have been observed in thermal waters. Radon is a chemically inert gas and its radiations have been shown to be carcinogenic at high doses. This element is present in natural and artificial grottoes or therapeutic caves and medical spas in addition to underground mine areas. Radioactive radon therapy exemplifies the model of therapeutic radiation hormesis. Results from clinical trials and molecular-level studies have shown beneficial rather than detrimental responses at low-level exposures (Zdrojewicz and Strzelczyk 2006). Small doses and dose rates of radiation often exhibit significant health hormetic effects based on observations in epidemiological and experimental studies (Sanders 2010) (also see Chapter 7). Radon therapy involves its intake through inhalation, or by transcutaneous absorption of radon dissolved in water. In Europe, therapeutic radon waters are widely available to treat inflammatory and painful diseases. A growing number of studies showed analgesic and anti-inflammatory effects for a variety of diseases including asthma, bronchitis, and psoriasis (Erickson 2007). Moreover, Falkenbach et al. (2005) described five trials of radon therapy for rheumatoid arthritis, three of which were double-blind, that showed beneficial effects. Radon therapy was also effective in treating dyslipidemia associated with cardiovascular disease (Iashina et al. 2011). Therefore, evidence consistent with the concept of hormesis shows an adaptive-healthy effect at low levels of radon radiation.

## 8.5 CONCLUSIONS

Several published studies reported therapeutic success in water environments through activation of prosurvival responses. Nevertheless, there are many studies, not well-controlled and with poor methodological quality and findings, which should be taken into careful consideration. Hydrothermal therapy is an adjunctive medical approach and it has been proven to be helpful in many ways. Despite this, there is still medical skepticism and controversy regarding integrating hydrothermal treatments with conventional treatments. Here, we described the potential impact of the hormesis concept as a unifying framework on the hydrothermal therapeutic field. Since the dose-response relationship is crucial for the hormetic model, specific effort to standardize protocols should be encouraged. In conclusion, it is unknown whether the benefits of sauna bathing extend to a generally healthy population. However, even though conclusive large-scale data are still lacking, studies continue to show that the low-intensity stress associated with thermal hydrotherapy may improve quality of life and many physiological processes, such as cardiovascular hemodynamic and various health markers.

## ACKNOWLEDGMENT

This work has been partially supported by funds from Fondazione per la Ricerca Scientifica Termale (FoRST), Rome.

## REFERENCES

Ardiç F, Őzgen M, Aybek H, Rota S, Cubukçu D, Gökgöz A. 2007. Effects of balneotherapy on serum IL-1, PGE2 and LTB4 levels in fibromyalgia patients. *Rheumatol Int* 27: 441–446.

Bellometti S, Galzigna L. 1998. Serum levels of a prostaglandin and a leukotriene after thermal mud pack therapy. *J Investig Med* 46: 140–145.

Bellometti S, Richelmi P, Tassoni T, Bertè F. 2005. Production of matrix metalloproteinases and their inhibitors in osteoarthritic patients undergoing mud bath therapy. *Int J Clin Pharmacol Res* 25: 77–94.

Bender T, Bálint G, Prohászka Z, Géher P, Tefner IK. 2013. Evidence-based hydro-and balneotherapy in Hungary—A systematic review and meta-analysis. *Int J Biometeorol* [Epub ahead of print].

Bender T, Bariska J, Vághy R, Gomez R, Imre Kovács. 2007. Effect of balneotherapy on the antioxidant system—A controlled pilot study. *Arch Med Res* 38: 86–89.

Bender T, Karagulle Z, Balint GP, Gutenbrunner C, Balint PV, Sukenik S. 2005. Hydrotherapy, balneotherapy, and spa treatment in pain management. *Rheumatol Int* 25: 220–224.

Benedetti S, Canino C, Tonti G, Medda V, Calcaterra P, Nappi G, Salaffi F, Canestrari F. 2010. Biomarkers of oxidation, inflammation and cartilage degradation in osteoarthritis patients undergoing sulfur-based spa therapies. *Clin Biochem* 43: 973–978.

Calabrese EJ, Blain RB. 2011. The hormesis database: The occurrence of hormetic dose responses in the toxicological literature. *Regul Toxicol Pharmacol* 61: 73–81.

Calabrese V, Cornelius C, Dinkova-Kostova AT, Calabrese EJ, Mattson MP. 2010. Cellular stress responses, the hormesis paradigm, and vitagenes: Novel targets for therapeutic intervention in neurodegenerative disorders. *Antioxid Redox Signal* 13: 1763–1811.

Calabrese V, Stella AM, Butterfield DA, Scapagnini G. 2004. Redox regulation in neurodegeneration and longevity: Role of the heme oxygenase and HSP70 systems in brain stress tolerance. *Antioxid Redox Signal* 6: 895–913.

Calderwood SK, Murshid A, Prince T. 2009. The shock of aging: Molecular chaperones and the heat shock response in longevity and aging—A mini-review. *Gerontology* 55: 550–558.

Calvert JW, Jha S, Gundewar S, Elrod JW, Ramachandran A, Pattillo CB, Kevil CG, Lefer DJ. 2009. Hydrogen sulfide mediates cardioprotection through Nrf2 signaling. *Circ Res* 105: 365–374.

Cozzi F, Lazzarin P, Todesco S, Cima L. 1995. Hypothalamic–pituitary–adrenal axis dysregulation in healthy subjects undergoing mud-bath applications. *Arthritis Rheum* 38: 724–726.

de Souto Araujo ZT, de Miranda Silva Nogueira PA, Cabral EE, de Paula Dos Santos L, da Silva IS, Ferreira GM. 2012. Effectiveness of low-intensity aquatic exercise on COPD: A randomized clinical trial. *Respir Med* 106: 1535–1543.

Dewey WC, Freeman ML, Raaphorst GP, Clark EP, Wong RSL, Highfield DP, Spiro IJ et al. 1980. Cell biology of hyperthermia and radiation. In: *Radiation Biology in Cancer Research*, Meyn RE and Withers HR (eds.), pp. 589–623. New York, Raven Press.

Dewey WC, Hopwood LE, Sapareto SA, Gerweck LE. 1977. Cellular responses to combinations of hyperthermia and radiation. *Radiology* 123: 463–474.

Dugué B, Leppänen E. 2000. Adaptation related to cytokines in man: Effects of regular swimming in ice-cold water. *Clin Physiol* 20: 114–121.

Erickson BE. 2007. The therapeutic use of radon: A biomedical treatment in Europe; an "alternative" remedy in the United States. *Dose Response* 5: 48–62.

Falkenbach A, Kovacs J, Franke A, Jörgens K, Ammer K. 2005. Radon therapy for the treatment of rheumatic diseases—Review and meta-analysis of controlled clinical trials. *Rheumatol Int* 25: 205–210.

Fioravanti A, Cantarini L, Bacarelli MR, de Lalla A, Ceccatelli L, Blardi P. 2011. Effects of spa therapy on serum leptin and adiponectin levels in patients with knee osteoarthritis. *Rheumatol Int* 31: 879–882.
Gething MJ, Sambrook J. 1992. Protein folding in the cell. *Nature* 355: 33–45.
Gutenbrunner C, Bender T, Cantista P, Karagülle Z. 2010. A proposal for a worldwide definition of health resort medicine, balneology, medical hydrology and climatology. *Int J Biometeorol* 54: 495–507.
Hahn GM. 1982. *Hyperthermia and Cancer*. New York, Plenum Publishing.
Iashina LM, Shatrova LE, Zhdanova KS, Kuznetsova TA. 2011. [The influence of radon baths on the lipid profile of patients with cardiovascular diseases and dyslipidemia]. *Vopr Kurortol Fizioter Lech Fiz Kult* March–April(2): 3–4.
Ikeda Y, Biro S, Kamogawa Y, Yoshifuku S, Eto H, Orihara K, Kihara T, Tei C. 2001. Repeated thermal therapy upregulates arterial endothelial nitric oxide synthase expression in Syrian golden hamsters. *Jpn Circ J* 65: 434–438.
Jackson R. 1990. Waters and spas in the classical world. *Med Hist Suppl* (10): 1–13.
Kendig EL, Le HH, Belcher SM. 2010. Defining hormesis: Evaluation of a complex concentration response phenomenon. *Int J Toxicol* 29: 235–246.
Kihara T, Biro S, Imamura M, Yoshifuku S, Takasaki K, Ikeda Y, Otuji Y, Minagoe S, Toyama Y, Tei C. 2002. Repeated sauna treatment improves vascular endothelial and cardiac function in patients with chronic heart failure. *J Am Coll Cardiol* 39: 754–759.
Kuennen M, Gillum T, Dokladny K, Bedrick E, Schneider S, Moseley P. 2011. Thermotolerance and heat acclimation may share a common mechanism in humans. *Am J Physiol Regul Integr Comp Physiol* 301: R524–R533.
Lagisz M, Hector KL, Nakagawa S. 2013. Life extension after heat shock exposure: Assessing meta-analytic evidence for hormesis. *Ageing Res Rev* 12: 653–660.
Liu W, Wang D, Liu K, Sun X. 2012. Nrf2 as a converging node for cellular signaling pathways of gasotransmitters. *Med Hypotheses* 79: 308–310.
Lyng H, Monges OR, Bohler PJ, Rofstad EK. 1991. Relationships between thermal dose and heat-induced tissue and vascular damage after thermoradiotherapy of locally advanced breast carcinoma. *Int J Hyperthermia* 7: 403–415.
Malhotra D, Portales-Casamar E, Singh A, Srivastava S, Arenillas D, Happel C, Shyr C et al. 2010. Global mapping of binding sites for Nrf2 identifies novel targets in cell survival response through ChIP-Seq profiling and network analysis. *Nucleic Acids Res* 38: 5718–5734.
Masuda A, Miyata M, Kihara T, Minagoe S, Tei C. 2004. Repeated sauna therapy reduces urinary 8-epi-prostaglandin F2α. *Jpn Heart J* 45: 297–303.
Matz H, Orion E, Wolf R. 2003. Balneotherapy in dermatology. *Dermatol Ther* 16: 132–140.
Michalsen A, Lüdtke R, Bühring M, Spahn G, Langhorst J, Dobos GJ. 2003. Thermal hydrotherapy improves quality of life and hemodynamic function in patients with chronic heart failure. *Am Heart J* 146: 728–733.
Mila-Kierzenkowska C, Woźniak A, Szpinda M, Boraczyński T, Woźniak B, Rajewski P, Sutkowy P. 2012. Effects of thermal stress on the activity of selected lysosomal enzymes in blood of experienced and novice winter swimmers. *Scand J Clin Lab Invest* 72: 635–641.
Nasermoaddeli A, Kagamimori S. 2005. Balneotherapy in medicine: A review. *Environ Health Prev Med* 10: 171–179.
Netzer WJ, Hartl FU. 1998. Protein folding in the cytosol: Chaperonin-dependent and -independent mechanisms. *Trends Biochem Sci* 23: 68–73.
Ohtsuka Y, Yabunaka N, Fujisawa H, Watanabe I, Agishi Y. 1994. Effect of thermal stress on glutathione metabolism in human erythrocytes. *Eur J Appl Physiol Occup Physiol* 68: 87–91.
Ohtsuka Y, Yabunaka N, Watanabe I, Noro H, Agishi Y. 1996. Balneotherapy and platelet glutathione metabolism in type II diabetic patients. *Int J Biometeorol* 39: 156–159.

Okada M, Hasebe N, Aizawa Y, Izawa K, Kawabe J, Kikuchi K. 2004. Thermal treatment attenuates neointimal thickening with enhanced expression of heat-shock protein 72 and suppression of oxidative stress. *Circulation* 109: 1763–1768.

Oláh M, Koncz A, Fehér J, Kálmánczhey J, Oláh C, Balogh S, Nagy G, Bender T. 2010. The effect of balneotherapy on C-reactive protein, serum cholesterol, triglyceride, total antioxidant status and HSP-60 levels. *Int J Biometeorol* 54: 249–254.

Perk J, Perk L, Bodén C. 1996. Cardiorespiratory adaptation of COPD patients to physical training on land and in water. *Eur Respir J* 9: 248–252.

Predmore BL, Lefer DJ, Gojon G. 2012. Hydrogen sulfide in biochemistry and medicine. *Antioxid Redox Signal* 17: 119–140.

Prentice WE. 1999. The science of therapeutic modalities. In: *Therapeutic Modalities in Sports Medicine*, Prentice WE (ed.). New York, McGraw-Hill.

Rattan SIS, Demirovic D. 2010. Hormesis can and does work in humans. *Dose Response* 8: 58–63.

Reigstad LJ, Jorgensen SL, Lauritzen SE, Schleper C, Urich T. 2011. Sulfur-oxidizing chemolithotrophic proteobacteria dominate the microbiota in high arctic thermal springs on Svalbard. *Astrobiology* 11: 665–678.

Ristow M, Zarse K, Oberbach A, Klöting N, Birringer M, Kiehntopf M, Stumvoll M, Kahn CR, Blüher M. 2009. Antioxidants prevent health-promoting effects of physical exercise in humans. *Proc Natl Acad Sci USA* 106: 8665–8670.

Ritter HTM. 1996. Instrumentation considerations. In: *Thermal Agents in Rehabilitation*, Michlovitz SL (ed.), pp. 62–63. Philadelphia, FA Davis.

Sanders CL. 2010. *Radiation Hormesis and the Linear-No-Threshold Assumption*. Germany, Springer-Verlag. pp. 214.

Sapareto SA, Dewey WC. 1984. Thermal dose determination in cancer therapy. *Int J Radiat Oncol Biol Phys* 10: 787–800.

Scapagnini G, Davinelli S, Sapere N, Zella D, Fortunati N. 2012. Heat-shock protein 70 is affected by thermal treatment. *Med Hydrol Balneol Environ Aspects* 66: 192–193.

Shirakura T, Kubota K. 1992. Balneotherapy in hematology and immunology. *Jpn J Biometeor* 29: 15–23.

Siewert C, Siewert H, Winterfeld HJ, Strangfeld D. 1994. Changes of central and peripheral hemodynamics during isometric and dynamic exercise in hypertensive patients before and after regular sauna therapy. *Z Kardiol* 83: 652–657.

Sodha NR, Clements RT, Feng J, Liu Y, Bianchi C, Horvath EM, Szabo C, Sellke FW. 2008. The effects of therapeutic sulfide on myocardial apoptosis in response to ischemia-reperfusion injury. *Eur J Cardiothorac Surg* 33: 906–913.

Sramek P, Simeckova M, Jansky L, Savlikova J, Vybiral S. 2000. Human physiological responses to immersion into water of different temperatures. *Eur J Appl Physiol* 81: 436–442.

Steiner FJF, Valkenburg HA, van de Stadt RJ, Stoyanova-Drenska M, Zant J. 1979. [Balneology treatment of patients with rheumatoid arthritis] [Dutch]. *Ned Tijdschr Geneeskd* 123: 661–664.

Szabó C. 2007. Hydrogen sulphide and its therapeutic potential. *Nat Rev Drug Discov* 6: 917–935.

Take H, Kubota K, Tamura K, Kurabayashi H, Shirakura T, Miyawaki S, Kobayashi I. 1996. Activation of circulating platelets by hyperthermal stress. *Eur J Med Res* 1: 562–564.

Tamura K, Kubota K, Kurabayashi H, Shirakura T. 1996. Effects of hyperthermal stress on the fibrinolytic system. *Int J Hyperthermia* 12: 31–36.

Tei C, Horikiri Y, Park JC, Jeong JW, Chang KS, Toyama Y, Tanaka N. 1995. Acute hemodynamic improvement by thermal vasodilation in congestive heart failure. *Circulation* 91: 2582–2590.

Tei C, Tanaka N. 1996. Thermal vasodilation as a treatment of congestive heart failure: A novel approach. *J Cardiol* 27: 29–30.
Valitutti S, Castellino F, Musiani P. 1990. Effect of sulfurous (thermal) water on T lymphocyte proliferative response. *Ann Allergy* 65: 463–468.
van der Zee J. 2002. Heating the patient: A promising approach? *Ann Oncol* 13: 1173–1184.
van Tubergen A, van der Linden S. 2002. A brief history of spa therapy. *Ann Rheum Dis* 61: 273–275.
Wang R. 2010. Hydrogen sulfide: The third gasotransmitter in biology and medicine. *Antioxid Redox Signal* 12: 1061–1064.
Yamashita N, Hoshida S, Taniguchi N, Kuzuya T, Hori M. 1998. Whole-body hyperthermia provides biphasic cardioprotection against ischemia/reperfusion injury in the rat. *Circulation* 98: 1414–1421.
Zdrojewicz Z, Strzelczyk JJ. 2006. Radon treatment controversy. *Dose Response* 4: 106–118.

# 9 Cardiac Ischemic Preconditioning and the Ischemia/Reperfusion Injury

*Andreas Simm and Rüdiger Horstkorte*

## CONTENTS

9.1 Introduction .................................................................................................. 167
9.2 Reactive Oxygen Species–Induced Cell Death ........................................... 169
9.3 Reactive Oxygen Species and Neurodegenerative Autoimmune
    Diseases ........................................................................................................ 170
9.4 Ischemia/Reperfusion Injury and Mechanisms of Cardiac Protection ........ 171
9.5 Short- and Long-Term Protection of Ischemic Preconditioning ................. 176
9.6 Hormesis: The Concept of Cell Protection .................................................. 176
9.7 Conclusion ................................................................................................... 178
References ............................................................................................................ 178

## 9.1 INTRODUCTION

Until the nineteenth century, the spectrum of human diseases, life span and causes of death had largely remained unchanged. In the twentieth century, best-performance life expectancy unexpectedly started to increase by a quarter of a year per year (Oeppen and Vaupel 2002). Reasons for these advances were improved environmental hygiene, an increased productivity of agriculture, and a breakthrough in biomedical sciences in the last decades of the nineteenth century and throughout the twentieth century (De Flora et al. 2005). In parallel, a shift from infectious to chronic degenerative diseases as predominant causes of death took place. As our lifespan is still increasing and age is a major risk factor for most of these chronic diseases, these diseases will continue to play a central role as the dominant cause of death. The ongoing demographic change is associated with more patients having serious comorbidities.

Within the degenerative diseases, heart failure (HF) is, beside neurodegenerative disorders, such as dementia or multiple sclerosis (MS), of major interest. HF is primarily a disease of the elderly, with an annual incidence of 1% after age 65 years, which nearly doubles every decade thereafter (Rosamond et al. 2007). Bleumink et al. (2004) from the Rotterdam Study reported that participants older than 70 years accounted for 88% of new HF cases. HF is presently the most frequent reason for inpatient hospital admission in Germany (Neumann et al. 2009). The importance of cardiovascular diseases (CVDs) for the health of all nations was mentioned by the World Health Organization in 2013 on their Web site (http://www.who.int/cardiovascular_diseases/en/): (1) an estimated 17.3 million people died from CVDs in 2008, representing 30% of all global deaths; (2) over 80% of CVD deaths take place in low- and middle-income countries; and (3) by 2030 more than 23 million people will die annually from CVDs. Within CVDs, coronary heart disease (CHD) is especially important as it is the leading single cause of morbidity and mortality worldwide (Lloyd-Jones et al. 2009). Ischemic heart disease results from an occlusion of a major coronary artery and is the main cause of acute myocardial infarction (AMI), which can lead to pump failure of the heart and arrhythmias often leading to sudden death.

Despite improved medical therapy, patients with CHD still suffer significant morbidity and mortality. Therefore, with the aging of the general population, increasing numbers of elderly patients will need interventional cardiac treatments, such as cardiac surgery, angioplasty, and thrombolysis. However, the results of these interventions in the elderly are inferior to those in the young (Rosenfeldt et al. 2002). Analyzing the risk profiles of elderly patients by means of data sets from all cardiac surgical centers in Germany for the year 2007 (a total of 47,881 operations), patients over age 75 have (1) significantly more prognosis-determining comorbidities and risk factors and (2) higher complication rates and lethality, for example, a 3.7-fold risk elevation for inhospital lethality compared to patients under age 65 (Friedrich et al. 2009).

Degenerative diseases like CHD are mainly based on a lifelong accumulation of molecular damage within molecules, cells, and tissues. As long as this damage is not reflected by a phenotype or a disease, most people summarize this as a physiological aging process. On the other hand, at some stages, the aggrieved molecules, for example, mutated DNA, oxidized lipids, or glycated proteins, will lead to reduced cell and organ function and at the end, to diseases and death. This was mentioned by Lakatta (2002): "age-associated changes in cardiac and vascular properties alter the substrate upon which cardiovascular disease is superimposed." During aging, many organs, including the heart, still retain their basic function but come into trouble if the organism gets stressed. It was shown, for example, that the stroke volume index as well as the left ventricular ejection fraction is preserved in older persons at rest, whereas the acute cardiac output is reduced during exhaustive exercise (Lakatta 2002). The accumulation of damage seems to be responsible for the reduction in organ function, especially the functional reserve, which is needed when the organ is stressed. This is to some point reflected by the New York Heart Association functional classification of the heart, a classification system of the extent of HF. It starts with limitations only at high physical activity, goes to shortness of breath and/or angina during ordinary activity, and ends up with severe limitations and symptoms at rest, indicating increasing reduction in functional reserve. Indeed, reduced organ function is one of the hallmarks in human aging and is one explanation of the increased risk during cardiac surgery in the elderly as well.

## 9.2 REACTIVE OXYGEN SPECIES–INDUCED CELL DEATH

Oxygen-dependent radicals or reactive oxygen species (ROS) are one basic cause of the aging process per se as well as of cell death in different organs due to necrosis and apoptosis (Simm et al. 2008). ROS are well-known as critical mediators of cardiac injury and death of cardiac myocytes. A free radical is, per definition, a molecule with an unpaired, highly reactive electron. The free radical theory of aging (Harman 1956) proposed that the cell and tissue damage due to the production and reactivity of intracellular ROS is the major determinant of lifespan. Intracellular ROS are primarily generated by the mitochondrial respiratory chain, and are mainly due to electron leakage naturally occurring at complexes I and III (Balaban et al. 2005). Mitochondria generate approximately 90% of cellular energy via oxidative phosphorylation and thus consume most of the oxygen within a eukaryotic cell. The metabolic rate of a cell is therefore related to mitochondrial function and, due to radical leakage, is a driving force for the aging process (Szibor and Holtz 2003). Besides mitochondria, the NADPH oxidases (NOX) seem to play a significant role in ROS production within cells as well. Interestingly, NOX1 and NOX4 mediated cellular senescence induced by resveratrol in human endothelial cells (Schilder et al. 2009).

But how does oxidative stress cause damage? When trying to find a partner for its lone electron, the free radical captures an electron from another molecule, which in turn becomes unstable and reacts with further molecules. Harmful molecules can arise that damage proteins, membranes, and nucleic acids, particularly DNA, including mitochondrial DNA. One major marker of oxidative damage of nucleic acids is the 8-hydroxy-29-deoxyguanosine (8-OHdG) adduct. It is formed when ROS, especially the hydroxyl radical, act on deoxyguanine in DNA (Ravanat et al. 2000). 8-OHdG can alter gene expression as it inhibits methylation. This indeed influences the epigenetic regulation of the genome. In parallel, it is mutagenic because it can be paired with adenosine rather than cytosine during DNA replication, leading to GC-to-AT conversion (the most frequent type of spontaneous mutation). Oxidation of the DNA induces in most cases long-term effects, for example, by exchanged gene expression.

Whereas for a long time, research on radical damage was mostly focused on DNA, it is now clear that proteins seem to be the major targets for ROS, mainly because of their high overall abundance. It has been estimated that proteins can scavenge the majority (50%–75%) of reactive species generated (Davies et al. 1999). The reaction with ROS may alter in consequence the protein structure, thereby changing in a straight line protein function. A major marker of severe protein oxidation is the formation of carbonyl groups. Protein carbonyls may be generated by the oxidation of several amino acid side chains (mainly Lys, Arg, Pro, and Thr). Protein modification can change directly and quickly the function of the molecules or induce fast-occurring damage. Dasgupta et al. (2012) showed a causal relationship between carbonylation, protein aggregation, and apoptosis of neurons undergoing oxidative damage. Baraibar and Friguet (2013) summarized the potential deleterious effects of protein irreversible oxidative modifications in key cellular pathways during aging and in the pathogenesis of age-related diseases. Beside the carbonylation of proteins, lipid peroxidation causes particularly destructive effects on cell membranes. If the lipid peroxidation by ROS is severe, it causes myocardial necrosis (Tavazzi et al. 1998).

In summary, ROS play a role in both apoptotic and necrotic cell death. In general, moderate oxidative stress induces apoptosis, whereas necrotic cell death is triggered when cells have a higher exposure to ROS (Saito et al. 2006). For this reason, it was believed for a long time that the inhibition of ROS by antioxidants was the best way to protect the tissue.

## 9.3 REACTIVE OXYGEN SPECIES AND NEURODEGENERATIVE AUTOIMMUNE DISEASES

A typical target of ROS-induced tissue damage is, especially in combination with a deregulated immune system, the nerve system. Inflammation is a crucial element of the immune response. During the inflammatory process, leucocytes leave the bloodstream and migrate into the tissue, where they secrete degrading enzymes or ROS to fight microorganisms or damaged tissue. Dysregulation of inflammation by autoimmune processes causes severe neurodegenerative diseases, including MS (Hafler and Weiner 1987).

The classical view of MS pathogenesis is a T-cell-receptor-mediated recognition of the myelin sheet by autoreactive $T_H$-lymphocytes (Martin and McFarland 1995). These $T_H$-lymphocytes then cause a cascade by activation of further immune cells, secretion of cytokines, or proteolytic enzymes, finally leading to degradation of the myelin sheet. ROS are heavily involved in the tissue damage during MS pathogenesis (van Horssen et al. 2011). MS patients suffer from increased concentrations of all oxidized products from lipids in the plasma and the cerebrospinal fluid. In particular, macrophages and glia cells of MS patients secrete high levels of TNF-α, which then stimulate the production of superoxide anions and peroxides that can be found in plasma and cerebrospinal fluid (Klebanoff et al. 1986). Superoxide anions react together with NO to peroxynitrite, a very toxic agent for neurons and glia cells. Furthermore, most MS patients have decreased levels of many components of the antioxidative defense system, such as thioredoxin (Trx) or antioxidants, and increased expression of the inducible NO synthase. This leads to increased concentrations of toxic peroxynitrits, which can be followed by 3-nitrotyrosine, a typical product of peroxynitrite in demyelinated regions.

Interestingly, it has been demonstrated that Trx itself is neuroprotective and promotes neurite outgrowth (Horstkorte et al. 2010). Moreover, also in other neurodegenerating diseases, such as Morbus Alzheimer, Trx expression is reduced (Patenaude et al. 2005). Taken together, ROS and the ROS defense system play an important role during pathogenesis of MS. However, there are recent reports promoting a change in the dogma that ROS are the major player in the progression of neurodegenerating diseases. For example, ROS protect mice from bacterial infections (Pizzolla et al. 2012) and ROS produced by NOX2 are beneficial in autoimmune conditions (Hultqvist et al. 2009). Therefore, there is a need for further research regarding whether by oxidative shielding it could be possible to treat neurodegenerating diseases (Naviaux 2012). Whereas the discussion about the role of ROS started in the nervous system, much more is known about the conflict of ROS as a destroying or protecting agent in the cardiovascular system, especially during ischemia/reperfusion.

## 9.4 ISCHEMIA/REPERFUSION INJURY AND MECHANISMS OF CARDIAC PROTECTION

It is well-known that early reperfusion of the heart is the best method of salvaging previously ischemic myocardium. On the other hand, reperfusion by itself, especially when reperfusion therapy is almost delayed, can be a double-edged sword because it may lead to an accelerated and additional myocardial injury. This is known as reperfusion injury (Braunwald and Kloner 1985).

Many mechanisms may contribute to this phenomenon. During ischemia, especially older hearts exhibit an excessive accumulation of cytosolic calcium, which impairs functional recovery after reperfusion (Ataka et al. 1992). This age-related alteration in the ion homeostasis may be caused by the reduced expression of the sarcoplasmic reticulum $Ca^{2+}$ ATPase, which uptakes $Ca_i^{2+}$ into the sarcoplasmic reticulum (Cain et al. 1998; Schmidt et al. 2000). In addition, there is a metabolic mismatch between the oxidation of glucose and free fatty acids in the ischemic heart. During normal perfusion, mitochondria generate adenosine triphosphate (ATP), consume large amounts of $O_2$, and contribute, for example, due to the expression of the Mn superoxide dismutase (Mn-SOD), to a balanced generation and scavenging of ROS. During ischemia, the lack of $O_2$ inhibits electron flow and myocardial ATP utilization becomes inefficient. The energy (ATP) needed is now produced primarily by anaerobic glycolysis, which is less effective in comparison to the aerobic pathway and thereby generates much more hydrogen ions within the cell. If ischemia is prolonged, $Na^+/K^+$ ATPase gets inhibited (because of the drop in ATP levels) and the intracellular acidification (induced by lactate production and the hydrolysis of ATP) activates the $Na^+/H^+$ exchanger (NHE), that is, the cell tries to restore the intracellular pH (Perrelli et al. 2011) (see Figure 9.1).

The NHE, a ubiquitous, plasma-membrane protein present in all mammalian cell types, exists in six isoforms (Donowitz et al. 2013). The myocardial NHE system consists predominantly of the NHE1 isoform of the NHE multigene family (Fliegel et al. 1993). Although this potent physiological system reduces cell acidosis by extruding protons from the cell, it does so by exchanging $H^+$ for $Na^+$, leading, in turn, to $Na_i^+$ overload, cell swelling, and subsequent $Na_i^+$ exchange for $Ca^{2+}$, along

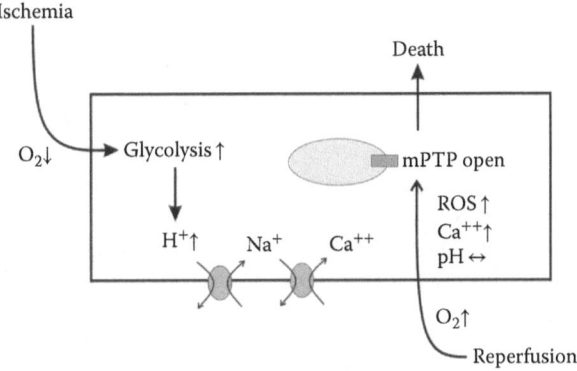

**FIGURE 9.1** Ischemia/reperfusion injury.

with further $Ca_i^{2+}$ accumulation and overload (Karmazyn et al. 1999). Within the first minutes of reperfusion and oxygen supply, a large amount of ROS is formed as the electron transport in the respiratory chain of the mitochondria is still reduced. ROS, the accumulated $Ca_i$, and the normalization of the intracellular pH open the mitochondrial permeability transition pores (mPTP). This leads to the decline of the mitochondrial membrane potential, reduced ATP formation, breakdown of plasma membrane gradients and stability, and cell death (see Figure 9.1) (Di Lisa et al. 2011; Javadov and Karmazyn 2007).

Ischemic preconditioning (IPC) is the most widely recognized method of achieving cell protection against an ischemia/reperfusion injury. Several brief ischemia/reperfusion cycles applied in advance of a prolonged ischemic insult resulted into protection, first recognized by Murry and coworkers in 1986. Interestingly, a lot of different signaling pathways were described, which all were able to protect the heart. For example, ligand-mediated activation of a variety of cell surface receptors for agonists such as adenosine or angiotensin has been shown to be effective in cardioprotection, as has direct activation of their downstream mediator enzymes, for example, protein kinase (PK)A, PKB/AKT, PKC, and PKG (Juhaszova et al. 2009). Sollott and coworkers (2009) postulated that all these different pathways need an integration point at the level of the mPTP, with the most likely candidate being GSK-3β. GSK-3β is a serine/threonine kinase which is involved in many cellular functions such as cellular proliferation, migration, glucose regulation, and cell death. It got its name as a kinase that phosphorylates and inhibits glycogen synthase (Embi et al. 1980). Beside its role in glycogen metabolism, GSK-3β was shown to have more than 40 substrates and plays a role in many intracellular processes. An unexpected difference from many other kinases is the observation that GSK is active in the unstimulated cell (as an unphosphorylated enzyme), thereby inhibiting downstream signals. Phosphorylation of GSK-3β thus inhibits the kinase and activates downstream pathways (Juhaszova et al. 2009). Survival signals stimulating the upstream AKT kinase inhibit GSK-3β. Therefore, it was believed that knocking out GSK should be protective against cell death. In contrast, during fetal development, mouse knockouts that do not express GSK-3β die due to apoptosis of the liver (Hoeflich et al. 2000), implicating that the regulatory network is more complex.

IPC results in increased phosphorylation of glycogen GSK-3β, downstream of the PI3-kinase-Akt-pathway. Mimicking GSK phosphorylation by direct inhibition of this kinase using the inhibitor SB 216763 before ischemia reduces the ischemia/reperfusion injury (Tong et al. 2002). Inactivation of the signaling kinase GSK-3β has been shown to strongly inhibit opening of the mPTP in cardiomyocytes (Juhaszova et al. 2004). Interestingly, many different agents and pathways inducing myocardial protection against ischemia/reperfusion injury involve GSK-3β inactivation. Examples beside IPC are treatments with erythropoietin (Nishihara et al. 2006), bradykinin (Park et al. 2006), isofluorane (Feng et al. 2005), or $Zn^{2+}$ (Chanoit et al. 2008).

One can distinguish between two types of cell protection: On the one hand, a direct and short-lived pathway exists, which normally does not involve gene expression but may directly lead to the inactivation of the GSK. This is often called the *trigger phase* because after the primary stimuli (trigger) the heart stays protected for about an hour even if the trigger substance is no longer present (Downey et al. 2007). On the other hand, there is protection with a memory effect. This kind of protection, for example, induced by IPC, leads to a long-lasting protection that exists for hours after the primary stimuli. IPC activates complex kinase cascades as well as $K^+$ influx into the mitochondria via the mito$K_{ATP}$ channel. This leads to slight mitochondrial swelling and a moderate activation of respiration and local ROS production (Juhaszova et al. 2009). Intracellular ROS can mediate a variety of effects on signaling pathways and protein activities that may be particularly important for cardiomyocyte function including PKA (Brennan et al. 2006), PKG (Burgoyne et al. 2007), the ryanodine receptor (Xu et al. 1998), the small G protein Ras (Kuster et al. 2006), class II histone deacetylases (Ago et al. 2008), and the metabolic enzyme glyceraldehyde-3-phosphate dehydrogenase (Eaton et al. 2002). In IPC, ROS activates the ε isoform of PKC, which in turn phosphorylates and inactivates GSK (Juhaszova et al. 2004). GSK would be responsible for the long-lasting activation/opening of the mPTP and the following cell death (see Figure 9.2).

mPTP was originally observed during experiments on $Ca^{2+}$-induced swelling and permeability of mitochondria. The permeability change was reversible by chelating $Ca^{2+}$ using ethylenediaminetetraacetic acid. The pore has no real specificity as charged and neutral molecules can pass it (Hunter and Haworth 1979). Opening of the mPTP increases the permeability of the inner membrane for molecules up to 1500 Da, which leads to mitochondrial swelling and depolarization. This permeabilization of the inner mitochondrial membrane is called mitochondrial permeability transition. Long-lasting opening of the mPTP has disastrous consequences for the mitochondria. They rapidly swell; this swelling occurs because the high concentration of proteins in the matrix space, which cannot pass the mPTP, exerts a large colloidal osmotic pressure. The cristae unfold as the matrix swells so that the outer membrane ruptures. The unspecific loss of outer membrane integrity also causes a spillage of mitochondrial death factors, like cytochrome c, from the inter-membrane space to the cytosol (Kinnally et al. 2011). Finally, the collapse of the membrane potential combined with the leakage of pyridine nucleotides, for example, NADH, can lead to the

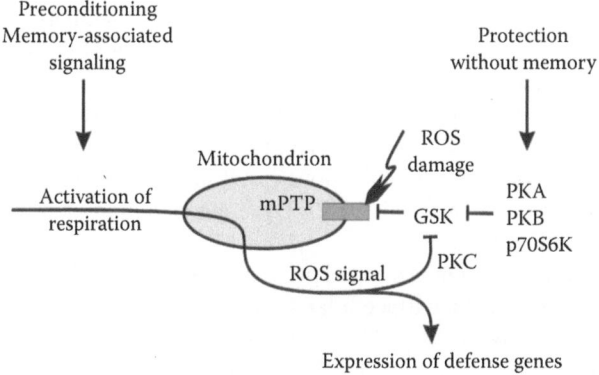

**FIGURE 9.2** Model of cell protection. (Adapted from Juhaszova, M. et al., *Cardiovasc Res*, 66, 2, 233–44, 2005. With permission.)

generation of ROS via the direct transfer of electrons to molecular oxygen (Bernardi et al. 2006; Halestrap and Pasdois 2009). As all these events in combination with a fast drop in ATP production will lead to necrosis, it is generally accepted that this pore is only central to necrosis (Halestrap 2005). On the other hand, opening of the mPTP can lead via cytochrome c release to apoptosis as well. In the context of ischemia, it is important to know that the opening probability of mPTP of the de-energized mitochondria is pH dependent. The opening is hugely reduced below pH 7.4, a condition occurring due to increased glycolysis during sufficiently prolonged ischemia (Nicolli et al. 1993). Whereas a lot is known about the regulation and activity of this pore, the molecular identity of the protein(s) forming this pore is still unknown.

The integration of ROS within the protective IPC signaling was proposed as IPC's protection can be mimicked by transient exposure to an oxygen radical generating system, and, conversely, an ROS scavenger can abolish protection from IPC (Baines et al. 1997; Tritto et al. 1997). In chick cardiomyocytes, the addition of $H_2O_2$ (15 μmol/L) under normoxic conditions also induced preconditioning-like protection (Vanden Hoek et al. 1998). In mouse and rat cardiac cells, the treatment with Maillard products from nutritional extracts leads to a modest intracellular ROS increase and protects against following ischemia-induced cell death (Ruhs et al. 2010). Because a prominent target of ROS in redox signaling is the PKC (see Section 9.4 and Figure 9.2), a correlation between PKC and preconditioning was proposed. Infusion of hypoxanthine together with xanthine oxidase in rabbit hearts induced protection only when combined (due to free radical production). Polymyxin B, an inhibitor of PKC, abolished this protection (Baines et al. 1997; Yellon and Downey 2003). Evidence exists that ROS-activated PKC will not only protect the heart during IPC but also during postischemic conditioning as a persistent activation of PKC is needed for protection under both conditions (Pain et al. 2000; Penna et al. 2006). These results would explain the observation that a PKC activator could rescue hearts experiencing acidic and hypoxic reperfusion (Cohen et al. 2008). Indeed, it has been reported that ROS can activate PKC in vitro by reacting with the zinc finger region of the molecule (Korichneva 2006).

To understand the ROS regulation of molecules within the signaling pathways, the mechanism of how ROS alters protein function should be briefly addressed. In most cases, the oxidative interface mainly consists of the redox regulation of redox-reactive cysteine (Cys) residues on proteins by ROS. Oxidation of these residues forms reactive sulfenic acid (–SOH) that can form disulfide bonds with nearby cysteines (–S–S–) or undergo further oxidation to sulfinic (–SO2H) or sulfonic (–SO3H) acid; if nearby nitrogen is available, sulfenic acid may also form a sulfonamide (Wagener et al. 2013). These oxidative modifications result in changes in the structure and/or function of the protein (Ray et al. 2012). Different signaling pathways are directly affected by oxidative stress.

One is part of the mitogen-activated protein kinase (MAPK) cascades consisting of four major MAPKs: the extracellular signal-related kinases (Erk1/2), the c-Jun N-terminal kinases (JNK), the p38 kinase (p38), and the big MAPK1 (BMK1/Erk5). These kinases play essential roles in cellular responses to a wide variety of signals, including oxidative stressors (Ray et al. 2012). The apoptosis signal-regulated kinase 1 (ASK1), an upstream kinase that regulates the JNK and p38MAPK pathways, is activated under stress conditions including oxidative stress leading to apoptosis.

In ASK1(–/–) embryonic fibroblasts, oxidative stress-induced sustained activations of JNK and p38 are lost, and ASK1(–/–) cells are resistant to $H_2O_2$-induced apoptosis (Ichijo et al. 1997; Tobiume et al. 2001). In parallel, inhibition of nuclear MAPK phosphatase MKP-1 by ROS has also been shown to activate JNK and p38MAPK activity and signaling (Liu et al. 2010). In a study on adiponectin-induced cell proliferation of adult hippocampal neural stem/progenitor cells, it was shown that p38MAPK is important for the phosphorylation of GSK-3β on Ser-389, a key inhibitory site (Zhang et al. 2011).

Another signaling pathway that plays a key role in cell survival in response to stress is the phosphoinositide 3-kinase (PI3K) pathway. PI3Ks function early in intracellular signal transduction pathways and affect many biological functions. A further level of complexity derives from the existence of eight PI3K isoforms, which are divided into class I, class II, and class III PI3Ks. PI3K signaling has been implicated in metabolic control, immunity, angiogenesis, and cardiovascular homeostasis, and is one of the most frequently deregulated pathways in cancer (Vanhaesebroeck et al. 2010). PI3K catalyzes the synthesis of the second messenger-like membrane-bound phosphatidylinositol 3,4,5 triphosphates (PtdIns(3,4,5)P3) (PIP3), which serves as a signaling molecule to recruit proteins such as PKB (AKT) serine/threonine kinase that function as survival kinase (Cantrell 2001). The synthesis of PIP3 is negatively regulated by the phosphatase and tensin homology phosphatase (PTEN), which dephosphorylates PIP3 back to PIP2. Inactivation of these lipid phosphatases has been implicated in disease. These include lipid phosphatases for PtdIns(3,4,5)P3, such as PTEN, a tumor suppressor that is frequently inactivated in cancer so that the PI3K pathway is constitutively active (Salmena et al. 2008). Through PTEN, the PI3K pathway is as well subject to reversible redox regulation by ROS. $H_2O_2$ was shown to oxidize and inactivate human PTEN through disulfide bond formation between the catalytic domain Cys-124 and Cys-71 residues (Kwon et al. 2004). It was also demonstrated that endogenously generated ROS causes oxidation of PTEN leading to the activation of the PI3K pathway (Seo et al. 2005). An important downstream target of PI3K and AKT is again the GSK-3β, which is essential for mPTP regulation (see Section 9.4 and Figure 9.2).

At the end, besides direct phosphorylation of target proteins by these stress/survival kinases, an effect on gene transcription can also be detected. To prevent long-term oxidative stress, the cell must respond to ROS by inducing the intracellular antioxidant defense system, including enzymes like the superoxide dismutase (SOD). Antioxidative enzymes play a major role in reducing ROS levels; therefore, redox regulation of transcription factors will affect cellular response to oxidative stress. Redox factor-1 (Ref-1) is a multifunctional protein that regulates transcription factor activities, for example, the activator protein 1, p53, nuclear factor kappa B, and hypoxia inducible factor 1 (Tell et al. 2009). Cys-65 and Cys-93 of Ref-1 appear to be the redox active sites required for the modification of the DNA binding of targeted transcription factors (Walker et al. 1993).

Whereas the results described above seem to fit to the model of tissue protections, there are still some problems regarding the age of the sample population. Most of the experiments were done in cells or young animals. On the other side, the patients who have to be treated are, in most cases, old. This has to be solved before positive results can be expected in the human system.

Analyzing right atrial trabeculae of patients at different ages (from 40 to 80 years) after simulated ischemia (hypoxic substrate-free superfusion) with and without 5 minutes of ischemic pretreatment, only tissue from young donors displayed a positive postischemic contractile recovery due to IPC. Interestingly, the IPC conserves the expression of cardioprotective determinants in aged hearts despite limited functional recovery (Bartling et al. 2003a). Young/adult myocardium showed PKC up-regulation after IP along with improved postischemic contractility. Although there was no functional benefit, PKC increased in older myocardium as well. Subsequent mRNA analyses demonstrated that IP stabilizes the mRNA expression of protective proteins (Hsp70, Bcl-2/-xL, IAPs) in both aging groups (Bartling et al. 2003b). It seems that there are additional factors that may change with age, which are not clarified until now.

## 9.5 SHORT- AND LONG-TERM PROTECTION OF ISCHEMIC PRECONDITIONING

Whereas primary studies on IPC focus on a first window of protection (SWOP) within the first 2 hours, it was discovered that a *second window* of protection occurred in response to the same IP protocols (Pagliaro et al. 2001). The (SWOP) appears about 24 hours after completion of the primary IP protocol and persists for 3–4 days. In contrast to the first protective phase within the first 1 or 2 hours, SWOP likely works through the induction of gene expression and/or new protein synthesis. Many candidates have been proposed including heat shock proteins, (inducible nitric oxide synthetase) iNOS, antioxidant proteins, and growth factors (Vander Heide 2011). It was proposed that both these windows can be explained on the level of the mitochondria. Memory-associated protection acts on mitochondrial targets associated with an increased $K^+$ influx into, or retention by, mitochondria. Consequent regulatory mitochondrial matrix swelling and increased respiration produce a redox signal resulting from moderately increased ROS production, which results in local PKC activation and GSK-3β inhibition. On the one hand, activated PKC can signal in a positive-feedback loop to mitoK-ATP, producing a system memory. On the other hand, a redox signal can induce intracellular signaling described above to induce protective gene expression, for example, the increased expression of the SODs as well as the catalase. For example, Mn-SOD protein expression was increased in the heart in mice after 30-minute coronary ligation followed by 24-hour reperfusion (Li et al. 2011). Memory-lacking protection can be triggered by cell surface receptor ligands or pharmacological agents causing activation of diverse signaling pathways that can bypass the mitochondrial volume-regulatory mechanism and converge on GSK-3β (Juhaszova et al. 2009).

## 9.6 HORMESIS: THE CONCEPT OF CELL PROTECTION

Ischemia/reperfusion injury is characterized by necrotic as well as apoptotic cell death. Previously it was believed that the time of ischemia was the most important cause of cell disruption. In contrast, it is now clear that ROS-induced damage at the time of reperfusion is a major burden. ROS is indeed an important cause of cell death.

Experimental studies based on the pathogenic role of ROS in myocardial damage following ischemia/reperfusion events have given promising results for antioxidant cardioprotection. Therefore, it was expected that treatments with exogenous antioxidant agents could protect the heart against lethal reperfusion injury in clinical models (Rodrigo et al. 2013a). However, although a number of strategies have been devised to reduce this injury, the beneficial effects in clinical settings have been disappointing to date (Rodrigo et al. 2013b). Clinical trials designed to study cardioprotection by long-term administration of vitamins C and E have failed to demonstrate beneficial effects (Rapola et al. 1997; Rodrigo et al. 2013a; Sesso et al. 2008). Therefore, an ROS signal seems to be important for long-term protection of the heart: moderate ROS increase seems to be an important stressor that activates an adaptive stress response, raising the resistance against high doses of following ROS. In most cases, this is associated with an altered gene expression.

IPC represents this adaptive phenomenon, which is known as *hormesis*. It describes the biological response to *harmless* doses of stressors. If this is true, *in vivo ischemic preconditioning* should have an impact on the outcome of patients. Indeed, there are some studies in favor of this hypothesis. Angina pectoris is a temporary chest pain, due to an inadequate supply of oxygen to the heart muscle. This can be compared to conditions resembled by experimental IPC. Patients presenting angina 1 week before AMI were less likely than those without angina to experience inhospital death, HF, or the combined endpoint of inhospital death and HF. Left ventricular function was more frequently depressed in elderly patients without preinfarction angina than with, and the incidence of arrhythmias was greater in the former group (Jimenez-Navarro et al. 2001). This effect seems to be age-dependent, as other studies demonstrated that the protection afforded by angina in adult patients seems to be lost in old patients. In one study on 990 patients who underwent coronary angiography within 12 hours after the onset of AMI, prodromal angina in the 24 hours before infarction was associated in patients younger than 70 years with a smaller infarct size and better short- and long-term survival. However, such a beneficial effect was not observed in elderly patients above 70 years (Ishihara et al. 2000). In line with these results, inhospital death, congestive HF, and shock were more frequent in adult patients younger than 65 years (with myocardial infarction) than without in those with previous angina. In contrast, the presence or absence of previous angina before AMI in elderly patients above 65 seems not to influence the respective outcomes (Abete et al. 1997). Interestingly, a follow-up study of the same group could demonstrate that the protective role of preinfarction angina on inhospital mortality was present again in elderly patients showing a high level of physical activity (Abete et al. 2001).

It seems to be clear that the hormetic response on IPC can be influenced by the aging process on the one hand, but may be not lost in total as shown by the impact of exercise on the angina effect in the elderly. Follow-up studies have to clarify the mechanisms behind these age-related changes. Another possibility would be the development of hormesis-stimulating compounds, which can initiate a comparable adaptive stress response that renders the heart resistant against a following ischemia (infarction). The hormesis response should allow stressed cardiac myocytes to avoid cell death until the interrupted blood flow is restored after reopening the blockade within the coronary arteries by reperfusion therapies like thrombolysis or angioplasty.

## 9.7 CONCLUSION

Whereas many mechanisms were discussed to reduce cell stress and cell death in pathophysiological situations, for example, the ischemia/reperfusion injury, the new concept of the ischemia/reperfusion injury as a hormetic response should lead to new therapeutic options. Inducing moderate oxidative stress, for example, by nutritional extracts (Ruhs et al. 2010) may be the basis for new concepts of the treatment of patients in advance of open-heart surgery to avoid the ischemia/reperfusion damage associated with the operation procedure per se.

## REFERENCES

Abete, P., N. Ferrara, F. Cacciatore, A. Madrid, S. Bianco, C. Calabrese, C. Napoli et al. 1997. Angina-induced protection against myocardial infarction in adult and elderly patients: A loss of preconditioning mechanism in the aging heart? *J Am Coll Cardiol* 30(4): 947–54.

Abete, P., N. Ferrara, F. Cacciatore, E. Sagnelli, M. Manzi, V. Carnovale, C. Calabrese et al. 2001. High level of physical activity preserves the cardioprotective effect of preinfarction angina in elderly patients. *J Am Coll Cardiol* 38(5): 1357–65.

Ago, T., T. Liu, P. Zhai, W. Chen, H. Li, J. D. Molkentin, S. F. Vatner, and J. Sadoshima. 2008. A redox-dependent pathway for regulating class II HDACs and cardiac hypertrophy. *Cell* 133(6): 978–93.

Ataka, K., D. Chen, S. Levitsky, E. Jimenez, and H. Feinberg. 1992. Effect of aging on intracellular Ca2+, pHi, and contractility during ischemia and reperfusion. *Circulation* 86(5 Suppl): II371–6.

Baines, C. P., M. Goto, and J. M. Downey. 1997. Oxygen radicals released during ischemic preconditioning contribute to cardioprotection in the rabbit myocardium. *J Mol Cell Cardiol* 29(1): 207–16.

Balaban, R. S., S. Nemoto, and T. Finkel. 2005. Mitochondria, oxidants, and aging. *Cell* 120(4): 483–95.

Baraibar, M. A., and B. Friguet. 2013. Oxidative proteome modifications target specific cellular pathways during oxidative stress, cellular senescence and aging. *Exp Gerontol* 48(7): 620–5.

Bartling, B., I. Friedrich, R. E. Silber, and A. Simm. 2003a. Ischemic preconditioning is not cardioprotective in senescent human myocardium. *Ann Thorac Surg* 76(1): 105–11.

Bartling, B., C. Hilgefort, I. Friedrich, R. E. Silber, and A. Simm. 2003b. Cardio-protective determinants are conserved in aged human myocardium after ischemic preconditioning. *FEBS Lett* 555(3): 539–44.

Bernardi, P., A. Krauskopf, E. Basso, V. Petronilli, E. Blachly-Dyson, F. Di Lisa, and M. A. Forte. 2006. The mitochondrial permeability transition from in vitro artifact to disease target. *FEBS J* 273(10): 2077–99.

Bleumink, G. S., A. M. Knetsch, M. C. Sturkenboom, S. M. Straus, A. Hofman, J. W. Deckers, J. C. Witteman, and B. H. Stricker. 2004. Quantifying the heart failure epidemic: Prevalence, incidence rate, lifetime risk and prognosis of heart failure The Rotterdam Study. *Eur Heart J* 25(18): 1614–9.

Braunwald, E., and R. A. Kloner. 1985. Myocardial reperfusion: A double-edged sword? *J Clin Invest* 76(5): 1713–9.

Brennan, J. P., S. C. Bardswell, J. R. Burgoyne, W. Fuller, E. Schroder, R. Wait, S. Begum, J. C. Kentish, and P. Eaton. 2006. Oxidant-induced activation of type I protein kinase A is mediated by RI subunit interprotein disulfide bond formation. *J Biol Chem* 281(31): 21827–36.

Burgoyne, J. R., M. Madhani, F. Cuello, R. L. Charles, J. P. Brennan, E. Schroder, D. D. Browning, and P. Eaton. 2007. Cysteine redox sensor in PKGIa enables oxidant-induced activation. *Science* 317(5843): 1393–7.

Cain, B. S., D. R. Meldrum, K. S. Joo, J. F. Wang, X. Meng, J. C. Cleveland, Jr., A. Banerjee, and A. H. Harken. 1998. Human SERCA2a levels correlate inversely with age in senescent human myocardium. *J Am Coll Cardiol* 32(2): 458–67.

Cantrell, D. A. 2001. Phosphoinositide 3-kinase signalling pathways. *J Cell Sci* 114(Pt 8): 1439–45.

Chanoit, G., S. Lee, J. Xi, M. Zhu, R. A. McIntosh, R. A. Mueller, E. A. Norfleet, and Z. Xu. 2008. Exogenous zinc protects cardiac cells from reperfusion injury by targeting mitochondrial permeability transition pore through inactivation of glycogen synthase kinase-3beta. *Am J Physiol Heart Circ Physiol* 295(3): H1227–33.

Cohen, M. V., X. M. Yang, and J. M. Downey. 2008. Acidosis, oxygen, and interference with mitochondrial permeability transition pore formation in the early minutes of reperfusion are critical to postconditioning's success. *Basic Res Cardiol* 103(5): 464–71.

Dasgupta, A., J. Zheng, and O. A. Bizzozero. 2012. Protein carbonylation and aggregation precede neuronal apoptosis induced by partial glutathione depletion. *ASN Neuro.* **2012** Apr 10;4(3). pii: e00084. doi: 10.1042/AN20110064.

Davies, M. J., S. Fu, H. Wang, and R. T. Dean. 1999. Stable markers of oxidant damage to proteins and their application in the study of human disease. *Free Radic Biol Med* 27(11–12): 1151–63.

De Flora, S., A. Quaglia, C. Bennicelli, and M. Vercelli. 2005. The epidemiological revolution of the 20th century. *FASEB J* 19(8): 892–7.

Di Lisa, F., M. Canton, A. Carpi, N. Kaludercic, R. Menabo, S. Menazza, and M. Semenzato. 2011. Mitochondrial injury and protection in ischemic pre- and postconditioning. *Antioxid Redox Signal* 14(5): 881–91.

Donowitz, M., C. Ming Tse, and D. Fuster. 2013. SLC9/NHE gene family, a plasma membrane and organellar family of Na(+)/H(+) exchangers. *Mol Aspects Med* 34(2–3): 236–51.

Downey, J. M., A. M. Davis, and M. V. Cohen. 2007. Signaling pathways in ischemic preconditioning. *Heart Fail Rev* 12(3–4): 181–8.

Eaton, P., N. Wright, D. J. Hearse, and M. J. Shattock. 2002. Glyceraldehyde phosphate dehydrogenase oxidation during cardiac ischemia and reperfusion. *J Mol Cell Cardiol* 34(11): 1549–60.

Embi, N., D. B. Rylatt, and P. Cohen. 1980. Glycogen synthase kinase-3 from rabbit skeletal muscle. Separation from cyclic-AMP-dependent protein kinase and phosphorylase kinase. *Eur J Biochem* 107(2): 519–27.

Feng, J., E. Lucchinetti, P. Ahuja, T. Pasch, J. C. Perriard, and M. Zaugg. 2005. Isoflurane postconditioning prevents opening of the mitochondrial permeability transition pore through inhibition of glycogen synthase kinase 3beta. *Anesthesiology* 103(5): 987–95.

Fliegel, L., J. R. Dyck, H. Wang, C. Fong, and R. S. Haworth. 1993. Cloning and analysis of the human myocardial Na+/H+ exchanger. *Mol Cell Biochem* 125(2): 137–43.

Friedrich, I., A. Simm, J. Kotting, F. Tholen, B. Fischer, and R. E. Silber. 2009. Cardiac surgery in the elderly patient. *Dtsch Arztebl Int* 106(25): 416–22.

Hafler, D. A., and H. L. Weiner. 1987. T cells in multiple sclerosis and inflammatory central nervous system diseases. *Immunol Rev* 100: 307–32.

Halestrap, A. 2005. Biochemistry: A pore way to die. *Nature* 434(7033): 578–9.

Halestrap, A. P., and P. Pasdois. 2009. The role of the mitochondrial permeability transition pore in heart disease. *Biochim Biophys Acta* 1787(11): 1402–15.

Harman, D. 1956. Aging: A theory based on free radical and radiation chemistry. *J Gerontol* 11(3): 298–300.

Hoeflich, K. P., J. Luo, E. A. Rubie, M. S. Tsao, O. Jin, and J. R. Woodgett. 2000. Requirement for glycogen synthase kinase-3beta in cell survival and NF-kappaB activation. *Nature* 406(6791): 86–90.

Horstkorte, R., S. Reinke, C. Bauer, W. Reutter, and M. Kontou. 2010. N-Propionylmannosamine-induced over-expression and secretion of thioredoxin leads to neurite outgrowth of PC12 cells. *Biochem Biophys Res Commun* 395(3): 296–300.

Hultqvist, M., L. M. Olsson, K. A. Gelderman, and R. Holmdahl. 2009. The protective role of ROS in autoimmune disease. *Trends Immunol* 30(5): 201–8.

Hunter, D. R., and R. A. Haworth. 1979. The Ca2+-induced membrane transition in mitochondria. I. The protective mechanisms. *Arch Biochem Biophys* 195(2): 453–9.

Ichijo, H., E. Nishida, K. Irie, P. ten Dijke, M. Saitoh, T. Moriguchi, M. Takagi, K. Matsumoto, K. Miyazono, and Y. Gotoh. 1997. Induction of apoptosis by ASK1, a mammalian MAPKKK that activates SAPK/JNK and p38 signaling pathways. *Science* 275(5296): 90–4.

Ishihara, M., H. Sato, H. Tateishi, T. Kawagoe, Y. Shimatani, K. Ueda, K. Noma, A. Yumoto, and K. Nishioka. 2000. Beneficial effect of prodromal angina pectoris is lost in elderly patients with acute myocardial infarction. *Am Heart J* 139(5): 881–8.

Javadov, S., and M. Karmazyn. 2007. Mitochondrial permeability transition pore opening as an endpoint to initiate cell death and as a putative target for cardioprotection. *Cell Physiol Biochem* 20(1–4): 1–22.

Jimenez-Navarro, M., J. J. Gomez-Doblas, J. Alonso-Briales, J. M. Hernandez Garcia, G. Gomez, A. G. Alcantara, I. Rodriguez-Bailon et al. 2001. Does angina the week before protect against first myocardial infarction in elderly patients? *Am J Cardiol* 87(1): 11–5.

Juhaszova, M., C. Rabuel, D. B. Zorov, E. G. Lakatta, and S. J. Sollott. 2005. Protection in the aged heart: Preventing the heart-break of old age? *Cardiovasc Res* 66(2): 233–44.

Juhaszova, M., D. B. Zorov, S. H. Kim, S. Pepe, Q. Fu, K. W. Fishbein, B. D. Ziman et al. 2004. Glycogen synthase kinase-3beta mediates convergence of protection signaling to inhibit the mitochondrial permeability transition pore. *J Clin Invest* 113(11): 1535–49.

Juhaszova, M., D. B. Zorov, Y. Yaniv, H. B. Nuss, S. Wang, and S. J. Sollott. 2009. Role of glycogen synthase kinase-3beta in cardioprotection. *Circ Res* 104(11): 1240–52.

Karmazyn, M., X. T. Gan, R. A. Humphreys, H. Yoshida, and K. Kusumoto. 1999. The myocardial Na(+)-H(+) exchange: Structure, regulation, and its role in heart disease. *Circ Res* 85(9): 777–86.

Kinnally, K. W., P. M. Peixoto, S. Y. Ryu, and L. M. Dejean. 2011. Is mPTP the gatekeeper for necrosis, apoptosis, or both? *Biochim Biophys Acta* 1813(4): 616–22.

Klebanoff, S. J., M. A. Vadas, J. M. Harlan, L. H. Sparks, J. R. Gamble, J. M. Agosti, and A. M. Waltersdorph. 1986. Stimulation of neutrophils by tumor necrosis factor. *J Immunol* 136(11): 4220–5.

Korichneva, I. 2006. Zinc dynamics in the myocardial redox signaling network. *Antioxid Redox Signal* 8(9–10): 1707–21.

Kuster, G. M., D. A. Siwik, D. R. Pimentel, and W. S. Colucci. 2006. Role of reversible, thioredoxin-sensitive oxidative protein modifications in cardiac myocytes. *Antioxid Redox Signal* 8(11–12): 2153–9.

Kwon, J., S. R. Lee, K. S. Yang, Y. Ahn, Y. J. Kim, E. R. Stadtman, and S. G. Rhee. 2004. Reversible oxidation and inactivation of the tumor suppressor PTEN in cells stimulated with peptide growth factors. *Proc Natl Acad Sci USA* 101(47): 16419–24.

Lakatta, E. G. 2002. Age-associated cardiovascular changes in health: Impact on cardiovascular disease in older persons. *Heart Fail Rev* 7(1): 29–49.

Li, Y., M. Cai, Y. Xu, H. M. Swartz, and G. He. 2011. Late phase ischemic preconditioning preserves mitochondrial oxygen metabolism and attenuates post-ischemic myocardial tissue hyperoxygenation. *Life Sci* 88(1–2): 57–64.

Liu, R. M., J. Choi, J. H. Wu, K. A. Gaston Pravia, K. M. Lewis, J. D. Brand, N. S. Mochel et al. 2010. Oxidative modification of nuclear mitogen-activated protein kinase phosphatase 1 is involved in transforming growth factor beta1-induced expression of plasminogen activator inhibitor 1 in fibroblasts. *J Biol Chem* 285(21): 16239–47.

Lloyd-Jones, D., R. Adams, M. Carnethon, G. De Simone, T. B. Ferguson, K. Flegal, E. Ford et al. 2009. Heart disease and stroke statistics—2009 update: A report from the American Heart Association Statistics Committee and Stroke Statistics Subcommittee. *Circulation* 119(3): e21–181.

Martin, R., and H. F. McFarland. 1995. Immunological aspects of experimental allergic encephalomyelitis and multiple sclerosis. *Crit Rev Clin Lab Sci* 32(2): 121–82.

Murry, C. E., R. B. Jennings, and K. A. Reimer. 1986. Preconditioning with ischemia: A delay of lethal cell injury in ischemic myocardium. *Circulation* 74(5):1124–36.

Naviaux, R. K. 2012. Oxidative shielding or oxidative stress? *J Pharmacol Exp Ther* 342(3): 608–18.

Neumann, T., J. Biermann, R. Erbel, A. Neumann, J. Wasem, G. Ertl, and R. Dietz. 2009. Heart failure: The commonest reason for hospital admission in Germany: Medical and economic perspectives. *Dtsch Arztebl Int* 106(16): 269–75.

Nicolli, A., V. Petronilli, and P. Bernardi. 1993. Modulation of the mitochondrial cyclosporin A-sensitive permeability transition pore by matrix pH. Evidence that the pore open-closed probability is regulated by reversible histidine protonation. *Biochemistry* 32(16): 4461–5.

Nishihara, M., T. Miura, T. Miki, J. Sakamoto, M. Tanno, H. Kobayashi, Y. Ikeda, K. Ohori, A. Takahashi, and K. Shimamoto. 2006. Erythropoietin affords additional cardioprotection to preconditioned hearts by enhanced phosphorylation of glycogen synthase kinase-3 beta. *Am J Physiol Heart Circ Physiol* 291(2): H748–55.

Oeppen, J., and J. W. Vaupel. 2002. Demography. Broken limits to life expectancy. *Science* 296(5570): 1029–31.

Pagliaro, P., D. Gattullo, R. Rastaldo, and G. Losano. 2001. Ischemic preconditioning: From the first to the second window of protection. *Life Sci* 69(1): 1–15.

Pain, T., X. M. Yang, S. D. Critz, Y. Yue, A. Nakano, G. S. Liu, G. Heusch, M. V. Cohen, and J. M. Downey. 2000. Opening of mitochondrial K(ATP) channels triggers the preconditioned state by generating free radicals. *Circ Res* 87(6): 460–6.

Park, S. S., H. Zhao, R. A. Mueller, and Z. Xu. 2006. Bradykinin prevents reperfusion injury by targeting mitochondrial permeability transition pore through glycogen synthase kinase 3beta. *J Mol Cell Cardiol* 40(5): 708–16.

Patenaude, A., M. R. Murthy, and M. E. Mirault. 2005. Emerging roles of thioredoxin cycle enzymes in the central nervous system. *Cell Mol Life Sci* 62(10): 1063–80.

Penna, C., R. Rastaldo, D. Mancardi, S. Raimondo, S. Cappello, D. Gattullo, G. Losano, and P. Pagliaro. 2006. Post-conditioning induced cardioprotection requires signaling through a redox-sensitive mechanism, mitochondrial ATP-sensitive K+ channel and protein kinase C activation. *Basic Res Cardiol* 101(2): 180–9.

Perrelli, M.-G., P. Pagliaro, and C. Penna. 2011. Ischemia/reperfusion injury and cardioprotective mechanisms: Role of mitochondria and reactive oxygen species. *World J Cardiol* 3(6): 186–200.

Pizzolla, A., M. Hultqvist, B. Nilson, M. J. Grimm, T. Eneljung, I. M. Jonsson, M. Verdrengh et al. 2012. Reactive oxygen species produced by the NADPH oxidase 2 complex in monocytes protect mice from bacterial infections. *J Immunol* 188(10): 5003–11.

Rapola, J. M., J. Virtamo, S. Ripatti, J. K. Huttunen, D. Albanes, P. R. Taylor, and O. P. Heinonen. 1997. Randomised trial of alpha-tocopherol and beta-carotene supplements on incidence of major coronary events in men with previous myocardial infarction. *Lancet* 349(9067): 1715–20.

Ravanat, J. L., P. Di Mascio, G. R. Martinez, M. H. Medeiros, and J. Cadet. 2000. Singlet oxygen induces oxidation of cellular DNA. *J Biol Chem* 275(51): 40601–4.

Ray, P. D., B. W. Huang, and Y. Tsuji. 2012. Reactive oxygen species (ROS) homeostasis and redox regulation in cellular signaling. *Cell Signal* 24(5): 981–90.

Rodrigo, R., M. Libuy, F. Feliu, and D. Hasson. 2013a. Molecular basis of cardioprotective effect of antioxidant vitamins in myocardial infarction. *Biomed Res Int.* **2013**;2013: 437613. doi: 10.1155/**2013**/437613. Epub **2013** Jul 14.

Rodrigo, R., J. C. Prieto, and R. Castillo. 2013b. Cardioprotection against ischaemia/reperfusion by vitamins C and E plus n-3 fatty acids: Molecular mechanisms and potential clinical applications. *Clin Sci (Lond)* 124(1): 1–15.

Rosamond, W., K. Flegal, G. Friday, K. Furie, A. Go, K. Greenlund, N. Haase et al. 2007. Heart disease and stroke statistics—2007 update: A report from the American Heart Association Statistics Committee and Stroke Statistics Subcommittee. *Circulation* 115(5): e69–171.

Rosenfeldt, F. L., S. Pepe, A. Linnane, P. Nagley, M. Rowland, R. Ou, S. Marasco, and W. Lyon. 2002. The effects of ageing on the response to cardiac surgery: Protective strategies for the ageing myocardium. *Biogerontology* 3(1–2): 37–40.

Ruhs, S., N. Nass, B. Bartling, H. J. Bromme, B. Leuner, V. Somoza, U. Friess, R. E. Silber, and A. Simm. 2010. Preconditioning with Maillard reaction products improves antioxidant defence leading to increased stress tolerance in cardiac cells. *Exp Gerontol* 45(10): 752–62.

Saito, Y., K. Nishio, Y. Ogawa, J. Kimata, T. Kinumi, Y. Yoshida, N. Noguchi, and E. Niki. 2006. Turning point in apoptosis/necrosis induced by hydrogen peroxide. *Free Radic Res* 40(6): 619–30.

Salmena, L., A. Carracedo, and P. P. Pandolfi. 2008. Tenets of PTEN tumor suppression. *Cell* 133(3): 403–14.

Schilder, Y. D., E. H. Heiss, D. Schachner, J. Ziegler, G. Reznicek, D. Sorescu, and V. M. Dirsch. 2009. NADPH oxidases 1 and 4 mediate cellular senescence induced by resveratrol in human endothelial cells. *Free Radic Biol Med* 46(12): 1598–606.

Schmidt, U., F. del Monte, M. I. Miyamoto, T. Matsui, J. K. Gwathmey, A. Rosenzweig, and R. J. Hajjar. 2000. Restoration of diastolic function in senescent rat hearts through adenoviral gene transfer of sarcoplasmic reticulum Ca(2+)-ATPase. *Circulation* 101(7): 790–6.

Seo, J. H., Y. Ahn, S. R. Lee, C. Yeol Yeo, and K. Chung Hur. 2005. The major target of the endogenously generated reactive oxygen species in response to insulin stimulation is phosphatase and tensin homolog and not phosphoinositide-3 kinase (PI-3 kinase) in the PI-3 kinase/Akt pathway. *Mol Biol Cell* 16(1): 348–57.

Sesso, H. D., J. E. Buring, W. G. Christen, T. Kurth, C. Belanger, J. MacFadyen, V. Bubes, J. E. Manson, R. J. Glynn, and J. M. Gaziano. 2008. Vitamins E and C in the prevention of cardiovascular disease in men: The Physicians' Health Study II randomized controlled trial. *JAMA* 300(18): 2123–33.

Simm, A., N. Nass, B. Bartling, B. Hofmann, R. E. Silber, and A. Navarrete Santos. 2008. Potential biomarkers of ageing. *Biol Chem* 389(3): 257–65.

Szibor, M., and J. Holtz. 2003. Mitochondrial ageing. *Basic Res Cardiol* 98(4): 210–8.

Tavazzi, B., D. Di Pierro, M. Bartolini, M. Marino, S. Distefano, M. Galvano, C. Villani, B. Giardina, and G. Lazzarino. 1998. Lipid peroxidation, tissue necrosis, and metabolic and mechanical recovery of isolated reperfused rat heart as a function of increasing ischemia. *Free Radic Res* 28(1): 25–37.

Tell, G., F. Quadrifoglio, C. Tiribelli, and M. R. Kelley. 2009. The many functions of APE1/Ref-1: Not only a DNA repair enzyme. *Antioxid Redox Signal* 11(3): 601–20.

Tobiume, K., A. Matsuzawa, T. Takahashi, H. Nishitoh, K. Morita, K. Takeda, O. Minowa, K. Miyazono, T. Noda, and H. Ichijo. 2001. ASK1 is required for sustained activations of JNK/p38 MAP kinases and apoptosis. *EMBO Rep* 2(3): 222–8.

Tong, H., K. Imahashi, C. Steenbergen, and E. Murphy. 2002. Phosphorylation of glycogen synthase kinase-3beta during preconditioning through a phosphatidylinositol-3-kinase— Dependent pathway is cardioprotective. *Circ Res* 90(4): 377–9.

Tritto, I., D. D'Andrea, N. Eramo, A. Scognamiglio, C. De Simone, A. Violante, A. Esposito, M. Chiariello, and G. Ambrosio. 1997. Oxygen radicals can induce preconditioning in rabbit hearts. *Circ Res* 80(5): 743–8.

van Horssen, J., M. E. Witte, G. Schreibelt, and H. E. de Vries. 2011. Radical changes in multiple sclerosis pathogenesis. *Biochim Biophys Acta* 1812(2): 141–50.

Vanden Hoek, T. L., L. B. Becker, Z. Shao, C. Li, and P. T. Schumacker. 1998. Reactive oxygen species released from mitochondria during brief hypoxia induce preconditioning in cardiomyocytes. *J Biol Chem* 273(29): 18092–8.

Vander Heide, R. 2011. Clinically useful cardioprotection: Ischemic preconditioning then and now. *J Cardiovasc Pharmacol Ther* 16(3–4): 251–4.

Vanhaesebroeck, B., J. Guillermet-Guibert, M. Graupera, and B. Bilanges. 2010. The emerging mechanisms of isoform-specific PI3K signalling. *Nat Rev Mol Cell Biol* 11(5): 329–41.

Wagener, F. A., C. E. Carels, and D. M. Lundvig. 2013. Targeting the redox balance in inflammatory skin conditions. *Int J Mol Sci* 14(5): 9126–67.

Walker, L. J., C. N. Robson, E. Black, D. Gillespie, and I. D. Hickson. 1993. Identification of residues in the human DNA repair enzyme HAP1 (Ref-1) that are essential for redox regulation of Jun DNA binding. *Mol Cell Biol* 13(9): 5370–6.

Xu, L., J. P. Eu, G. Meissner, and J. S. Stamler. 1998. Activation of the cardiac calcium release channel (ryanodine receptor) by poly-S-nitrosylation. *Science* 279(5348): 234–7.

Yellon, D. M., and J. M. Downey. 2003. Preconditioning the myocardium: From cellular physiology to clinical cardiology. *Physiol Rev* 83(4): 1113–51.

Zhang, D., M. Guo, W. Zhang, and X. Y. Lu. 2011. Adiponectin stimulates proliferation of adult hippocampal neural stem/progenitor cells through activation of p38 mitogen-activated protein kinase (p38MAPK)/glycogen synthase kinase 3β (GSK-3β)/β-catenin signaling cascade. *J Biol Chem* 286(52): 44913–20.

# 10 Cerebral Ischemia

*Yannick Béjot and Philippe Garnier*

## CONTENTS

10.1 Introduction .................................................................................................. 185
10.2 Transient Ischemic Attack: A Clinical Model of
       Preconditioning? ........................................................................................... 186
       10.2.1 Transient Ischemic Attack: A Risk Factor of Stroke ....................... 186
       10.2.2 Effect of Transient Ischemic Attack on Outcome after
              Subsequent Ischemic Stroke ............................................................. 186
10.3 Possible Explanation of the Observed Neuroprotection Conferred by
       Transient Ischemic Attack: Cerebral Ischemic Tolerance ........................... 193
       10.3.1 Definition of Ischemic Tolerance ..................................................... 193
       10.3.2 Tolerance Paradigms ........................................................................ 193
       10.3.3 Time Windows for Cerebral Ischemic Tolerance ............................ 194
       10.3.4 Overview of Endogenous Mechanisms of Cerebral Ischemic
              Tolerance ........................................................................................... 195
10.4 Clinical Application of Ischemic Tolerance ................................................ 195
10.5 Conclusion .................................................................................................... 197
References ............................................................................................................... 197

## 10.1 INTRODUCTION

Stroke is a devastating disease that affects more than 15 million individuals and is responsible for approximately 6 million deaths each year worldwide (Strong et al. 2007). Besides its deleterious consequences on survival, stroke resulted in more than 50 million lost disability-adjusted life years around the world in 2005, accounting for 13% of the global burden of disease in the population aged over 60, and 38% of the burden of cardiovascular diseases in the same age group (Strong et al. 2007). Therefore, a major challenge for the coming years will be to develop strategies to reduce the harmful consequences of stroke.

In this chapter, we discuss the interest of ischemic preconditioning in reducing poststroke disability, from clinical observations to pathophysiological support, and its potential applications in terms of therapeutic strategies.

## 10.2 TRANSIENT ISCHEMIC ATTACK: A CLINICAL MODEL OF PRECONDITIONING?

### 10.2.1 Transient Ischemic Attack: A Risk Factor of Stroke

Transient ischemic attack (TIA) is defined by the World Health Organization (WHO) as "rapidly developing clinical signs of focal (at times global) disturbance of cerebral function, lasting less than 24 hours or leading to death with no apparent cause other than that of vascular origin." This is a frequent condition that affects 30–60 individuals per 100,000 annually (Rothwell et al. 2004; Béjot et al. 2007; Cancelli et al. 2011; Fonseca et al. 2012). TIA is considered a risk factor of ischemic stroke. Indeed, 7%–25% patients with ischemic stroke have a history of TIA, and in almost half of them, TIA occurs in the week preceding stroke (Albers et al. 2002; Rothwell and Warlow 2005; Rothwell et al. 2007; Béjot et al. 2008). Similarly, the risk of stroke after TIA is up to 5% in the first 48 hours and can reach 10%–20% at 90 days (Albers et al. 2002). Therefore, TIA is not benign, but should be regarded as a warning sign, and patients must be managed similarly to ischemic stroke patients in terms of diagnostic procedures and secondary prevention.

This assumption is also supported by the fact that more recently, with the emergence of modern imaging techniques, in particular magnetic resonance imaging (MRI), it has been demonstrated that brain ischemic lesions can be found in TIA patients, especially when the duration of symptoms is long (Engelter et al. 1999; Kidwell et al. 1999). In fact, although the time window for the classical TIA definition is large (24 hours), in the great majority of TIA patients symptoms persist for less than 1 hour and even less than 30 minutes (Pessin et al. 1977; Weisberg 1991). Similarly, only 15% of patients with neurological symptoms lasting more than 1 hour recover within the first 24 hours (Levy 1988). These data suggest that there is a clear continuum between TIA and ischemic stroke, and this led the TIA Working Group to redefine TIA as a brief episode of neurological dysfunction caused by focal brain or retinal ischemia, with clinical symptoms typically lasting less than 1 hour, and without evidence of acute infarction (Albers et al. 2002). Although this definition is not applicable for epidemiological research purposes because patients with suspected TIA would require systematic MRI assessment, which is clearly impossible in population-based studies (Béjot and Giroud 2009), it is nonetheless of interest because it underlines the importance of the duration of TIA symptoms on the consequences of ischemic damage.

### 10.2.2 Effect of Transient Ischemic Attack on Outcome after Subsequent Ischemic Stroke

Although patients with TIA have a greatly increased risk of stroke, the concept that previous TIA could have a potential neuroprotective effect in patients with subsequent ischemic stroke recently came to light in clinical research. Indeed, several observational studies aimed to assess whether patients with prior TIA presenting with a subsequent ischemic stroke had a better prognosis than patients without TIA. Among the 12 studies published in the literature (Table 10.1), 8 identified prior TIA as a protective factor for handicap after subsequent ischemic stroke (Weih et al. 1999;

### TABLE 10.1
### Studies That Evaluated the Effect of Prior TIA on Outcome in Patients with Subsequent Ischemic Stroke

| Authors (Year of Publication) | Number of Patients | Study Location | Study Design | Outcomes Measured | Results |
|---|---|---|---|---|---|
| Aboa-Eboulé et al. (2011) | 3015 | Dijon, France | Prospective population-based stroke registry | 1-month and 1-year all-cause mortality | Recent (<4 weeks) and brief (<30 minutes) TIA is associated with a decrease in the risk of severe handicap in patients with subsequent cardioembolic IS or nonlacunar noncardioembolic IS, and a decrease in the risk of death at 1 year in patients with nonlacunar noncardioembolic IS |
| | | | All patients with a first-ever IS in Dijon, France between 1985 and 2008 | Handicap at discharge: none to moderate (m-Rankin 0–3) vs. severe (m-Rankin >3) | |
| | | | IS classified into three subtypes: lacunar IS, cardioembolic IS, nonlacunar noncardioembolic IS | | |
| Della-Morte et al. (2008) | 203 | Naples, Italy | Retrospective analysis of medical charts of consecutive patients aged more than 65 admitted for nonlacunar first-ever IS between 2000 and 2005 | Clinical severity on admission measured by the NIHSS score | No difference in either NIHSS score on admission, or in m-Rankin on admission and at discharge between IS patients with and without history of TIA |
| | | | | Handicap on admission and at discharge measured by the m-Rankin Scale | |

*(Continued)*

**TABLE 10.1 (Continued)**
**Studies That Evaluated the Effect of Prior TIA on Outcome in Patients with Subsequent Ischemic Stroke**

| Authors (Year of Publication) | Number of Patients | Study Location | Study Design | Outcomes Measured | Results |
|---|---|---|---|---|---|
| Fu et al. (2008) | 120 | Shanghai, China | Retrospective analysis of medical charts of consecutive patients admitted for atherothrombotic IS between 2002 and 2007 with and without prior TIA in the same vascular territory | Difference between clinical severity on admission and at discharge measured by the NIHSS score | Lower severity on discharge observed in IS patients with a prodromal ipsilateral TIA lasting <4 minutes, or with two prior TIA events  No difference for IS patients with a TIA for which the interval between the TIA event and IS was 1–7 days |
| Zsuga et al. (2008) | 2874 | Debrecen, Hungary | Prospective analysis of all consecutive patients admitted for hemispheric IS between 2002 and 2004 | Intrahospital mortality | Previous TIA is associated a lower risk of inhospital death |
| Deplanque et al. (2006) | 362 | Lille, France | Prospective analysis of all consecutive patients admitted for hemispheric IS between 2002 and 2004 | Clinical severity on admission measured by the NIHSS score (NIHSS 0–5 vs. >5)  Handicap at 8 days measured by the m-Rankin Scale and the Barthel Index | Previous TIA is associated with lesser severity on admission  No difference in handicap at 8 days between IS patients with and without prior TIA |
| Schaler (2005) | 130 | Berne, Switzerland | Retrospective analysis of medical charts of consecutive IS patients with vascular occlusion treated by intra-arterial thrombolysis with urokinase between 2000 and 2002, with or without prior ispsilateral TIA lasting less than 60 minutes. Exclusion of patients with previous TIA lasting more than 60 minutes. | Clinical severity at discharge measured by the NIHSS score (NIHSS 0–10 vs. >10) | Previous ipsilateral TIA lasting 10–20 minutes or recent (<7 days) is associated with lesser severity and handicap at discharge |

| Study | Location | N | Design | Measures | Findings |
|---|---|---|---|---|---|
| Wegener et al. (2004) | Düsseldorf, Hamburg, Heidelberg, and Berlin, Germany | 65 | Retrospective analysis of consecutive nonlacunar IS patients with an MRI performed within the first 12 hours after stroke onset, between 1997 and 2001 | Handicap at discharge measured by the m-Rankin Scale<br><br>Recanalization rates measured by Thrombolysis in Myocardial Infarction (TIMI) score<br><br>Clinical severity on admission measured by the NIHSS score<br><br>Handicap on admission and at discharge measured by the m-Rankin Scale<br><br>Infarct volume on MRI (perfusion and diffusion-weighted imaging)<br><br>Recanalization rates measured by TIMI score on MR angiography | Better recanalization is observed in IS patients with either prior short (<20 minutes) ipsilateral TIA, 2 or 3 prior ipsilateral TIA, or recent (<7 days) prior ipsilateral TIA<br><br>Lower severity on admission and handicap both on admission and at discharge observed in IS patients with prior TIA<br><br>Smaller final infarct volume in IS patients with prior TIA<br><br>No difference in recanalization rates between IS patients with and without prior TIA |
| Arboix et al. (2004) | Barcelona, Spain | 1753 | Prospective analysis of all consecutive patients admitted for IS (classified as lacunar and nonlacunar IS) between 1986 and 1997 | Handicap at discharge measured by the m-Rankin Scale (score ≤2 vs. >3) | Previous TIA is associated with better functional outcome at discharge in nonlacunar IS patients |

*(Continued)*

**TABLE 10.1 (Continued)**
Studies That Evaluated the Effect of Prior TIA on Outcome in Patients with Subsequent Ischemic Stroke

| Authors (Year of Publication) | Number of Patients | Study Location | Study Design | Outcomes Measured | Results |
|---|---|---|---|---|---|
| Johnston (2004) | 180 | Northern California | Retrospective analysis of all consecutive TIA patients admitted for IS between 1997 and 1998 (n = 1707). Analysis restricted to patients with subsequent IS within 3 months of follow-up (n = 180). | Handicap at discharge measured by the m-Rankin Scale (score ≤2 vs. >3) | No difference in handicap among patients with IS occurring within 1 day, 1–7 days, and 7–90 days after the TIA |
| Sitzer et al. (2004) | 4797 | State of Hesse, Germany | Prospective analysis of all consecutive patients admitted for anterior circulation IS between 1998 and 2000 | Handicap at discharge measured by the m-Rankin Scale (0–1 vs. ≥2) and the Barthel Index (90–100 vs. <90) | Previous TIA is associated with better functional outcome at discharge in IS patients |
| Moncayo et al. (2000) | 2490 | Lausanne, Switzerland | Prospective analysis of all consecutive patients admitted for anterior circulation IS between 1979 and 1997 | Handicap at one-month measured by a five-level scale | Brief (10–20 minutes) TIA is associated with a better functional outcome at discharge in IS patients. Better effect observed for recent TIA (<7 days vs. 7 days to 1 month vs. >1 month) |

| Weih et al. (1999) | 148 | Berlin, Germany | Retrospective analysis of medical charts of consecutive IS patients admitted between 1994 and 1998 | Clinical severity on admission measured by the CNS<br><br>Handicap after at least 3 months of follow-up measured by the m-Rankin Scale (0–1 vs. ≥2) | Previous TIA is associated with lesser severity on admission and better functional outcome at follow-up in IS patients |

CNS, Canadian Neurological Scale; IS, ischemic stroke; MRI, magnetic resonance imaging; NIHSS, National Institute of Health Stroke Scale; TIA, transient ischemic attack; Thrombolysis in Myocardial Infarction (TIMI) score.

Lacunar IS were defined as small (<15 mm in diameter) subcortical infarcts resulting from occlusion of a penetrating artery; cardioembolic IS were defined as infarcts involving an arterial territory because of a thrombus from a cardiac source, mainly atrial fibrillation; nonlacunar noncardioembolic IS were defined as infarcts involving an arterial territory without a cardiac source of embolism.

The modified Rankin Scale (m-Rankin Scale) is a commonly used scale for measuring poststroke disability. The scale runs from 0 (no symptom) to 6 (death). Patients with a score of 0–1 or 0–2 are usually considered to have a good outcome.

The Barthel Index is a scale used to assess disability in patients with various diseases, including stroke. It evaluates a patient's self-care abilities in 10 areas. Total score ranges from 0 to 100. Patients with a score of 90 or more are usually considered to be independent.

Moncayo et al. 2000; Arboix et al. 2004; Sitzer et al. 2004; Wegener et al. 2004; Schaller 2005; Fu et al. 2008; Aboa-Eboulé et al. 2011). In addition, two studies concluded that prior TIA was associated with a reduced risk of either in-hospital or 1-year mortality (Zsuga et al. 2008; Aboa-Eboulé et al. 2011). In contrast, three studies failed to demonstrate any effect of prior TIA on either survival or the functional prognosis in ischemic stroke patients (Johnston 2004; Deplanque et al. 2006; Della-Morte et al. 2008).

These studies are difficult to compare because of methodological differences. First of all, only one study was population based (Aboa-Eboulé et al. 2011), and it is well-known that a selection bias may accompany hospital-based cohorts because they usually include patients with more severe stroke. Second, even though all studies used the standard WHO definition of TIA, except one that excluded patients with symptoms lasting more than 60 minutes, marked differences were observed concerning the duration of TIA, the time between the TIA and ischemic stroke, or the type of ischemic stroke concerned. Finally, the outcomes measured also differed from one study to another in terms of the scales used to quantify handicap or duration of follow-up.

However, despite these methodological limitations, studies which concluded that prior TIA had a positive effect on reducing the risk of handicap after subsequent ischemic stroke presented several common features. First, the beneficial effect of previous TIA was essentially observed in cases of nonlacunar ischemic stroke. One explanation could be that lacunar strokes, which are defined as small (<15 mm in diameter) subcortical infarcts resulting from occlusion of a penetrating artery, are generally associated with a good prognosis, thus making it difficult to demonstrate a statistical difference between patients with and those without prior TIA in such cases. Second, a better functional outcome was observed for recent prestroke TIA (from 1 day to 1 month preceding the ischemic stroke, depending on the study), whereas TIA that occurred long before the stroke did not seem to confer such a beneficial effect. Third, the duration of the TIA also appeared to be an important factor. Actually, in studies that considered this parameter, brief TIA (lasting up to 30 minutes), but not TIA with symptoms lasting longer, was associated with a better prognosis after subsequent ischemic stroke. Finally, an important observation was that the beneficial effect of prior TIA on functional outcome was noted no matter the side of the TIA (ipsilateral or contralateral to the stroke).

From these observations, it could be speculated that prestroke TIA may have a neuroprotective effect on handicap and mortality. The fact that several studies found a lesser degree of clinical severity on admission in patients with than in those without prior TIA supports this hypothesis (Weih et al. 1999; Wegener et al. 2004; Deplanque et al. 2006). Another argument is that in an imaging-based study, smaller infarct volumes were observed in ischemic stroke patients with prior TIA (Wegener et al. 2004). However, beyond this neuroprotective effect, it cannot be excluded that prior TIA may also be beneficial in terms of functional outcome by promoting poststroke recovery.

Although appealing, these results from observational clinical studies must be interpreted with caution. Possible residual confounding factors may have contributed to the observed findings. For example, ischemic stroke patients with a history of TIA are somewhat different from those without TIA in terms of prestroke medication. Indeed, after a TIA, secondary prevention based on antithrombotic agents, antihypertensive drugs, and statins is usually recommended. Consequently, because previous works have suggested that the premorbid use of aspirin or statins may be associated with a better prognosis in ischemic stroke patients (Kalra et al. 2000; Marti-Fabregas et al. 2004; Lampl et al. 2005; Moonis et al. 2005; Deplanque et al. 2006; Karlikaya et al. 2006; Arboix et al. 2012; Aboa-Eboulé et al. 2013; Béjot et al. 2013), the potential protective effect of TIA alone should be interpreted with caution. Of note, few of the studies listed in Table 10.1 took prestroke medication into account in their multivariable analyses (Wegener et al. 2004; Deplanque et al. 2006; Aboa-Eboulé et al. 2011).

## 10.3 POSSIBLE EXPLANATION OF THE OBSERVED NEUROPROTECTION CONFERRED BY TRANSIENT ISCHEMIC ATTACK: CEREBRAL ISCHEMIC TOLERANCE

### 10.3.1 DEFINITION OF ISCHEMIC TOLERANCE

Ischemic tolerance is defined by the phenomenon that takes place when a preconditioning insult near to, but below the threshold of damage induces protection against a subsequent ischemic insult. This phenomenon was initially described in the myocardium (Murry et al. 1986). More recently, evidence that supports cerebral ischemic tolerance as an active process emerged from the literature.

### 10.3.2 TOLERANCE PARADIGMS

Different models have been used to induce ischemic tolerance. Ischemic tolerance models have been the most widely studied, both in vitro and in vivo (Table 10.2). In vivo models have been developed in the rodent brain and rely on two primary models: global ischemia and focal ischemia. Both models have been shown to induce ischemic tolerance, though brief occlusion of the middle cerebral artery probably mimics a better clinical presentation of prodromal TIA with subsequent stroke

**TABLE 10.2**
**Ischemic Tolerance Models Used in the Literature**

*In Vivo Models*

*Global ischemia*
  Permanent occlusion of the vertebral arteries combined with brief occlusion of bilateral common carotid arteries
  Bilateral carotid artery occlusion combined with systemic hypotension

*Focal ischemia*
  Transient occlusion of the middle cerebral artery
  Permanent occlusion of the middle cerebral artery

*In Vitro Models*

  Neuron cultures
  Hippocampal slice preparations

(Kitagawa et al. 1990; Kato et al. 1991; Simon et al. 1993; Ueda and Nowak 2005). Of note, it has been shown that global ischemic preconditioning can also induce ischemic tolerance against focal ischemia and vice versa (Simon et al. 1993). In vitro models, based on the generation of hypoxic conditions via oxygen–glucose deprivation or by using chemical ischemia that models the reversible energy failure occurring during transient ischemia in vivo applied to neuron cultures or hippocampal slice preparations, have also induced ischemic tolerance (Grabb and Choi 1999; Garnier et al. 2003; Tanaka et al. 2004). These models offer a better opportunity to investigate reproducible molecular and biochemical responses before studying them in more complex systems.

Beyond these ischemic tolerance models, other types of preconditioning have recently been developed, suggesting that the options for inducing tolerance are not specific to either the location or the type of injury, which may have major implications for the potential clinical applications (Dirnagl et al. 2009). Remote preconditioning is defined by the preconditioning of one organ or system that leads to the protection of another organ. Hence, according to this concept, ischemia occurring in one organ protects against a subsequent prolonged ischemia in another distant organ. For example, in rats, the induction of short episodes of ischemia in a limb protects against focal ischemia by reducing infarct size (Ren et al. 2008). Cross-conditioning refers to prior exposure to nonischemic noxious stress that confers ischemic tolerance. Several cross-conditioning stimuli have been associated with the induction of cerebral ischemic tolerance, including temperature changes (both hyperthermia [Nowak et al. 1990] and hypothermia [Yunoki et al. 2002]), cortical spreading depression (Matsushima et al. 1996), inflammation (Ohtsuki et al. 1996; Liu et al. 2000; Furuya et al. 2005), pharmacological interventions (Ohtsuki et al. 1992; Kis et al. 2003), epilepsy (Plamondon et al. 1999), and traumatic brain injury (Otori et al. 2004). In addition, it has been shown that some anesthetics can trigger preconditioning (Kitano et al. 2007; Wei et al. 2007).

### 10.3.3 Time Windows for Cerebral Ischemic Tolerance

As initially demonstrated for the myocardium, two time windows have been identified for the cerebral ischemic tolerance conferred by preconditioning: a rapid window and a delayed window. The rapid window of ischemic tolerance appears within the first minutes after the preconditioning stimulus and confers short, transient protection, whereas the delayed window usually starts 24 hours after preconditioning, and lasts several days with a peak observed at 48–72 hours (Dirnagl et al. 2003). Of note, delayed preconditioning is considered more robust for the brain than that conferred early and is generally associated with the requirement of de novo protein synthesis (Kirino 2002). Another important consideration is that after a delay of 7–14 days, no ischemic tolerance is observed. This result is of interest when considering the observations from clinical observational studies on TIA. Indeed, only a recent TIA confers protection against subsequent ischemic stroke, which seems to confirm the importance of the time window for cerebral ischemic tolerance.

More recently, the concept of postconditioning has been described. It has been demonstrated that sublethal stimuli or insults performed immediately or up to two days after the cerebral ischemia can reduce both neuron death and the final infarct volume and improve neurological recovery in focal, global, and in vitro models. Several stimuli, including intermittent occlusion or reperfusion, brief ischemia or hypoxia, and the application of a neurotoxic agent, have been described (Zhao 2009).

### 10.3.4 Overview of Endogenous Mechanisms of Cerebral Ischemic Tolerance

Numerous studies, based on both in vivo and in vitro models, have investigated the pathophysiological processes involved in preconditioning and cerebral ischemic tolerance. However, the findings are still not fully understood, and a unified view on the phenomenon is still impossible, despite many review articles published on this topic (Kirino 2002; Dirnagl et al. 2009). Schematically, it has been suggested that potential mechanisms could be divided into at least two categories: a cellular defense function against ischemia and a cellular stress response and the synthesis of stress proteins (Kirino 2002).

Historically, the cellular stress response was shown to be essential to ensure cell survival. This phenomenon relies on the synthesis of various stress proteins belonging to the family of heat shock proteins (Glazier et al. 1994; Kato et al. 1994). These proteins work as cellular "chaperones" by unfolding misfolded cellular proteins so as to cope with their accumulation and aggregation, and to fight against apoptosis.

In addition, different mechanisms of various natures, including protein phosphorylation, nitric oxide production, immediate early gene expression, inhibition of apoptosis, production of cytokines, inflammatory-mediated protection, glial cell proliferation, activation of antioxidant defenses, preservation of energy content, reduction of organelle stress, posttranscriptional regulation through the alteration of cerebral miRNA, and genomic reprogramming, have been shown to be implicated in cellular defenses. Concerning the latter, the role of hypoxia-inducible factor-1, which transactivates about 100 genes implicated in oxygen transport, angiogenesis, vasomotor control, cell survival, pH regulation, and energy metabolism, has been underlined (Dirnagl et al. 2009). Whatever the exact mechanisms involved, the cascade of events leads to the reinforcement of factors that contribute to cell survival and inhibit cell death.

## 10.4 CLINICAL APPLICATION OF ISCHEMIC TOLERANCE

On the basis of observations of the potential beneficial effect of prior TIA on recovery after subsequent ischemic stroke, and the animal and in vitro models, the reality of a preconditioning phenomenon to induce brain protection against ischemia is becoming more and more evident. The development of therapeutic strategies that aim to stimulate preconditioning thus appears to be promising and could be an exciting avenue for research on stroke treatment. However, there are still many questions to resolve and difficulties to overcome before translating this knowledge into effective treatment for patients.

First of all, as ischemic tolerance induced by preconditioning stimuli in animal models of stroke has been shown to confer protection over a short time window and because ischemic stroke is a sudden and unpredictable event, anticipating preconditioning strategies before stroke onset is totally impossible. Hence, clinical trials should enroll a large number of patients because of the overall low rates of stroke incidence in the general population. As a consequence, the potential expected beneficial effects would be outweighed by the harmful effects observed in individuals exposed to a treatment that they do not actually need. An alternative method would be to identify individuals with a high risk of ischemic stroke so as to reduce sample size. Up to now, no such clinical trials on preconditioning have been conducted. However, a small number of trials on postconditioning strategies, in which the conditioning treatment is given just after ischemic stroke onset, are ongoing (Mergenthaler and Dirnagl 2011). An important additional remark is that, by analogy, conventional neuroprotection trials, in which neuroprotective pharmacological or nonpharmacological interventions are also provided in the first hours following the ischemic event, have failed to show any efficacy (Ginsberg 2008).

A second and major issue to be dealt with before designing and implementing clinical trials is that ischemic preconditioning may necessarily involve brain damage that might have been missed in previous animal studies that were short in duration with limited periods of follow-up (Sommer 2008). This brain damage could lead to important deleterious consequences in terms of functional and cognitive impairment. In line with this assumption, a recent population-based clinical study showed that although a TIA before subsequent ischemic stroke was associated with a better functional and vital prognosis, it was also accompanied by an increased risk of early poststroke dementia (Jacquin et al. 2012). In addition, whether or not preconditioning mimetics are safe is still unclear. For these reasons, it has been proposed that preconditioning effectors could be a safer way to induce cerebral ischemic tolerance than preconditioning mimetics or ischemic preconditioning itself (Dirnagl et al. 2009).

For these reasons, the development of preconditioning strategies appears to be very interesting in situations of anticipated high risk of ischemic stroke. An illustrative example is the use of remote ischemic preconditioning for cerebral and cardiac protection in patients undergoing carotid endarterectomy. In a pilot study in which patients were preconditioned using 10 minutes of lower limb ischemia–reperfusion, remote ischemic preconditioning appeared to be safe, and further large clinical trials are now required to determine whether this procedure could confer clinical benefits (Walsh et al. 2010). Another example is the interest of hyperbaric oxygen preconditioning. Several animal studies have shown that this procedure may be associated with the induction of cerebral ischemic tolerance, and may be as effective as hypoxic models, with the advantage of being safer (Schäbitz et al. 2004; Freiberger et al. 2006). In a randomized controlled clinical trial, repeated hyperbaric oxygen preconditioning for 70 minutes per day for the 5 consecutive days before on-pump coronary artery bypass graft surgery was associated with a reduced length of stay in the intensive care unit and a decrease in the use of inotropic drugs (Li et al. 2011). Interestingly, in patients who underwent hyperbaric oxygen preconditioning, the serum levels of the S100B protein measured 1 hour after cross-clamp removal and of the neuron-specific enolase measured from 6 hours to 24 hours postoperatively, which are two potential neurobiochemical markers of brain injury, were significantly lower than those observed in the control group. Of note, a similar result was found for troponin I. These findings suggested that repeated hyperbaric oxygen might promote both myocardial and cerebral protection. Larger trials are required to confirm this assumption. Similarly, another study that aims to evaluate the neuroprotective effect of electroacupuncture pretreatment in patients undergoing cardiac surgery is currently recruiting patients (clinical trial identifier: NCT01020266).

Finally, in the general population, easy-to-implement preventive interventions may promote cerebral preconditioning. Indeed, several animal studies have shown that physical exercise may have a neuroprotective role in case of subsequent cerebral ischemia by involving various mechanisms, although there is still controversy about both the intensity and the duration of the exercise that are needed to induce brain ischemic tolerance (Zhang et al. 2011). However, the impact of regular physical activity in humans remains to be investigated in randomized controlled trials.

## 10.5 CONCLUSION

Although our knowledge is still incomplete, preconditioning by the induction of ischemic tolerance could be a promising avenue of research to reduce disability after ischemic stroke. High-quality preclinical studies are needed to provide better understanding of the endogenous mechanisms of ischemic tolerance, and to guide the design and the implementation of future randomized controlled trials.

## REFERENCES

Aboa-Eboulé C, Béjot Y, Osseby GV, Rouaud O, Binquet C, Marie C, Cottin Y, Giroud M, Bonithon-Kopp C. 2011. Influence of prior transient ischaemic attack on stroke prognosis. *J Neurol Neurosurg Psychiatry* 82: 993–1000.

Aboa-Eboulé C, Binquet C, Jacquin A, Hervieu M, Bonithon-Kopp C, Durier J, Giroud M, Béjot Y. 2013. Effect of previous statin therapy on severity and outcome in ischemic stroke patients: A population-based study. *J Neurol* 260: 30–7.

Albers GW, Caplan LR, Easton JD, Fayad PB, Mohr JP, Saver JL, Sherman DG; TIA Working Group. 2002. Transient ischemic attack—Proposal for a new definition. *N Engl J Med* 347: 1713–6.

Arboix A, Cabeza N, García-Eroles L, Massons J, Oliveres M, Targa C, Balcells M. 2004. Relevance of transient ischemic attack to early neurological recovery after nonlacunar ischemic stroke. *Cerebrovasc Dis* 18: 304–11.

Arboix A, Garcia-Eroles L, Oliveres M, Targa C, Balcells M, Massons J. 2012. Pretreatment with statins improves early outcome in patients with first-ever ischaemic stroke: A pleiotropic effect of statins or a beneficial effect of hypercholesterolemia? *BMC Neurol* 10: 47.

Béjot Y, Aboa-Eboulé C, de Maistre E, Jacquin A, Troisgros O, Hervieu M, Osseby GV, Rouaud O, Giroud M. 2013. Prestroke antiplatelet therapy and early prognosis in stroke patients: The Dijon Stroke Registry. *Eur J Neurol* 20: 879–90.

Béjot Y, Caillier M, Ben Salem D, Couvreur G, Rouaud O, Osseby GV, Durier J, Marie C, Moreau T, Giroud M. 2008. Ischemic stroke subtypes and associated risk factors: A French population-based study. *J Neurol Neurosurg Psychiatry* 79: 1344–8.

Béjot Y, Giroud M. 2009. Epidemiological implications of the new definition of transient ischemic attack. *Neuroepidemiology* 33: 358.

Béjot Y, Rouaud O, Benatru I, Durier J, Caillier M, Couvreur G, Fromont A et al. 2007. Trends in the incidence of transient ischemic attacks, premorbid risk factors and the use of preventive treatments in the population of Dijon, France from 1985 to 2004. *Cerebrovasc Dis* 23: 126–31.

Cancelli I, Janes F, Gigli GL, Perelli A, Zanchettin B, Canal G, D'Anna L, Russo V, Barbone F, Valente M. 2011. Incidence of transient ischemic attack and early stroke risk: Validation of the ABCD2 score in an Italian population-based study. *Stroke* 42: 2751–7.

Della-Morte D, Abete P, Gallucci F, Scaglione A, D'Ambrosio D, Gargiulo G, De Rosa G et al. 2008. Transient ischemic attack before nonlacunar ischemic stroke in the elderly. *J Stroke Cerebrovasc Dis* 17: 257–62.

Deplanque D, Masse I, Lefebvre C, Libersa C, Leys D, Bordet R. 2006. Prior TIA, lipid-lowering drug use, and physical activity decrease ischemic stroke severity. *Neurology* 67: 1403–10.

Dirnagl U, Becker K, Meisel A. 2009. Preconditioning and tolerance against cerebral ischaemia: From experimental strategies to clinical use. *Lancet Neurol* 8: 398–412.

Dirnagl U, Simon RP, Hallenbeck JM. 2003. Ischemic tolerance and endogenous neuroprotection. *Trends Neurosci* 26: 248–54.

Engelter ST, Provenzale JM, Petrella JR, Alberts MJ. 1999. Diffusion MR imaging and transient ischemic attacks. *Stroke* 30: 2762–3.

Fonseca PG, Weiss PA, Harger R, Moro CH, Longo AL, Gonçalves AR, Whiteley WN, Cabral NL. 2012. Transient ischemic attack incidence in Joinville, Brazil, 2010: A population-based study. *Stroke* 43: 1159–62.

Freiberger JJ, Suliman HB, Sheng H, McAdoo J, Piantadosi CA, Warner DS. 2006. A comparison of hyperbaric oxygen versus hypoxic cerebral preconditioning in neonatal rats. *Brain Res* 1075: 213–22.

Fu Y, Sun JL, Ma JF, Geng X, Sun J, Liu JR, Song YJ, Chen SD. 2008. The neuroprotection of prodromal transient ischaemic attack on cerebral infarction. *Eur J Neurol* 15: 797–801.

Furuya K, Zhu L, Kawahara N, Abe O, Kirino T. 2005. Differences in infarct evolution between lipopolysaccharide-induced tolerant and nontolerant conditions to focal cerebral ischemia. *J Neurosurg* 103: 715–23.

Garnier P, Ying W, Swanson RA. 2003. Ischemic preconditioning by caspase cleavage of poly(ADP-ribose) polymerase-1. *J Neurosci* 23: 7967–73.

Ginsberg MD. 2008. Neuroprotection for ischemic stroke: Past, present and future. *Neuropharmacology* 55: 363–89.

Glazier SS, O'Rourke DM, Graham DI, Welsh FA. 1994. Induction of ischemic tolerance following brief focal ischemia in rat brain. *J Cereb Blood Flow Metab* 14: 545–53.

Grabb MC, Choi DW. 1999. Ischemic tolerance in murine cortical cell culture: Critical role for NMDA receptors. *J Neurosci* 19: 1657–62.

Jacquin A, Aboa-Eboulé C, Rouaud O, Osseby GV, Binquet C, Durier J, Moreau T, Bonithon-Kopp C, Giroud M, Béjot Y. 2012. Prior transient ischemic attack and dementia after subsequent ischemic stroke. *Alzheimer Dis Assoc Disord* 26: 307–13.

Johnston SC. 2004. Ischemic preconditioning from transient ischemic attacks? Data from the Northern California TIA Study. *Stroke* 35: 2680–2.

Kalra L, Perez I, Smithard DG, Sulch D. 2000. Does prior use of aspirin affect outcome in ischemic stroke? *Am J Med* 108: 205–9.

Karlikaya G, Varlbas F, Demirkaya M, Orken C, Tireli H. 2006. Does prior aspirin use reduce stroke mortality? *Neurologist* 12: 263–7.

Kato H, Liu Y, Araki T, Kogure K. 1991. Temporal profile of the effects of pretreatment with brief cerebral ischemia on the neuronal damage following secondary ischemic insult in the gerbil: Cumulative damage and protective effects. *Brain Res* 553: 238–42.

Kato H, Liu Y, Kogure K, Kato K. 1994. Induction of 27-kDa heat shock protein following cerebral ischemia in a rat model of ischemic tolerance. *Brain Res* 634: 235–44.

Kidwell CS, Alger JR, Di Salle F, Starkman S, Villablanca P, Bentson J, Saver JL. 1999. Diffusion MRI in patients with transient ischemic attacks. *Stroke* 30: 1174–80.

Kirino T. Ischemic tolerance. 2002. *J Cereb Blood Flow Metab* 22: 1283–96.

Kis B, Rajapakse NC, Snipes JA, Nagy K, Horiguchi T, Busija DW. 2003. Diazoxide induces delayed pre-conditioning in cultured rat cortical neurons. *J Neurochem* 87: 969–80.

Kitagawa K, Matsumoto M, Tagaya M, Kuwabara K, Hata R, Handa N, Fukunaga R, Kimura K, Kamada T. 1990. "Ischemic tolerance" phenomenon found in the brain. *Brain Res* 528: 21–4.

Kitano H, Young JM, Cheng J, Wang L, Hurn PD, Murphy SJ. 2007. Gender-specific response to isoflurane preconditioning in focal cerebral ischemia. *J Cereb Blood Flow Metab* 27: 1377–86.

Lampl Y, Boaz M, Sadeh M. 2005. The significance of prestroke aspirin dosage in fatal outcome of acute stroke. *Clin Neuropharmacol* 28: 55–9.

Levy DE. 1988. How transient are transient ischemic attacks? *Neurology* 38: 674–7.

Li Y, Dong H, Chen M, Liu J, Yang L, Chen S, Xiong L. 2011. Preconditioning with repeated hyperbaric oxygen induces myocardial and cerebral protection in patients undergoing coronary artery bypass graft surgery: A prospective, randomized, controlled clinical trial. *J Cardiothorac Vasc Anesth* 25: 908–16.

Liu J, Ginis I, Spatz M, Hallenbeck JM. 2000. Hypoxic preconditioning protects cultured neurons against hypoxic stress via TNF-alpha and ceramide. *Am J Physiol Cell Physiol* 278: C144–53.

Marti-Fabregas J, Gomis M, Arboix A, Aleu A, Pagonabarraga J, Belvis R, Cocho D et al. 2004. Favorable outcome of ischemic stroke in patients pretreated with statins. *Stroke* 35: 1117–21.

Matsushima K, Hogan MJ, Hakim AM. 1996. Cortical spreading depression protects against subsequent focal cerebral ischemia in rats. *J Cereb Blood Flow Metab* 16: 221–6.

Mergenthaler P, Dirnagl U. 2011. Protective conditioning of the brain: expressway or roadblock? *J Physiol* 589: 4147–55.

Moncayo J, de Freitas GR, Bogousslavsky J, Altieri M, van Melle G. 2000. Do transient ischemic attacks have a neuroprotective effect? *Neurology* 54: 2089–94.

Moonis M, Kane K, Schwiderski U, Sandage BW, Fisher M. 2005. HMG-CoA reductase inhibitors improve acute ischemic stroke outcome. *Stroke* 36: 1298–1300.

Murry CE, Jennings RB, Reimer KA. 1986. Preconditioning with ischemia: A delay of lethal cell injury in ischemic myocardium. *Circulation* 74: 1124–36.

Nowak TS Jr, Bond U, Schlesinger MJ. 1990. Heat shock RNA levels in brain and other tissues after hyperthermia and transient ischemia. *J Neurochem* 54: 451–8.

Ohtsuki T, Matsumoto M, Kuwabara K, Kitagawa K, Suzuki K, Taniguchi N, Kamada T. 1992. Influence of oxidative stress on induced tolerance to ischemia in gerbil hippocampal neurons. *Brain Res* 599: 246–52.

Ohtsuki T, Ruetzler CA, Tasaki K, Hallenbeck JM. 1996. Interleukin-1 mediates induction of tolerance to global ischemia in gerbil hippocampal CA1 neurons. *J Cereb Blood Flow Metab* 16: 1137–42.

Otori T, Friedland JC, Sinson G, McIntosh TK, Raghupathi R, Welsh FA. 2004. Traumatic brain injury elevates glycogen and induces tolerance to ischemia in rat brain. *J Neurotrauma* 21: 707–18.

Pessin MS, Duncan GW, Mohr JP, Poskanzer DC. 1977. Clinical and angiographic features of carotid transient ischemic attacks. *N Engl J Med* 296: 358–62.

Plamondon H, Blondeau N, Heurteaux C, Lazdunski M. 1999. Mutually protective actions of kainic acid epileptic preconditioning and sublethal global ischemia on hippocampal neuronal death: Involvement of adenosine A1 receptors and K(ATP) channels. *J Cereb Blood Flow Metab* 19: 1296–308.

Ren C, Gao X, Steinberg GK, Zhao H. 2008. Limb remote-preconditioning protects against focal ischemia in rats and contradicts the dogma of therapeutic time windows for preconditioning. *Neuroscience* 151: 1099–1103.

Rothwell PM, Coull AJ, Giles MF, Howard SC, Silver LE, Bull LM, Gutnikov SA et al. 2004. Change in stroke incidence, mortality, case-fatality, severity, and risk factors in Oxfordshire, UK from 1981 to 2004 (Oxford Vascular Study). *Lancet* 363: 1925–33.

Rothwell PM, Giles MF, Chandratheva A, Marquardt L, Geraghty O, Redgrave JN, Lovelock CE et al. 2007. Effect of urgent treatment of transient ischaemic attack and minor stroke on early recurrent stroke (EXPRESS study): A prospective population-based sequential comparison. *Lancet* 370: 1432–42.

Rothwell PM, Warlow CP. 2005. Timing of TIAs preceding stroke: Time widow for prevention is very short. *Neurology* 64: 817–20.

Schäbitz WR, Schade H, Heiland S, Kollmar R, Bardutzky J, Henninger N, Müller H et al. 2004. Neuroprotection by hyperbaric oxygenation after experimental focal cerebral ischemia monitored by MRI. *Stroke* 35: 1175–9.

Schaller B. 2005. Ischemic preconditioning as induction of ischemic tolerance after transient ischemic attacks in human brain: Its clinical relevance. *Neurosci Lett* 377: 206–11.

Simon RP, Niiro M, Gwinn R. 1993. Prior ischemic stress protects against experimental stroke. *Neurosci Lett* 163: 135–7.

Sitzer M, Foerch C, Neumann-Haefelin T, Steinmetz H, Misselwitz B, Kugler C, Back T. 2004. Transient ischaemic attack preceding anterior circulation infarction is independently associated with favourable outcome. *J Neurol Neurosurg Psychiatry* 75: 659–60.

Sommer C. 2008. Ischemic preconditioning: Postischemic structural changes in the brain. *J Neuropathol Exp Neurol* 67: 85–92.

Strong K, Mathers C, Bonita R. 2007. Preventing stroke: Saving lives around the world. *Lancet Neurol* 6: 182–7.

Tanaka H, Yokota H, Jover T, Cappuccio I, Calderone A, Simionescu M, Bennett MV, Zukin RS. 2004. Ischemic preconditioning: Neuronal survival in the face of caspase-3 activation. *J Neurosci* 24: 2750–9.

Ueda M, Nowak TS, Jr. 2005. Protective preconditioning by transient global ischemia in the rat: Components of delayed injury progression and lasting protection distinguished by comparisons of depolarization thresholds for cell loss at long survival times. *J Cereb Blood Flow Metab* 25: 949–58.

Walsh SR, Nouraei SA, Tang TY, Sadat U, Carpenter RH, Gaunt ME. 2010. Remote ischemic preconditioning for cerebral and cardiac protection during carotid endarterectomy: Results from a pilot randomized clinical trial. *Vasc Endovascular Surg* 44: 434–9.

Wegener S, Gottschalk B, Jovanovic V, Knab R, Fiebach JB, Schellinger PD, Kucinski T et al. 2004. Transient ischemic attacks before ischemic stroke: Preconditioning the human brain? A multicenter magnetic resonance imaging study. *Stroke* 35: 616–21.

Wei H, Liang G, Yang H. 2007. Isoflurane preconditioning inhibited isoflurane-induced neurotoxicity. *Neurosci Lett* 425: 59–62.

Weih M, Kallenberg K, Bergk A, Dirnagl U, Harms L, Wernecke KD, Einhäupl KM. 1999. Attenuated stroke severity after prodromal TIA: A role for ischemic tolerance in the brain? *Stroke* 30: 1851–4.

Weisberg LA. 1991. Clinical characteristics of transient ischemic attacks in black patients. *Neurology* 41: 1410–4.

Yunoki M, Nishio S, Ukita N, Anzivino MJ, Lee KS. 2002. Characteristics of hypothermic preconditioning influencing the induction of delayed ischemic tolerance. *J Neurosurg* 97: 650–7.

Zhang F, Wu Y, Jia J. 2011. Exercise preconditioning and brain ischemic tolerance. *Neuroscience* 177: 170–6.

Zhao H. 2009. Ischemic postconditioning as a novel avenue to protect against brain injury after stroke. *J Cereb Blood Flow Metab* 29: 873–85.

Zsuga J, Gesztelyi R, Juhasz B, Kemeny-Beke A, Fekete I, Csiba L, Bereczki D. 2008. Prior transient ischemic attack is independently associated with lesser in-hospital case fatality in acute stroke. *Psychiatry Clin Neurosci* 62: 705–12.

# 11 Optimal Stress, Psychological Resilience, and the Sandpile Model

*Martha Stark*

## CONTENTS

11.1 Optimal Frustration and Optimal Challenge ............................................. 202
11.2 Environmental Stressors ........................................................................... 203
11.3 Stressful Stuff Happens............................................................................. 204
11.4 Web of Life................................................................................................ 204
11.5 Optimization of Flow ................................................................................ 205
11.6 Hormetic Dose-Response Relationship..................................................... 206
11.7 MindBodyMatrix: A Chaotic System ....................................................... 207
11.8 Healing Cycles of Defensive Collapse and Adaptive Recovery................ 208
11.9 Self-Organizing Systems Resist Perturbation ........................................... 209
11.10 Challenge versus Support of the Patient's Defensive Structures............... 209
11.11 Traumatic versus Optimal Stress .............................................................. 210
11.12 Creation of Cognitive Dissonance............................................................. 211
11.13 From Gain to Pain ..................................................................................... 212
11.14 Relentless Hope: The Refusal to Grieve.................................................... 213
11.15 From Relentless Hope to Sober Acceptance ............................................. 213
11.16 Grieving Process ....................................................................................... 214
11.17 Belated Processing of Unmastered Experience ........................................ 216
11.18 Sandpile Model and the Paradoxical Impact of Stress.............................. 216
11.19 Difference between a Poison and a Medication ....................................... 217
11.20 Resilience: Continuous Adjustment to Instability .................................... 218
11.21 Optimization of Mental Potential.............................................................. 219
11.22 Dis-Order and Dis-Ease ............................................................................ 220
11.23 Conclusion ................................................................................................ 220
References............................................................................................................. 221

> The world breaks everyone and afterward many are strong at the broken places.
>
> **Hemingway, 1929**

As a psychoanalyst and holistic psychiatrist, I have long been interested in understanding how exactly it is that my patients get better—in other words, what exactly it is that allows them to reverse underlying dysfunctionality so that they can advance from illness to wellness. Over the course of the years, I have come to appreciate something that is at once both completely obvious and quite profound, namely, that it will be input from the outside and the patient's capacity to process, integrate, and adapt to its impact that will ultimately enable the patient to get better. In other words, there must be both environmental input (which will constitute the dose) and capacity of the system to manage that input (which will constitute the response).

As it happens, however, more often than not it will actually be *stressful input from the outside* and the patient's *capacity to process, integrate, and adapt to the impact of this stress* that will provoke recovery. In other words, more often than not it will be not so much gratification but rather frustration against a backdrop of gratification—to which the psychodynamic literature refers as *optimal frustration* (Kohut 1971)—that will provide the therapeutic leverage needed to provoke, after initial disruption, eventual revitalization of the system at a higher level of functionality and adaptive capacity. Expressed in somewhat different terms, more often than not it will be not so much support but rather challenge against a backdrop of support (i.e., optimal challenge) that, by virtue of the anxiety thereby elicited, will then provide the impetus needed to transform dysfunctional defense into more functional adaptation.

Admittedly, defenses have a component that is adaptive and adaptations have a component that is defensive. But defenses involve lower-level processing (and therefore a lower level of complexity), whereas adaptations involve higher-level processing (and therefore a higher level of complexity). By the same token, defenses are automatic and are mobilized almost immediately, whereas adaptations emerge only over time and are more evolved. Defenses are reflexive; adaptations are more reflective. Psychoanalysts speak of the *need to defend* and the *capacity to adapt*; so, too, they speak of *defensive need* and *adaptive capacity* to highlight the distinction between unhealthy and healthy (Stark 1994a,b).

## 11.1 OPTIMAL FRUSTRATION AND OPTIMAL CHALLENGE

If the therapist offers only gratification and support—and neither frustration of primitive desire nor challenge to the patient's maladaptive patterns of acting, reacting, and interacting—then there will be nothing that needs to be mastered and therefore little incentive for transformation and growth. Therapeutic input, however, that provides an optimal level of stress and anxiety (in the form of interventions that offer just the right balance of frustration and gratification and just the right combination of challenge and support) can ultimately provoke not only reversal of underlying dysfunctionality but also optimization of functionality by tapping into the patient's innate striving toward health and inborn ability to self-correct in the face of environmental perturbation.

It is to highlight the clinical usefulness of optimal stress to provoke healing and revitalization of resilience that I am so boldly advancing the idea that if the patient is provided with only gratification and support, then there will be insufficient impetus for transformation and growth. Let me now qualify that rather bold assertion by saying that, in most instances, gratification and support alone will not be enough to reverse chronic dysfunctionality, forestall age-related decline, or promote optimal health. In other words, direct support is necessary but not always sufficient. Therefore a more accurate rendering would be the following: Usually reversal of underlying dysfunctionality and fine-tuning of functionality require therapeutic input that provides not only direct support but also optimal challenge. Whereas the therapeutic effectiveness of direct support is intuitively obvious, the therapeutic effectiveness of a combination of support and challenge is more counterintuitive. In truth, direct support and optimal challenge work in concert. Whereas optimal challenge provokes recovery and revitalization by prompting the system to adapt, direct support facilitates healing by reinforcing the system's underlying adaptability and restoring its adaptation reserves—thereby honing the system's ability to adapt to, and benefit from, ongoing stressful environmental input.

As I will later hope to demonstrate, support—in the form of lightening the system's load (to correct for toxicities) and replenishing the system's reserves (to correct for deficiencies)—will facilitate the flow of information and energy throughout the system's vast network of interconnected channels. The more seamless that flow, the better able will the system be to process, integrate, and adapt to the impact of the myriad of environmental stressors to which it is being continuously exposed. The more effective the processing, integrating, and adapting, the more resilient will the system be and the better able not only to manage the impact of stress but also to benefit from the stress, by using that stressful input to provide the impetus for reversal of underlying dysfunctionality and revitalization of resilience.

## 11.2 ENVIRONMENTAL STRESSORS

Of note is the fact that I am here using the term *environment* to encompass the living and the nonliving; the external and the internal; and the psychological, the physiological, and the energetic. Environmental stressors, therefore, may take the form of such challenges as an anxiety-provoking but ultimately health-promoting interpretation; the accumulation of metabolic waste products in the body; an interpersonal disappointment; exposure to the aluminum found in antiperspirants or the mercury found in dental amalgams; psychological, physical, or sexual abuse; contact with a carcinogenic pesticide (such as the insect repellent DEET); the loss of a parent; or the ingestion of endocrine disruptors (such as the phthalates found in plastic bottles).

In my own psychoanalytic writings, I have found it clinically useful to think in terms of psychological stressors as involving both the *presence of bad* and the *absence of good* in the early-on parent–child relationship; both *too much that was bad* between parent and child and *not enough that was good*; trauma and abuse on the one hand, deprivation and neglect on the other—toxicities and deficiencies that were internally recorded and structuralized as psychic scars in the developing mind of the young child. By the same token, I have found it clinically useful to think, more generally, in terms of environmental stressors as involving both *too much bad* (environmental toxicities) and *not enough good* (environmental deficiencies). Too much rejection by the caregiver, not enough love and support. Too much oxidative stress from electron-scavenging free radicals, not enough electron-donating (and therefore neutralizing) antioxidants. Too much criticism, not enough acceptance. Too many antibiotics altering the balance of healthy flora in the gastrointestinal tract, not enough probiotics (or beneficial bacteria) to restore that balance. Too many anxiety-provoking interpretations, not enough anxiety-assuaging empathic interventions. In other words, both *too much bad* and *too little good* are environmental stressors, the cumulative impact of which will potentially compromise the health and vitality of the individual.

## 11.3 STRESSFUL STUFF HAPPENS

Stressful stuff happens. But it will be how well the individual is able to process and integrate its impact—psychologically, physiologically, and energetically—that will make it either a growth-disrupting event (when the impact of the stress is simply too much to be processed, integrated, and adapted to) or a growth-promoting opportunity (when the impact of the stress, although initially destabilizing, is ultimately able to provoke restabilization of the system at a more evolved level of functionality and mature capacity). As we will see, the villain in our piece will be traumatic stress—here defined as stressful input that overwhelms and disrupts because it is simply too much to be handled; the heroine in our piece will be optimal stress—here defined as (ongoing) stressful input that ultimately strengthens by triggering healing cycles of first disruption and then repair, first destabilization and then restabilization at ever-higher levels of complexity, integration, and adaptive capacity.

In essence, whether the primary target is mind or body and the clinical manifestation therefore psychiatric or medical, the critical issue will be the ability of the living system (to which I will also be referring as the MindBodyMatrix) to manage stress through adaptation. The MindBodyMatrix is a term that I have coined to highlight the complex interdependence of mind and body; it also reflects a keen appreciation for the intimate and precise relationship between the health and vitality of the mind and that of the body.

## 11.4 WEB OF LIFE

Whether described as the *ground regulation system* (Pischinger 1991), the *biological terrain* (Bernard 1865, 1957), the *connective tissue matrix* (Oschman 2000), the *extracellular matrix* (Rea and Patel 2010), the *living matrix* (Sheldrake 1988; McTaggart 2001; Braden 2007; Lipton 2009), or the MindBodyMatrix (Stark 2008, 2012), the living system is an intricate web of interdependent living tissue that extends from the surface of the body to its innermost recesses, ultimately penetrating every cell in the body. This vast regulatory network of complex and interwoven pathways allows for the high-speed transmission of regulatory information and vibratory energy throughout the body's expanse and is therefore ultimately responsible for the body's intrinsic ability to regulate and repair itself.

It is a ground regulation system that includes the connective tissue or extracellular matrix, the cytoskeleton or cellular matrix, the nuclear matrix, and the molecular structures linking these matrices (Rea and Patel 2010). More specifically, this web of life comprises a continuous meshwork of connective tissue fibers (glycoproteins that are either structural, such as collagen and elastin, or cross-linking, such as fibronectin and laminin) dispersed throughout an amorphous ground substance (a colloidal gel consisting primarily of large sugar–protein complexes). These sugar–protein complexes—proteoglycan/glycosaminoglycan (PG/GAG) macromolecules—contain a positively charged, core protein backbone (PG) to which negatively charged, highly polymerized glycan side chains (GAGs) are attached. These GAG side chains stand out from the PG backbone in a bristle-like fashion and are surrounded by, and tightly bound to, polarized water molecules.

In the language of solid-state physics, the branch of physics that studies highly ordered systems (such as crystalline structures), we now understand this ground regulation system to be a liquid crystal with semiconducting properties (Oschman 2000). In other words, because the living matrix is a highly ordered array of molecules (primarily glycoproteins, PG/GAGs, and water) closely packed and tightly organized in a crystal-like lattice structure, it has the semiconducting properties of a crystal and, as such, allows for the near-instantaneous flow of regulatory information and vibratory energy throughout the entire fabric of the body. It is this crystallinity that enables the living matrix to conduct *energy quanta* or *biophotons* (i.e., units of information and energy) at almost the speed of light, transmitting both information (in much the way that a phone line transmits information) and energy (in much the way that the wire to a toaster transmits energy). This lightning-fast dissemination of information and energy throughout the body's expanse makes the living matrix the ideal vehicle for the maintenance of homeostatic balance within the body.

When the concept of homeostasis was first introduced by Cannon (1932), his focus was on regulatory organ systems, most especially the autonomic nervous and endocrine systems. In a little-known but brilliant volume first published more than 20 years ago and entitled *Matrix and Matrix Regulation: Basis for a Holistic Theory in Medicine*, however, Pischinger (1991) presents a fairly compelling case for the idea that maintenance of balance within the body is accomplished by way of the ground regulation system. His contention is that nerve endings and capillary networks are embedded within this matrix; but nowhere do they have direct contact with functioning cells. And so it is not simply regulatory organ systems but the ground regulation system itself that makes possible the maintenance of balance within the body. It is therefore crucial that this matrix operates as effectively as possible because the health and vitality of the body's cells are entirely dependent on it.

## 11.5  OPTIMIZATION OF FLOW

To optimize the flow of information and energy throughout the living system, it is essential that the body be kept as uncongested, nutrient-rich, well-oxygenated, alkaline, electron-rich, negatively charged, energetically unblocked, well-balanced, relaxed, well-rested, and structurally aligned as possible. Of course, regular exercise will also be critical for maintaining the system's resilience and adaptability. Most effective is high-intensity interval training, an exercise strategy that alternates periods of short intense anaerobic exercise with less intense aerobic recovery periods. As with psychotherapeutic interventions that alternately challenge and support, here too cycles of anaerobic activity alternating with aerobic activity appear to be associated with fine-tuning the matrix and optimizing functionality.

Furthermore, to regulate gene expression and protein function, it will be important that there be adequate methylation of the various components of the ground regulation system, easily enough obtained by way of nutrient supplementation with such methyl donors as folic acid, vitamin B12, trimethylglycine, dimethylglycine, S-adenosylmethionine, and dimethylaminoethanol. Particularly critical for maintenance of a healthy and vital living matrix will be water, the presence of which allows the matrix to function as a liquid crystal, thereby making possible the high-speed body-wide propagation of information and energy necessary to ensure the resilience of the matrix, its adaptability, and its capacity to reconstitute at ever-higher levels of order, coherence, and integration.

## 11.6 HORMETIC DOSE-RESPONSE RELATIONSHIP

If the MindBodyMatrix cannot process, integrate, and adapt to stressful input, then over time the health of the system will become increasingly compromised; but if the MindBodyMatrix is able to process, integrate, and ultimately adapt to the potentially devastating impact of all the environmental stressors to which it is being continuously exposed, then over time the system will evolve, through iterative cycles of destabilization and restabilization, to ever-higher levels of health not just *in spite of* the stressful input but *by way of* that input—this benefit is the result of tapping into the *wisdom of the body* (Cannon 1932) and jumpstarting the system's innate ability to self-correct in the face of optimal challenge.

How might we understand the seemingly paradoxical response of the MindBodyMatrix to stress? And why is it that high doses of a stressor would be toxic whereas lower doses of that same stressor might be therapeutic? Hormesis is the term used to describe this biphasic dose-response relationship, whereby the response to a stressor will shift from stimulatory to inhibitory at some point along the dose-response (or stress–response) curve, that is, low-dose stimulation but higher dose inhibition (or, depending upon the context, higher dose overstimulation). In other words, because of the hormetic (biphasic) effect, there is not always a simple linear relationship between the dose of a stressor and the clinical response.

A good example of the hormetic effect in psychopharmacology is provided by one of the most impressive and most often cited clinical trials on the efficacy of ω-3 fatty acids in the treatment of unipolar depression. In an elegantly designed, 12-week, double-blind, placebo-controlled study conducted by Peet and Horrobin (2002), patients receiving eicosapentaenoic acid (EPA)—a heart-healthy, memory-boosting supplement also used to treat depression—were divided into three dosage groups (1, 2, and 4 g/day). All the participants had experienced persistent depression despite treatment with standard antidepressants at adequate dosages. The findings from this study showed that the patients receiving 1 g EPA per day had the best outcome (with more than half achieving a 50% reduction in depressive symptomatology on the

Hamilton Depression Rating Scale). Interestingly, the authors of the study found that the group receiving 2 and 4 g/day showed little evidence of efficacy. There are many ways to interpret this surprising finding, but one possible explanation involves the hormetic (biphasic) effect of low-dose stimulation and higher dose inhibition (or, in this instance, perhaps overstimulation).

Although long marginalized in the toxicological literature (which has historically embraced a linear no-threshold model whereby toxins are thought to be toxic at whatever their dose and to become ever more toxic at ever-higher doses), the concept of hormesis, which literally means *to excite*, is now slowly gaining acceptance in some academic circles, largely through the extraordinary research efforts of Edward Calabrese (Calabrese and Baldwin 2003; Mattson and Calabrese 2010; also, see 1), an avant-garde toxicologist who believes that the hormetic effect is an almost universal biological phenomenon (whatever the stressor, whatever the biological system, whatever the endpoint being measured). Calabrese's contention (2011)—and one supported by Suresh Rattan, Éric Le Bourg, and all the contributors to this seminal volume—is that a biphasic dose-response is a manifestation of the system's adaptive response to stress, a modest overcompensation, Calabrese suggests, in the face of threatened disruption to homeostatic balance. My chapter addresses the paradoxical impact of stress on, most especially, the mind (here conceptualized as an open, self-organizing chaotic system) and develops the idea that an optimal dose of stressful input (i.e., an optimal challenge), by tapping into the system's resilience and intrinsic ability to heal itself, can indeed provoke modest overcompensation and a strengthening at the broken places.

## 11.7 MINDBODYMATRIX: A CHAOTIC SYSTEM

The MindBodyMatrix is what complexity (or chaos) theorists would describe as an open, complex adaptive, nonlinear dynamical, self-organizing chaotic system. Because of this, progression from illness to wellness—necessary to reverse underlying dysfunctionality—will never be simple, straightforward, or linear. Rather, evolution from poor health to good health will generally be a much more protracted and sometimes wildly unpredictable process involving multiple stops and starts, downs and ups, backward and forward movements, regressions and progressions, disruptions and repairs. Briefly, in the language of complexity theory (Strogatz 1994; Kauffman 1995; Ho 1998; Buchanan 2000), an open system is

1. Chaotic, which speaks to the system's underlying order despite its apparent randomness—an order that will emerge as the system evolves
2. Complex, which speaks to the intricate interdependence of the system's constituent components
3. Adaptive, which speaks to the system's capacity to benefit from experience
4. Nonlinear, which speaks to the totally unpredictable but deeply patterned evolution of the system over time and in response to input from the outside
5. Dynamical, which speaks to the emergence of novel structural configurations involving both repetition and innovation

6. Self-organizing, which speaks to the spontaneous emergence of system-wide patterns arising solely from the complex interplay of the system's components—these global patterns are an emergent property of the system itself, not a property imposed upon the system by an external influence

Indeed, complexity theory has it that the internal structure of a self-organizing (or chaotic) system is intrinsically such that, in response to input from the environment, order will ultimately emerge from chaos (Prigogine 1984; Strogatz 2003). This process of self-organization shows nonlinearity, with erratic, often dramatic, and sometimes catastrophic transitions from one state of complexity to another whenever some critical threshold (the timing for which is never knowable in advance) is reached.

A quintessential self-organizing (chaotic) system—whereby order emerges from chaos as the system evolves over time—is crystallization (i.e., the spontaneous emergence of beautifully patterned crystals from solutions of randomly moving molecules). Other examples of self-organizing systems include the assemblage of rippled dunes from grains of sand and the generation of swirling spiral patterns in hurricanes—the rippled dunes and the swirling spirals are emergent properties of self-organizing systems. Or consider the phenomenon whereby female roommates will, over time, begin to menstruate on the same cycle—this synchronization of their menstrual cycles is an emergent property of a self-organizing system. Or consider the phenomenon whereby thousands of fireflies gathered in trees at night and flashing on and off randomly will, after a while, begin to flash in unison—their synchronization is an emergent property of a self-organizing system and a dramatic illustration of the phase-locking of biorhythms. Or consider the phenomenon (in the inanimate realm) whereby a number of grandfather clocks with their pendulums initially swinging randomly will eventually entrain, such that all the pendulums will be swinging in precise synchrony (Bentov 1977).

## 11.8 HEALING CYCLES OF DEFENSIVE COLLAPSE AND ADAPTIVE RECOVERY

What patterns will emerge as the patient, here conceptualized as a self-organizing system, advances from chaos to coherence and from disorder to orderedness? As I have been suggesting, over time and by way of input that either challenges or supports, healing cycles of defensive collapse and adaptive reconstitution at ever-higher levels of complexity and capacity will be induced as the patient responds – either defensively (prompting collapse) or adaptively (prompting reconstitution) – to ongoing stressful input. The therapist, by way of her stressful input, will precipitate rupture in order to trigger repair; and she will do this repeatedly. Indeed, a patient's journey from illness to wellness involves progression through these iterative cycles of disruption and repair as the patient evolves from chaos and dysfunctionality to coherence and functionality.

Decades ago, Cannon (1932) was speaking to the wondrousness of the ongoing remodeling process that takes place on a microscopic level, moment by moment, within the living system (to which Cannon referred as the *fluid matrix*) when he wrote: "the system is open, engaging in free exchange with the outer world ... the structure itself is not permanent but is being continuously broken down by the wear and tear of action, and as continuously built up again by processes of repair." On a more macroscopic level and in the poignant words of James Baldwin (1961):

> Any real change implies the breakup of the world as one has always known it, the loss of all that gave one an identity, the end of safety. And at such a moment, unable to see and not daring to imagine what the future will now bring forth, one clings to what one knew, or thought one knew; to what one possessed or dreamed that one possessed. Yet, it is only when one is able, without bitterness or self-pity, to surrender a dream one has long cherished or a privilege one has long possessed that one is set free—one has set oneself free—for higher dreams, for greater privileges. All have gone through this, go through it, each according to one's degree, throughout their lives. It is one of the irreducible facts of life.

Finally, the remodeling process that is continuously taking place in all self-organizing systems is captured in the eloquent words of Deepak Chopra (2012): "All great changes are preceded by chaos."

## 11.9 SELF-ORGANIZING SYSTEMS RESIST PERTURBATION

With that said, it is important to understand that self-organizing (chaotic) systems resist perturbation (Krebs 2006). No matter how compromised they might be, self-organizing systems—fueled as they are by their homeostatic tendency to remain constant over time—are inherently resistant to change; they have an inertia that must be overcome if the system is ever to evolve from impaired capacity to more robust capacity.

Let me show this fundamental principle by speaking to the almost universal resistance to change manifested by psychiatric patients.

Consider a patient who is clinging tenaciously to dysfunctional defenses that had once served her but that have long since outlived their usefulness. As a self-organizing system, the patient must be sufficiently perturbed (i.e., impacted) by input from the outside (i.e., the therapist's interventions) that there will be impetus (i.e., force needed to bring about change) for her to relinquish her attachment to these maladaptive patterns of acting, reacting, and interacting. In essence, the therapist's interventions must have enough stressful impact that they will be able to challenge the homeostatic balance (i.e., the status quo) of the patient's dysfunctional defenses. By the same token, the therapist's interventions must also provide enough support that this input, in combination with the patient's innate striving toward health (i.e., her resilience), will prompt the patient to evolve to a higher level of functionality and adaptive capacity.

## 11.10 CHALLENGE VERSUS SUPPORT OF THE PATIENT'S DEFENSIVE STRUCTURES

Psychotherapists are therefore ever-busy formulating interventions that will either challenge or support—challenge the patient by directing her attention to where she is not (but where the therapist would want the patient to be) or support the patient by resonating with where she is (and where the patient would seem to need to be). On the basis of his/her moment-by-moment assessment of what the patient can tolerate (Stark 1994a,b), the therapist will therefore either challenge (by way of anxiety-provoking interpretive statements that call into question defenses to which the patient has long clung to preserve her psychological equilibrium) or support (by way of anxiety-assuaging empathic statements that honor these self-protective defenses—a therapeutic stance often referred to as *going with the resistance*).

Back and forth, back and forth. Challenge then support, challenge then support. The therapist will alternately challenge (e.g., by highlighting the price the patient is coming to recognize that she pays for refusing to let go of her dysfunctional defenses) and support (e.g., by resonating empathically with the patient's investment in holding on to those defenses)—all with an eye to creating tension within the patient between her dawning awareness of just how costly her defenses have become and her new-found understanding of just how much she feels the need for them and, thus, her intense attachment to them. In fact, as we sit with our patients, there is ever tension within us as well—between, on the one hand, our vision of who we think the patient could be (were she but able to make healthier choices) and, on the other hand, our respect for the reality of who she is (and for the choices, no matter how unhealthy, that she is continuously making). And, on some level, we are therefore always struggling to find an optimal balance between challenging the patient with our vision of the choices that we believe she could be making and supporting her by honoring the choices that she is making. Our interventions that challenge will almost inevitably increase the patient's anxiety; our interventions that support will almost inevitably decrease it (Stark 1999).

If the therapist's interventions make the patient too anxious, the patient may get defensive and then, because she is too overwhelmed to process and integrate the impact of the stress occasioned by the therapist's input, be unable to take in—or benefit from—that input. But if the anxiety elicited by the therapist's interventions is more manageable, the patient may then be able to process and integrate the stressful input and ultimately adapt to it by reconstituting at a higher level of self-awareness and complex understanding.

The dose-response relationship between the therapist's input and the patient's response clearly shows the hormetic (biphasic) effect: If the stressful impact is too much (i.e., overly stimulating), the patient will react defensively; but if the stressful impact is more modest (i.e., optimally stimulating), the patient will be able to respond adaptively. In essence, the capacity to cope with stress is a story not about defense but about adaptation.

## 11.11 TRAUMATIC VERSUS OPTIMAL STRESS

In truth, the psychiatric patient can respond in any one of the three ways to the therapist's stressful input:

1. Too much challenge, too much anxiety, too much stimulation, too much stress will be too overwhelming for the patient to process and integrate,

triggering instead defensive collapse and at least temporary derailment of the therapeutic process.
2. Too little challenge, too little anxiety, too little stimulation, too little stress will provide too little impetus for transformation and growth because there will be nothing that needs to be mastered; too little challenge will serve simply to reinforce the status quo.
3. But just the right amount of challenge, just the right amount of anxiety, just the right amount of stimulation, just the right amount of stress—to which the father of stress, Hans Selye (1974, 1978), referred as *eustress* and to which others (Scott 2009) and I refer as *optimal stress*—will offer just the right combination of challenge and support needed to optimize the potential for transformation and growth at the cutting edge.

Striving always to strike just the right balance, the therapist will therefore alternately challenge (when possible) and support (when necessary). The therapist's intent is always to generate an optimal level of anxiety within the patient, anxiety that will then provide the therapeutic leverage needed to propel the patient to ever-higher levels of integration, balance, and harmony.

How do we know, for sure, what the optimal level of anxiety will be? How do we know, for sure, when to challenge and when to support? We do not—no more than we can predict, with certainty, when there will be buying opportunities in the stock market because it is reaching a level of support and when there will be selling opportunities because it is reaching a level of resistance. We just do the best we can to buy low and sell high and then make whatever adjustments we need to optimize the return on our portfolio.

## 11.12 CREATION OF COGNITIVE DISSONANCE

So, too, with respect to the stressful (i.e., anxiety-provoking but ultimately insight-promoting) therapeutic interventions that we offer our patients, where we simply do the best we can to (1) challenge the patient by highlighting what she is coming to appreciate as the price she pays for holding on to her dysfunctionality and then, once she has been made sufficiently anxious, (2) support her as she, tapping into her resilience and innate striving toward health, begins to accept the reality that she needs to let go of the dysfunctionality. For example, as the patient comes to recognize that her (defensive) holding back for fear of being judged and found inadequate has seriously interfered with her ability to advance herself professionally, she begins to understand that—if she is ever to move forward in her career and, more generally, in her life—she must ultimately relinquish her attachment to keeping herself hidden and not seen.

And so it is that we, with our finger ever on the pulse of the patient's level of anxiety and capacity to tolerate further challenge, alternately challenge (when possible) and support (when necessary) to create tension within the patient between her dawning awareness of both the price she is paying for clinging to her defenses and her investment in holding on to them (even so)—tension that will ultimately become the fulcrum for therapeutic change.

You know that eventually you will need to confront—and grieve—the reality that your mother will never apologize for what she did and that you won't get better until you let go of your hope that someday she will, but you're not quite yet ready to do that because of your fear that you might not survive the heartbreak you would feel were you to be forced to face that devastatingly painful truth.

Even though you know that someday you're going to have to deal with the despair you feel about having developed fibromyalgia, in the moment you find yourself hoping that maybe it will simply go away. You know that you should be taking more responsibility for living a healthier lifestyle; but, right now, it all feels so overwhelming—and you're not convinced that it would make all that much difference anyway.

You know, on some level, that you are consigning yourself to a lifetime of chronic frustration and heartache as long as you cling to your hope that maybe someday, somehow, some way, Tony will change his mind and come back to you; but, in the moment, all you can think about is how good it had felt to be loved so deeply and so passionately.

You know that your need to keep what really matters to you hidden, incommunicado, private means that you will never really be able to have deep connection or real intimacy with anyone; but you have been betrayed so many times in the past that you are not now sure that you will ever again dare to open your heart.

You're coming to understand that your anger can put people off, but you tell yourself that you have a right to be as angry as you want because of how much you have had to suffer over the years.

You know that there is an element of choice in living your life in the self-defeating ways that you do and that eventually you will need to understand why you are so invested in sabotaging yourself; but, for now, you can't imagine giving up your various self-indulgences because they are what enable you to get through your day.

You know that, if you are ever to get on with your life, you will have to let go of your conviction that your childhood scarred you for life, but it's hard not to feel like damaged goods when you grew up with an abusive father who was always calling you a loser.

As the therapist repeatedly juxtaposes the patient's *knowledge* of reality with the patient's *experience* of it, the patient will be forced to see ever more clearly—even if reluctantly—the discrepancy between the sobering realities with which she is being confronted (be it the price she has paid for refusing to let go of her dysfunctional defenses, disillusioning truths about the object of her desire, accountability for her dysfunctional choices, or the therapeutic work she has yet to do to evolve to a higher level of emotional maturity) and her self-protective need to defend herself against having to confront those stressful truths.

## 11.13  FROM GAIN TO PAIN

Why must the patient be made anxious before she will relinquish her dysfunctional defenses? Earlier, and by way of answering this intriguing question, I had offered the explanation that self-organizing systems resist perturbation. I would now like to add, more specifically, that we must also never lose sight of the fact that the patient has an intensely ambivalent attachment to her defenses; no matter how dysfunctional or how costly those defenses might be, over time they have also served her, enabling her to survive. We must therefore always appreciate, and be respectful of, the patient's conflicted attachment to these defenses—to which Freud (1923) aptly referred as the *adhesiveness of the id*. Indeed, as long as the *gain* is greater than the *pain* and holding on to the defense provides more benefit than cost, then the patient will *maintain* the defense and *remain* entrenched in her dysfunction.

But once the *pain* becomes greater than the *gain* and holding on to the defense becomes more costly (i.e., ego-dystonic) than beneficial (i.e., ego-syntonic), then it is that the stress and *strain* created within the patient by the *cognitive dissonance* (Festinger 1957) between the pain and the gain will compel her to relinquish her attachment to the unhealthy defense in favor of a healthier adaptation. In other words, as the patient becomes ever more aware of the fact that she is holding on to two cognitions that are at odds with each other, she will become more and more uncomfortable—and this ever-increasing discomfort will ultimately prompt her to surrender her attachment to the cognition that is the more untenable of the two.

## 11.14 RELENTLESS HOPE: THE REFUSAL TO GRIEVE

By way of a more specific clinical illustration, I would like now to highlight a particular defense that I believe figures prominently in the psychology of many patients. I have written elsewhere of this self-protective, but ultimately self-sabotaging, defense as *relentless hope* (Stark 1994a,b, 1999) and as speaking to the patient's refusal to confront—and grieve—certain intolerably painful realities about her objects (a somewhat peculiar term used in the psychodynamic literature to signify important people in a patient's life).

Relentless hope is a defense to which the patient clings in order not to have to feel the pain of her disappointment in the object, the hope a defense ultimately against grieving. The patient's refusal to deal with the pain of her grief about the object (be it an infantile, a contemporary, or a transference object) fuels the relentlessness with which she pursues it—both the relentlessness of her hope that she might yet be able to make the object over into what she would want it to be and the relentlessness of the outrage she experiences in those moments of dawning recognition that, despite her best efforts and most fervent desire, she might never be able to make that actually happen. But, even more fundamentally, what fuels the intensity of the patient's pursuit is the fact of the object's existence as separate from hers, as outside the sphere of her omnipotence, and as therefore unable to be either possessed or controlled. In truth, it is this very immutability of the object—the fact that the object cannot be forced to change—that provides the propulsive fuel for the patient's relentless pursuit.

## 11.15 FROM RELENTLESS HOPE TO SOBER ACCEPTANCE

In what follows, I will be speaking to the manner in which the therapist, by alternately challenging and then supporting, can facilitate transformation of the patient's relentless hope (a dysfunctional defense) into sober acceptance (a more functional adaptation). Expressed in somewhat different terms, the clinical material below shows transformation of the defensive need to hold on into the adaptive capacity to let go, once illusions of omnipotent control of the object (a defense) become transformed into humble acceptance of the limits of one's power to possess and control the object (an adaptation).

Many a psychiatric patient, as a child, has suffered great heartache at the hands of a misguided, even if well-intentioned, parent, be it in the form of psychological trauma and abuse (*too much bad*) or emotional deprivation and neglect (*not enough good*). Such a patient may never have had occasion to confront the pain of her grief about the parent's unwitting, but devastating, betrayal of her. Instead, she has defended herself against the pain of her heartache by pushing it out of her awareness and clinging instead to the illusion of her parent (or a stand-in for her parent) as good and as ultimately forthcoming if she (the patient) could but get it right. Under the sway of her repetition compulsion, the patient—once she goes out into the world—may well find herself, there too, delivering into each new relationship her desperate hope that perhaps this time, were she to be but good enough, want it badly enough, or suffer deeply enough, she might yet be able to make her new partner over into the perfect parent she never had, but should have had, early on.

If the patient is ever to transform the chaotic energy of her unmastered grief about what was bad into the coherent structure of thoughtful appreciation for what was good (i.e., transform defense into adaptation), then she must someday dare to let herself first remember the anguish of just how heartbreaking it really was—both the parental errors of commission (*presence of bad*) and the parental errors of omission (*absence of good*)—and then confront, and mourn, the pain of her grief about the parent, grief against which she has spent a lifetime unconsciously defending herself.

The therapist's goal will be to facilitate transformation of the patient's need to defend against the pain of her grief by clinging to her relentless hope into the capacity to adapt by confronting the pain of her grief and coming to a place of serene acceptance with respect to what the object cannot do and appreciation for what it can. The bad news, of course, will be the sadness the patient experiences as she begins to accept the sobering reality that disappointment is an inevitable and necessary aspect of relationships. The good news, however, will be the wisdom she acquires as she comes to appreciate ever more profoundly the subtleties and nuances of relationship and begins to make her peace with the harsh reality of life's imperfections.

## 11.16  GRIEVING PROCESS

To facilitate the patient's grieving, the therapist's interventions will first challenge the patient by directing her attention to where she is not (i.e., to acknowledgement of the reality of the object's limitations, separateness, and immutability) and then support the patient by resonating empathically with where she is (i.e., with the heartbreak the patient is experiencing as she begins to confront that painfully disillusioning reality).

> As you begin to confront the reality that John is not who you had thought he was, it just breaks your heart and you wonder how you will be able to go on.
>
> In those moments when you let yourself remember just how limited your mother really is and just how defensive she can get whenever you try to hold her accountable, the pain goes so deep.
>
> As you begin to admit to yourself that probably your father will never recognize what it took for you to be able to accomplish all that you have done in your life, the disappointment that you feel is almost unbearable—especially because so much of what you did was to please him.

Again and again, the therapist will first increase the patient's anxiety by highlighting what the patient, at least on some level, really does know to be the devastating truth about the object and then decrease the patient's anxiety by resonating empathically with the grief the patient experiences as she begins to face that sobering reality. The therapist, ever attuned to the patient's affective experience in the moment, wants the patient to be able to feel that she (the patient) is not alone with the pain of her grief.

> As you begin to recognize that your mother, despite all that you've done in an effort to make her understand, is still off in her own self-absorbed world and almost totally oblivious to your appeals to be heard, you find yourself feeling absolutely devastated and overwhelmed with despair.
>
> As you become increasingly aware of the price you've paid for clinging to the hope that someday you might be able to get your father to love you, your heart breaks and you find yourself going to a very dark and desperate place.

Back and forth, back and forth—challenging then supporting, challenging then supporting—the therapist first directing the patient's attention to disillusioning truths and then supporting the patient as the patient sits with and at last feels, to the depths of her soul, her devastating disappointment. It is hoped that the patient, within the context of safety provided by the relationship with her therapist, will be finally able to access the reservoir of tears that have accumulated inside of her over time. And it is hoped that the patient will be able to feel secure enough that she can experience the sadness, the anguish, the torment, the fury, and the impotent rage that have been pent-up inside her for so long.

The patient must come to accept the reality that she is ultimately powerless to do anything to make her objects, both past and present, different. She can, and should, do things to change herself, but she cannot change her objects—and she will have to come to terms with that sobering truth. Such is the work of grieving—and mastering the experience of loss, disappointment, heartbreak, and defeat; such is the work of making one's peace with reality and moving on. As Kopp (1969) has written, "Genuine grief is the sobbing and wailing which express the acceptance of our helplessness to do anything about losses."

The grieving process—like any evolutionary process—will involve these recursive cycles of challenge and then support, increasing the patient's anxiety and then decreasing it, precipitating disruption by reminding the patient of painful truths and then facilitating repair by *holding* the patient as she confronts the pain of her grief, first prompting defensive collapse and then triggering adaptive reconstitution at everhigher levels of nuanced understanding and sober acceptance.

It will be only once the patient has been able to process and integrate the dissociated grief that she will be able to relinquish her relentless pursuit of the unattainable. The patient will have transformed unhealthy defense into healthier adaptation once she has grieved and, in the process, developed a more refined awareness of the limitations inherent in relationships and a more evolved capacity to accept that which she cannot change.

Relevant here is the Serenity Prayer: "God grant me the serenity to accept the things I cannot change; courage to change the things I can; and wisdom to know the difference." Perhaps as we get older, we become sadder; but we also become more aware, more accepting, and more accountable—as we transform the defensive need to have reality be a certain way into the adaptive capacity to accept it as it is. More specifically, as the defensive need to hold on becomes transformed into the adaptive capacity to let go, realistic (mature) hope replaces relentless (infantile) hope. Along these same lines, Searles (1996) has suggested that realistic hope arises in the context of surviving disappointment. I am here reminded of the New Yorker cartoon in which a gentleman, seated in a restaurant named The Disillusionment Café, is awaiting the arrival of his order. The waiter returns to his table and announces, "Your order is not ready, nor will it ever be."

## 11.17 BELATED PROCESSING OF UNMASTERED EXPERIENCE

In essence, psychotherapy affords the patient an opportunity, often long after the fact, to manage experience that had once been overwhelming (and therefore defended against) but that can now, with enough support from the outside, be processed, integrated, and adapted to. Psychotherapy is therefore a story about the belated processing of unmastered experience and—in the face of optimal challenge (provided by stressful therapeutic interventions that alternately challenge and support) and by way of tapping into the patient's innate capacity to self-repair—reconstitution at ever-higher levels of awareness, acceptance, and accountability (Stark 1999).

In other words, psychotherapy offers patients the opportunity to process the cumulative impact of a myriad of environmental stressors—psychological, physiological, and energetic—to which they have been exposed over the course of their lives and ultimately to reintegrate at a higher level of functionality and complexity. As noted repeatedly throughout this chapter, it will be ongoing exposure to environmental impingement (in the form of the therapist's stressful interventions) that will, counterintuitively, provide the therapeutic leverage—the impetus—for such transformation.

More generally, the ever-evolving psychotherapeutic process can be conceptualized as a story about transforming unhealthy defense into healthier adaptation, whether it be the transformation of (1) *resistance* into *awareness* (in the language of classical psychoanalysis); (2) *relentlessness* into *acceptance* (in the language of those psychological theories that focus on the internal *absence of good*); or (3) *reenactment* into *accountability* (in the language of those psychological theories that focus on the internal *presence of bad*) (Stark 1999).

## 11.18 SANDPILE MODEL AND THE PARADOXICAL IMPACT OF STRESS

Let us now look a little more closely at the way in which chaos theorists conceptualize the evolution, over time, of complex adaptive systems in response to environmental stressors. Complexity theory states that simple rules can lead to complex behaviors, that is, to an infinite number of potential outcomes but, ultimately, the emergence of certain underlying patterns, properties, and structures.

Consider the following very simple set of two rules: (1) grains of sand are being added one at a time to a sandpile; and (2) whenever a grain of sand lands right on top of another one, there is a good chance that the sandpile will topple in one direction or another.

Long intriguing to chaos theorists is this sandpile model (Bak et al. 1987; Bak 1996), which is believed to offer a dramatic depiction of the cumulative impact, over time, of environmental impingement on open systems. Evolution of the sandpile is governed by some complex mathematical formulas and is well-known in many scientific circles; but the sandpile model is rarely applied to living systems and has never been used to show either the adaptability and resilience of the living system or the paradoxical impact of stress on it. My contention, however, is that this simulation model provides an elegant visual metaphor for how the MindBodyMatrix is continuously refashioning itself at ever-higher levels of complexity and integration—not just *in spite of* stressful input from the outside but *by way of* that input.

Consider a sandpile (here representing the patient) to which grains of sand (here representing environmental stressors) are being continuously added. As the sandpile evolves, an underlying pattern will begin to emerge, characterized by recursive cycles of collapse and recovery, disruption and repair, and defensive disorganization and adaptive reorganization, the sandpile growing ever larger even as it is becoming ever more unstable and ever more precariously balanced at the edge of chaos—in the words of Bak (1996), "perpetually out-of-balance, but organized in a poised state."

Amazingly enough, the grains of sand being steadily added to the gradually evolving sandpile are the occasion for both its disruption and its repair. Not only do the grains of sand being added precipitate partial collapse of the sandpile, but also they become the means by which the sandpile is able to build itself back up—each time at a new level of homeostasis (i.e., a new allostatic set point). The system is therefore able not only to manage the impact of the stressful input but also to benefit from that impact.

## 11.19  DIFFERENCE BETWEEN A POISON AND A MEDICATION

The noted sixteenth-century Swiss physician Paracelsus is credited with having written that the difference between a poison and a medication is the dosage thereof (Paracelsus 2004). And, I would add, the system's capacity—a function of its underlying resilience—to process, integrate, and ultimately adapt to the impact of that stressor (Szent-Gyorgyi 1960). Stressful input, therefore, is inherently neither bad (poison) nor good (medication), which is to say that the therapist's interventions are inherently neither toxic (poison) nor therapeutic (medication). Rather, the dosage of the stressor, the underlying resilience and adaptability of the system, and the *intimate edge* (Ehrenberg 1992) between stressor and system will determine whether the system, in response to the environmental input, defends and devolves to ever-greater disorganization or adapts and evolves, by way of a series of healing cycles, to ever more complex levels of organization and dynamic balance.

In other words, if the interface between stressor and system is such that the stressor is able to provoke recovery within the system, then what would have been poison becomes medication, what would have constituted toxic input becomes therapeutic input, what would have been deemed traumatic stress becomes optimal stress, and what would have overwhelmed becomes transformative. I am, of course, speaking here to the therapeutic use of stress to provoke recovery by activating the body's innate ability to repair itself (Cannon 1932; Sapolsky 1994; McEwen 1998; Bland 1999; McEwen 2002). In essence, what does not kill you makes you stronger.

## 11.20 RESILIENCE: CONTINUOUS ADJUSTMENT TO INSTABILITY

To repeat: How well the MindBodyMatrix is able to process and integrate environmental perturbations will be a reflection of the system's underlying health, more specifically, a reflection of its resilience, its regulatory capacity, and its ability to adapt. Even more fundamentally, the system's ability to manage stress will be a story about its underlying orderedness and fluidity. I am of course speaking to the system's capacity to recover in the aftermath of disruption occasioned by environmental challenge. In fact, the hallmark of a system's resilience will be this capacity to adaptively reconstitute (i.e., to self-organize) in response to regulatory input from the outside. The health of the matrix is therefore a story about its capacity to adapt, to self-regulate, and to restore its homeostatic balance in the face of challenge.

Of note is the fact that implicit in this conceptualization of self-regulation is the compelling idea that a living system will be able to preserve its stability only by way of continuous adjustment to instability. Indeed, in 1963, two obstetricians made an intriguing discovery about the paradoxical relationship between regularity of the fetal heart rate and fetal mortality. They observed that the more metronome-like the heartbeat, the less likely the fetus would be to survive. In other words, the regularity of the fetal heart rate was found to be highly correlated with fetal death. By the same token, the obstetricians observed that the greater the heart rate variability (i.e., the more variable the heart's beat-to-beat intervals), the more likely the fetus would be to thrive (Hon and Lee 1963). This regulatory capacity, critically important for the maintenance of health, has been described as "the ability to survive change by changing" (Meadows 1972).

Richet (1900), a French physiologist, was also addressing this seeming paradox when he made the following observation about the living system:

> The living being is stable. It must be so in order not to be destroyed, dissolved or disintegrated by the colossal forces, often adverse, which surround it. By an apparent contradiction it maintains its stability only if it is excitable and capable of modifying itself according to external stimuli and adjusting its response to the stimulation. In a sense it is stable because it is modifiable—the slight instability is the necessary condition for the true stability of the organism.

In essence, health speaks to the capacity to continuously adjust to ongoing environmental perturbation and to adaptively reorganize at ever-newer homeostatic (allostatic) set points.

## 11.21 OPTIMIZATION OF MENTAL POTENTIAL

Although my focus throughout this chapter has been on the therapeutic use of stress to restore the functionality of a dysfunctional system, it must also be noted that optimal stress can be used to fine-tune the functionality of an already well-functioning system and to slow the progression of age-related decline in functionality. As an example, optimal challenge of the brain will serve to sharpen mental acuity, to decelerate cognitive decline, and to combat the effects of aging on the brain. Just as athletes can improve their physical fitness by optimally challenging their bodies with physical exercise (e.g., by way of interval training), so too people can maintain or even improve their mental fitness by optimally challenging their minds with brain teasers (e.g., by way of mathematical puzzles, word games, crossword puzzles, logic problems, and memory challenges). In fact, any mental exercise requiring deliberate and concentrated effort (e.g., active repetition, focused attention, learning a new skill or a new language, reflection, or meditation) will promote mental agility and delay age-related decline in mental capacity.

In essence, the brain is a muscle that needs to be exercised, and, when it is subjected over time to optimal levels of stress, among other things, there will be significant increases in the level of brain-derived neurotrophic factor (BDNF)—a neuroprotective growth factor that is produced in the basal forebrain, frontal cortex, and hippocampus, all of which are critically important for learning, memory (especially long term), and higher thinking (Zoladz and Pilc 2010). More specifically, BDNF supports the survival of existing neurons, fosters the generation of new neurons from neural stem cells (i.e., neurogenesis), and encourages the development of new and more flexible connections between neurons (i.e., synaptic plasticity). BDNF has been likened to fertilizer for the brain's nerve cells, prompting them to grow more rapidly and to develop stronger connections. In essence, it is brain food—or, in the words of Ratey (2008), a "Miracle-Gro" for the brain. By tweaking the brain's neural circuitry and thereby revitalizing the brain's neural connectivity, BDNF makes the processing and integrating of neural messages more efficient and more accurate. In other words, optimal challenge of the mind promotes neuroplasticity, that is, the brain's amazing ability to reorganize, repair, and restructure itself.

In addition to puzzles and games, our brains will be stimulated when we are exposed to situations that are new, unusual, different, novel, or unexpected; when our daily routines are disrupted; or when we combine two senses (such as listening to music and smelling flowers or watching a sunset and tapping our fingers). Exercising more than one sense at the same time is a form of cross-training for the brain because it taps into the brain's inherent tendency to form associations between different types of information. Whereas routine activity (i.e., doing the same thing day in and day out) can deaden the brain, spicing things up by introducing variety into one's daily routines can provide the optimally stressful challenge needed to activate underused neural pathways and connections, thereby making the brain more fit and flexible.

In sum, all mental exercises that challenge our cognitive skills would appear to enhance brain fitness by tapping into the system's innate striving toward integration, balance, and harmony in the face of optimal challenge. But a word of caution:

Improvement in functionality of the brain (as of the body) in response to optimal stimulation will be possible only to a point, because eventually the overtrained system (be it the body or the brain) will become overstimulated and/or overloaded, its adaptation reserves depleted, its recovery capacity exceeded. At that point, the system's ability to tolerate further challenge will become limited and its functionality will be compromised.

## 11.22　DIS-ORDER AND DIS-EASE

I am proposing that the capacity to tolerate stress (a hallmark of optimal functionality and a prerequisite for the reversal of underlying chronic dysfunctionality) is ultimately a story about the system's ability to process and integrate the impact of environmental impingement, which in turn will be a reflection of the underlying orderedness of the liquid crystalline matrix and the ease therefore with which information and energy can be propagated throughout its expanse—lack of order manifesting as "dis-order" and disrupted ease of flow manifesting as "dis-ease." And so it is that we speak of psychiatric disorders and diseases and of medical disorders and diseases.

The word dis-order is a particularly auspicious term inasmuch as it speaks to the idea not only of randomness (the less than optimally ordered distribution of molecules in the matrix) but also of ill health, both mental and physical. Indeed, psychiatric and medical disorders are about the dis-ordered distribution of molecules in the crystalline array that constitutes the living matrix and, because of this dis-order, disruption to the ease of flow of information and energy throughout its vast network—dis-ease.

Whether the primary involvement is of mind or body, dis-order (disrupted orderedness within the MindBodyMatrix) and dis-ease (disrupted ease of flow within the MindBodyMatrix) are implicated in the generation of both psychiatric and medical conditions. Optimal health (both mental and physical) is therefore a story about orderedness and ease of flow within the matrix, bad health a story about dis-order and dis-ease, that is, disrupted order and therefore disrupted ease of flow within the matrix.

## 11.23　CONCLUSION

Again, bad stuff happens. But it will be how well an individual is able to deal with the impingement (i.e., to process and integrate it) that will make of it either a growth-disrupting (sandpile-destabilizing) event or a growth-promoting (sandpile-restabilizing) opportunity. In other words, it will be how well the MindBodyMatrix is able to manage the cumulative impact, over time, of environmental stressors that will either hasten a compromised system's deterioration or support a more resilient system's evolution toward increasing complexity and functionality.

So whether the primary target is mind or body and the clinical manifestation therefore psychiatric or medical, the critical issue will be the ability of the MindBodyMatrix to handle stress through adaptation. Again, too much stress will overwhelm and prompt defense; too little stress will offer too little opportunity for transformation and growth; but just the right amount of stress—optimal stress—will provide just the right amount of therapeutic leverage needed to induce, after initial disruption, adaptive reconstitution at ever-higher levels of complexity, integration, mastery, and adaptive capacity.

And I wonder—is there not a certain beauty in brokenness, a beauty never achieved by things unbroken? If a bone is fractured and then heals, the area of the break will be stronger than the surrounding bone and will not easily break again. Are we too not stronger at our broken places? Is there not a certain beauty in brokenness, a quiet strength we acquire from surviving adversity and hardship and mastering the experience of disappointment, heartbreak, and devastation? And, then, when we finally rise above it, don't we rise up in quiet triumph, even if only we notice?

## REFERENCES

Bak P. 1996. *How Nature Works: The Science of Self-Organized Criticality*. Springer-Verlag, New York.
Bak P, Tang C, and Wiesenfeld K. 1987. Self-organized criticality: An explanation of 1/f noise. *Phys Rev Lett* 59(4): 381–384.
Baldwin J. 1961. *Nobody Knows My Name: More Notes of a Native Son*. Dial, New York.
Bentov I. 1977. *Stalking the Wild Pendulum: On the Mechanics of Consciousness*. Destiny Books, Rochester, VT.
Bernard C. 1865. *An Introduction to the Study of Experimental Medicine*. Paris, France
Bernard C. 1957. *An Introduction to the Study of Experimental Medicine*. Dover Publications, New York.
Bland J. 1999. *Genetic Nutritioneering*. McGraw-Hill, New York.
Braden G. 2007. *The Divine Matrix: Bridging Time, Space, Miracles, and Belief*. Hay House, Carlsbad, CA.
Buchanan M. 2000. *Ubiquity: Why Catastrophes Happen*. Three Rivers Press, New York.
Calabrese EJ. 2011. Toxicology rewrites its history and rethinks its future: Giving equal focus to both harmful and beneficial effects. *Environ Toxicol Chem* 30(12): 2658–2673.
Calabrese EJ and Baldwin LA. 2003. The hormetic dose–response model is more common than the threshold model in toxicology. *Toxicol Sci* 71(2): 246–250.
Cannon WB. 1932. *The Wisdom of the Body*. W. W. Norton, New York.
Chopra D. 2012. *Spiritual Solutions: Answers to Life's Greatest Challenges*. Harmony Books, New York.
Ehrenberg D. 1992. *The Intimate Edge: Extending the Reach of Psychoanalytic Interaction*. W. W. Norton, New York.
Festinger L. 1957. *A Theory of Cognitive Dissonance*. Row, Peterson and Company, Evanston, IL.
Freud S. 1923. *The Ego and the Id*. W. W. Norton, New York.
Gladwell M. 2002. *The Tipping Point: How Little Things Can Make a Big Difference*. Little, Brown and Company, New York.
Hemingway E. 1929. *A Farewell to Arms*. Charles Scribner's Sons, New York.
Ho MW. 1998. *The Rainbow and the Worm: The Physics of Organisms*. World Scientific, River Edge, NJ.
Hon EH and Lee ST. 1963. Electronic evaluation of the fetal heart rate. VIII. Patterns preceding fetal death, further observations. *Am J Obstet Gynecol* 87: 814–826.
Huang EJ and Reichardt LF. 2001. Neurotrophins: Roles in neuronal development and function. *Annu Rev Neurosci* 24: 677–736.

Kauffman S. 1995. *At Home in the Universe: The Search for the Laws of Self-Organization and Complexity*. Oxford University Press, New York.

Kohut H. 1971. *The Analysis of the Self: A Systematic Approach to the Psychoanalytic Treatment of Narcissistic Personality Disorders*. University of Chicago Press, New York.

Kopp S. 1969. The refusal to mourn. *Voices*, Spring: 30–35.

Krebs C. 2006. *Nutrition for the Brain: Feeding Your Brain for Optimum Performance*. Michelle Anderson Publishing, South Yarra, VIC, Australia.

Lipton B. 2009. *Spontaneous Evolution: Our Positive Future*. Hay House, Carlsbad, CA.

McEwen BS. 1998. Stress, adaptation, and disease. Allostasis and allostatic load. *Ann NY Acad Sci* 840: 33–44.

McEwen BS. 2002. *The End of Stress As We Know It*. Joseph Henry Press, Washington, DC.

McTaggart L. 2001. *The Field: The Quest for the Secret Force of the Universe*. HarperCollins, London.

Mattson MP and Calabrese EJ (eds). 2010. *Hormesis: A Revolution in Biology, Toxicology and Medicine*. Springer-Verlag, New York.

Meadows D. 1972. *The Limits to Growth*. Chelsea Green Publishing, White River Junction, VT.

Oschman JL. 2000. *Energy Medicine: The Scientific Basis*. Churchill Livingstone, New York.

Paracelsus T. 2004. *The Archidoxes of Magic*. Turner R (trans). Ibis Publishing, Temecula, CA.

Peet M and Horrobin DF. 2002. A dose-ranging study of the effects of ethyl-eicosapentaenoate in patients with ongoing depression despite apparently adequate treatment with standard drugs. *Arch Gen Psychiatry* 59(10): 913–919.

Pischinger A. 1991. *Matrix and Matrix Regulation: Basis for a Holistic Theory in Medicine*. Heine H (ed). Haug International, Brussels, Belgium.

Prigogine I. 1984. *Order Out of Chaos: Man's New Dialogue with Nature*. Bantam Books, New York.

Ratey J. 2008. *Spark: The Revolutionary New Science of Exercise and the Brain*. Little, Brown and Company, New York.

Rea WJ and Patel K. 2010. *Reversibility of Chronic Degenerative Disease and Hypersensitivity: Regulating Mechanisms of Chemical Sensitivity*, Vol 1. CRC Press, Boca Raton, FL.

Richet C. 1900. Functions of defense. *Dictionnaire de physiologie*, Vol. 4. Félix Alcan, Paris, France.

Sapolsky RM. 1994. *Why Zebras Don't Get Ulcers*. W. H. Freeman, New York.

Scott C. 2009. *Optimal Stress: Living in Your Best Stress Zone*. Wiley, Hoboken, NJ.

Searles H. 1996. *Countertransference and Related Subjects: Selected Papers*. International Universities Press, New York.

Selye H. 1974. *Stress without Distress*. Harper & Row, New York.

Selye H. 1978. *The Stress of Life*. McGraw-Hill, New York.

Sheldrake R. 1988. *The Presence of the Past: Morphic Resonance and the Habits of Nature*. Times Books, New York.

Stark M. 1994a. *Working with Resistance*. Jason Aronson, Northvale, NJ.

Stark M. 1994b. *A Primer on Working with Resistance*. Jason Aronson, Northvale, NJ.

Stark M. 1999. *Modes of Therapeutic Action: Enhancement of Knowledge, Provision of Experience, and Engagement in Relationship*. Jason Aronson, Northvale, NJ.

Stark M. 2008. Hormesis, adaptation, and the sandpile model. *Crit Rev Toxicol* 38(7): 641–644.

Stark M. 2012. The sandpile model: Optimal stress and hormesis. *Dose Response* 10(1): 66–74.

Sterling P. 2004. Principles of allostasis: Optimal design, predictive regulation, pathophysiology, and rational therapeutics. In: *Allostasis, Homeostasis, and the Costs of Physiological Adaptation*, Schulkin J (ed), pp. 17–64. Cambridge University Press, New York.

Strogatz S. 1994. *Nonlinear Dynamics and Chaos: With Applications to Physics, Biology, Chemistry, and Engineering*. Perseus Books, Cambridge, MA.

Strogatz S. 2003. *Sync: How Order Emerges from Chaos in the Universe, Nature, and Daily Life*. Hyperion, New York.
Szent-Gyorgyi A. 1960. *Introduction to a Submolecular Biology*. Academic Press, New York.
Williams RJ. 1956. *Biochemical Individuality: The Basis for the Genetotrophic Concept*. Keats Publishing, New Canaan, CT.
Zoladz JA and Pilc A. 2010. The effect of physical activity on the brain derived neurotrophic factor: From animal to human studies. *J Physiol Pharmacol* 61(5): 533–541.

# Section III

## Molecular Mechanisms of Hormesis

# 12 Molecular Stress Response Pathways as the Basis of Hormesis

*Dino Demirovic, Irene Martinez de Toda, and Suresh I.S. Rattan*

## CONTENTS

12.1 Introduction ..................................................................................................227
12.2 Heat Shock Response....................................................................................228
12.3 Oxidative Stress Response............................................................................231
12.4 Nutritional Stress Response..........................................................................232
12.5 Energy Stress Response................................................................................233
12.6 DNA Damage Response...............................................................................234
12.7 Unfolded Protein Response..........................................................................235
12.8 Inflammatory Response................................................................................236
12.9 Conclusions and Perspectives.......................................................................237
Acknowledgment....................................................................................................238
References..............................................................................................................238

## 12.1 INTRODUCTION

The biological processes of stress detection and response are crucial with respect to the physiological outcome as being hormetic, harmful, or lethal. In this context, the term *stress* is defined as a signal generated by any physical, chemical, or biological factor (stressor), which in a living system initiates a series of events both as an immediate stress response (SR) that can last from a few seconds to a few hours and as a delayed SR lasting for much longer before returning to near-basal levels (Rattan 2008, 2012b; Demirovic and Rattan 2013). The immediate and late responses to external and internal stressors effectively determine the molecular, biochemical, and physiological stability in a dynamic and interactive manner. There are three main aspects of SR: (1) immediate SR involving receptors and intracellular signaling during the period of disturbance and exposure to stressors; (2) delayed SR involving sensors and modulators in the presence of stressors or after the removal of stressors; and (3) downstream effectors for counteracting the effects of disturbance and for reestablishing homeodynamics. At present, it is not completely understood how these three steps at the molecular level are maintained interactively in terms of kinetics and intensity, and how these may alter during growth, development, and aging.

Initiation of one or more SR pathways in response to a stressor is the first step toward hormesis. The key conceptual features of hormesis are the stress-induced disruption of homeodynamics, activation of maintenance and repair processes, modest overcompensation, reestablishment of homeodynamics, and the adaptive nature of the whole process (Rattan 2008; Calabrese et al. 2012). In this chapter, we give a brief overview of the major molecular level SR pathways that are well-identified in human and other eukaryotic systems, and that form the mechanistic basis of hormesis. Table 12.1 lists the molecular SR pathways along with their known stressors, specific proteins, and signaling molecules involved in the immediate response, and the proteins and organelles involved as effectors in the delayed response. Based on the involvement of one or more molecular SR pathways, higher order (cellular, organ level, and body level) SR is manifested, which include apoptosis, inflammation, thermoregulation, and hyperadrenocorticism.

## 12.2 HEAT SHOCK RESPONSE

The so-called heat shock response (HSR) is the most studied of the cellular SR, and is also one of the most conserved and universal SRs. It is named after the original discovery of the heat-induced preferential synthesis of a variety of novel proteins, termed *heat shock proteins* (HSPs) or stress proteins, which are essential for the survival of cells and organisms in the event of severe stress (Verbeke et al. 2001; Powers and Workman 2007; Kaarniranta et al. 2009). Other inducers of HSR include toxins, heavy metals, antibiotics, infections, ethanol, and inflammation. Even cold stress leads to the induction of synthesis of several, but not all, HSPs, as reported for insects (Burton et al. 1988; Colinet et al. 2010). The HSP family consists of a number of chaperones that are involved in protein refolding and remodeling, prevention of aggregation, and stimulation of proteolysis. The nomenclature for HSP has traditionally been in accordance with their molecular weights, for example, HSP90, HSP70, and HSP27, but now there is a modified nomenclature for HSP genes and proteins consistent with the HUGO Gene Nomenclature Committee and used in the National Centre of Biotechnology Information Entrez Gene Database (Kampinga et al. 2009). HSPs are present during both stressful and normal conditions, and in the latter situation they serve as sensors of the cellular redox status (Ahn and Thiele 2003).

The induction of HSR is mediated by transcription factors called heat shock factors (HSFs). There are at least four HSFs, depending on the system, but of these, HSF1 is the most abundant and most studied (Verbeke et al. 2001; Anckar and Sistonen 2011). Under normal conditions, HSF1 is present in the cytoplasm in an inactivated form in a dynamic complex with HSP90. When cells are exposed to heat or any other stressor that can cause protein denaturation, protein unfolding increases and nonnative proteins begin to accumulate. These nonnative proteins compete with HSF1 for binding with HSP90, resulting in the release of the transcription factor. The transformation of HSF1 from its inactive to active form is a multistep process: (1) conversion of HSF1 from monomer to a homotrimer in a reaction driven by hydrophobic interactions between LZ1 repeat sequences; (2) secondary changes in HSF conformation such that certain domains believed to be involved in the transcription-promoting activities of

## TABLE 12.1
## Molecular Stress Response Pathways and Their Stressors and Effectors Involved in the Immediate and Late Responses

| Response Name | Stressors | Immediate Response | Late Response | Effector Molecules |
|---|---|---|---|---|
| HSR | Heat, heavy metals, antibiotics, denatured proteins | Homotrimerization of HSF1, nuclear translocation and binding to HSE sequences in promoters of HSP genes | Induction of transcription and translation of HSP | Chaperones, co-chaperones, proteasome, and other proteases |
| OSR | Free radicals, ROS, pro-oxidants, low levels of oxygen | Translocation of Nrf2, FOXO, or HIF1 to the nucleus | HO-1, VEGF, iNOS | HO-1, FOXO |
| NSR | Food limitation, hypoxia, damaged organelles | Induction of AP, altered LC3-I and LC3-II, formation of autophagosomes | Increased number of lysosomes | Lysosomal proteases |
| ESR or sirtuin response | Energy depletion | Enhanced NAD$^+$ concentration | Sirtuins protein level; altered NAD$^+$/NADH ratio | Sirtuins |
| DDR | Radiation, oxidants, free radicals, ROS | Recruitment of ATM (via MRN complex) to double-strand breaks and ATRIP (via ATRIP) to single-strand DNA regions | Increased levels of checkpoint proteins p53, p16, p21, mortalin | DNA repair enzymes |
| UPR or ER stress response | Unfolded and misfolded proteins in ER | Cleavage of ATF-6α and binding to ERSE sequences in promoters of ER chaperones genes | Synthesis of ER stress chaperones | Chaperones and co-chaperones |
| ISR | Pathogens, allergens, damaged macromolecules | Activation of STATs by tyrosine phosphorylation | Increased level of NF-κB and inflammatory interleukins | Cytokines, NOS, COX-2 |

HSR, heat shock response; HSF1, heat shock factors 1; HSE, heat shock element; HSP, heat shock proteins; OSR, oxidative stress response; ROS, reactive oxygen species; Nrf2, nuclear factor (erythroid-derived 2); FOXO, forkhead box; HIF1, hypoxia-inducible factors 1; HO-1, heme oxygenase-1; VEGF, vascular endothelial growth factor; iNOS, inducible nitric oxide synthase; NSR, nutritional stress response; AP, autophagy; LC3, microtubule-associated protein 1 light chain 3; ESR, energy stress response; NAD$^+$, nicotinamide adenine dinucleotide; NADH, nicotinamide adenine dinucleotide (reduced form); DDR, DNA damage response; ATM, ataxia-telangiectasia mutated; MRN complex, MRE11–RAD50–NBS1 complex; UPR, unfolded protein response; ER, endoplasmic reticulum; ATF-6α, activating transcription factor 6; ERSE, ER stress-response element; ISR, inflammatory stress response; STATs, signal transducers and activators of transcription; NF-κB, nuclear factor-kappa B; NOS, nitric oxide synthase; COX-2, cyclooxygenase-2.

HSF1 become exposed; and (3) stress-induced hyperphosphorylation of the transcription factor. HSF1 then gets translocated into the nucleus, and binds to the heat shock element (HSE) sequences in the promoter regions of the genes that encode HSP. The homotrimerization of HSF1 has been shown to be essential for HSF to bind with high affinity to the HSE sequences found upstream of all HSP genes. However, the trimerization, although essential for DNA binding activity, is not enough for HSF1 to become a transcriptionally active factor, and requires hyperphosphorylation. It has also been proposed that for mammalian and *Drosophila* cells translocation of HSF1 from the cytoplasm to the nucleus is a step that regulates HSF1 activity. On the other hand, other studies have localized the unstressed form of HSF1 in the nucleus, suggesting that HSF1 is predominantly a nuclear protein prior to being exposed to stress and that the translocation is not part of the multistep process in HSF1 activation (Verbeke et al. 2001; Anckar and Sistonen 2011).

The translocation of the HSF1 trimers and their binding to the HSE results in a de novo RNA transcription of the HSP genes, followed by the preferential synthesis of several HSPs as repair and maintenance molecules. The major functions of the HSP within a cell are protein–protein interactions to ensure proper folding and to establish the correct protein conformation. Through their role in stabilization of the unfolded or misfolded proteins, HSPs also assist in transportation of proteins across membranes. Some HSPs can also be excreted out of the cells and can go into other parts of the body, including other tissues, organs, and body fluids. In those cases, the systemic functions of HSPs range from their role in the transportation of proteins across membranes to their function as immuno-modulators, which include stimulation of both innate and adaptive immune responses.

Optimal HSR is essential for the survival of cells. However, the nature of the stressor, which may vary in both intensity and duration, will determine whether the cell system can tolerate multiple rounds of exposure. Furthermore, although the general mechanisms of severe HSR are well-understood (Feder and Hofmann 1999; Verbeke et al. 2000; Sun and MacRae 2005), it is not clear whether there are any significant differences between mild heat stress with hormetic effects, and severe heat stress, repeated exposure to which has deleterious effects (Park et al. 2005). It is likely that the physiological cost of stress in terms of energy utilization, molecular damage overload, and metabolic shift determines the difference between the outcome of mild and severe stress. Also, it is yet to be understood how the transient appearance of HSP leads to biologically amplified hormetic effects at various other levels of cellular functioning, such as improved proteasome activity, enhanced resistance to other stresses, maintenance of the cytoskeletal integrity, and others, including extension of life span.

Numerous papers have reported the health- and longevity-promoting effects of various HSR-inducing hormetins in a wide range of experimental systems, such as yeast, nematodes, fruit flies, rodents, and human cells in culture (Rattan 2005, 2008). Such HSR-mediating hormetins include physical exercise, thermal stress, and some plant extracts, for example, the Chinese herb Sanchi (Rattan et al. 2013). Although some questions have been raised about the wider application of heat as a hormetin (Lagisz et al. 2013), thermal hormesis for humans through HSR induction is also being reported (see Chapter 8 in this book).

## 12.3 OXIDATIVE STRESS RESPONSE

A major SR pathway in the cellular defense against oxidative or electrophilic stress is the so-called antioxidant response, which should correctly be called *oxidative stress response* (OSR). Most commonly, ORS is mediated by Nrf2 (nuclear factor [erythroid-derived 2]-like 2, also known as NFE2L2). Nrf2 belongs to the family of redox-sensitive Nrf transcription factors and is retained in the cytoplasm by Kelch-like ECH-associated protein 1 (Keap1), which also facilitates its degradation by Cullin-3 (Cul3)-dependent ubiquitination, giving Nrf2 a half-life of 15 to 20 minutes (Motohashi and Yamamoto 2004; Zhang and Gordon 2004). The activation of Nrf2 in terms of its nuclear translocation can occur both by electrophile-modulated changes in the Keap1 structure leading to the dissociation of Nrf2–Keap1 complex and by the modification of three critical cystein residues in Keap1 disrupting the Keap1–Cul3 ubiquitination system and enhancing Nrf2 stability.

Translocation of Nrf2 into the nucleus leads to its heterodimerization with macrophage-activating factor protein, followed by binding to the antioxidant response element or electrophile response element sequences in the promoter region of several antioxidative and phase II detoxifying genes. Nrf2 thus controls the expression of hundreds of genes whose protein products are involved in the detoxification and elimination of reactive oxidants and other electrophilic agents through conjugative reactions and by enhancing cellular antioxidant capacity (Nguyen et al. 2009). Nrf2 is ubiquitously expressed in all tissues, with the highest concentrations in the kidney, muscle, lung, heart, liver, and brain.

Some of the major antioxidative gene products induced by Nrf2 activation are glutathione S-transferase that catalyzes the conjugation of glutathione with endogenous and exogenous electrophiles, (such as reduced nicotinamide adenine dinucleotide (NAD[P]H) oxidoreductase that catalyzes the reduction and detoxification of highly reactive quinones), and heme oxygenase-1 (HO-1), also known as HSP32, which catalyzes the breakdown of heme into the antioxidant biliverdin (Nguyen et al. 2009). Nrf2-initiated synthesis of numerous antioxidant gene products strengthens the antioxidative ability of cells and leads to several beneficial effects. This end result of achieving antioxidative effects masks the original pro-oxidative action of the compounds that caused the activation of Nrf2 in the first place. Numerous so-called antioxidants, including various polyphenols, flavonoids, spices, herbal extracts, and other natural and synthetic molecules are actually pro-oxidants, which act as hormetins by activating stress-induced defense pathways (Barone et al. 2009).

Another side of the oxidative SR is hypoxia, which is due to lack of oxygen. In mammals, the primary transcriptional response to hypoxic stress is mediated by the hypoxia-inducible factors (HIFs) (Warnecke et al. 2003; Maxwell 2005). HIFs are obligate heterodimers consisting of an $O_2$-labile $\alpha$-subunit and a stable $\beta$-subunit. Mammals possess three isoforms of HIF$\alpha$. HIF-1$\alpha$ is expressed ubiquitously in all cells, whereas HIF-2$\alpha$ and HIF-3$\alpha$ are selectively expressed in certain tissues, including vascular endothelial cells, type II pneumocytes, renal interstitial cells, liver parenchymal cells, and cells of the myeloid lineage. HIF-1 is regulated through the stability of the $\alpha$-subunit, which depends on the oxygen availability. The $\beta$-subunit is constitutively expressed and translated, and its levels are maintained in a constant way. In contrast, the HIF-1α protein has a short half-life of 5 minutes given that its levels are regulated by proteasomal degradation (Huang et al. 1998).

During normoxia, two proline residues (Pro402 and Pro564) of HIF-1α are hydroxylated and recognized by the von Hippel-Lindau (pVHL) ubiquitin E3 ligase complex, leading to its degradation by the proteasome (Majmundar et al. 2010). But under hypoxic conditions, the activity of prolyl hydroxylases is inhibited (given that they need oxygen as substrate), so the residues are not hydroxylated and consequently HIF-1α is not degraded. It becomes stable and translocates into the nucleus, where it dimerizes with HIF-1β and binds to HIF-responsive elements (HRE sequences) of genes to enhance the synthesis of proteins required for survival in low oxygen conditions, such as glycolytic enzymes, vascular endothelial growth factor (VEGF), inducible nitric oxide synthase (iNOS), HO-1, and erythropoietin (Fedele et al. 2002).

Exercise, as a hormetic agent, involves both HSR and hypoxia responses. The contraction of muscles requires large amounts of oxygen to be consumed during the oxidative phosphorylation in the mitochondria. This results in excess production of reactive oxygen species (ROS) that will trigger the OSR in the cells (Radak et al. 2005, 2008a,b). Exercise results in an increased availability and bioactivity of nitric oxide (NO), a free radical derived from the endothelium. There is a special relationship between NO and ROS, where NO serves as an antioxidant at lower oxygen levels, whereas at higher concentrations it becomes highly reactive and cytotoxic by interacting with superoxide, resulting in formation of peroxynitrite ($ONOO^-$). Hormetic effects of hypoxia and ischemia are discussed separately in this book (see Chapters 9 and 10).

## 12.4 NUTRITIONAL STRESS RESPONSE

Lack of sufficient nutrients is a major stress factor for biological systems, and induces the so-called nutritional stress response(s) (NSR) to survive through the period of deprivation. At the organismic level, NSR also includes reduced metabolism and reproduction cessation, but at the cellular level autophagy (AP) is a major SR pathway (Yang and Klionsky 2010; Scherz-Shouval and Elazar 2011; Choi et al. 2013). AP is a regulatory "self-eating process" that involves cytoplasmic degradation of proteins, macromolecules, and other large compartments of the cell, such as the mitochondria. Under normal conditions, AP acts, together with the lysosomal and proteasomal pathways, as one of the mechanisms for the degradation and turnover of damaged proteins and organelles (Shintani and Klionsky 2004; Singh et al. 2009). This is also known as macroautophagy, which operates at a basal level constitutively. However, under conditions of limitation or deprivation of amino acids, growth factors, and other nutrients, or when macromolecules become damaged, aggregated, fibrillated, or in some other way modified and not used by the cells, or the cell lacks nutrients or energy, AP is enhanced.

The AP flux machinery is initiated by the inhibition of mammalian target of rapamycin (mTOR), which in its protein complex form mTORC1 is a Ser/Thr kinase (Pous and Codogno 2011). Although the exact mechanisms by which energy limitation and nutritional deprivation inactivate mTOR are still unknown, the inactivation of mTORC1 kinase activity results in the formation of a cytoplasmic double membrane or phagophore, probably originating from the endoplasmic reticulum (ER). This process is regulated by numerous proteins, such as Vps34, that interact with Beclin1, Ambra, and Bcl-2 (Maiuri et al. 2010; Scherz-Shouval and Elazar 2011). A complex of Atg-5, Atg-12, and Atg-16 is responsible for the formation of an isolation double membrane and recruitment of microtubule-associated protein 1 light chain 3 (LC3) into the membrane. LC3 is a cytoplasmic and constitutively expressed protein, which, during the formation of the autophagic isolation membrane, becomes conjugated with phosphatidylethanolamine, forming LC3-II, and gets incorporated into the autophagosomal double membrane. The double membrane then expands and forms a closed sphere called the mature autophagosome, which is recognized by the lysosomal associated membrane proteins, and the contents of the autophagosome, including LC3-II, are degraded by the lysosomal enzymes. The appearance and disappearance of the LC3-II can be used as a marker for the autophagic flux (Tanida and Waguri 2010; Weidberg et al. 2010).

In the context of hormesis, NSR is considered to be the basis for explaining the well-established health-beneficial and longevity-promoting effects of calorie restriction (Masoro 2007). Induction of AP and enhanced basal levels of autophagic flux on chronic or intermittent and periodic CR and fasting is the earliest and consistent molecular mechanism for clearing up the intracellular debris (Bergamini et al. 2003; Cavallini et al. 2008; Cuervo 2008; Blagosklonny 2013; Choi et al. 2013).

## 12.5 ENERGY STRESS RESPONSE

The cellular response to reduced levels of energy, as indicated by increased ratios of adenosine monophosphate (AMP)/adenosine triphosphate and nicotinamide adenine dinucleotide ($NAD^+$)/NADH may be termed the *energy stress response* (ESR). The AMP-activated protein kinase (AMPK) is a heterotrimer and is a crucial integrator of signals that monitor energy balance and regulate several biochemical pathways in the cell (Steinberg and Kemp 2009). Energy deficiency may result from enhanced biochemical demands during exercise, food limitation, glucose uptake impairment, and enhanced fatty acid oxidation. AMPK is involved in the regulation of energy metabolism by directly affecting the transcription of several genes for key metabolic pathways, especially carbohydrate and lipid metabolism, protein biosynthesis, and mitochondrial biogenesis. Another major player in the ESR is the family of seven sirtuins (SIRT) in humans (Hipkiss 2008; Morris 2013). Of these, SIRT1 and SIRT6 are localized in the nucleus, SIRT2 is located in the cytoplasm, SIRT3, 4 and 5 are mitochondrial sirtuins, and SIRT7 is localized in the nucleolus. Each SIRT has a characteristic enzymatic activity, such as SIRT1 and SIRT2 being $NAD^+$-dependent deacetylases, and SIRT4 and SIRT6 mostly mono-adenosine diphosphate -ribosyl transferases (Morris 2013).

When cellular $NAD^+$ levels increase as a result of, for example, exercise, fasting or calorie restriction, SIRT1 is activated and causes deacetylation of its target proteins, such as histones, forkhead box transcription factor, cyclic-AMP response element-binding protein-regulated transcription coactivator-2, sterol regulatory element-binding protein, and transcriptional coactivator peroxisome proliferator-activated receptor γ coactivator-1α. SIRTs are considered general stress sensors, protecting the cells from apoptosis and premature senescence. ESR mediated via SIRTs affects a wide range of cellular, biochemical, and metabolic processes, such as enhanced AP, increased gluconeogenesis, increased fatty acid oxidation, increased insulin secretion, increased glucose uptake, and decreased inflammation. These effects have various health implications with respect to the prevention, treatment, or management of several diseases, including diabetes, sarcopenia, osteoporosis, inflammatory arthropathies, neurodegenerative diseases, and cancers.

Thus, efficient ESR involves both AMPK and SIRTs, which are mutually reinforcing and interdependent, with varying time kinetics (Steinberg and Kemp 2009; Morris 2013). For example, whereas activation of AMPK occurs within minutes of energy deficiency, SIRT activation is a delayed response that may occur after a few hours. Therefore, potential hormetins that induce ESR may have either AMPK or SIRTs as their main mode of action. Some of the well-known ESR-inducing chemical and nutritional hormetins are resveratrol, quercetin, rolipram, spices, and other polyphenols and synthetic SIRT-activating and AMPK-activating compounds (Balstad et al. 2011; Morris 2013; Pallauf and Rimbach 2013).

## 12.6 DNA DAMAGE RESPONSE

DNA damage response (DDR) is another important SR involved in realizing the hormetic effects of radiation and other DNA damaging agents. DNA lesions, which induce DDR, include formation of abasic sites, base modifications, singe-strand breaks (SSBs) and double-strand breaks (DSBs), crosslinks, and other adducts. DDR is primarily the induction and activation of a complex series of repair and removal pathways, which involve more than 200 genes and their products (Jackson and Bartek 2009; Moskalev et al. 2013). These pathways include base excision repair (BER), nucleotide excision repair, non-homologous end-joining, homologous recombination, and ataxia-telangiectasia mutated (ATM)-mediated and ataxia-telangiectasia-mutated and Rad3-related (ATR)-mediated DDR signaling. Each of these pathways is well-understood in detailed molecular terms (Jackson and Bartek 2009; Gredilla et al. 2012; Moskalev et al. 2013).

The key DR-initiating or signaling proteins in mammalian cells are the proteins ATM and ATR, which are phosphatidyl inositol 3-kinase-like kinases (Shiloh 2003; Abraham 2004). ATM is activated by DSBs, whereas ATR is recruited to single-stranded DNA regions, which arise at stalled replication forks or during the processing of bulky lesions such as UV photoproducts (Zou and Elledge 2003; Zou et al. 2003). ATM, which exists as an inactive dimer, undergoes autophosphorylation, triggering monomerisation and activation, and is recruited to DSB by the MRE11–RAD50–NBS1 complex (Bakkenist and Kastan 2004; Falck et al. 2005). ATR is also recruited after DNA damage, in this case to replication protein A–coated SSBs via its interacting partner, ATR-interacting protein. Once these kinases are recruited to DNA damage sites, they phosphorylate substrates such as p53, Brca1, Chk1, Chk2, and Rad17, which collectively inhibit DNA replication, cell-cycle progression, and mitosis, thus promoting DNA repair, recombination, or apoptosis (Zou and Elledge 2003). ATM/ATR signaling also enhances repair by inducing the transcription and translation of DNA-repair proteins, by recruiting repair factors to the damage sites, and by activating repair proteins by posttranslational modifications. Another important SSB- and DSB-signaling protein is poly(ADP)-ribose polymerase, which binds SSB and BER intermediates and promotes their repair (Bürkle 2000; Shall and de Murcia 2000; von Zglinicki et al. 2001; Yelamos et al. 2011).

DDR in the context of hormesis in human health and disease is mostly described by the mild and/or background radiation effects (see Chapter 7). Although health-beneficial hormetic effects of various DNA damaging agents, especially radiation exposure to UV-, X-, and γ-rays, have been reported, the details of the DDR have not been well-studied (Calabrese and Baldwin 2000; Vaiserman 2008). An efficient and optimal DDR is considered to be crucial in diverse biological settings, such as generating immune receptor diversity, maintaining telomeres, appropriate cell differentiation, cancer suppression, prevention of neuronal degeneration, and other metabolic syndromes (Jackson and Bartek 2009).

## 12.7 UNFOLDED PROTEIN RESPONSE

Unfolded or misfolded proteins in the ER lead to the activation of several intracellular signal transduction pathways, collectively known as the unfolded protein response (UPR) or endoplasmic SR (Schroder and Kaufman 2005a,b; Walter and Ron 2011). Several types of stressors or proteotoxic agents can induce UPR, for example, oxidants, hypoxia, hormones, viral infection, and toxic chemicals, by causing protein unfolding or misfolding in the ER lumen. UPR activation leads to the expansion of the ER membrane and incorporation of the ER space with protein-folding machinery (Walter and Ron 2011).

Three different classes of ER stress sensors or signal transducers have been identified: inositol-requiring enzyme-1 (IRE1), activating transcription factor-6 (ATF6), and protein kinase RNA-like ER kinase (PERK). In each case, an integral membrane protein senses the protein-folding status in the ER lumen and transmits this information across the ER membrane to the cytosol. IRE1α and IRE1β are ER membrane-bound endoribonucleases that initiate spliceosome-independent mRNA splicing of X-box binding protein 1 (XBP1) in response to ER stress (Patil and Walter 2001). Activation of XBP1 requires multiple steps, such as transcription, splicing, and then translation to produce an active form of XBP1. Only the spliced form of XBP1 functions as a potent transcription factor and binds directly to the endoplasmic reticulum stress element (ERSE) and thus can activate transcription of ER chaperone genes, but also can activate transcription of the ER-associated degradation machinery (Yoshida 2007). Moreover, XBP1 carries a functional ERSE sequence in its promoter allowing it to transactivate its own transcription. Thus, once produced, XBP1 can function in a more sustained fashion, as the XBP1 activation cycle continues as long as IRE1 is activated or as long as unfolded proteins are present in the ER (Yoshida et al. 2000).

ATF6 is the other crucial initiator of UPR since its activation is immediate because it is achieved by cleavage of a preexisting protein (Yoshida et al. 2003). ATF6 is normally located as an ER-transmembrane protein, which upon accumulation of unfolded proteins is packaged into transport vesicles and delivered to the Golgi apparatus. ATF6 then encounters site-1 and site-2 protease, which sequentially remove the luminal domain and the transmembrane anchor (Walter and Ron 2011). The N-terminal cytosolic fragment, ATF6(N), then moves into the nucleus and activates UPR target genes, such as the proteins involved in protein folding BiP (a chaperone of HSP70 family), glucose-regulated protein 94 (a chaperone of HSP90 family), and protein disulfide isomerase.

The third major ER stress signal transducer, PERK, is a ubiquitously expressed ER membrane-bound protein kinase, which phosphorylates itself and the $\alpha$-subunit of the eukaryotic initiation factor-2 (eIF2$\alpha$). Phosphorylation of eIF2 by PERK inactivates eIF2 and inhibits mRNA translation. This stops or slows down the flux of protein entry into the ER and reduces the load of ER stress (Harding et al. 1999). However, preferential translation of some mRNAs, which contain short open reading frames in their 5'-untranslated regions, continues even in the absence of eIF2 activity. The transcription factor ATF4 is one such product, activating the transcription of CHOP and GADD34, which have important roles in the regulation of apoptosis and cell cycle arrest.

ER stress is considered hormetic in the sense that UPR induced by mild ER stress improves the functioning of cells by increasing chaperone levels, enhancing antioxidant capacity, and clearing up molecular debris (Salminen and Kaarniranta 2010). Several natural and synthetic hormetins work through UPR, for example, various plant extracts and phytochemicals such as curcumin, barberine, and resveratrol (Salminen and Kaarniranta 2010).

## 12.8 INFLAMMATORY RESPONSE

Inflammation is a complex and localized immune response with systemic consequences, and which destroys, reduces, or sequesters both the harmful agent and the wounded tissue (Candore et al. 2010). Whereas low-level and intermittent inflammatory response has protective and health supportive effects, chronic inflammation has several harmful effects, including the emergence of several diseases, acceleration of aging, and shortening of life span. Inflammation involves many cell types (macrophages, neutrophils, monocytes, and others), and their molecular products such as cytokines, chemokines, prostaglandins, angiogenic factors, and nuclear factor-kappa B (NF-κB). The key factor in early inflammatory SR is the family of NF-κB transcription factors. NF-κB is also considered a downstream tuner connected with PI3K/Akt and mitogen-activated protein kinases signaling networks (Chirumbolo 2012).

There are numerous activators of the cellular inflammatory response, including microbial invasion, ROS, and activation through interaction with specific cell surface receptors by inflammatory cytokines, of which the most important is the tumor necrosis factor (TNF). During noninflammatory conditions, NF-κB exists as a dimer sequestered in the cytoplasm by the inhibitor of kappa B (IκB) and prevents it from being translocated to the nucleus. The activation of NF-κB requires a degradation of IκB proteins facilitated by IκB kinase, which phosphorylates two serine residues on the IκB and thereby modifies the IκB by ubiquitination, making it degradable by the proteasome. This results in the translocation of the NF-κB to the nucleus, which then heterodimerizes with the protein RelB, to form a ternary complex with DNA, promoting gene transcription. There are more than 150 NF-κB-target genes (Pahl 1999), leading to the de novo synthesis of proteins that facilitate tumor progression, inflammation (e.g., TNF, interleukin-1 [IL-1]), cell survival (e.g., B-cell lymphoma-extra large, calf intestinal alkaline phosphatase), angiogenesis (e.g., VEGF, IL-1, IL-8), proliferation (e.g., TNF, IL-1, IL-6, cyclin D1), and tumor promotion (e.g., cyclooxygenase-2, iNOS, MMP-9). The role of inflammatory response in hormesis in health and disease is discussed in detail in Chapter 13.

## 12.9 CONCLUSIONS AND PERSPECTIVES

The SR pathways discussed in this chapter constitute a critical arm of the cellular homeodynamic space, which is defined as the ability of living systems to counteract stress by repair, removal, or adaptation (Rattan 2012b). Not all pathways of SR respond to every stressor, and although there may be some overlap, generally SR pathways are quite specific. The specificity of SR is mostly determined by the nature of the disturbance or the damage induced by the stressor and the variety of downstream effectors involved. Furthermore, an induction of a specific SR pathway as the first response (immediate response) does not rule out the induction of one or more other SR pathways later on (delayed response). A complete and successful SR for effective homeodynamics and for the maintenance of the homeodynamic space requires both immediate and delayed SR in an optimal manner. It is therefore important that all SR pathways are analyzed simultaneously and a complete SR profile is established under a given condition, such as age, health, and disease status, and during and after exposure to single or multiple stressors (Demirovic and Rattan 2013). Furthermore, being able to map the kinetics and amplitude of different SRs, and their effects on each other, can form the basis to evaluate the health status of an individual and to develop effective means of aging modulators and maintainers of homeodynamic space (Rattan 2013).

Discovering novel hormetic agents by putting potential candidates through a screening process for their ability to induce one or more SR pathways in cells and organisms can be a successful strategy (Rattan 2012a; Rattan et al. 2013). Such agents have been termed *hormetins*, which refer to all such conditions that bring about biologically beneficial effects by initially causing low-level molecular damage, leading to the activation of one or more SR pathways, and eventually strengthening the homeodynamics (Rattan 2012a; Rattan et al. 2013). Hormetins are further categorized as (1) physical hormetins, such as exercise, thermal shock, and irradiation; (2) mental hormetins, such as mental challenge and focused attention or meditation; and (3) biological and nutritional hormetins, such as infections, gut flora, micronutrients, spices, and other sources of plant and animal origin (Rattan 2012a).

A frequent observation regarding the health applications of hormesis is that a single hormetin, such as exercise, heat shock, or caloric restriction, can strengthen the overall performance of cells and enhance other abilities, including the physiological and psychological ones. This happens by hormetin-induced initiation of a cascade of processes resulting in a biological amplification of beneficial effects (Rattan and Demirovic 2010; Rattan 2012a; Rattan et al. 2013). It should also be pointed out that at the molecular level the immediate and singular effects are generally quantitatively quite small. However, it is the repeated exposure to hormetins that eventually brings about large health-beneficial effects. Finally, since identical biological end results could be achieved by activating mechanistically very different molecular pathways, it is very important to know the earliest steps in the mode of action of a potential hormetin for establishing its uniqueness, specificity, and applicability.

## ACKNOWLEDGMENT

Laboratory of Cellular Ageing is financially partially supported by a research grant from LVMH Recherche, Saint Jean de Braye, France.

## REFERENCES

Abraham, R. T. 2004. PI 3-kinase related kinases: "Big" players in stress-induced signaling pathways. *DNA Repair (Amst)* 3: 883–887.

Ahn, S. G. and D. J. Thiele. 2003. Redox regulation of mammalian heat shock factor 1 is essential for Hsp gene activation and protection from stress. *Genes Dev* 17: 516–528.

Anckar, J. and L. Sistonen. 2011. Regulation of HSF1 function in the heat stress response: Implications in aging and disease. *Annu Rev Biochem* 80: 1089–1115.

Bakkenist, C. J. and M. B. Kastan. 2004. Initiating cellular stress responses. *Cell* 118: 9–17.

Balstad, T. R., H. Carlsen, M. C. Myhrstad, M. Kolberg, H. Reiersen, L. Gilen, K. Ebihara, I. Paur, and R. Blomhoff. 2011. Coffee, broccoli and spices are strong inducers of electrophile response element-dependent transcription in vitro and in vivo—studies in electrophile response element transgenic mice. *Mol Nutr Food Res* 55: 185–197.

Barone, E., V. Calabrese, and C. Mancuso. 2009. Ferulic acid and its therapeutic potential as a hormetin for age-related diseases. *Biogerontology* 10: 97–108.

Bergamini, E., G. Cavallini, A. Donati, and Z. Gori. 2003. The anti-ageing effects of calorie restriction may involve stimulation of macroautophagy and lysosomal degradation, and can be intensified pharmacologically. *Biomed Pharmacol* 57: 203–208.

Blagosklonny, M. V. 2013. M(o)TOR of aging: MTOR as a universal molecular hypothalamus. *Aging (Albany NY)* 5: 490–494.

Bürkle, A. 2000. Poly(ADP-ribosyl)ation: A posttranslational protein modification linked with genome protection and mammalian longevity. *Biogerontology* 1: 41–46.

Burton, V., H. K. Mitchell, P. Young, and N. S. Petersen. 1988. Heat shock protection against cold stress of Drosophila melanogaster. *Mol Cell Biol* 8: 3550–3552.

Calabrese, E., I. Iavicoli, and V. Calabrese. 2013. Hormesis: Its impact on medicine and health. *Hum Exp Toxicol* 32: 120–152.

Calabrese, E. J. and L. A. Baldwin. 2000. The effects of gamma rays on longevity. *Biogerontology* 1: 309–319.

Candore, G., C. Caruso, and G. Colonna-Romano. 2010. Inflammation, genetic background, and longevity. *Biogerontology* 11: 565–573.

Cavallini, G., A. Donati, Z. Gori, and E. Bergamini. 2008. Towards an understanding of the anti-aging mechanism of calorie restriction. *Curr Aging Sci* 1: 4–9.

Chirumbolo, S. 2012. Possible role of NF-kappaB in hormesis during ageing. *Biogerontology* 13: 637–646.

Choi, A. M., S. W. Ryter, and B. Levine. 2013. Autophagy in human health and disease. *N Engl J Med* 368: 651–662.

Colinet, H., S. F. Lee, and A. Hoffmann. 2010. Temporal expression of heat shock genes during cold stress and recovery from chill coma in adult Drosophila melanogaster. *FEBS J* 277: 174–185.

Cuervo, A. M. 2008. Calorie restriction and aging: The ultimate "cleansing diet." *J Gerontol A Biol Sci Med Sci* 63: 547–549.

Demirovic, D. and S. I. Rattan. 2013. Establishing cellular stress response profiles as biomarkers of homeodynamics, health and hormesis. *Exp Gerontol* 48: 94–98.
Falck, J., J. Coates, and S. P. Jackson. 2005. Conserved modes of recruitment of ATM, ATR and DNA-PKcs to sites of DNA damage. *Nature* 434: 605–611.
Fedele, A. O., M. L. Whitelaw, and D. J. Peet. 2002. Regulation of gene expression by the hypoxia-inducible factors. *Mol Interv* 2: 229–243.
Feder, M. E. and G. E. Hofmann. 1999. Heat-shock proteins, molecular chaperones, and the stress response. *Annu Rev Physiol* 61: 243–282.
Gredilla, R., C. Garm, and T. Stevnsner. 2012. Nuclear and mitochondrial DNA repair in selected eukaryotic aging model systems. *Oxid Med Cell Longev* 2012: 282438.
Harding, H. P., Y. Zhang, and D. Ron. 1999. Protein translation and folding are coupled by an endoplasmic-reticulum-resident kinase. *Nature* 397: 271–274.
Hipkiss, A. 2008. Energy metabolism, altered proteins, sirtuins and ageing: Converging mechanisms? *Biogerontology* 9: 49–55.
Huang, L. E., J. Gu, M. Schau, and H. F. Bunn. 1998. Regulation of hypoxia-inducible factor 1alpha is mediated by an O2-dependent degradation domain via the ubiquitin-proteasome pathway. *Proc Natl Acad Sci USA* 95: 7987–7992.
Jackson, S. P. and J. Bartek. 2009. The DNA-damage response in human biology and disease. *Nature* 461: 1071–1078.
Kaarniranta, K., A. Salminen, E. L. Eskelinen, and J. Kopitz. 2009. Heat shock proteins as gatekeepers of proteolytic pathways—implications for age-related macular degeneration (AMD). *Ageing Res Rev* 8: 128–139.
Kampinga, H. H., J. Hageman, M. J. Vos, H. Kubota, and R. M. Tanguay. 2009. Guidelines for the nomenclature of the human heat shock proteins. *Cell Stress Chaperones* 14: 105–111.
Lagisz, M., K. L. Hector, and S. Nakagawa. 2013. Life extension after heat shock exposure: Assessing meta-analytic evidence for hormesis. *Ageing Res Rev* 12: 653–660.
Maiuri, M. C., A. Criollo, and G. Kroemer. 2010. Crosstalk between apoptosis and autophagy within the Beclin 1 interactome. *EMBO J* 29: 515–516.
Majmundar, A. J., W. J. Wong, and M. C. Simon. 2010. Hypoxia-inducible factors and the response to hypoxic stress. *Mol Cell* 40: 294–309.
Masoro, E. J. 2007. The role of hormesis in life extension by dietary restriction. *Interdiscip Top Gerontol* 35: 1–17.
Maxwell, P. H. 2005. Hypoxia-inducible factor as a physiological regulator. *Exp Physiol* 90: 791–797.
Morris, B. J. 2013. Seven sirtuins for seven deadly diseases of aging. *Free Radic Biol Med* 56: 133–171.
Moskalev, A. A., M. V. Shaposhnikov, E. N. Plyusnina, A. Zhavoronkov, A. Budovsky, H. Yanai, and V. E. Fraifeld. 2013. The role of DNA damage and repair in aging through the prism of Koch-like criteria. *Ageing Res Rev* 12: 661–684.
Motohashi, H. and M. Yamamoto. 2004. Nrf2-Keap1 defines a physiologically important stress response mechanism. *Trends Mol Med* 10: 549–557.
Nguyen, T., P. Nioi, and C. B. Pickett. 2009. The Nrf2-antioxidant response element signaling pathway and its activation by oxidative stress. *J Biol Chem* 284: 13291–13295.
Pahl, H. L. 1999. Activators and target genes of Rel/NF-kappaB transcription factors. *Oncogene* 18: 6853–6866.
Pallauf, K. and G. Rimbach. 2013. Autophagy, polyphenols, and healthy ageing. *Ageing Res Rev* 12: 237–252.
Park, H. G., S. I. Han, S. Y. Oh, and H. S. Kang. 2005. Cellular responses to mild heat stress. *Cell Mol Life Sci* 62: 10–23.
Patil, C. and P. Walter. 2001. Intracellular signaling from the endoplasmic reticulum to the nucleus: The unfolded protein response in yeast and mammals. *Curr Opin Cell Biol* 13: 349–355.

Pous, C. and P. Codogno. 2011. Lysosome positioning coordinates mTORC1 activity and autophagy. *Nat Cell Biol* 13: 342–344.
Powers, M. V. and P. Workman. 2007. Inhibitors of the heat shock response: Biology and pharmacology. *FEBS Lett* 581: 3758–3769.
Radak, Z., H. Y. Chung, and S. Goto. 2005. Exercise and hormesis: Oxidative stress-related adaptation for successful aging. *Biogerontology* 6: 71–75.
Radak, Z., H. Y. Chung, and S. Goto. 2008a. Systemic adaptation to oxidative challenge induced by regular exercise. *Free Radic Biol Med* 44: 153–159.
Radak, Z., H. Y. Chung, E. Koltai, A. W. Taylor, and S. Goto. 2008b. Exercise, oxidative stress, and hormesis. *Ageing Res Rev* 7: 34–42.
Rattan, S. I. S. 2005. Hormetic modulation of aging and longevity by mild heat stress. *Dose-Response* 3: 533–546.
Rattan, S. I. S. 2008. Hormesis in aging. *Ageing Res Rev* 7: 63–78.
Rattan, S. I. S. 2012a. Rationale and methods of discovering hormetins as drugs for healthy ageing. *Expert Opin Drug Discov* 7: 439–448.
Rattan, S. I. S. 2012b. Biogerontology: From here to where? The Lord Cohen Medal Lecture, 2011. *Biogerontology* 13: 83–91.
Rattan, S. I. S. 2013. Healthy ageing, but what is health? *Biogerontology* 14: 673–677.
Rattan, S. I. S. and D. Demirovic. 2010. Hormesis as a mechanism for the anti-aging effects of calorie restriction. In: *Calorie Restriction, Aging and Longevity*, Everitte, A. V., S. I. S. Rattan, D. G. Le Couteur, and R. de Cabo (eds.). Springer, Dordrecht, pp. 233–245.
Rattan, S. I. S., V. Kryzch, S. Schnebert, E. Perrier, and C. Carine Nizard. 2013. Hormesis-based anti-aging products: A case study of a novel cosmetic. *Dose Response* 11: 99–108.
Salminen, A. and K. Kaarniranta. 2010. ER stress and hormetic regulation of the aging process. *Ageing Res Rev* 9: 211–217.
Scherz-Shouval, R. and Z. Elazar. 2011. Regulation of autophagy by ROS: Physiology and pathology. *Trends Biochem Sci* 36: 30–38.
Schroder, M. and R. J. Kaufman. 2005a. ER stress and the unfolded protein response. *Mutat Res* 569: 29–63.
Schroder, M. and R. J. Kaufman. 2005b. The mammalian unfolded protein response. *Annu Rev Biochem* 74: 739–789.
Shall, S. and G. de Murcia. 2000. Poly(ADP-ribose) polymerase-1: What have we learned from the deficient mouse model? *Mutat Res* 460: 1–15.
Shiloh, Y. 2003. ATM and related protein kinases: Safeguarding genome integrity. *Nat Rev Cancer* 3: 155–168.
Shintani, T. and D. J. Klionsky. 2004. Autophagy in health and disease: A double-edged sword. *Science* 306: 990–995.
Singh, R., S. Kaushik, Y. Wang, Y. Xiang, I. Novak, M. Komatsu, K. Tanaka, A. M. Cuervo, and M. J. Czaja. 2009. Autophagy regulates lipid metabolism. *Nature* 458: 1131–1135.
Steinberg, G. R. and B. E. Kemp. 2009. AMPK in health and disease. *Physiol Rev* 89: 1025–1078.
Sun, Y. and T. H. MacRae. 2005. The small heat shock proteins and their role in human disease. *FEBS J* 272: 2613–2627.
Tanida, I. and S. Waguri. 2010. Measurement of autophagy in cells and tissues. *Methods Mol Biol* 648: 193–214.
Vaiserman, A. M. 2008. Irradiation and hormesis. In: *Mild Stress and Healthy Aging: Applying Hormesis in Aging Research and Interventions*, Le Bourg, E. and S. I. S. Rattan (eds.). Springer, Dordrecht, pp. 21–41.
Verbeke, P., B. F. C. Clark, and S. I. S. Rattan. 2000. Modulating cellular aging in vitro: Hormetic effects of repeated mild heat stress on protein oxidation and glycation. *Exp Gerontol* 35: 787–794.

Verbeke, P., J. Fonager, B. F. C. Clark, and S. I. S. Rattan. 2001. Heat shock response and ageing: Mechanisms and applications. *Cell Biol Int* 25: 845–857.

von Zglinicki, T., A. Bürkle, and T. B. L. Kirkwood. 2001. Stress, DNA damage, and ageing—an integrative approach. *Exp Gerontol* 36: 1049–1062.

Walter, P. and D. Ron. 2011. The unfolded protein response: From stress pathway to homeostatic regulation. *Science* 334: 1081–1086.

Warnecke, C., W. Griethe, A. Weidemann, J. S. Jurgensen, C. Willam, S. Bachmann, Y. Ivashchenko et al. 2003. Activation of the hypoxia-inducible factor-pathway and stimulation of angiogenesis by application of prolyl hydroxylase inhibitors. *FASEB J* 17: 1186–1188.

Weidberg, H., E. Shvets, T. Shpilka, F. Shimron, V. Shinder, and Z. Elazar. 2010. LC3 and GATE-16/GABARAP subfamilies are both essential yet act differently in autophagosome biogenesis. *EMBO J* 29: 1792–1802.

Yang, Z. and D. J. Klionsky. 2010. Eaten alive: A history of macroautophagy. *Nat Cell Biol* 12: 814–822.

Yelamos, J., J. Farres, L. Llacuna, C. Ampurdanes, and J. Martin-Caballero. 2011. PARP-1 and PARP-2: New players in tumour development. *Am J Cancer Res* 1: 328–346.

Yoshida, H. 2007. ER stress and diseases. *FEBS J* 274: 630–658.

Yoshida, H., T. Matsui, N. Hosokawa, R. J. Kaufman, K. Nagata, and K. Mori. 2003. A time-dependent phase shift in the mammalian unfolded protein response. *Dev Cell* 4: 265–271.

Yoshida, H., T. Okada, K. Haze, H. Yanagi, T. Yura, M. Negishi, and K. Mori. 2000. ATF6 activated by proteolysis binds in the presence of NF-Y (CBF) directly to the cis-acting element responsible for the mammalian unfolded protein response. *Mol Cell Biol* 20: 6755–6767.

Zhang, Y. and G. B. Gordon. 2004. A strategy for cancer prevention: Stimulation of the Nrf2-ARE signaling pathway. *Mol Cancer Ther* 3: 885–893.

Zou, L. and S. J. Elledge. 2003. Sensing DNA damage through ATRIP recognition of RPA-ssDNA complexes. *Science* 300: 1542–1548.

Zou, Y., S. M. Shell, C. D. Utzat, C. Luo, Z. Yang, N. E. Geacintov, and A. K. Basu. 2003. Effects of DNA adduct structure and sequence context on strand opening of repair intermediates and incision by UvrABC nuclease. *Biochemistry* 42: 12654–12661.

# 13 Inflammatory Pathways

*Salvatore Chirumbolo*

## CONTENTS

13.1 Introduction: Inflammation in the Network and in Chaotic Perspective ...... 243
13.2 PI3K/Akt Pathway ............................................................................................ 247
    13.2.1 Introduction ........................................................................................ 247
    13.2.2 PI3K/Akt Signaling in Inflammation and Hormesis ....................... 250
    13.2.3 Targeting the PTEN/PI3K/Akt/mTOR Signaling Pathway to Face at Inflammation and Cancer ...................................................... 252
13.3 MAP Kinase/ERK/JNK Pathway ...................................................................... 254
    13.3.1 Introduction ........................................................................................ 254
    13.3.2 MAPK/ERK Signaling in Toxicology, Inflammation, and Hormesis ....................................................................................... 257
13.4 NF-κB Signaling Pathway ................................................................................. 260
    13.4.1 Introduction ........................................................................................ 260
    13.4.2 Role of NF-κB in Inflammation and Hormesis ................................ 262
13.5 Chaotic Perspective of the Inflammatory Pathways in Hormesis ................. 265
13.6 Conclusions ........................................................................................................ 269
References .................................................................................................................. 269

## 13.1 INTRODUCTION: INFLAMMATION IN THE NETWORK AND IN CHAOTIC PERSPECTIVE

Inflammation is often depicted as a complex response to noxious agents or foreign molecules to maintain self-integrity and individuality. In this simplistic and quite reductive description, mechanisms underlying inflammation are often described as linear phenomena; they have a certain predictable behavior, mainly due to the current knowledge of cell immunity, and are organized, likewise-oriented arrows sharing determined secondary ways with other systems to build up a comprehensible related information network. The major purpose of an inflammatory response is to "reply" to insults and put an end to acute immune reactions, to restore a previous homeostatic balance.

In a more general way, this hallmark should turn inflammation into a linear vectorial event toward homeostasis, but, actually, homeostatic balance cannot be merely defined as coming back to a static equilibrium, without pulsed, cyclic, or oscillating dynamics: inflammatory pathways, which rule the inflammatory response at a cellular and molecular level, are highly susceptible to the chaotic rules pertaining to all biological complex systems, particularly the immune system.

Throughout the whole biological evolutionary pathway, systems unpredictable physical events in the environment, such as the immune system, evolved as complex chaotic systems or complex devices. A full comprehension of the many signaling pathways involved in the molecular response elicited by inflammatory agents needs, probably, a new approach dealing with complexity and bifurcations, despite the many difficulties related to the mathematics of chaos.

The immune system as a whole is a complex system and its relationship with external stimuli includes also a chaotic behavior. Although the best picture of inflammatory pathways is a cascade-related network, where few main triggering agents elicit a downstream interrelated branch of kinases and phosphatases oriented toward gene expression, many of them have typical chaotic oscillation patterns. Their interrelationship creates bifurcations that are very sensitive entries for cell decision by different agonists; the switch tuned by bifurcations may even lead to dramatic changes only by using the intrinsic energy of the chaotic oscillatory mechanism, with less power expenditure and yet by harnessing stressors to ameliorate the chaotic system. In this context, the exact role of the immune system as a chaotic attractor in the organism has not yet been fully elucidated, despite the many contributions to the field (Weisbuch 1990; Bernardes and Zorzenon Don Santos 1997; Dalgleish 1999; Caravagna et al. 2010, 2012).

In the network model by Jerne (1974), each individual immune repertoire will be different depending on the lymphocytes that participate in the connected network. Using a very simple cellular automata model of the immune repertoire dynamics, Bernardes and Zorzenon Don Santos showed that although the usual regimes (stable and chaotic) attained by this automata were not interesting from the biological point of view, the transition region, at the edge of chaos, was very appropriate to describe such dynamics: in this region, they obtained a functional connected network involving 10%–20% of the lymphocytes available in the repertoire, as suggested by Jerne (1974) and Bernardes and Zorzenon Don Santos (1997). This should mean that a major relationship in a complex network occurs in bifurcation/transition regions. However, the network hypothesis of many complex mechanisms underlying cell biology has deeply influenced the comprehension and description of inflammatory signaling pathways (Buchman 2002), thus hindering for a long time a full approach to the immune system through a chaotic perspective.

Quite recently, the so-called network paradigm, which can be considered as an application of graph theory to biology, has proven to be a powerful approach to gain insights into biological complexity and catalyzed the advancement of systems biology, chaos theory, and possibly new suggestions toward the comprehension of hormesis (Tieri et al. 2010; Calabrese et al. 2011; Cornelius et al. 2013). In this perspective, the concept of degeneracy, which is highly intertwined with another recently proposed organizational principle, that is, the "bow tie" architecture (Li et al. 2012), appeared as a hallmark of many complex systems defined as the ability exerted by structurally different elements to perform the same function. The consideration of concepts such as degeneracy, bow tie architecture, and network resulted in a powerful new interpretative tool that could take into account the constructive role of noise (stochastic fluctuations) and is able to grasp the major characteristics of biological complexity, that is, the capacity to turn an apparently chaotic and highly dynamic set of signals into functional information.

The chaotic system appears as a biological strategy to address subtle and continuous changes within the space of events, although mechanisms underlying chaos may generate systems that operate as a short circuit. In a chaotic system, a "frustration" event occurs when the global structure is such that local connectivity patterns responsible for stable behavior are intertwined, leading to mutually competing attractors and chaotic itinerancy among brief appearance of these attractors. Frustration appears as a short circuit in the complex dynamics of a chaotic system, at the minimum level of Boolean relationships among three elements within a network. In some cases, the presence of frustration has been positively assessed as enlarging the repertoire of equilibrium configurations (Amit 1989; Sherrington 1990), while in some others the same phenomenon has been perceived as something to prevent due to the instability it induces in the network.

According to some authors, (Bersini and Calenbuhr 1997), frustration destabilizes the network and provokes an unpredictable "wavering" among the stable dynamic regimes, which characterizes the same network when it is interconnected in a nonfrustrated way. An example is provided by the immune idiotypic network in which the prevailing behavior is oscillatory. Common to all the signaling networks is the description of frustrated chaos as a succession of attempts to relax the network into one of the oscillatory regimes given by a weaker and nonfrustrated connectivity, which appears to be an impossible achievement, making the dynamics rambling over brief but repelling orbits. Frustration is systematically lower in transcriptional networks, as modeled at a functional level, than in signaling and metabolic networks (modeled at the stoichiometry level) (Iacono and Altafini 2010). Signaling networks lack energetic barriers and this promotes global order; in few words, if an average overall possible perturbations is met, it is possible to observe that unlike for transcriptional networks, in the signaling/metabolic networks, the probability of finding the system in its least-frustrated configuration tends to be high also in correspondence of a moderate energetic regime, meaning that, in spite of the higher frustration, these networks can achieve a globally ordered response to perturbations even for moderate values of the strength of the interactions (Iacono and Altafini 2010). This bulk of considerations and evidence should suggest that the complex mechanism of signaling in inflammation is a possible ground for hormesis occurring.

In this hypothetical and mathematical scenario, some authors have developed a comparable approach, called the "sandpile model" of hormesis. The sandpile model was actually developed by chaos theorists and represents a visual metaphor for the cumulative impact of environmental stressors on complex adaptive systems; complex adaptive systems are continuously modifying and refashioning themselves, presumably at ever-higher levels of complexity and integration, not just in spite of stressful or noxious inputs coming from the outside but because of them and through their intrinsic complexity (Stark 2008, 2012). In this perspective, stressful input is therefore inherently neither bad nor good, but its effect depends on the chaotic state of the targeted system. Although it is rarely explicitly stated, the traditional view of the immune system is that its normal state is regulated by a homeostatic equilibrium. In this context, the standard response to a pathogen should be represented by a rapid transient expansion, followed by a return to the healthy steady state once the infection is cleared. If the pathogen cannot be eliminated, resulting in chronic infection, the immune system ends up in a different steady state, which may perhaps slowly drift in time, as occurs in human immunodeficiency virus infection. This largely static picture implicitly underlies the majority of the experimental analysis in immunology and the static interactive representation of molecular inflammatory pathways. However, there are a growing number of examples where the immune system undergoes complex temporal oscillations (Stark et al. 2007). These oscillations represent a chaotic behavior that can also occur through intracellular mechanisms, such as $Ca^{2+}$ concentrations in T cells (Dolmetsch and Lewis 1994; Launay et al. 2004). Interestingly, these oscillations appear to have a functional role, for instance, with their frequency being related to the type of T-cell response.

Steady-state and chaotic oscillations probably represent the molecular engine of the intrinsic ordered and network-fashioned language used by molecular signaling to organize a cellular response. This has been reported also for cytoskeleton, for example (Enculescu et al. 2008). This behavior, which is a hallmark of the model near the second transition, appeared similar to that of a system undergoing a so-called canard explosion (Krupa and Szmolyan 2001). The actin/cytoskeleton model further exhibited bistability between stationary states and limit cycles (Enculescu et al. 2008).

In a chaotic perspective, inflammation is probably regulated by oscillating chaotic systems leading to bifurcations, which in turn render the whole molecular network able to be modulated by weak signals (Kumar et al. 2004), and oscillating systems such as intracellular calcium and cytoskeleton participate in the overall complexity. (Dolmetsch and Lewis, 2013)

Inflammation is a very complex mechanism involving panoply of signaling molecules and interactive networks. The time course and intensity of responses by resident and circulating cells may be regulated by various inflammatory signaling, including Src family kinases, protein kinase C (PKC), growth factor tyrosine kinase receptors, nicotinamide adenine dinucleotide phosphate, reactive oxygen species (ROS), phosphoinositide-3-kinase (PI3K)/Akt, mitogen-activated protein kinases (MAPKs), nuclear factor-kappa B (NF-κB), activator protein-1, and other signaling molecules. These signaling molecules and their network regulate both key inflammatory signaling transduction pathways and target proteins involved in inflammation. Many of them exhibit a chaotic behavior, as indicated in Section 13.5 in this chapter.

The approach suggested by system biology to focus on the language and "logics" used by cells to activate a successful response to stressors might give interesting insights on the molecular mechanisms underlying hormesis. A major concern is to highlight which kind of behavior lies beneath the complex interactions, which make up the visual network of a signaling cascade, which are often represented as a logical, understandable scheme made of arrows, hubs, cut lines, boxes, and nets. Inflammatory pathways are often indicated as interconnected functional relationships within a dynamic network. This fashion may hinder the comprehension of the oscillatory/chaotic behavior of many signaling dynamics. Nevertheless, this simplistic approach allows the reader to highlight hubs as possible sensitive spots where molecular function gets a chaotic behavior and origins bifurcations. This depends mainly on interacting chaotic oscillators.

In this chapter, the first part deals with three major signaling pathways, as described by current molecular biology and hormesis, namely, the PI3K/Akt signaling pathway, the MAPK/extracellular signal–regulated kinase (ERK)/c-Jun-N-terminal kinase (JNK) pathway, and the NF-κB signaling pathway.

## 13.2 PI3K/AKT PATHWAY

### 13.2.1 INTRODUCTION

Figure 13.1 describes the PI3K/Akt signaling pathway. From a molecular point of view, PI3Ks are a family of related intracellular signal transducer enzymes capable of phosphorylating the 3-position hydroxyl group of the inositol ring of phosphatidylinositol (PtdIns). They are also known as phosphatidylinositol-3-kinases. The various 3-phosphorylated phosphoinositides that are produced by PI3K, namely, PtdIns3P, PtdIns(3,4)P2, PtdIns(3,5)P2, and PtdIns(3,4,5)P3, function in a mechanism by which an assorted group of signaling proteins, containing a phosphoinositide-binding (PX) structural domain, pleckstrin homology (PH) domains, FYVE domains, namely the four Cys-rich domain from Fab 1, YOTB (yeast orthologue of PIKfyve), Vac 1 (vesicle transport protein), and EEA1 (Early Endosome Antigen 1), and other phosphoinositide-binding domains, are recruited to various cellular membranes. The PH domain of Akt binds directly to PtdIns(3,4,5)P3 and PtdIns(3,4)P2, which are produced by activated PI3K (Franke et al. 1997; Krakauer 2012). Since PtdIns(3,4,5)P3 and PtdIns(3,4)P2 are restricted to the plasma membrane, this results in translocation of Akt to the plasma membrane. Likewise, the phosphoinositide-dependent kinase-1 (PDK1 or, rarely referred to as PDPK1) also contains a PH domain that binds directly to PtdIns(3,4,5)P3 and PtdIns(3,4)P2, causing it to also translocate to the plasma membrane upon activation of PI3K. Therefore, the co-localization of activated PDK1 and Akt allows the latter to become phosphorylated by PDK1 on Thr308, leading to partial activation of Akt.

Full activation of Akt occurs upon phosphorylation of Ser473 by the TORC2 complex of the mammalian target of rapamycin (mTOR) protein kinase. In this context, the nomenclature can be confusing, as PDK1 also refers to the unrelated enzyme pyruvate dehydrogenase kinase, isozyme 1. Similarly, TORC2 also refers to the unrelated transcription factor Transducer of Regulated CREB activity 2, which has recently been renamed CREB-regulated transcription coactivator 2 to reduce the confusion. Many other proteins have been identified that are regulated by PtdIns(3,4,5)P3, including Bruton's tyrosine kinase, General Receptor for Phosphoinositides-1 (GRP1), and the O-linked N-acetylglucosamine transferase. The PI3K/Akt inflammatory pathway constitutes an important network in the intracellular receptor–mediated response, which regulates the signaling of multiple biological processes such as apoptosis, metabolism, inflammation, stress response, cell proliferation, and cell growth.

Akt, also known as protein kinase B (PKB), is a serine/threonine-specific protein kinase that in mammals comprises three highly homologous members known as PKBα (Akt1), PKBβ (Akt2), and PKBγ (Akt3). PKB/Akt is activated in cells exposed to diverse stimuli such as hormones, growth factors, and extracellular matrix components. The activation mechanism remains to be fully characterized, but occurs downstream of PI3K. PI3K generates phosphatidylinositol-3,4,5-trisphosphate (PIP3), a lipid second messenger essential for the translocation of PKB/Akt to the plasma membrane, where it is phosphorylated and activated by PDK-1 and possibly other kinases. PKB/Akt phosphorylates and regulates the function of many cellular proteins involved in processes that include metabolism, apoptosis, and proliferation. Recent evidence indicates that PKB/Akt is frequently constitutively active

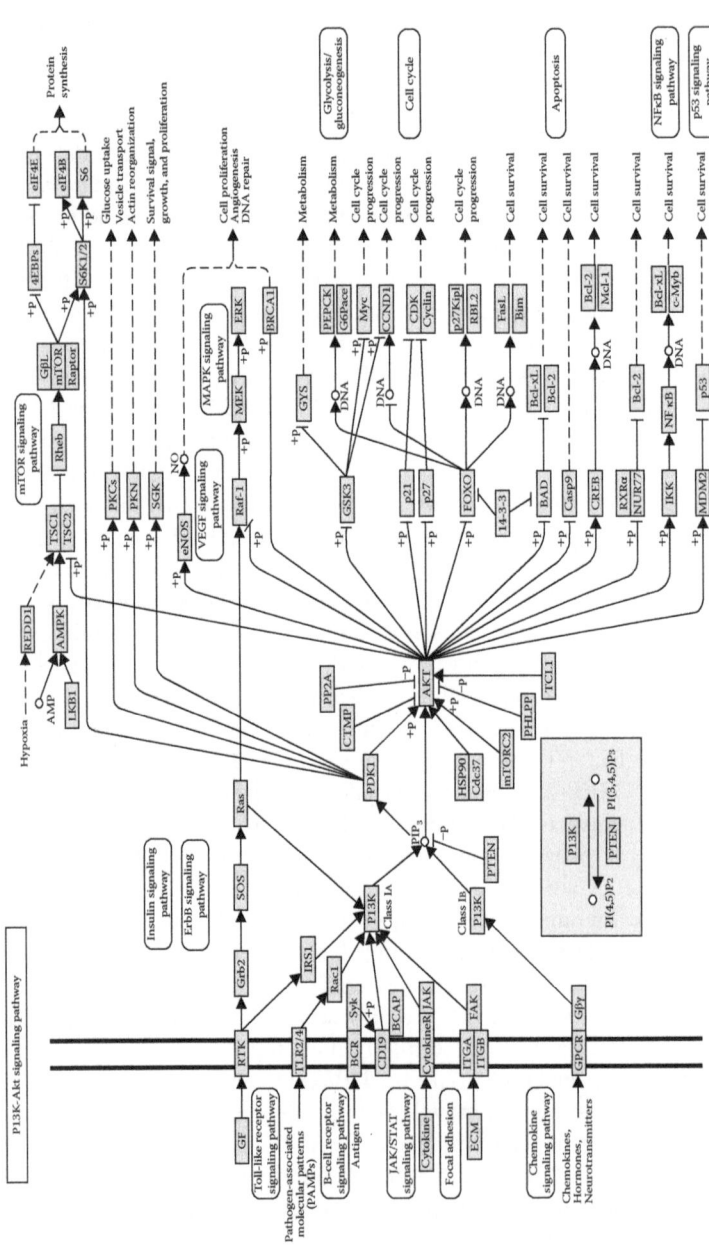

**FIGURE 13.1** Signaling network of the phosphatidylinositol 3′-kinase (PI3K)/Akt signaling pathway. This pathway is activated by many types of cellular stimuli and inflammatory and toxic insults and regulates fundamental cellular functions such as gene transcription, protein translation, proliferation, growth, and survival. The binding of growth factors to their receptor tyrosine kinase (RTK) or G protein-coupled receptors (GPCR) stimulates class Ia and Ib PI3K isoforms, respectively. PI3K catalyzes the production of phosphatidylinositol-3,4,5-triphosphate (PIP3) at the cell membrane. The pathway is activated by immune activation including Toll-like receptors, cytokine, and B-cell receptor signaling. PIP3 in turn serves as a second messenger that helps to activate Akt. Once active, Akt can control key cellular processes by phosphorylating substrates involved in apoptosis, protein synthesis, metabolism, and cell cycle. Pathway hsa04151 from Kyoto Encyclopaedia of Genes and Genomes.

in many types of human cancer. Constitutive PKB/Akt activation can occur due to amplification of PKB/Akt genes or as a result of mutations in components of the signaling pathway that activates PKB/Akt. Although the mechanisms have not yet been fully characterized, constitutive PKB/Akt signaling is believed to promote proliferation and increased cell survival and thereby contribute to cancer progression.

Akt1 is implicated in cell metabolism, survival migration, and gene expression; however, little is known about the role of other specific Akt isoforms during inflammation in vivo. The loss of Akt1 did not affect leukocyte functions in vitro, and bone marrow transplant experiments suggest that host Akt1 regulates leukocyte migration into inflamed tissues. An example of inflammation is paw-induced edema. Carrageenan-induced edema and the direct pro-permeability actions of bradykinin and histamine were reduced dramatically in $Akt1^{-/-}$ knockout versus wild-type mice (Di Lorenzo et al. 2009). These findings are supported by in vitro experiments showing that Akt1 deficiency or blockade of nitric oxide synthase markedly reduces histamine-stimulated changes in trans-endothelial electrical resistance of microvascular endothelial cells. Akt1 is necessary for acute inflammation and exerts its actions primarily through regulation of vascular permeability, leading to edema and leukocyte extravasation (Di Lorenzo et al. 2009)—this may explain why the axis PI3K/Akt is frequently targeted by antiallergic hormetins.

The network originating from Akt signaling may lead to NF-κB, p53, and mTOR signaling pathways and is inhibited by PTEN (Georgescu et al. 2010). Phosphatase and tensin homolog or phosphatase and tensin homolog deleted from chromosome 10 (PTEN) is a dual protein/lipid phosphatase, the main substrate of which is PIP3, the product of PI3K (Zhang et al. 2012). The regulation and function ascribed to PTEN have become more diverse since its discovery as a putative phosphatase mutated in many human tumors. PTEN function is positively and negatively regulated at the transcriptional level as well as posttranslationally by phosphorylation, oxidation, and acetylation. Deregulation of PTEN is implicated in other human diseases in addition to cancers, including diabetes and obesity. Actually, modulation of PTEN level has widespread therapeutic applications to those tumorigenesis and nontumor diseases (Leslie and Downes 2004; Tamguney and Stokoe 2007). As PTEN exerts enzymatic activity as a $PIP_3$ phosphatase, thus opposing the activity of PI3K, the concerted action to increase the availability of $PIP_3$ in cancer cells, relying either on other phosphoinositide enzymes or on the intrinsic regulation of PTEN activity by other molecules, is a major concern to comprehend the role of PTEN in an inflammatory event. In particular, the synergy between PTEN and the circle of its direct interacting proteins will be brought forth in an attempt to understand both the activation of the PI3K/Akt pathway and the connections with other parallel oncogenic pathways.

The understanding of the interplay between the modulators of the PI3K/Akt pathway in cancer should eventually lead to the design of therapeutic approaches with increased efficacy in the clinic (Ghayad and Cohen 2010; Almhanna et al. 2011). Increase in PIP3 recruits Akt to the membrane where it is activated by other kinases also dependent on PIP3. During inflammation or an allergic response, due to allergen challenge in mouse models, PI3K activity increases significantly, while PTEN protein expression and PTEN activity decreased during allergic inflammation, such as in ovalbumin (OVA)-induced asthma. Immunoreactive PTEN localized in epithelial layers around the bronchioles in control mice; anyway, intratracheal administration of PI3K inhibitors or adaptor of PTEN remarkably reduced bronchial inflammation and airway hyperresponsiveness (Kwak et al. 2003). Many components of this pathway have been described as causal forces in inflammation, particularly in cancer.

### 13.2.2 PI3K/Akt Signaling in Inflammation and Hormesis

Many phytochemicals may ameliorate cell disorders through a hormetic mechanism. PI3K was invoked to explain the hormetic role of Z-ligustilide in PC12 cells against oxygen glucose deprivation (OGD)-induced neuronal cell death. Z-ligustilide not only triggered stress response by causing ROS formation and transient GSH depletion but also activated survival-promoting signals through cross-talking of PI3K and NF-E2-related factor 2 (Nrf2) pathways. A key finding was that Z-ligustilide preconditioning protected PC12 cells from OGD-induced injury either at a low concentration for a prolonged period of time or at a high concentration for a short period of time. Presumably, mild preconditioning stimulated moderate ROS production, but effectively activated hormetic signals and induced stress responsive genes. In contrast, higher concentrations of Z-ligustilide could be toxic over a prolonged period of time due to massive ROS production (Qi et al. 2012).

An approach to prevent inflammation-derived insults by targeting PI3K activity has been recently attempted. PI3K is known to induce the activation of NF-κB; therefore, synthetic inhibitors of PI3K activation would demonstrate anti-inflammatory potential. Some authors have assayed a preferential p110α/γ PI3K inhibitor (compound 8C; PIK-75) in inflammation-based assays, with promising results (Dagia et al. 2010), an evidence reported also by current research on PI3K synthetic inhibitors (Bender et al. 2011; Opel et al. 2011; Guenther et al. 2013). Furthermore, natural inhibitors targeting the PI3K/Akt pathway have been also thoroughly investigated (Lee et al. 2011; Adhami et al. 2012; Cheng et al. 2013; Hyam et al. 2013). Recent investigation of the downstream effector pathways for specific peptide hormones, growth factors, and cytokines with cardioprotective properties has identified molecules involved in the progression of cardiac hypertrophy and heart failure, including PI3K/Akt signals and mTOR. Using genetically modified transgenic or knockout mice and adenoviral targeting to manipulate expression or function in experimental models of heart failure, several investigators have demonstrated that the PI3K/Akt pathway regulates cardiomyocyte size, survival, angiogenesis, and inflammation in both physiological and pathological cardiac hypertrophy (Rigor et al. 2009; Damilano et al. 2010; Aoyagi and Matsui 2011; Ghigo et al. 2011).

Approaches to prevent PI3K-related cardiac dysfunction have been recently attempted (He et al. 2012; Gao et al. 2013a). Recent evidence has shown that repetitive exposure of diabetic mice to low-dose radiation (LDR) at 25 mGy could significantly attenuate diabetes-induced renal inflammation, oxidative damage, remodeling, and dysfunction, for which the underlying mechanism yet remains unknown. A measure of 75 mGy of X-rays can stimulate the Akt signaling pathway and upregulate Nrf2 expression and function in diabetic kidneys; single exposure of 25 mGy did not, but three exposures to 25 mGy of X-rays could offer a similar effect as single exposure to 75 mGy on the stimulation of Akt phosphorylation and the upregulation of

Nrf2 expression and transcription function (Xing et al. 2012). The involvement of the PI3K/Akt signaling in radiation-induced hormesis might explain why non-ionizing radiation, inflammation, and oxidative stress response are closely related (Schonfeld et al. 2010; Xia et al. 2010). Further evidence reported that LDR (single doses from 0.3 to 1.0 Gy) in clinical practice, which is mostly used to treat patients with several inflammatory diseases and painful degenerative disorders, exert anti-inflammatory effects (Gaipl et al. 2009). However, the molecular and cellular mechanisms are not fully analyzed and most of the observed effects are based on empirical studies. A biphasic appearance of cell death in irradiated polymorphonuclear (PMN) cells was observed, displaying a relative maximum at 0.3 Gy and minimum at 0.5 Gy, respectively. This biphasic course of cell death was coincident with the protein level of total cellular Akt, suggesting the role of PI3K/Akt signaling network in radiation-induced inflammatory response (Gaipl et al. 2009).

Notwithstanding, although Akt triggering leads to the activation of various pathways related to cell survival, the roles of Akt in modulating cellular responses induced by ionizing radiation (IR) in normal human cells remain unclear. Probably, Akt activation inhibits cell death during radiation-induced apoptosis through the regulation of phosphorylation and expression of pancreas and liver cells. The Akt pathway functions as one of the important regulatory mechanisms required for modulating IR sensitivity (Park et al. 2009). This evidence hampers true comprehension of the role of Akt targeting for anti-inflammatory therapy. Experiments performed by co-immunoprecipitation showed a complex formation of activated Akt and DNA-dependent protein kinase catalytic subunit (DNA-PKcs), supporting the assumption that Akt plays an important regulatory role in the activation of DNA-PKcs in irradiated cells; this assumptions leads to the suggestion that targeting Akt enhances radiation sensitivity of lung cancer cell lines A549 and H460, most likely through specific inhibition of DNA-PKcs-dependent DNA-double strand breaks repair, but not through enhancement of radiation-induced apoptosis (Toulany et al. 2008a).

The role exerted by the PI3K/Akt against inflammation and oxidative stress caused by irradiation is intertwined with MAPKs. IR-induced X-ray repair cross-complementing group 1 protein (XRCC1) expression is dependent on the expression level of DNA-PKcs and basal activity status of PI3K/Akt signaling. Likewise, potential of IR-induced XRCC1 expression depends on its basal expression level. XRCC1 expression induced by irradiation, however, was independent of PI3K/Akt signaling, but dependent on MAPK-ERK1/2 (Toulany et al. 2008b). On the other hand, low doses of irradiation may be therapeutically improving (Wang et al. 2008). For example, multiple exposures to low doses of radiation significantly suppress diabetes-induced systemic and renal inflammatory response and renal oxidative damage, resulting in a prevention of renal dysfunction and fibrosis (Zhang et al. 2009).

Natural hormetins play also an utmost role in PI3K/Akt-mediated hormesis (Chirumbolo 2010, 2013; Kao et al. 2010; Yang et al. 2011). In this context, depending on little changes in the flavone backbone and on subtle mechanisms of cell behavior and responsiveness, flavonoids can have a modulating, biphasic, and regulatory action on immunity and inflammation by acting on several signaling pathways, particularly by affecting PI3K and NF-κB-dependent networks (Chirumbolo 2010); in this field, the number of flavones and flavonols that have been assayed is still increasing, mainly because of their chemical similarity with quercetin, although the evidence reported in the literature about the action of flavonoids is limited to a restricted group of molecules. Many of the effects reported about flavonoids regard quercetin as probably the most diffused and known nature-derived flavonol. Like many other molecules sharing a flavone ring, quercetin affects immunity and inflammation by acting mainly on leukocytes and targeting many intracellular signaling kinases and phosphatases and enzymes and membrane proteins, particularly PI3K due to its interaction with PI3K catalytic site (Walker et al. 2000; Chirumbolo et al. 2010), often crucial for a cellular-specific function.

### 13.2.3 Targeting the PTEN/PI3K/Akt/mTOR Signaling Pathway to Face at Inflammation and Cancer

The PI3K signaling pathway is activated in a broad spectrum of human cancers, either directly by genetic mutation or indirectly through activation of receptor tyrosine kinases or inactivation of the PTEN tumor suppressor. The key nodes of this pathway have emerged as important therapeutic targets for the treatment of cancer. For example, epigallocatechin-3-gallate (EGCG), a major component of green tea, is an ATP-competitive inhibitor of both PI3K and mTOR at relatively low doses, with $K_i$ values of 380 and 320 nM, respectively. The potency of EGCG against PI3K and mTOR is within physiologically relevant concentrations. In addition, EGCG inhibits cell proliferation and Akt phosphorylation at Ser473 in MDA-MB-231 and A549 cells. Molecular docking studies showed that EGCG binds well to the PI3K kinase domain active site, agreeing with the finding that EGCG competes for ATP binding (Van Aller et al. 2011).

Agents such as chronic infections, obesity, alcohol, tobacco, radiation, environmental pollutants, and a high-calorie diet have been recognized as major risk factors for the most common types of pathologies involving inflammatory response and immunity. All these risk factors are also linked to cancer through inflammation. Although it is well known that acute inflammation, which persists for the short term, mediates host defenses against infections, chronic inflammation that lasts for the long term can predispose the host to various chronic illnesses, including cancer (Aggarwal et al. 2009). While targeting PI3K/Akt signaling proteins would promise new expectancy in cancer therapy and inflammatory diseases (Hernandez-Aya and Gonzalez-Angulo 2011), some contradictory evidence was also reported, albeit together with encouraging results (Ning et al. 2007; Maxwell et al. 2012). For example, PI3K inhibition by LY294002 can synergistically enhance radiation efficacy through dephosphorylation of Akt in cervical cancer, and PI3K inhibition alone can also suppress tumor re-growth evidence that may provide novel therapeutic opportunities to enhance the effect of radiotherapy against bladder cancer cell lines and cervical cancer (Gupta et al. 2003; Liu et al. 2011). Moreover, the PI3K/Akt signaling pathway may lead to an anti-inflammatory action, through an interleukin 10-mediated response (Tapia-Abellan et al. 2013); this evidence should suggest a more complex modulating role of PI3K/Akt signaling in innate inflammation, besides an activation of NF-κB function.

In experimental autoimmune encephalomyelitis (EAE), the administration of the estrogen receptor-beta (ERβ) ligand 2,3-bis(4-hydroxyphenyl)-propionitrile (DPN) decreased clinical disease, was neuroprotective, stimulated endogenous myelination, and improved axon conduction without altering peripheral cytokine production or reducing central nervous system inflammation (Kumar et al. 2013). DPN treatment of EAE animals resulted in phosphorylated ERβ and activated the PI3K/serine-threonine-specific protein kinase (Akt)/mTOR signaling pathway, a pathway required for oligodendrocyte survival and axon myelination. The PI3K/Akt/mTOR pathway participates also in staphylococcal superantigen-induced toxicity and updates potential therapeutics against superantigens (Krakauer 2010; Krakauer 2012). Basal-like breast cancer is frequently associated with an increased activity of the PI3K pathway, which is critical for cell growth, survival, and angiogenesis. The use of Akt (MK-2206) and mTOR (MK-8669) inhibitors, together with PTEN knockdown, resulted in tumor growth inhibition; compared to control (GFP) knockdown, PTEN knockdown led to a more dramatic reduction in cell proliferation and tumor growth inhibition in response to MK-8669 and MK-2206 both in vitro and in vivo (Xu et al. 2013).

The PI3K/Akt/mTOR network plays a key regulatory function in cell survival, proliferation, migration, metabolism, angiogenesis, and apoptosis. Genetic aberrations found at different levels, with either activation of oncogenes or inactivation of tumor suppressors, make this pathway one of the most commonly disrupted in human breast cancer. The PI3K-dependent phosphorylation and activation of the serine/threonine kinase Akt is a key activator of cell survival mechanisms. The activation of the oncogene PIK3CA and the loss of regulators of Akt, including the tumor suppressor gene PTEN, are mutations commonly found in breast tumors. Akt relieves the negative regulation of mTOR to activate protein synthesis and cell proliferation through S6K and 4EBP1. The common activation of the PI3K pathway in breast cancer has led to the development of compounds targeting the effector mechanisms of the pathway, including selective and pan-PI3K/pan-Akt inhibitors (Ghayad and Cohen 2010), rapamycin analogs for mTOR inhibition, and TOR-catalytic subunit inhibitors. The influences of other oncogenic pathways such as Ras-Raf-Mek on the PI3K pathway and the known feedback mechanisms of activation have prompted the use of compounds with broader effects at multiple levels and rational combination strategies to obtain a more potent antitumor activity and possibly a meaningful clinical effect (Xu et al. 2004; Martelli et al. 2011; Barrett et al. 2012).

Recently published evidence reported an involvement of Akt in the anti-apoptotic effect induced by LiCl (Suganthi et al. 2012). Breast cancer cells respond in a diverse manner to LiCl, that is, at lower concentrations, namely 1, 5, and 10 mM, LiCl induces cell survival by inhibiting apoptosis through regulation of GSK-3β, caspase-2, Bax, and cleaved caspase-7 and by activating antiapoptotic proteins, such as Akt, β-catenin, Bcl-2, and cyclin D1. In contrast, at high concentrations (50 and 100 mM), it induces apoptosis by reversing these effects (Suganthi et al. 2012). This relationship between toxicology and genotoxic/cancerogenic substances in Akt-mediated hormesis has been reported also for BDE-47, a congener of polybrominated diphenyl ethers. After BDE-47 treatment at low concentration, the expression of proliferating cell nuclear antigen, cyclin D1, DNA-PKcs, and phosphorylated PKB (p-Akt) in the $HepG_2$ cells was markedly upregulated. However, in DNA-PKcs-inhibited cells, the promotion effect on cell proliferation was significantly suppressed, suggesting that BDE-47 had a hormesis effect in $HepG_2$ cells and DNA-PKcs/Akt pathway might be involved in regulation of cell proliferation and apoptosis (Wang et al. 2012).

Targeting PI3K/Akt signaling in hormesis still represents a challenge for current research, because of the complex pattern eliciting the hormetic response. If hormesis arises from stress response as an adaptation mechanism, the concept of adaptation across generations, by natural selection, equates to the (game theoretic) maximization of fitness (the success with which one individual produces more individuals), while self-organizing-based adaptation, within generations, equates to energetic efficiency and the matching of intake and biosynthesis they need. Emlen (1998) discusses implications of the attractor hypothesis for a wide variety of genetic and physiological phenomena, including genetic architecture, directed mutation, genetic imprinting, paramutation, hormesis, plasticity, optimality theory, genotype-phenotype linkage, and puncuated equilibrium, presenting suggestions for tests of the hypothesis. If the rate of random changes in a biological system is proportional, even only roughly, to the amount of environmental stress, a virtual force is created, acting in the direction of stress relief (Shimansky 2010): this may be the only case in which molecular targeting of PI3K or Akt/mTOR signaling lead to beneficial effects on inflammation and cancer.

## 13.3 MAP KINASE/ERK/JNK PATHWAY

### 13.3.1 Introduction

The MAPK/ERK signaling cascade, which was thoroughly described in the early 1990s and recently reviewed and updated (Seger and Krebs 1995; Kyriakis and Avruch 2012), is mainly activated by stress and inflammation (de Dios et al., 2010). Figure 13.2 describes the main network of the MAPK/ERK/JNK signaling pathway. Acute as well as chronic inflammation is thought to be central to the pathogenesis of many diseases, such as rheumatoid arthritis, asthma, chronic obstructive pulmonary disease, and acute respiratory distress syndrome. The site and specific characteristics of the inflammatory responses might be different in each of these diseases but all are characterized by the recruitment of leukocytes to the site of tissue injury and activation of immune and inflammatory cells.

The process of tissue injury is complex and requires intercellular communication between infiltrating leukocytes, endothelium, resident epithelium, alveolar macrophages, and smooth muscle cells. Migration and activation of leukocytes is initiated either by physical injury or infection, or a local immune response, or both, and requires a series of orchestrated signals. One of the many signaling pathways used is the MAPK pathway (Herlaar and Brown 1999). The participation of p38/MAPK and other components of the MAPK signaling pathway on the cellular inflammatory response have suggested in the past that MAPK targeting would be a therapeutic tool to prevent and treat inflammation (Wang et al. 1998; Lee et al. 1999, 2000; Kumar et al. 2003; Saklatvala 2004; Gao et al. 2013b). Recently, intra-coronary administration of tacrolimus in an acute myocardial infarction mini-pig model by ligating the left anterior descending coronary artery attenuated inflammation and MAPK signaling, limited infarct size, and preserved left ventricular function (Yang et al. 2013).

MAPKs and G-proteins are implicated in the cardiomyocyte inflammatory response to lipopolysaccharide (LPS) as well as cross talk through COX-2-generated $PGE_2$ (Frazier et al. 2012). Besides their involvement in inflammation, MAPKs orchestrate the recruitment of gene transcription, protein biosynthesis, cell cycle

# Inflammatory Pathways 255

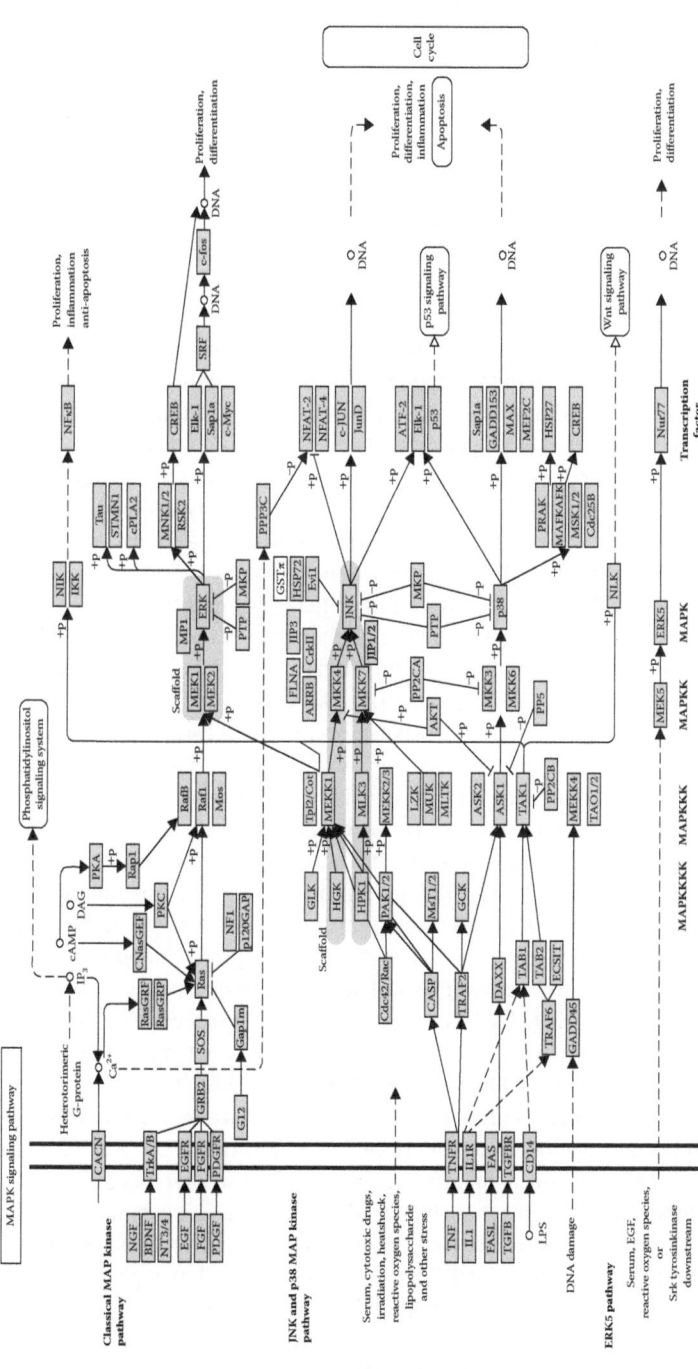

**FIGURE 13.2** Signaling network of mitogen-activated protein kinase (MAPK). The MAPK cascade is a highly conserved module that is involved in various cellular functions, including cell proliferation, differentiation, and migration. Several factors inducing a cellular stress response (serum antigens/allergens, cytotoxic drugs, irradiation, heat shock, oxygen or nitrogen reactive species, bacterial and viral products, cytokines, etc.) are active factors able to elicit a MAPK-mediated response. Mammals express at least four distinctly regulated groups of MAPKs, extracellular signal-related kinases (ERK)-1/2, c-Jun-N-terminal kinases (JNK)1/2/3, p38 proteins (p38alpha/beta/gamma/delta), and ERK5, which are activated by specific MAPKKs: MEK1/2 for ERK1/2, MKK3/6 for p38, MKK4/7 (JNKK1/2) for JNKs, and MEK5 for ERK5. Each MAPKK, however, can be activated by more than one MAPKKK, increasing the complexity and diversity of MAPK signaling. Presumably each MAPKKK confers responsiveness to distinct stimuli. For example, activation of ERK1/2 by growth factors depends on the MAPKKK c-Raf, but other MAPKKKs may activate ERK1/2 in response to pro-inflammatory stimuli. Pathway hsa04010 from Kyoto Encyclopaedia of Genes and Genomes.

control, apoptosis, and differentiation. MAPKs are phosphorylated and activated through simultaneous Tyr and Thr phosphorylation within a distinct and evolutionarily conserved Thr-X-Tyr motif in the kinase subdomain VIII activation loop. This phosphorylation is catalyzed by members of the dual-specificity MAPK kinases (MAP2Ks, also called MEKs or MKKs). The MAP2Ks are activated by Ser/Thr phosphorylation, again, within a conserved motif in kinase subdomain VIII. This phosphorylation is catalyzed by a bewilderingly termed group of protein kinase families referred to collectively as MAPK kinase kinases (MAP3Ks).

MAPK core signaling modules are themselves regulated by a poorly understood array of upstream components (Chang and Karin 2001; Keshet and Seger 2010). MAPKs are evolutionarily conserved regulators that mediate signal transduction and play essential roles in various physiological processes. There are three main families of MAPKs in mammals, whose functions are regulated by activators, inhibitors, substrates, and scaffolds, which together form delicate signaling cascades in response to different extracellular or intracellular stimulation. MAPK signaling is tightly regulated so that optimal biological activities are achieved and health is maintained (Imajo et al. 2006; Zhang and Dong 2007). However, how the specificity of the signaling flow along each cascade is achieved is still relatively unclear.

From a molecular point of view, mammals express multiple MAPK pathways. The majority of these are, along with NF-κB pathway (see discussion in Section 13.4), recruited by stress and inflammatory stimuli rather than by mitogens (Gantke et al. 2012). At least in mice, NF-κB-mediated inflammation may be exacerbated by MAPK activity. NF-κB, activated by inhibitor of κB (IκB) kinase (IKK), is a key regulator of inflammation, innate immunity, and tissue integrity. NF-κB and one of its main activators and transcriptional targets, tumor necrosis factor-alpha (TNF-α), are upregulated in many inflammatory diseases that are accompanied by tissue destruction; inflammation, which is driven by massive TNF-α production, requires additional activation of p38, ERK and MAPKs (Guma et al. 2011). Accordingly, these pathways represent a substantial trove of potentially important targets for novel anti-inflammatory drugs.

The ERKs (ERK-1 and ERK-2; MAPK-3 and MAPK-1) were the first mammalian MAPKs to be identified. The ERKs are most familiar as insulin-activated protein kinase and MAPK recruited by agonists that engage the Ras proto-oncoprotein. Ras, in turn, recruits the Raf family of MAP3Ks. The Rafs activate two ERK-specific MAP2Ks: MEK1 and MEK2 (MAP2K-1 and MAP2K-2, respectively). The biochemistry, biology, and regulation of Ras-dependent ERKs activation have been extensively reviewed (McKay and Morrison 2007; Raman et al. 2007).

The ERK1/2 pathway of mammals is probably the best-characterized MAPK system. The most important upstream activators of this pathway are the Raf proteins (A-Raf, B-Raf, or c-Raf), the key mediators of response to growth factors (EGF, FGF, PDGF, etc.), but other MAP3 kinases such as c-Mos and Tpl2/Cot can also play the same role. All these enzymes phosphorylate and thus activate MKK1 and/or MKK2 kinases, which are highly specific for ERK1 and ERK2. The latter phosphorylate a number of substrates important for cell proliferation and cell cycle progression (RSKs, Elk-1 transcription factor, etc.) (Kyriakis and Avruch 2012).

In contrast to the relatively well-insulated ERK1/2 pathway, mammalian p38 and JNKs have most of their activators shared at the MAP3K level (MEKK1, MEKK4,

ASK1, TAK1, MLK3, TAOK1, etc.). In addition, some MAP2K enzymes may activate both p38 and JNK (MKK4), while others are more specific for either JNK (MKK7) or p38 (MKK3 and MKK6). Due to these interlocks, there are very few, if any, stimuli that can elicit JNK activation without simultaneously activating or reversing p38 (Cargnello and Roux 2011). Both JNK and p38 signaling pathways are responsive to stress stimuli, such as cytokines, ultraviolet irradiation, heat shock, and osmotic shock, besides being involved in cell differentiation and apoptosis. JNKs have a number of dedicated substrates that only they can phosphorylate (c-Jun, NFAT4, etc.), while p38s also have some unique targets (e.g., the MAPKAP (MAP kinase-activated protein) kinases MK2 and MK3), ensuring the need for both to respond to stressful stimuli.

ERK5 is part of a fairly well-separated pathway in mammals. Its sole-specific upstream activator MKK5 is turned on in response to the MAP3 kinases MEKK2 and MEKK3. The specificity of these interactions is provided by the unique architecture of MKK5 and MEKK2/3, both containing N-terminal PB1 domains, enabling direct heterodimerization with each other (Nakamura et al. 2003). Although the details are poorly known, MEKK2 and MEKK3 respond to mechanical as well as other stimuli to direct endothelia formation and cardiac morphogenesis. While also implicated in brain development, the embryonic lethality of ERK5 inactivation due to cardiac abnormalities underlines its central role in mammalian vasculogenesis (Regan et al. 2002). It is notable that conditional knockout of ERK5 in adult animals is also lethal, due to the widespread disruption of endothelial barriers (Regan et al. 2002). Mutations in the upstream components of the ERK5 pathway are thought to underlie cerebral cavernous malformations (the CCM complex) in humans. In many instances, the ERKs can also be activated, in a manner independent of Ras, by pro-inflammatory stimuli, including cytokines of the TNF family, by pathogen-associated molecular pattern molecules (PAMPs), such as LPS produced by invading microbial pathogens, and by damage-associated molecular pattern molecules (DAMPs), endogenously produced danger signals such as oxidized low-density lipoprotein in atherosclerosis and crystalline uric acid in gout. PAMPs/DAMPs engage Pattern recognition receptor (PRRs), including those of the IL-1 receptor/Toll-like receptor (TLR) family. These mechanisms of ERK activation play an unexpectedly important role in innate immunity and inflammation. The ERKs are encoded by two genes, *ERK1* and *ERK2*. As with most conventional MAPKs, the ERKs require dual Tyr and Thr phosphorylation at two closely spaced residues in the activation loop of subdomain VIII: Thr185-Glu-Tyr187 (ERK2) or Thr203-Glu-Tyr205 (ERK1) (Raman et al. 2007).

### 13.3.2 MAPK/ERK Signaling in Toxicology, Inflammation, and Hormesis

The MAPK pathway seems to be fully involved also in hormetic mechanisms (Wetzker and Rubio 2012). As a matter of fact, stress has been defined as a complex response of biological systems to potentially harmful factors, so-called stressors, which typically include environmental factors such as toxins, irradiation, dietary restriction, or infectious agents. The binding of a stressor to a specific receptor in the target cell may mediate different outcomes at different doses by acting solely through that receptor. This may be true, for example, for receptors that have been shown to be involved in regeneration as well as degeneration in the target cell. For example, the PI3K/Akt signaling pathway is known to suppress apoptosis and induce proliferation (Vanhaesebroeck et al. 2010) or, contrary, to induce apoptosis (Schulthess et al. 2009). Context-dependent feedback loops, as suggested for the alternative regulatory functions of MAPK following EGF or NGF stimulation, could contribute to such alternative functions of signaling networks (Santos et al. 2007). The MAPK pathway is involved in stress following radiative exposition. For example, the activation of several members in the MAPK/ERK signaling pathway, including c-Raf, MEK, and ERK, was observed in rat mesenchymal stem cells (MSCs) exposed to 75 mGy X-rays, thus showing that hormesis induced by low-dose ionizing radiation stimulates MSC proliferations in the in vitro condition through the activation of the MAPK/ERK pathway (Liang et al. 2011). The MAPK/ERK/JNK signal axis seems to be involved also in the hormetic induction of cell growth and repair by $CO_2$ laser radiation in cultured oral cells and periodontal fibroblasts (Iwasaka et al. 2011a,b).

MAPKs are involved also in the hormetic effect exerted by metal toxicology, particularly with $CdCl_2$ and $HgCl_2$ (Hao and Hao 2011a,b). Many toxicological surveys have shown that $Cd^{2+}$ and $Hg^{2+}$ may induce cell proliferation and apoptosis through a biphasic dose-response relationship in human cells, although the mechanisms underlying this phenomenon are still unknown. However, $Cd^{2+}$ and $Hg^{2+}$ can stimulate cell proliferation at lower concentrations (0.05 and 0.5 µM), but inhibit it at higher concentrations (50 and 500 µM). Apoptosis increases at higher concentrations (50 and 500 µM) of $Cd^{2+}$ and $Hg^{2+}$. While 0.5 µM $Cd^{2+}$ and $Hg^{2+}$ decrease the JNK phosphorylation, 50 µM $Cd^{2+}$ and $Hg^{2+}$ increase the JNK and p38 phosphorylation (Hao and Hao 2009). The activation of the MAPK pathway may be involved in the biphasic effect induced by $Cd^{2+}$ and $Hg^{2+}$. Biphasic effects with low doses of metals actually involve the participation of the MAPK/ERK/JNK signaling axis (He et al. 2007; Jung et al. 2008; Jiang et al. 2009; Mantha and Jumarie 2010).

Dietary phytochemicals, such as kaempferol and genistein, can modulate MAPK signaling through a hormetic mechanism (Gopalakrishnan et al. 2006; Gopalakrishnan and Tony Kong 2008; Calabrese et al. 2010; Birringer 2011; Murakami and Ohnishi 2012). Schisandrin B, probably the most abundant dibenzocyclooctadiene isolated from the fruit of *Schisandra chinensis* (Turcz) Baillon or Wu-Wei-Zi (trans-literally meaning "the fruit of five tastes" in Chinese), is an antioxidant molecule that causes a dose-dependent and sustained increase in ROS production as well as a time-dependent activation of MAPK, particularly ERK1/2 (Lam and Ko 2012). In this context, the activation of MAPK was followed by an enhanced translocation of Nrf2 to the nucleus and the eliciting of a glutathione-dependent antioxidant response in cultured hepatocytes and cardiomyocytes. This evidence is of great importance, because Nrf2 functions as a redox-sensitive transcription factor that is translocated from the cytosol to the nucleus upon phosphorylation by upstream kinases and then binds to the antioxidant response element consensus sequence to induce a glutathione antioxidant response through the expression of antioxidant proteins such as glutathione reductase, glutathione peroxidase, and glutathione transferase.

The eliciting of adaptive responses to oxidative stress often requires one or more members of the MAPK cascade, suggesting therefore the possible involvement of oxidative stress-sensitive JNK and p38/MAPK to prevent neurodegeneration or promote a cardioprotective action. Many natural-derived polyphenols act as neuroprotective, cardioprotective, and chemopreventive agents through the involvement of a stress response by MAPK signaling (Speciale et al. 2011; Vauzour 2012) but a wider activity has been reported for many other inflammatory mechanisms, including cancer. Among these natural plant-derived components, special attention has been recently devoted to hesperetin from citrus fruit in PC12 cells (Hwang et al. 2012), to quercetin, rutin, myricetin, chrysin, EGCG, epicatechin, catechin, resveratrol, and xanthohumol in colorectal cancer cell lines (Araújo et al. 2011), to pinocembrin in rat hepatocarcinogenesis (Punvittayagul et al. 2012). Flavonoids act through an MAPK/ERK-mediated response to oxidative and chemicophysical or mechanical stress and elicit a neuroprotective (Karmarkar et al. 2011; Dai et al. 2013; Zhao et al. 2013), cardiovascular protective (Chae et al. 2007; Pan et al. 2008; Kim et al. 2011; Aggeli et al. 2013), and hepatoprotective action (Weng et al. 2011).

During inflammation, the role of MAPK in endothelial cells may be enhanced by natural phytochemicals in hormetic doses. LPS stimulation of endothelial cells induces the expression of intercellular adhesion molecule-1 (ICAM-1), a critical adhesion molecule involved in the adhesive interaction between leukocytes and endothelial cells in shock and inflammation. Although there is little literature about the role of p38/MAPK in ICAM-1 protein expression of LPS-induced endothelial cells, it is still not defined whether gene transcription is regulated by p38/MAPK in ICAM-1 expression of LPS-induced endothelial cells. Recent evidence has reported that the upregulation of ICAM-1 expression of LPS-induced endothelial cells in vitro and in vivo is mediated by p38/MAPK pathway at the level of gene transcription. The ICAM-1 expression of LPS-induced endothelial cells is characteristic of time dependence and dose dependence and tolerates chronic LPS stimulation, and the inhibition of the p38/MAPK signal pathway may be used as an approach to attenuate ICAM-1 production in the treatment of septic shock (Yan et al. 2002). While stress response occurs, MAPKs and NF-κB have to be considered as two major regulators of gene transcription and metabolism in response to oxidative, energetic, and mechanical stress in skeletal muscle. Chronic activation of these signaling pathways has been implicated in the development and perpetuation of various pathologies such as diabetes and cachexia. However, both MAPK and NF-κB are also stimulated by exercise, which promotes improvements in fuel homeostasis and can prevent skeletal muscle atrophy (Kramer and Goodyear 2007). Although limited, there is additional evidence to suggest cross talk between signaling mediated by MAPK and NF-κB with exercise.

A puzzling relationship exists between innate response through TLRs and MAPK involvement in skeletal muscle cells (Zbinden-Foncea et al. 2012). Actually TLRs are transmembrane proteins that detect a variety of molecular components mostly derived from microorganisms, but furthermore TLR2 and TLR4 are among others present in liver, adipose tissue, and skeletal muscle. Extracellular long-chain fatty acids bind TLR2 and TLR4 and induce downstream signaling cascades implicated in cellular stress and inflammatory processes (Francaux 2009). Evidence indicates that TLR activation by non esterified fatty acids (NEFAs) may participate in the development of metabolic syndrome through insulin resistance. Exercise seems to induce a downregulation of TLR expression in various tissues, a mechanism that may take part in the protective effect of exercise against insulin resistance. Moreover, TLRs seem to mediate the activation of p38/MAPK and JNK by extracellular NEFAs during endurance exercise. The activity of MAPKs should be explained in the context of stress response due to oxidative stressors derived from muscular efforts; p38α/βMAPK are involved in mediating oxidative stress–induced autophagy-related genes, suggesting that p38α/βMAPK regulates both the ubiquitin–proteasome and the autophagy–lysosome systems in the case of muscle wasting (McClung et al. 2010). Hormesis related to muscular exercise, therefore, involves fundamentally MAPK/ERK signaling (Ji et al. 2006, 2007; Kefaloyianni et al. 2006; Ji 2007).

Toxic or poisonous substances may activate MAPK-mediated signaling by involving other actors within the cell. Arsenic, for example, activates all three unfolded protein response (UPR) regulatory proteins in the skin. Arsenic induces IRE1 phosphorylation that resulted in augmented splicing of X-box binding protein 1 (XBP-1) leading to its migration to the nucleus, and also enhances transcriptional activation of downstream target proteins (Li et al. 2011). Hyperphosphorylation of PERK that induces eukaryotic translation initial factor 2α (eIF2α) in a phosphorylation-dependent manner enhances translation of ATF4, in addition to augmenting proteolytic activation of ATF6 in arsenic-treated skin. A similar increase in the expression of C/EBP homologous protein (CHOP) was also observed (Li et al. 2011). Arsenic enhanced also XBP-1s, ATF4, and ATF6 regulated downstream chaperones GRP94 and GRP78 and induced inflammation-related p38/MAPKAPK-2 MAPK signaling and alterations in Th-1/Th-2/Th-17 cytokines/chemokines and their receptors; in this context, antioxidant N-acetyl cysteine blocked arsenic-induced ROS, with a concomitant attenuation of UPR and MAPK signaling and proinflammatory cytokine/chemokine signatures (Li et al. 2011).

## 13.4 NF-κB SIGNALING PATHWAY

### 13.4.1 Introduction

NF-κB was discovered in 1986 by Sen and Baltimore (1996) through its interaction with an 11–base pair sequence in the immunoglobulin light-chain enhancer in B cells. The transcription factor NF-κB is activated by numerous stimuli. Once NF-κB is fully activated, it participates in the regulation of various target genes in different cells to exert its biological functions, most of them dealing with cell survival and inflammation. NF-κB has often been referred to as a central mediator of the immune response, since a large variety of bacteria and viruses can lead to the activation of NF-κB, which in turn controls the expression of many inflammatory cytokines, chemokines, immune receptors, and cell surface adhesion molecules (Li and Stark 2002; Bouwmeester et al. 2004; Tergaonkar 2006).

Recent studies have shown that the transcription factor NF-κB may function more generally as a central regulator of stress responses, since different stressful conditions, including physical stress, oxidative stress, and exposure to certain chemicals, also lead to NF-κB activation. Furthermore, NF-κB blocks cell apoptosis in several cell types. Taken together, these findings make it clear that NF-κB plays an important

# Inflammatory Pathways

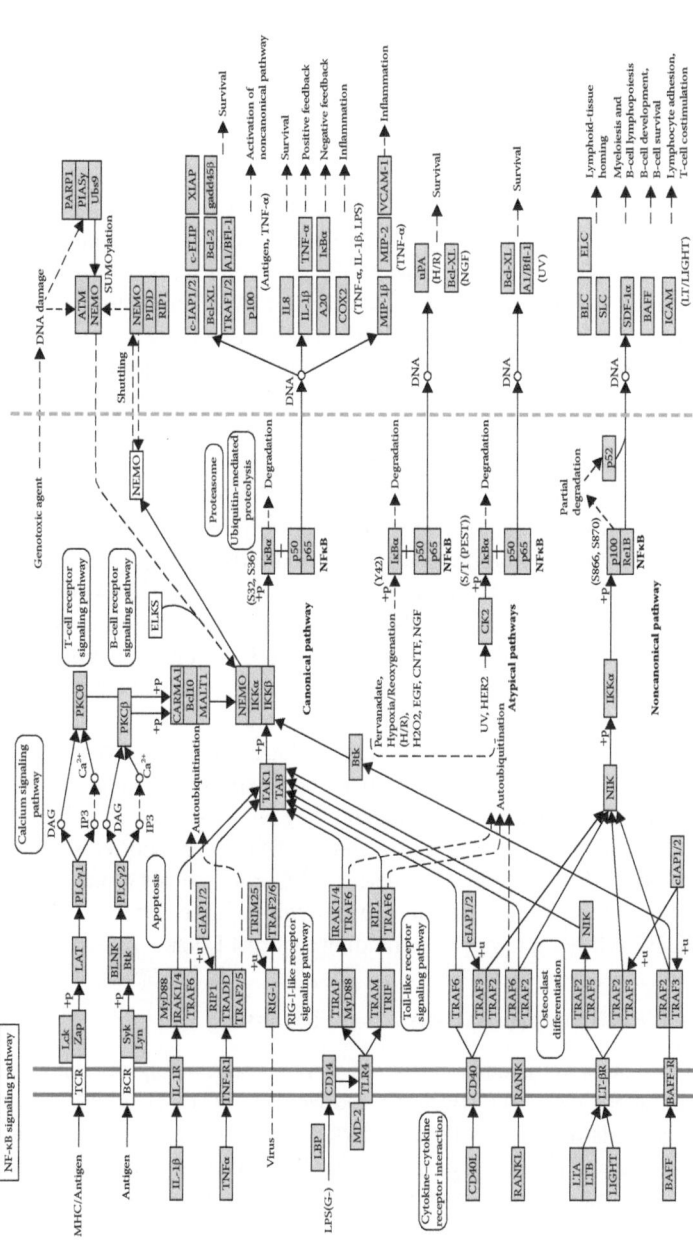

**FIGURE 13.3** Signaling network of nuclear factor-kappa B (NF-κB). The pathway is the generic name of a family of transcription factors that function as dimers and regulate genes involved in immunity, inflammation, and cell survival. There are several pathways leading to NF-κB-activation. The canonical pathway is induced by tumor necrosis factor-alpha (TNF-alpha), interleukin-1 (IL-1), or by-products of bacterial and viral infections. This pathway relies on IκB kinase (IKK)-mediated IκappaB-alpha phosphorylation on Ser32 and 36, leading to its degradation, which allows the p50/p65 NF-κB dimer to enter the nucleus and activate gene transcription. Atypical pathways are IKK-independent and rely on phosphorylation of IκappaB-alpha on Tyr42 or on Ser residues in IκappaB-alpha PEST domain. The noncanonical pathway is triggered by particular members of the TNFR = Tumor necrosis factor receptor; NIK = NF-κB inducing kinase superfamily, such as lymphotoxin-beta (LT-beta) or BAFF. It involves NIK and IKK-alpha-mediated p100 phosphorylation and processing to p52, resulting in nuclear translocation of p52/RelB heterodimers. Pathway has04064 from Kyoto Encyclopaedia of Genes and Genomes.

role in cell proliferation and differentiation. Figure 13.3 describes the main signaling network of NF-κB. The prototypical NF-κB functions as a heterodimer of p50 and p65 subunits. NF-κB is present in the cytoplasm of all cells as an inactive factor in complex with a member of the IκB inhibitor protein family. Diverse NF-κB-inducing stimuli lead to the activation of the IKK complex. IKK is a large multisubunit complex that specifically phosphorylates a pair of serine amino acid side chains in the amino-terminal region of NF-κB complex-associated IκB. Purification of IKK from cytokine-induced HeLa cells revealed that it is composed of three subunits (Häcker and Karin 2006; Huxford et al. 2011). These are referred to as IKKα (IKK1), IKKβ (IKK2), and IKKγ (NEMO, FIP3). Although IKKα and IKKβ are highly conserved protein subunits, they differ significantly in their cellular function. For example, the IKKβ subunit has been shown to be responsible for activating NF-κB in response to inflammatory stimuli by catalyzing the attachment of two phosphates near the amino-terminus of the classical IκB proteins. Furthermore, IKKβ itself is subject to phosphorylation-dependent regulation of its own catalytic activity. Once phosphorylated, IκB is recognized by a specific E3 ubiquitin–protein ligase complex, leading to its polyubiquitination. The 26S proteasome can then recognize and proteolyze IκB. Removal of IκB renders NF-κB active. Then, it rapidly translocates from the cytoplasm to the nucleus, where it binds specifically to DNA elements within the promoter regions of target genes and activates their transcription.

The classical NF-κB inhibitor proteins, IκBα, IκBβ, and IκBε, function primarily in the cell cytoplasm by masking NF-κB nuclear localization signals and blocking DNA binding. However, two additional classes of IκB proteins are also integral to NF-κB regulation. The proteins p105 and p100 play a dual role as IκB proteins and precursors of the mature NF-κB p50 and p52 subunits, respectively. The identification of a third general class of IκB proteins that function exclusively in the nucleus has been made recently (Trinh et al. 2008). The nuclear IκB proteins include Bcl-3, IκBζ (MAIL), and IκBNS. These proteins all show similar properties: their expression is regulated by NF-κB; they rapidly accumulate in the nucleus; and they have modulatory effects on NF-κB-dependent expression of specific target genes. Evidence was reported showing that in contrast to classical IκB proteins, the nuclear IκBζ protein binds preferentially to the NF-kB p50 homodimer. The formation of this protein–protein complex does not actually remove the NF-κB homodimer from binding to target DNA.

### 13.4.2 ROLE OF NF-κB IN INFLAMMATION AND HORMESIS

NF-κB transcription factors are evolutionarily conserved, coordinating regulators of immune and inflammatory responses (Baeuerle and Henkel, 1994). They also play a pivotal role in oncogenesis and metabolic disorders. Several studies during the past two decades have highlighted the key role of the IKK/NF-κB pathway in the induction and maintenance of the state of inflammation that underlies metabolic diseases such as obesity and type 2 diabetes. Recent reports, however, reveal an even more intimate connection between NF-κB and metabolism (Tornatore et al. 2012). The close relationship between inflammation and energetic balance is a hallmark of NF-κB signaling (Piva et al. 2006; Moretti et al. 2012). Due to its major role in inflammation, many studies have approached NF-κB targeting to prevent inflammation, cancer onset and development, and cardiovascular and degenerative diseases (Makarov 2000; Orlowski and Baldwin 2002; Lee et al. 2007; O'Sullivan et al. 2007; Sarkar et al. 2008; Yan and Greer 2008; Verhelst et al. 2013). The noncanonical IKKs IKKε and TANK-binding kinase 1 are induced in liver and fat by NF-κB activation upon high-fat diet feeding and in turn initiate a program of counter-inflammation that preserves energy storage. Amlexanox, an approved small-molecule therapeutic presently used in the clinic to treat aphthous ulcers and asthma, is an inhibitor of these kinases. Treatment of obese mice with amlexanox elevates energy expenditure through increased thermogenesis, producing weight loss, improved insulin sensitivity, and decreased steatosis (Reilly et al. 2013).

Actually, IKKε regulates energy balance in obese mice (Chiang et al. 2009). IKKε (-/-) mice had reduced body weight and leptin levels as well as higher insulin sensitivity when kept on chow diet. This fact did not correlate- with inflammatory parameters, measured in liver, adipose tissue, and plasma, as they were either unaltered or showed a trend toward upregulation (liver NF-κB activity, TNF-α, and IL-1β expression) (Scheja et al. 2011). IKKε might represent, therefore, a molecular bridge between obesity, metabolic syndrome, and inflammation (Olefsky 2009). This noncanonical component of the NF-κB signaling switches to different hubs' in the metabolic and energetic balance to inflammation (Calay and Hotamisligil 2013). Metabolic inflammation or "metaflammation" features the activation of several inflammatory kinases such as JNK, IKK, and protein kinase R. This has led to the working model that insulin resistance is mediated through stress-activated kinases, which may have evolved to counteract the effects of insulin, thereby diverting energy toward fighting off infection. The evidence shows that NF-κB appears to be more complex than expected. Moreover, the modulatory activity of the NF-κB pathway in survival and inflammation is further complicated by the role of micro-RNAs (Li et al. 2012b). For example, miR-34a functions as an important tumor suppressor during the process of carcinogenesis. NF-κB could elevate miR-34a expression levels through directly binding to its promoter. Wild-type p53 is responsible for NF-κB-mediated miR-34a transcriptional activity but not for NF-κB binding. Therefore, miR-34a abnormality generates human malignancies, opening new perspectives for the roles of miR-34a and NF-κB in tumor progression.

During physical exercise, the activation of the NF-κB signaling cascade has been shown to enhance the gene expression of important enzymes, such as mitochondrial superoxide dismutase (MnSOD) and inducible nitric oxide synthase (iNOS); furthermore, MAPK activations are involved in a variety of cellular functions, including growth, proliferation, and adaptation. ROS produced by xanthine oxydase may be involved in MAPK signaling, because inhibition of ROS with alkaline phosphatase (ALP) attenuated phosphorylation of these enzymes. Whether weakened MAPK signaling was related to the lowered MnSOD and iNOS expression remains to be investigated. However, it is plausible that integrated inputs from both NF-κB and MAPK signaling pathways are required to mediate gene expression of MnSOD and/or iNOS in the muscle cells in response to exercise stress (Ji et al. 2006). The activity of NF-κB can be modulated by exercise, therefore, to promote health span (Goto et al. 2007). This activity involves also the MAPK/ERK/JNK pathway (McClung et al. 2010; Zbinden-Foncea et al. 2012). Stress response, elicited by ROS and oxygen consumption as well as by energy wasting, triggers the MAPK/ERK/JNK and NF-κB pathways (Mattson 2008; Balan and Locke 2011). NF-κB represents, therefore, a bottleneck in the survival/inflammation mechanism expressed by cells. This evidence can be demonstrated also for plant-derived hormetins: dietary polyphenols also act hormetically, displaying cytoprotective effects at low doses, while excessive nutritional supplementation (i.e., high doses) can have negative consequences through the generation of more reactive and harmful intermediates with pathological consequences (Calabrese et al. 2010).

The moderate induction of stressors may give health advantages. For example, the molecular mechanism responsible for obesity-associated insulin resistance has been partially clarified, as increased fatty acid levels in muscle fibers promote diacylglycerol synthesis, which activates certain isoforms of PKC. This in turn triggers a kinase cascade that activates both IκB kinase-beta (IκK-β) and JNK, each of which can phosphorylate a key serine residue in insulin receptor substrate-1 (IRS-1), rendering it a poor substrate for the activated insulin receptor. Heat shock proteins Hsp27 and Hsp72 have the potential to prevent the activation of IκK-β and JNK, respectively; this suggests that induction of heat shock proteins may blunt the adverse impact of fat overexposure on insulin function (McCarty 2006). Indeed, bimoclomol, a heat shock protein coinducer being developed for treatment of diabetic neuropathy, and lipoic acid, suspected to be a heat shock protein inducer, have each demonstrated favorable effects on the insulin sensitivity of obese rodents, and parenteral lipoic acid is reported to improve the insulin sensitivity of type 2 diabetes (McCarty 2006). In *Drosophila melanogaster*, immunoblotting with antibodies specific for components of the Toll pathway indicated that, compared to non irradiated control or flies irradiated at 10 Gy, protein levels of the NF-κB transcriptional activator Dorsal and the interleukin-1 receptor associated kinase (IRAK)-like protein kinase Pelle were significantly increased in flies irradiated at 0.2 Gy. However, LDR did not alter expression of the transmembrane receptor Toll and Cactus, a homologue of mammalian IκBs. Additionally, LDR had no effect on the expression of Relish, a central NF-κB transcription factor in the Imd pathway, so suggesting that the effect involved only the Toll pathway (Seong et al. 2012). Mild cold shock can increase resistance to fungal infections in *D. melanogaster* (Le Bourg et al. 2009; Le Bourg 2012). In this study, low-dose irradiation of flies significantly increased resistance against gram-positive and gram-negative bacterial infections as well as expression of several antimicrobial peptide genes. Additionally, low-dose irradiation also resulted in a specific increase in expression of key proteins of the Toll signaling pathway and phosphorylated forms of p38 and JNK, but involved NF-κB to a lesser extent.

Mammalian cells can respond to damage and stress by activating various repair and survival pathways. Pre-conditioning the cells to sublethal stress is known to induce a pro-survival response that prevents damage and death. During hormesis response, PI3K/Akt and PKC activate NF-κB (Luna-Lopez et al. 2013). By investigating an oxidative-hormetic model, which was previously established in the L929 cell line by subjecting the cells to a mild oxidative stress of 50μM $H_2O_2$ for 9 hours, Luna-Lopez and colleagues identified two different translational mechanisms that participate in the regulation of Bcl-2 expression during the hormetic response, and involving PI3K/Akt signaling. These mechanisms converged in activating NF-κB. Interestingly, the noncanonical p50 subunit of the NF-κB family was apparently the subunit that participated during the oxidative-hormetic response. Whether NF-κB can be considered an attraction basin or an attractor itself is yet to be discovered. Probably, its chaotic expression depends on the noise level of stressors.

According to some authors, the balance between the production and scavenging of ROS, which should lead to homeostasis in general, is somehow shifted toward the formation of free radicals, which results in accumulated cell damage in time. We know that antioxidants can attenuate the damaging effects of ROS at least in vitro and delay many events that contribute to cellular aging. The use of multivitamin/mineral supplements has grown rapidly over the past decades; however, some recent studies demonstrated no effect of antioxidant therapy and sometimes the intake of antioxidants even increased mortality (Bjelakovic and Gluud 2007). Oxidative stress is damaging and beneficial for the organism, as some ROS are signaling molecules in cellular signaling pathways. Lowering the levels of oxidative stress by antioxidant supplements is not beneficial in such cases. The balance between ROS and antioxidants is optimal, as both extremes, oxidative and antioxidative stress, are damaging; this means, therefore, there is a need for accurate determination of an individual's oxidative stress levels before prescribing supplement antioxidants (Poljsac et al. 2013).

## 13.5 CHAOTIC PERSPECTIVE OF THE INFLAMMATORY PATHWAYS IN HORMESIS

A chaotic perspective involving concepts likewise bifurcation points, attractors and fractals, may give interesting and promising insights into the elucidation of hormesis in signaling pathways involved in stress response and inflammation (Aldridge et al. 2006; Kholodenko 2006; Novák and Tyson 2008; Wang et al. 2010; Chirumbolo 2012). Mathematical models and computer simulation may give some important suggestions to comprehend the underlying logic that rules and controls intracellular signaling pathways and the related hubs within the signal network through which hormetins and stressors at low doses are able to elicit a protective or beneficial response.

Networks created by interactive proteins in a signaling cascade may not necessarily generate a chaotic system. Those networks containing only sign-consistent loops, such as positive feedforward and feedback loops, function as monotone systems (Ma'ayan et al. 2008). These monotone systems display well-ordered behavior that excludes the possibility for chaotic dynamics. Perturbations of such systems have unambiguous global effects and a predictability characteristic that confers robustness and adaptability. It is found that the three biological networks contain far more positive "sign-consistent" feedback and feedforward loops than negative loops. Negative loops can be "eliminated" from the real networks by the removal of fewer links as compared with the corresponding shuffled networks. The abundance of positive feedforward and feedback loops in real networks emerges from the presence of hubs that are enriched with either negative or positive links. These observations have suggested that intracellular regulatory networks are "close-to-monotone," a characteristic that could contribute to the dynamical stability observed in cellular behavior. However, many networks susceptible to be tuned by subtle stimuli behave as chaotic attractors.

Huang and Ferrel (1996) formulated the first mathematical model of the MAPK cascade in the late 1990s. In their mathematical model, they concluded that the MAPK cascade was predicted to exhibit ultrasensitivity, with the degree of ultrasensitivity increasing as the cascade is descended. This behavior was robustly predicted for a wide range of assumed concentrations and $K_m$ values for the cascade enzymes and reactions, although the exact extent of the predicted ultrasensitivity varies as the assumed values are switched (Huang and Ferrel 1996). Figure 13.4 describes the kinetic scheme of MAPK cascade, according to Kholodenko (2000). The functional organization of signal transduction into protein phosphorylation cascades in itself, such as the MAPK cascade, greatly enhances the sensitivity of cellular targets to external stimuli and therefore may suggest an arguable hypothetical model to deepen the role of hormesis in cell signaling. The sensitivity increases multiplicatively with the number of cascade levels, so that a tiny change in a stimulus results in a large change in the response, a phenomenon referred to as ultra-sensitivity (Kholodenko 2000). In a variety of cell types, the MAPK cascades are imbedded in long feedback loops, positive or negative, depending on whether the terminal kinase stimulates or inhibits the activation of the initial level. Kholodenko (2000) showed that a negative feedback loop, combined with intrinsic ultrasensitivity of the MAPK cascade, could bring about sustained oscillations in MAPK phosphorylation. Actually, based on kinetic data on the MAPK cascades, the author predicted that the period of oscillations can range from minutes to hours. The phosphorylation level can vary between the base level and almost 100% of the total protein, hence the oscillations of the phosphorylation cascades and slow protein diffusion in the cytoplasm could lead to intracellular waves of phosphoproteins (Mankevich et al. 2004).

The MAPK cascade is a highly conserved series of three protein kinases implicated in diverse biological processes. The cascade arrangement has unexpected consequences for the dynamics of MAPK signaling (Zumsande and Gross 2010). ERK components in MAPK signaling behave as bistable elements. One difference was observed in PC12 cells and in neurons, where this mechanism is thought to be caused by different ERK activation dynamics, that is, EGF inducing a transient response, whereas NGF leads to a sustained ERK activation. Santos and colleagues used a combination of recently developed mathematical methods and experimentation and suggested that the difference in ERK dynamics was achieved by distinct topologies of the protein networks involved in EGF and NGF signaling (Santos et al. 2007). In particular, the activation of ERK with EGF (mediated by the EGF receptor) leads to negative feedback regulation of Raf-1 and thereby transient ERK activation. In contrast, the response to NGF (through the TrkA receptor) involves positive feedback regulation of Raf-1. The positive feedback loop leads to bistable dynamics of ERK activation, which causes sustained activation and differentiation (Kholodenko 2007). The addition of an explicit negative feedback within the MAPK signal network leads to chaotic oscillations (Kholodenko 2000), but the combination of double phosphorylation and enzyme sequestration introduces an implicit feedback leading to bistability (Markevich et al. 2004) and oscillations (Qiao et al. 2007; Hadac et al. 2013).

The chaotic approach might explain the disturbance caused by external factors on the genetic/epigenetic balance. The so-called bystander effects have dramatic consequences for DNA damage-mutation-cancer initiation paradigms of radiation carcinogenesis that provide the mechanistic justification for low-dose risk estimates. MAPK participates in cell proliferation control (Yokoyama and Nakamura 2011). If carcinogenesis does not result from directly induced DNA mutations, then the carcinogenic initiation process may not simply relate to radiation dose, and modification

# Inflammatory Pathways

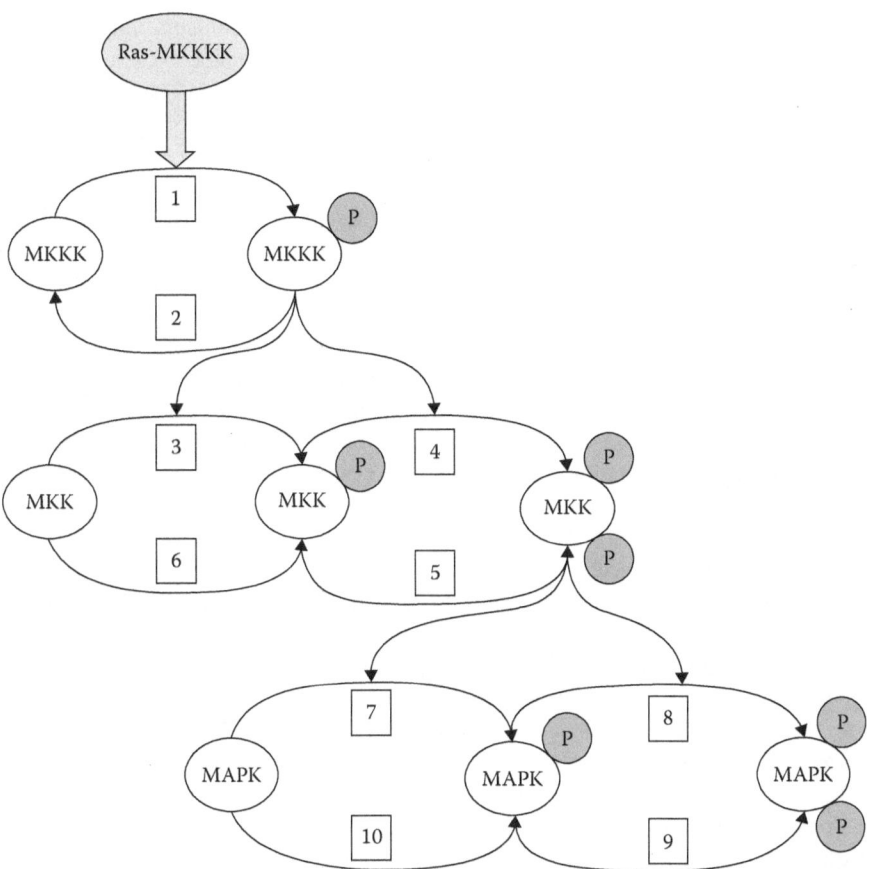

**FIGURE 13.4** Functional cascade of mitogen-activated protein kinase (MAPK) signaling according to Kholodenko. Each step is enumerated by its role in the oscillatory circle. Physicochemical models of signaling pathways are characterized by high levels of structural and parametric uncertainty, reflecting both incomplete knowledge about signal transduction and the intrinsic variability of cellular processes. As a result, these models try to predict the dynamics of systems with tens or even hundreds of free parameters, although here only a few are shown. At this level of uncertainty, model analysis should emphasize statistics of systems-level properties, rather than the detailed structure of solutions or boundaries separating different dynamic regimes. Based on the combination of random parameter search and continuation algorithms, a methodology for the statistical analysis of mechanistic signaling models was developed. In applying a methodology for the statistical analysis of mechanistic signaling models to the well-studied MAPK cascade model, a large region of oscillations explaining their emergence from single-stage bistability has been shown. The surprising abundance of strongly nonlinear (oscillatory and bistable) input/output maps revealed may be one of the reasons why the MAPK cascade in vivo is embedded in more complex regulatory structures. This type of analysis should accompany nonlinear multiparameter studies of stationary as well as transient features in network dynamics.

of the preclonal state through genetic and epigenetic mechanisms may actually occur. Some authors have proposed a chaotic or bifurcation model invoking autopoietic theory that could accommodate both beneficial (hormetic) and harmful effects of radiation at comparable doses (Mothersill and Seymour 2003).

The role of MAPK signaling appears to be relegated into the fence of stressors homeostasis. NF-κB might be the ultimate chance for the cell to reach a stress response stability. From this point of view, p38/MAPK are a group of serine/threonine protein kinases that together with ERK and JNK and other MAPKs act to convert different extracellular signals into specific cellular responses by interacting with and phosphorylating downstream targets. In contrast to the mitogenic ERK pathway, mammalian p38/MAPK family proteins (alpha, beta, gamma, and delta), with and without JNK participation, predominantly regulate inflammatory and stress response. Recent emerging evidence suggests that the p38 stress MAPK pathway may function as a tumor suppressor by regulating Ras-dependent and Ras-independent proliferation, transformation, invasion, and cell death by isoform-specific mechanisms (Loesch and Chen 2008). This is, effectively, a role quite different with respect to that of NF-κB (Schindler et al. 2007), addressed to pursue an anti-inflammatory response.

An oscillatory chaotic mechanism has been reported for the NF-κB signaling network. A sustained stimulation with an inflammatory cytokine, such as TNF-α, should induce substantial oscillations, observed at both the single cell and population levels, in the NF-κB signaling pathway (Wang et al. 2011). Although the mechanism has not yet been elucidated fully, a core system has been identified consisting of a negative feedback loop involving NF-κB (RelA:p50 heterodimer) and its inhibitor IκBα. Computational studies based on current chaotic-bifurcation models suggest that resonant interactions between periodic pulsatile forcing and the system's natural frequencies may become evident for sufficiently weak stimulation in NF-κB system. In addition, further simulations suggest that the nonlinearities of the NF-κB feedback oscillator mean that even sinusoidally modulated force can induce a rich variety of nonlinear interactions, thus leading to a bifurcated occurrence of events (survival/apoptosis, stress response/inflammation, and so forth).

Cellular rhythms are generated by complex interactions among genes, proteins, and metabolites. They are used to control every aspect of cell physiology, from signaling, motility, and development to growth, division, and death. Some authors have considered specific examples of oscillatory processes and discussed four general requirements for biochemical oscillations: negative feedback, time delay, sufficient "nonlinearity" of the reaction kinetics, and proper balancing of the timescales of opposing chemical reactions (Nelson et al. 2004). Positive feedback is one mechanism to delay the negative feedback signal. Biological oscillators can be classified according to the topology of the positive and negative feedback loops in the underlying regulatory mechanism. This oscillatory behavior, like MAPK signaling, may account for the regulation of gene expression and the genetic/epigenetic balance (Nelson et al. 2004).

NF-κB is one of the most important regulators in mammalian cells, involved in a large number of cellular responses. When the external stimulus is constant, the concentration of NF-κB inside the cell nucleus is either stationary or shows spiky oscillations. If the external stimulus is time-dependent, the response can be more intricate. The response to pulsed and periodic stimuli and the corresponding bifurcation diagram has been recently investigated (Fonslet et al. 2007). In the latter case, the time-dependent response can be chaotic with the NF-κB concentration converging to a strange attractor. The whole system constructed by a signaling network may be considered as consisted of some minimal sub-systems, represented by subsets of the whole system phase space, which are only loosely coupled to one another, each one no longer being reasonably subdivisible. For certain reasons, the dynamics of any such sub-system can be described as a trajectory in its corresponding phase subspace. Such a trajectory moves about until it encounters a region of that space called a basin of attraction, where feedback processes capture it, and then it remains there following a path referred to as an attractor, until eventually being shaken out by external forces. Self-organization may be considered as movement of the system into attractors in response to environmental changes (Emlen et al. 1998; Klonowski 2002).

## 13.6 CONCLUSIONS

Organisms are perpetually facing noxious insults but exhibit surprisingly diverse reaction patterns. Depending on the strength, frequency, and quality of the stress stimuli, biological systems may react with increased vitality, future stress resistance, or injury and degeneration. Whereas a multitude of such specific stress responses has been observed in diverse biological systems, the underlying molecular mechanisms are mainly unknown. Three main inflammation signaling pathways can be investigated in their chaotic perspective, working as chaotic oscillators able to generate bifurcation points or moving toward basins of attraction. Mathematics related to these models prevent researchers from expanding the debate about hormesis, also because of a traditional view about molecular signaling that takes into account the network hypothesis. These knowledge restrictions urge the exploration of specific molecular signaling reactions controlling the ambivalent responses of cells and organisms to noxious effects. The adaptive responses of signaling networks to defined stress stimuli need to be investigated in a time- and dose-resolved manner in cellular and organismic models by using the language of system biology. Anticipated results are expected to significantly advance the understanding of the molecular signatures of stress responses and may also promote ongoing efforts for the effective use of the organism's preventive and regenerative potentials in modern medicine.

## REFERENCES

Adhami VM, Syed DN, Khan N, Mukhtar H. Dietary flavonoid fisetin: A novel dual inhibitor of PI3K/Akt and mTOR for prostate cancer management. *Biochem Pharmacol*. 2012; 84(10): 1277–81.

Aggarwal BB, Vijayalekshmi RV, Sung B. Targeting inflammatory pathways for prevention and therapy of cancer: Short-term friend, long-term foe. *Clin Cancer Res*. 2009; 15(2): 425–30.

Aggeli IK, Koustas E, Gaitanaki C, Beis I. Curcumin acts as a pro-oxidant inducing apoptosis via JNKs in the isolated perfused Rana ridibunda heart. *J Exp Zool A Ecol Genet Physiol*. 2013; 19(6): 328–39.

Aldridge BB, Burke JM, Lauffenburger DA, Sorger PK. Physicochemical modelling of cell signalling pathways. *Nat Cell Biol.* 2006; 8(11): 1195–203.

Almhanna K, Strosberg J, Malafa M. Targeting AKT protein kinase in gastric cancer. *Anticancer Res.* 2011; 31(12): 4387–92.

Amit, DJ. *Modelling Brain Function: The World of Attractor Neural Networks.* Cambridge University Press: Cambridge; 1989.

Aoyagi T, Matsui T. Phosphoinositide-3 kinase signaling in cardiac hypertrophy and heart failure. *Curr Pharm Des.* 2011; 17(18): 1818–24.

Araújo JR, Gonçalves P, Martel F. Chemopreventive effect of dietary polyphenols in colorectal cancer cell lines. *Nutr Res.* 2011; 31(2): 77–87.

Baeuerle PA, Henkel T. Function and activation of NF-kappa B in the immune system. *Annu Rev Immunol.* 1994; 12: 141–79.

Balan M, Locke M. Acute exercise activates myocardial nuclear factor kappa B. *Cell Stress Chaperones.* 2011; 16(1): 105–11.

Barrett D, Brown VI, Grupp SA, Teachey DT. Targeting the PI3K/AKT/mTOR signaling axis in children with hematologic malignancies. *Paediatr Drugs.* 2012; 14(5): 299–316.

Bender A, Opel D, Naumann I, Kappler R, Friedman L, von Schweinitz D, Debatin KM, Fulda S. PI3K inhibitors prime neuroblastoma cells for chemotherapy by shifting the balance towards pro-apoptotic Bcl-2 proteins and enhanced mitochondrial apoptosis. *Oncogene.* 2011; 30(4): 494–503.

Bernardes AT, Zorzenon dos Santos RM. Immune network at the edge of chaos. *J Theor Biol.* 1997; 186: 173–187.

Bersini H, Calenbuhr V. Frustrated chaos in biological networks. *J Theor Biol.* 1997; 188(2): 187–200.

Birringer M. Hormetics: Dietary triggers of an adaptive stress response. *Pharm Res.* 2011; 28(11): 2680–94.

Bjelakovic G, Gluud C. Surviving antioxidant supplements. *J Natl Cancer Inst.* 2007; 99(10): 742–3.

Bouwmeester T, Bauch A, Ruffner H, Angrand PO, Bergamini G, Croughton K, Cruciat C, et al. A physical and functional map of the human TNF-alpha/NF-kappa B signal transduction pathway. *Nat Cell Biol.* 2004; 6(2): 97–105.

Buchman TG. The community of the self. *Nature.* 2002; 420: 246–51.

Calabrese V, Cornelius C, Cuzzocrea S, Iavicoli I, Rizzarelli E, Calabrese EJ. Hormesis, cellular stress response and vitagenes as critical determinants in aging and longevity. *Mol Aspects Med.* 2011; 32(4–6): 279–304.

Calabrese V, Cornelius C, Trovato A, Cavallaro M, Mancuso C, Di Rienzo L, Condorelli D, De Lorenzo A, Calabrese EJ. The hormetic role of dietary antioxidants in free radical-related diseases. *Curr Pharm Des.* 2010; 16(7): 877–83.

Calay ES, Hotamisligil GS. Turning off the inflammatory, but not the metabolic, flames. *Nat Med.* 2013; 19(3): 265–7.

Caravagna G, Barbuti R, d'Onofrio A. Fine-tuning anti-tumor immunotherapies via stochastic simulations. *BMC Bioinformatics.* 2012; 13 Suppl 4: S8

Caravagna G, d'Onofrio A, Milazzo P, Barbuti R. Tumour suppression by immune system through stochastic oscillations. *J Theor Biol.* 2010; 265(3): 336–45.

Cargnello M, Roux PP. Activation and function of the MAPKs and their substrates, the MAPK-activated protein kinases. *Microbiol Mol Biol Rev.* 2011; 75(1): 50–83.

Chae YJ, Kim CH, Ha TS, Hescheler J, Ahn HY, Sachinidis A. Epigallocatechin-3-O-gallate inhibits the angiotensin II-induced adhesion molecule expression in human umbilical vein endothelial cell via inhibition of MAPK pathways. *Cell Physiol Biochem.* 2007; 20(6): 859–66.

Chang L, Karin M. Mammalian MAP kinase signalling cascades. *Nature.* 2001; 410(6824): 37–40.

Cheng WY, Chiao MT, Liang YJ, Yang YC, Shen CC, Yang CY. Luteolin inhibits migration of human glioblastoma U-87 MG and T98G cells through downregulation of Cdc42 expression and PI3K/AKT activity. *Mol Biol Rep.* 2013; 40(9): 5315–26.

Chiang SH, Bazuine M, Lumeng CN, Geletka LM, Mowers J, White NM, Ma JT, et al. The protein kinase IKKepsilon regulates energy balance in obese mice. *Cell.* 2009; 138(5): 961–75.

Chirumbolo S. The role of quercetin, flavonols and flavones in modulating inflammatory cell function. *Inflamm Allergy Drug Targets.* 2010; 9(4): 263–85.

Chirumbolo S. Possible role of NF-κB in hormesis during ageing. *Biogerontology.* 2012; 13(6): 637–46.

Chirumbolo S. Quercetin in cancer prevention and therapy. *Integr Cancer Ther.* 2013; 12(2): 97–102.

Chirumbolo S, Marzotto M, Conforti A, Vella A, Ortolani R, Bellavite P. Bimodal action of the flavonoid quercetin on basophil function: An investigation of the putative biochemical targets. *Clin Mol Allergy.* 2010; 8: 13.

Cornelius C, Perrotta R, Graziano A, Calabrese EJ, Calabrese V. Stress responses, vitagenes and hormesis as critical determinants in aging and longevity: Mitochondria as a "chi." *Immun Ageing.* 2013; 10(1): 15.

Dagia NM, Agarwal G, Kamath DV, Chetrapal-Kunwar A, Gupte RD, Jadhav MG, Dadarkar SS, et al. A preferential p110alpha/gamma PI3K inhibitor attenuates experimental inflammation by suppressing the production of proinflammatory mediators in a NF-kappaB-dependent manner. *Am J Physiol Cell Physiol.* 2010; 298(4): C929–41.

Dai J, Chen L, Qiu YM, Li SQ, Xiong WH, Yin YH, Jia F, Jiang JY. Activations of GABAergic signaling, HSP70 and MAPK cascades are involved in baicalin's neuroprotection against gerbil global ischemia/reperfusion injury. *Brain Res Bull.* 2013; 90: 1–9.

Dalgleish A. The relevance of non-linear mathematics (chaos theory) to the treatment of cancer, the role of the immune response and the potential for vaccines. *QJM.* 1999; 92(6): 347–59.

Damilano F, Perino A, Hirsch E. PI3K kinase and scaffold functions in heart. *Ann N Y Acad Sci.* 2010; 1188: 39–45.

de Dios CH, Román E, Monge RA, Pla J. The role of MAPK signal transduction pathways in the response to oxidative stress in the fungal pathogen *Candida albicans*: Implications in virulence. *Curr Protein Pept Sci.* 2010; 11(8): 693–703.

Di Lorenzo A, Fernández-Hernando C, Cirino G, Sessa WC. Akt1 is critical for acute inflammation and histamine-mediated vascular leakage. *Proc Natl Acad Sci U S A.* 2009 Aug 25;106(34): 14552–7.

Dolmetsch RE, Lewis RS. Signaling between intracellular $Ca^{2+}$ stores and depletion-activated $Ca^{2+}$ channels generates $[Ca^{2+}]I$ oscillations in T lymphocytes. *J Gen Physiol.* 1994; 103: 365–88.

El Khattabi I, Sharma A. Preventing p38 MAPK-mediated MafA degradation ameliorates β cell dysfunction under oxidative stress. *Mol Endocrinol.* 2013; 27(7):1078–90.

Emlen JM, Freeman DC, Mills A, Graham JH. How organisms do the right thing: The attractor hypothesis. *Chaos.* 1998; 8(3): 717–26.

Enculescu M, Gholami A, Falcke M. Dynamic regimes and bifurcations in a model of actin-based motility. *Phys Rev E Stat Nonlin Soft Matter Phys.* 2008; 78(3 Pt 1): 031915.

Fonslet J, Rud-Petersen K, Krishna S, Jensen MH. Pulses and chaos: Dynamical response in a simple genetic oscillator. *Int J Modern Physics B.* 2007; 21: 4083.

Francaux M. Toll-like receptor signalling induced by endurance exercise. *Appl Physiol Nutr Metab.* 2009; 34(3): 454–8.

Franke TF, Kaplan DR, Cantley LC, Toker A. Direct regulation of the Akt proto-oncogene product by phosphatidylinositol-3,4-bisphosphate. *Science.* 1997; 275(5300): 665–8.

Frazier WJ, Xue J, Luce WA, Liu Y. MAPK signaling drives inflammation in LPS-stimulated cardiomyocytes: The route of crosstalk to G-protein-coupled receptors. *PLoS One*. 2012; 7(11): e50071.

Gaipl US, Meister S, Lödermann B, Rödel F, Fietkau R, Herrmann M, Kern PM, Frey B. Activation-induced cell death and total Akt content of granulocytes show a biphasic course after low-dose radiation. *Autoimmunity*. 2009; 42(4): 340–2.

Gantke T, Sriskantharajah S, Sadowski M, Ley SC. IκB kinase regulation of the TPL-2/ERK MAPK pathway. *Immunol Rev*. 2012; 246(1): 168–82.

Gao M, Ha T, Zhang X, Wang X, Liu L, Kalbfleisch J, Singh K, Williams D, Li C. The Toll-like receptor 9 ligand, CpG oligodeoxynucleotide, attenuates cardiac dysfunction in polymicrobial sepsis, involving activation of both phosphoinositide 3 kinase/Akt and extracellular-signal-related kinase signaling. *J Infect Dis*. 2013a; 207(9): 1471–9.

Gao JL, Lv GY, He BC, Zhang BQ, Zhang H, Wang N, Wang CZ, Du W, Yuan CS, He TC. Ginseng saponin metabolite 20(S)-protopanaxadiol inhibits tumor growth by targeting multiple cancer signaling pathways. *Oncol Rep*. 2013b; 30(1): 292–8.

Georgescu MM. PTEN tumor suppressor network in PI3K-Akt pathway control. *Genes Cancer*. 2010; 1(12): 1170–7.

Ghayad SE, Cohen PA. Inhibitors of the PI3K/Akt/mTOR pathway: New hope for breast cancer patients. *Recent Pat Anticancer Drug Discov*. 2010; 5(1): 29–57.

Ghigo A, Morello F, Perino A, Damilano F, Hirsch E. Specific PI3K isoform modulation in heart failure: Lessons from transgenic mice. *Curr Heart Fail Rep*. 2011; 8(3): 168–75.

Gopalakrishnan A, Tony Kong AN. Anticarcinogenesis by dietary phytochemicals: Cytoprotection by Nrf2 in normal cells and cytotoxicity by modulation of transcription factors NFkappaB and AP-1 in abnormal cancer cells. *Food Chem Toxicol*. 2008; 46: 1257–70.

Gopalakrishnan A, Xu CJ, Nair SS, Chen C, Hebbar V, Kong AN. Modulation of activator protein-1 (AP-1) and MAPK pathway by flavonoids in human prostate cancer PC3 cells. *Arch Pharm Res*. 2006; 8: 633–44.

Goto S, Naito H, Kaneko T, Chung HY, Radák Z. Hormetic effects of regular exercise in aging: Correlation with oxidative stress. *Appl Physiol Nutr Metab*. 2007; 32(5): 948–53.

Guenther M, Graab U, Fulda S. Synthetic lethal interaction between PI3K/Akt/mTOR and Ras/MEK/ERK pathway inhibition in rhabdomyosarcoma. *Cancer Lett*. 2013; 337(2): 200–9. doi:pii: S0304-3835(13)00379-0. 10.1016/j.canlet.2013.05.010.

Guma M, Stepniak D, Shaked H, Spehlmann ME, Shenouda S, Cheroutre H, Vicente-Suarez I, Eckmann L, Kagnoff MF, Karin M. Constitutive intestinal NF-κB does not trigger destructive inflammation unless accompanied by MAPK activation. *J Exp Med*. 2011; 208(9): 1889–900. doi:10.1084/jem.20110242. Epub 2011 Aug 8. Erratum in: *J Exp Med*. 2012; 209(10): 1901.

Gupta AK, Cerniglia GJ, Mick R, Ahmed MS, Bakanauskas VJ, Muschel RJ, McKenna WG. Radiation sensitization of human cancer cells in vivo by inhibiting the activity of PI3K using LY294002. *Int J Radiat Oncol Biol Phys*. 2003; 56(3): 846–53.

Häcker H, Karin M. Regulation and function of IKK and IKK-related kinases. *Sci STKE*. 2006; 2006(357): re13.

Hadač O, Schreiber I, Přibyl M. On the origin of bistability in the stage 2 of the Huang-Ferrell model of the MAPK signaling. *J Chem Phys*. 2013; 138(6): 065102.

Hao C, Hao W. Role of ERK in the hormesis induced by cadmium chloride in HEK293 cells. *[J Hygiene Research]*. 2011a; 40(4): 517–22.

Hao C, Hao W. The role of MAPK in the hormesis induced by $CdCl_2$ and $HgCl_2$ in RAW 264.7 cells. *J Health Toxicol*. 2011b; 03: 1–11

Hao C, Hao W, Wei X, Xing L, Jiang J, Shang L. The role of MAPK in the biphasic dose-response phenomenon induced by cadmium and mercury in HEK293 cells. *Toxicol In Vitro*. 2009; 23(4): 660–6.

He XQ, Chen R, Yang P, Li AP, Zhou JW, Liu QZ. Biphasic effect of arsenite on cell proliferation and apoptosis is associated with the activation of JNK and ERK1/2 in human embryo lung fibroblast cells. *Toxicol Appl Pharmacol.* 2007; 220(1): 18–24.

He D, Liu X, Pang Y, Liu L. Inhibitory effect of resveratrol on ischemia reperfusion-induced cardiocyte apoptosis and its relationship with PI3K-Akt signaling pathway. *[China Journal of Chinese Materia Medica].* 2012; 37(15): 2323–6.

Herlaar E, Brown Z. p38 MAPK signalling cascades in inflammatory disease. *Mol Med Today.* 1999; 5(10): 439–47.

Hernandez-Aya LF, Gonzalez-Angulo AM. Targeting the phosphatidylinositol 3-kinase signaling pathway in breast cancer. *Oncologist.* 2011; 16(4): 404–14.

Huang CY, Ferrell JE Jr. Ultrasensitivity in the mitogen-activated protein kinase cascade. *Proc Natl Acad Sci USA.* 1996; 93(19): 10078–83.

Huxford T, Hoffmann A, Ghosh G. Understanding the logic of IκB:NF-κB regulation in structural terms. *Curr Top Microbiol Immunol.* 2011; 349: 1–24.

Hwang SL, Lin JA, Shih PH, Yeh CT, Yen GC. Pro-cellular survival and neuroprotection of citrus flavonoid: The actions of hesperetin in PC12 cells. *Food Funct.* 2012; 3(10): 1082–90.

Hyam SR, Lee IA, Gu W, Kim KA, Jeong JJ, Jang SE, Han MJ, Kim DH. Arctigenin ameliorates inflammation in vitro and in vivo by inhibiting the PI3K/AKT pathway and polarizing M1 macrophages to M2-like macrophages. *Eur J Pharmacol.* 2013; 708(1–3): 21–9.

Iacono G, Altafini C. Monotonicity, frustration, and ordered response: An analysis of the energy landscape of perturbed large-scale biological networks. *BMC Syst Biol.* 2010; 4: 83.

Jerne NK. Towards a network theory of the immune system. *Ann Immunol Inst Pasteur.* 1974; 125C: 373–89.

Ji LL. Antioxidant signaling in skeletal muscle: A brief review. *Exp Gerontol.* 2007; 42(7): 582–93.

Ji LL, Gomez-Cabrera MC, Vina J. Exercise and hormesis: Activation of cellular antioxidant signaling pathway. *Ann N Y Acad Sci.* 2006; 1067: 425–35.

Ji LL, Gomez-Cabrera MC, Vina J. Role of nuclear factor kappaB and mitogen-activated protein kinase signaling in exercise-induced antioxidant enzyme adaptation. *Appl Physiol Nutr Metab.* 2007; 32(5): 930–5.

Jiang G, Duan W, Xu L, Song S, Zhu C, Wu L. Biphasic effect of cadmium on cell proliferation in human embryo lung fibroblast cells and its molecular mechanism. *Toxicol In Vitro.* 2009; 23(6): 973–8.

Jung YS, Jeong EM, Park EK, Kim YM, Sohn S, Lee SH, Baik EJ, Moon CH. Cadmium induces apoptotic cell death through p38 MAPK in brain microvessel endothelial cells. *Eur J Pharmacol.* 2008; 578(1): 11–8.

Kao TC, Shyu MH, Yen GC. Glycyrrhizic acid and 18beta-glycyrrhetinic acid inhibit inflammation via PI3K/Akt/GSK3beta signaling and glucocorticoid receptor activation. *J Agric Food Chem.* 2010; 58(15): 8623–9.

Karmarkar SW, Bottum KM, Krager SL, Tischkau SA. ERK/MAPK is essential for endogenous neuroprotection in SCN2.2 cells. *PLoS One.* 2011; 6(8): e23493.

Kefaloyianni E, Gaitanaki C, Beis I. ERK1/2 and p38-MAPK signalling pathways, through MSK1, are involved in NF-kappaB transactivation during oxidative stress in skeletal myoblasts. *Cell Signal.* 2006; 18(12): 2238–51.

Keshet Y, Seger R. The MAP kinase signaling cascades: A system of hundreds of components regulates a diverse array of physiological functions. *Methods Mol Biol.* 2010; 661: 3–38.

Kholodenko BN. Negative feedback and ultrasensitivity can bring about oscillations in the mitogen-activated protein kinase cascades. *Eur J Biochem.* 2000; 267(6): 1583–8.

Kholodenko BN. Cell-signalling dynamics in time and space. *Nat Rev Mol Cell Biol.* 2006; 7(3): 165–76.

Kholodenko BN. Untangling the signalling wires. *Nat Cell Biol.* 2007; 9(3): 247–9.

Kim SW, Kim CE, Kim MH. Flavonoids inhibit high glucose-induced up-regulation of ICAM-1 via the p38 MAPK pathway in human vein endothelial cells. *Biochem Biophys Res Commun.* 2011; 415(4): 602–7.

Klonowski W. Chaotic dynamics applied to signal complexity in phase space and time domain. *Chaos Solit Fract.* 2002; 14(9): 1379–87.

Krakauer T. Therapeutic down-modulators of staphylococcal superantigen-induced inflammation and toxic shock. *Toxins (Basel).* 2010; 2(8): 1963–83.

Krakauer T. PI3K/Akt/mTOR, a pathway less recognized for staphylococcal superantigen-induced toxicity. *Toxins (Basel).* 2012; 4(11): 1343–66.

Kramer HF, Goodyear LJ. Exercise, MAPK, and NF-kappaB signaling in skeletal muscle. *J Appl Physiol.* 2007; 103(1): 388–95.

Krupa M, Szmolyan P. Relaxation oscillation and canard explosion. *J Diff Equat.* 2001; 174(2): 312–68.

Kumar S, Boehm J, Lee JC. p38 MAP kinases: Key signalling molecules as therapeutic targets for inflammatory diseases. *Nat Rev Drug Discov.* 2003; 2(9): 717–26.

Kumar R, Clermont G, Vodovotz Y, Chow CC. The dynamics of acute inflammation. *J Theor Biol.* 2004; 230(2): 145–55.

Kumar S, Patel R, Moore S, Crawford DK, Suwanna N, Mangiardi M, Tiwari-Woodruff SK. Estrogen receptor β ligand therapy activates PI3K/Akt/mTOR signaling in oligodendrocytes and promotes remyelination in a mouse model of multiple sclerosis. *Neurobiol Dis.* 2013; 56C: 131–44.

Kwak YG, Song CH, Yi HK, Hwang PH, Kim JS, Lee KS, Lee YC. Involvement of PTEN in airway hyperresponsiveness and inflammation in bronchial asthma. *J Clin Invest.* 2003; 111(7): 1083–92.

Kyriakis JM, Avruch J. Mammalian MAPK signal transduction pathways activated by stress and inflammation: A 10-year update. *Physiol Rev.* 2012; 92(2): 689–737.

Imajo M, Tsuchiya Y, Nishida E. Regulatory mechanisms and functions of MAP kinase signaling pathways. *IUBMB Life.* 2006; 58(5–6): 312–7.

Iwasaka K, Hemmi E, Tomita K, Ishihara S, Katayama T, Sakagami H. Effect of $CO_2$ laser irradiation on hormesis induction in human pulp and periodontal ligament fibroblasts. *In Vivo.* 2011a; 25(5): 787–93.

Iwasaka K, Tomita K, Ozawa Y, Katayama T, Sakagami H. Effect of $CO_2$ laser irradiation on hormesis induction in cultured oral cells. *In Vivo.* 2011b; 25(1): 93–8.

Lam PY, Ko KM. Schisandrin B as a hormetic agent for preventing age-related neurodegenerative diseases. *Oxid Med Cell Longev.* 2012; 2012: 250825.

Launay P, Cheng H, Srivatsan S, Penner R, Fleig A, Kinet JP. TRPM4 regulates calcium oscillations after T cell activation. *Science.* 2004; 306(5700): 1374–7.

Le Bourg E. Combined effects of two mild stresses (cold and hypergravity) on longevity, behavioral aging, and resistance to severe stresses in *Drosophila melanogaster*. *Biogerontology.* 2012; 13(3): 313–28.

Le Bourg E, Massou I, Gobert V. Cold stress increases resistance to fungal infection throughout life in *Drosophila melanogaster*. *Biogerontology.* 2009; 10(5): 613–25.

Lee YC, Cheng TH, Lee JS, Chen JH, Liao YC, Fong Y, Wu CH, Shih YW. Nobiletin, a citrus flavonoid, suppresses invasion and migration involving FAK/PI3K/Akt and small GTPase signals in human gastric adenocarcinoma AGS cells. *Mol Cell Biochem.* 2011; 347(1–2): 103–15.

Lee CH, Jeon YT, Kim SH, Song YS. NF-kappaB as a potential molecular target for cancer therapy. *Biofactors.* 2007; 29(1): 19–35.

Lee JC, Kassis S, Kumar S, Badger A, Adams JL. p38 mitogen-activated protein kinase inhibitors—Mechanisms and therapeutic potentials. *Pharmacol Ther.* 1999; 82(2–3): 389–97.

Lee JC, Kumar S, Griswold DE, Underwood DC, Votta BJ, Adams JL. Inhibition of p38 MAP kinase as a therapeutic strategy. *Immunopharmacology*. 2000; 47(2–3): 185–201.

Leslie NR, Downes CP. PTEN function: How normal cells control it and tumour cells lose it. *Biochem J*. 2004; 382(Pt 1): 1–11.

Li J, Hua X, Haubrock M, Wang J, Wingender E. The architecture of the gene regulatory networks of different tissues. *Bioinformatics*. 2012; 28(18): i509–i514.

Li X, Stark GR. NFkappaB-dependent signaling pathways. *Exp Hematol*. 2002; 30(4): 285–96.

Li J, Wang K, Chen X, Meng H, Song M, Wang Y, Xu X, Bai Y. Transcriptional activation of microRNA-34a by NF-kappa B in human esophageal cancer cells. *BMC Mol Biol*. 2012b; 13: 4.

Li C, Xu J, Li F, Chaudhary SC, Weng Z, Wen J, Elmets CA, Ahsan H, Athar M. Unfolded protein response signaling and MAP kinase pathways underlie pathogenesis of arsenic-induced cutaneous inflammation. *Cancer Prev Res (Phila)*. 2011; 4(12): 2101–9.

Liang X, So YH, Cui J, Ma K, Xu X, Zhao Y, Cai L, Li W. The low-dose ionizing radiation stimulates cell proliferation via activation of the MAPK/ERK pathway in rat cultured mesenchymal stem cells. *J Radiat Res*. 2011; 52(3): 380–6.

Liu Y, Cui B, Qiao Y, Zhang Y, Tian Y, Jiang J, Ma D, Kong B. Phosphoinositide-3-kinase inhibition enhances radiosensitization of cervical cancer in vivo. *Int J Gynecol Cancer*. 2011; 21(1): 100–5.

Liu R, Wu CX, Zhou D, Yang F, Tian S, Zhang L, Zhang TT, Du GH. Pinocembrin protects against β-amyloid-induced toxicity in neurons through inhibiting receptor for advanced glycation end products (RAGE)-independent signaling pathways and regulating mitochondrion-mediated apoptosis. *BMC Med*. 2012; 10: 105.

Loesch M, Chen G. The p38 MAPK stress pathway as a tumor suppressor or more? *Front Biosci*. 2008; 13: 3581–93.

Luna-López A, González-Puertos VY, Romero-Ontiveros J, Ventura-Gallegos JL, Zentella A, Gomez-Quiroz LE, Königsberg M. A noncanonical NF-κB pathway through the p50 subunit regulates Bcl-2 overexpression during an oxidative-conditioning hormesis response. *Free Radic Biol Med*. 2013 Oct;63:41–50

Ma'ayan A, Lipshtat A, Iyengar R, Sontag ED. Proximity of intracellular regulatory networks to monotone systems. *IET Syst Biol*. 2008; 2(3): 103–12.

Makarov SS. NF-kappaB as a therapeutic target in chronic inflammation: Recent advances. *Mol Med Today*. 2000; 6(11): 441–8.

Mantha M, Jumarie C. Cadmium-induced hormetic effect in differentiated Caco-2 cells: ERK and p38 activation without cell proliferation stimulation. *J Cell Physiol*. 2010; 224(1): 250–61.

Markevich NI, Hoek JB, Kholodenko BN. Signaling switches and bistability arising from multisite phosphorylation in protein kinase cascades. *J Cell Biol*. 2004; 164(3): 353–9.

Martelli AM, Evangelisti C, Chappell W, Abrams SL, Bäsecke J, Stivala F, Donia M, et al. Targeting the translational apparatus to improve leukemia therapy: Roles of the PI3K/PTEN/Akt/mTOR pathway. *Leukemia*. 2011; 25(7): 1064–79.

Mattson MP. Hormesis and disease resistance: Activation of cellular stress response pathways. *Hum Exp Toxicol*. 2008; 27(2): 155–62.

Maxwell MJ, Tsantikos E, Kong AM, Vanhaesebroeck B, Tarlinton DM, Hibbs ML. Attenuation of phosphoinositide 3-kinase δ signaling restrains autoimmune disease. *J Autoimmun*. 2012; 38(4): 381–91.

McCarty MF. Induction of heat shock proteins may combat insulin resistance. *Med Hypotheses*. 2006; 66(3): 527–34.

McClung JM, Judge AR, Powers SK, Yan Z. p38 MAPK links oxidative stress to autophagy-related gene expression in cachectic muscle wasting. *Am J Physiol Cell Physiol*. 2010; 298(3): C542–9.

McKay MM, Morrison DK. Integrating signals from RTKs to ERK/MAPK. *Oncogene*. 2007; 26(22): 3113–21.

Moretti M, Bennett J, Tornatore L, Thotakura AK, Franzoso G. Cancer: NF-κB regulates energy metabolism. *Int J Biochem Cell Biol*. 2012; 44(12): 2238–43.

Mothersill C, Seymour C. Radiation-induced bystander effects, carcinogenesis and models. *Oncogene*. 2003; 22(45): 7028–33.

Murakami A, Ohnishi K. Target molecules of food phytochemicals: Food science bound for the next dimension. *Food Funct*. 2012; 3(5): 462–76.

Nakamura K, Johnson GL. PB1 domains of MEKK2 and MEKK3 interact with the MEK5 PB1 domain for activation of the ERK5 pathway. *J Biol Chem*. 2003; 278(39): 36989–92.

Nelson DE, Ihekwaba AE, Elliott M, Johnson JR, Gibney CA, Foreman BE, Nelson G, et al. Oscillations in NF-kappaB signaling control the dynamics of gene expression. *Science*. 2004; 306(5696): 704–8.

Ning S, Chen Z, Dirks A, Husbeck B, Hsu M, Bedogni B, O'Neill M, Powell MB, Knox SJ. Targeting integrins and PI3K/Akt-mediated signal transduction pathways enhances radiation-induced anti-angiogenesis. *Radiat Res*. 2007; 168(1): 125–33.

Novák B, Tyson JJ. Design principles of biochemical oscillators. *Nat Rev Mol Cell Biol*. 2008; 9(12): 981–91.

Olefsky JM. IKKepsilon: A bridge between obesity and inflammation. *Cell*. 2009; 138(5): 834–6.

Opel D, Naumann I, Schneider M, Bertele D, Debatin KM, Fulda S. Targeting aberrant PI3K/Akt activation by PI103 restores sensitivity to TRAIL-induced apoptosis in neuroblastoma. *Clin Cancer Res*. 2011 May 15;17(10): 3233–47.

Orlowski RZ, Baldwin AS Jr. NF-kappaB as a therapeutic target in cancer. *Trends Mol Med*. 2002; 8(8): 385–9.

O'Sullivan B, Thompson A, Thomas R. NF-kappa B as a therapeutic target in autoimmune disease. *Expert Opin Ther Targets*. 2007; 11(2): 111–22.

Pan W, Chang MJ, Booyse FM, Grenett HE, Bradley KM, Wolkowicz PE, Shang Q, Tabengwa EM. Quercetin induced tissue-type plasminogen activator expression is mediated through Sp1 and p38 mitogen-activated protein kinase in human endothelial cells. *J Thromb Haemost*. 2008; 6(6): 976–85.

Park HS, Yun Y, Kim CS, Yang KH, Jeong M, Ahn SK, Jin YW, Nam SY. A critical role for AKT activation in protecting cells from ionizing radiation-induced apoptosis and the regulation of acinus gene expression. *Eur J Cell Biol*. 2009; 88(10): 563–75.

Piva R, Belardo G, Santoro MG. NF-kappaB: A stress-regulated switch for cell survival. *Antioxid Redox Signal*. 2006; 8(3–4): 478–86.

Poljsac B, Šuput D, IrinaMilisav I. Achieving the balance between ROS and antioxidants: When to use the synthetic antioxidants. *Oxidat Med Cell Longev*. 2013; 2013: 956792. doi:10.1155/2013/956792.

Punvittayagul C, Pompimon W, Wanibuchi H, Fukushima S, Wongpoomchai R. Effects of pinocembrin on the initiation and promotion stages of rat hepatocarcinogenesis. *Asian Pac J Cancer Prev*. 2012; 13(5): 2257–61.

Qi H, Han Y, Rong J. Potential roles of PI3K/Akt and Nrf2-Keap1 pathways in regulating hormesis of Z-ligustilide in PC12 cells against oxygen and glucose deprivation. *Neuropharmacology*. 2012; 62(4): 1659–70.

Qiao L, Nachbar RB, Kevrekidis IG, Shvartsman SY. Bistability and oscillations in the Huang-Ferrell model of MAPK signaling. *PLoS Comput Biol*. 2007; 3(9): 1819–26.

Raman M, Chen W, Cobb MH. Differential regulation and properties of MAPKs. *Oncogene*. 2007; 26(22): 3100–12.

Regan CP, Li W, Boucher DM, Spatz S, Su MS, Kuida K. Erk5 null mice display multiple extraembryonic vascular and embryonic cardiovascular defects. *Proc Natl Acad Sci USA*. 2002; 99(14): 9248–53.

Reilly SM, Chiang SH, Decker SJ, Chang L, Uhm M, Larsen MJ, Rubin JR, et al. An inhibitor of the protein kinases TBK1 and IKK-ε improves obesity-related metabolic dysfunctions in mice. *Nat Med*. 2013; 19(3): 313–21.

Rigor DL, Bodyak N, Bae S, Choi JH, Zhang L, Ter-Ovanesyan D, He Z, et al. Phosphoinositide 3-kinase kt signaling pathway interacts with protein kinase Cbeta2 in the regulation of physiologic developmental hypertrophy and heart function. *Am J Physiol Heart Circ Physiol*. 2009; 296(3): H566–72.

Saklatvala J. The p38 MAP kinase pathway as a therapeutic target in inflammatory disease. *Curr Opin Pharmacol*. 2004; 4(4): 372–7.

Santos SD, Verveer PJ, Bastiaens PI. Growth factor-induced MAPK network topology shapes Erk response determining PC-12 cell fate. *Nat Cell Biol*. 2007; 9(3): 324–30.

Sarkar FH, Li Y, Wang Z, Kong D. NF-kappaB signaling pathway and its therapeutic implications in human diseases. *Int Rev Immunol*. 2008; 27(5): 293–319.

Scheja L, Heese B, Seedorf K. Beneficial effects of IKKε-deficiency on body weight and insulin sensitivity are lost in high fat diet-induced obesity in mice. *Biochem Biophys Res Commun*. 2011; 407(2): 288–94.

Schindler JF, Monahan JB, Smith WG. p38 pathway kinases as anti-inflammatory drug targets. *J Dent Res*. 2007 Sep; 86(9): 800–11.

Schonfeld SJ, Bhatti P, Brown EE, Linet MS, Simon SL, Weinstock RM, Hutchinson AA, et al. Polymorphisms in oxidative stress and inflammation pathway genes, low-dose ionizing radiation, and the risk of breast cancer among US radiologic technologists. *Cancer Causes Control*. 2010; 21(11): 1857–66.

Schulthess FT, Paroni F, Sauter NS, Shu L, Ribaux P, Haataja L, Strieter RM, Oberholzer J, King CC, Maedler K. CXCL10 impairs beta cell function and viability in diabetes through TLR4 signaling. *Cell Metab*. 2009; 9(2): 125–39.

Sherrington, D. Complexity due to disorder and frustration. In *Lectures in the Sciences of Complexity—SFI Studies in the Sciences of Complexity—Lect. Vol. II*, Ed. Erica Jen. Addison-Wesley, Santa Fe (NM) (USA) pp. 415–455; 1990.

Shimansky YP. Adaptive force produced by stress-induced regulation of random variation intensity. *Biol Cybern*. 2010; 103(2): 135–50.

Seger R, Krebs EG. The MAPK signaling cascade. *FASEB J*. 1995; 9(9): 726–35.

Sen R, Baltimore D. Multiple nuclear factors interact with the immunoglobulin enhancer sequences. *Cell*. 1986; 46(5): 705–16.

Seong KM, Kim CS, Lee BS, Nam SY, Yang KH, Kim JY, Park JJ, Min KJ, Jin YW. Low-dose radiation induces *Drosophila* innate immunity through Toll pathway activation. *J Radiat Res*. 2012; 53(2): 242–9.

Speciale A, Chirafisi J, Saija A, Cimino F. Nutritional antioxidants and adaptive cell responses: An update. *Curr Mol Med*. 2011; 11(9): 770–89.

Stark J, Chan C, George AJ. Oscillations in the immune system. *Immunol Rev*. 2007; 216: 213–31.

Stark M. Hormesis, adaptation, and the sandpile model. *Crit Rev Toxicol*. 2008; 38(7): 641–4.

Stark M. The sandpile model: Optimal stress and hormesis. *Dose Response*. 2012; 10(1): 66–74.

Suganthi M, Sangeetha G, Gayathri G, Ravi Sankar B. Biphasic dose-dependent effect of lithium chloride on survival of human hormone-dependent breast cancer cells (MCF-7). *Biol Trace Elem Res*. 2012; 150(1–3): 477–86.

Tamguney T, Stokoe D. New insights into PTEN. *J Cell Sci*. 2007; 120(Pt 23): 4071–9.

Tapia-Abellán A, Ruiz-Alcaraz AJ, Hernández-Caselles T, Such J, Francés R, García-Peñarrubia P, Martínez-Esparza M. Role of MAP kinases and PI3K-Akt on the cytokine inflammatory profile of peritoneal macrophages from the ascites of cirrhotic patients. *Liver Int*. 2013; 33(4): 552–60.

Tergaonkar V. NFkappaB pathway: A good signaling paradigm and therapeutic target. *Int J Biochem Cell Biol*. 2006; 38(10): 1647–53.

Tieri P, Grignolio A, Zaikin A, Mishto M, Remondini D, Castellani GC, Franceschi C. Network, degeneracy and bow tie integrating paradigms and architectures to grasp the complexity of the immune system. *Theor Biol Med Model*. 2010; 7: 32.

Tornatore L, Thotakura AK, Bennett J, Moretti M, Franzoso G. The nuclear factor kappa B signaling pathway: Integrating metabolism with inflammation. *Trends Cell Biol*. 2012; 22(11): 557–66.

Toulany M, Dittmann K, Fehrenbacher B, Schaller M, Baumann M, Rodemann HP. PI3K-Akt signaling regulates basal, but MAP-kinase signaling regulates radiation-induced XRCC1 expression in human tumor cells in vitro. *DNA Repair (Amst)*. 2008b; 7(10): 1746–56.

Toulany M, Kehlbach R, Florczak U, Sak A, Wang S, Chen J, Lobrich M, Rodemann HP. Targeting of AKT1 enhances radiation toxicity of human tumor cells by inhibiting DNA-PKcs-dependent DNA double-strand break repair. *Mol Cancer Ther*. 2008a; 7(7): 1772–81.

Trinh DV, Zhu N, Farhang G, Kim BJ, Huxford T. The nuclear I kappaB protein I kappaB zeta specifically binds NF-kappaB p50 homodimers and forms a ternary complex on kappaB DNA. *J Mol Biol*. 2008; 379(1): 122–35.

Van Aller GS, Carson JD, Tang W, Peng H, Zhao L, Copeland RA, Tummino PJ, Luo L. Epigallocatechin gallate (EGCG), a major component of green tea, is a dual phosphoinositide-3-kinase/mTOR inhibitor. *Biochem Biophys Res Commun*. 2011; 406(2): 194–9.

Vanhaesebroeck B, Guillermet-Guibert J, Graupera M, Bilanges B. The emerging mechanisms of isoform-specific PI3K signalling. *Nat Rev Mol Cell Biol*. 2010; 11(5): 329–41.

Vauzour D. Dietary polyphenols as modulators of brain functions: Biological actions and molecular mechanisms underpinning their beneficial effects. *Oxid Med Cell Longev*. 2012; 2012: 914273.

Verhelst K, Verstrepen L, Carpentier I, Beyaert R. IκB kinase ε (IKKε): A therapeutic target in inflammation and cancer. *Biochem Pharmacol*. 2013; 85(7): 873–80.

Walker EH, Pacold ME, Perisic O, Stephens L, Hawkins PT, Wymann MP, Williams RL. Structural determinants of phosphoinositide 3-kinase inhibition by wortmannin, LY294002, quercetin, myricetin, and staurosporine. *Mol Cell*. 2000; 6(4): 909–19.

Wang Z, Canagarajah BJ, Boehm JC, Kassisà S, Cobb MH, Young PR, Abdel-Meguid S, Adams JL, Goldsmith EJ. Structural basis of inhibitor selectivity in MAP kinases. *Structure*. 1998; 6(9): 1117–28.

Wang G, Krueger GR. Computational analysis of mTOR signaling pathway: Bifurcation, carcinogenesis, and drug discovery. *Anticancer Res*. 2010; 30(7): 2683–8.

Wang GJ, Li XK, Sakai K, Lu Cai. Low-dose radiation and its clinical implications: Diabetes. *Hum Exp Toxicol*. 2008; 27(2): 135–42.

Wang Y, Paszek P, Horton CA, Kell DB, White MR, Broomhead DS, Muldoon MR. Interactions among oscillatory pathways in NF-kappa B signaling. *BMC Syst Biol*. 2011; 5: 23.

Wang L, Zou W, Zhong Y, An J, Zhang X, Wu M, Yu Z. The hormesis effect of BDE-47 in HepG2 cells and the potential molecular mechanism. *Toxicol Lett*. 2012; 209(2): 193–201.

Weisbuch G. A shape space approach to the dynamics of the immune system. *J Theor Biol*. 1990; 143(4): 507–22.

Weng CJ, Chen MJ, Yeh CT, Yen GC. Hepatoprotection of quercetin against oxidative stress by induction of metallothionein expression through activating MAPK and PI3K pathways and enhancing Nrf2 DNA-binding activity. *N Biotechnol*. 2011; 28(6): 767–77.

Wetzker R, Rubio I. Hormetic signaling patterns. *Dose Response*. 2012; 10(1): 83–90.

Xia S, Zhao Y, Yu S, Zhang M. Activated PI3K/Akt/COX-2 pathway induces resistance to radiation in human cervical cancer HeLa cells. *Cancer Biother Radiopharm*. 2010; 25(3): 317–23.

Xing X, Zhang C, Shao M, Tong Q, Zhang G, Li C, Cheng J, Jin S, Ma J, Wang G, Li X, Cai L. Low-dose radiation activates Akt and Nrf2 in the kidney of diabetic mice: A potential mechanism to prevent diabetic nephropathy. *Oxid Med Cell Longev*. 2012; 2012: 291087.

Xu S, Li S, Guo Z, Luo J, Elllis MJ, Ma CX. Combined targeting of mTOR and AKT is an effective strategy for basal-like breast cancer in patient-derived xenograft models. *Mol Cancer Ther*. 2013; 12(8): 1665–75.

Xu G, Zhang W, Bertram P, Zheng XF, McLeod H. Pharmacogenomic profiling of the PI3K/PTEN-AKT-mTOR pathway in common human tumors. *Int J Oncol*. 2004; 24(4): 893–900.

Yan J, Greer JM. NF-kappa B, a potential therapeutic target for the treatment of multiple sclerosis. *CNS Neurol Disord Drug Targets*. 2008; 7(6): 536–57.

Yan W, Zhao K, Jiang Y, Huang Q, Wang J, Kan W, Wang S. Role of p38 MAPK in ICAM-1 expression of vascular endothelial cells induced by lipopolysaccharide. *Shock*. 2002; 17(5): 433–8.

Yang CH, Sheu JJ, Tsai TH, Chua S, Chang LT, Chang HW, Lee FY, et al. Effect of tacrolimus on myocardial infarction is associated with inflammation, ROS, MAP kinase and Akt pathways in mini-pigs. *J Atheroscler Thromb*. 2013; 20(1): 9–22.

Yang HJ, Youn H, Seong KM, Yun YJ, Kim W, Kim YH, Lee JY, Kim CS, Jin YW, Youn B. Psoralidin, a dual inhibitor of COX-2 and 5-LOX, regulates ionizing radiation (IR)-induced pulmonary inflammation. *Biochem Pharmacol*. 2011; 82(5): 524–34.

Yokoyama T, Nakamura T. Tribbles in disease: Signaling pathways important for cellular function and neoplastic transformation. *Cancer Sci*. 2011; 102(6): 1115–22.

Zbinden-Foncea H, Raymackers JM, Deldicque L, Renard P, Francaux M. TLR2 and TLR4 activate p38 MAPK and JNK during endurance exercise in skeletal muscle. *Med Sci Sports Exerc*. 2012; 44(8): 1463–72.

Zhang P, Chen JH, Guo XL. New insights into PTEN regulation mechanisms and its potential function in targeted therapies. *Biomed Pharmacother*. 2012; 66(7): 485–90.

Zhang Y, Dong C. Regulatory mechanisms of mitogen-activated kinase signaling. *Cell Mol Life Sci*. 2007; 64(21): 2771–89.

Zhang C, Tan Y, Guo W, Li C, Ji S, Li X, Cai L. Attenuation of diabetes-induced renal dysfunction by multiple exposures to low-dose radiation is associated with the suppression of systemic and renal inflammation. *Am J Physiol Endocrinol Metab*. 2009; 297(6): E1366–77.

Zhao L, Wang JL, Wang YR, Fa XZ. Apigenin attenuates copper-mediated β-amyloid neurotoxicity through antioxidation, mitochondrion protection and MAPK signal inactivation in an AD cell model. *Brain Res*. 2013; 1492: 33–45.

Zumsande M, Gross T. Bifurcations and chaos in the MAPK signaling cascade. *J Theor Biol*. 2010; 265(3): 481–91.

# 14 Oxidative Stress Response Pathways
## *Role of Redox Signaling in Hormesis*

*Li Li Ji*

## CONTENTS

14.1 Hormesis and Redox Signaling .................................................................. 282
14.2 General Mechanisms of Redox-Sensitive Signaling Pathways .................... 283
14.3 Antioxidant Gene Expression Controlled by Redox Signaling ................... 283
    14.3.1 Major Signaling Pathways ............................................................... 285
        14.3.1.1 NF-κB ................................................................................ 285
        14.3.1.2 MAPK ................................................................................ 286
        14.3.1.3 AP-1 ................................................................................... 288
    14.3.2 Exercise Activation of the Redox-Sensitive Signaling Pathway ....... 288
    14.3.3 Redox Regulation of Antioxidant Enzymes by Muscle Contraction .................................................................................... 289
14.4 Role of PGC-1α in Mitochondrial Biogenesis and Redox Signaling ................................................................................................... 290
    14.4.1 Mitochondrial Adaptation to Functional Needs .............................. 290
    14.4.2 Roles of PGC-1α in Mitochondrial Adaptations ............................. 291
        14.4.2.1 PGC-1α: A Master Regulator for Mitochondrial Biogenesis ......................................................................... 291
        14.4.2.2 PGC-1α Is Required for Antioxidant Enzyme Expression ........................................................................... 292
        14.4.2.3 PGC-1α as a Suppressor of Inflammation ........................ 292
        14.4.2.4 Mechanism of Exercise Activation of PGC-1α Signaling ............................................................................. 292
14.5 Mitochondrial Remodeling by Redox Signaling ........................................ 294
    14.5.1 Mitochondrial Fusion and Fission ................................................... 294
    14.5.2 Effect of Exercise on Mitochondrial Remodeling ........................... 295
    14.5.3 Role of UCP in Mitochondrial Antioxidant Defense ...................... 296
14.6 Conclusion ................................................................................................... 297
References ............................................................................................................. 298

## 14.1 HORMESIS AND REDOX SIGNALING

The term *hormesis* has been adopted to explain how a mild stress can result in favorable adaptations that protect the body against more severe stresses and disorders derived from physical stress or other reasons. Generation of reactive oxygen species (ROS) is a ubiquitous phenomenon associated with aerobic life. If not adequately kept in check, ROS generation may result in oxidative stress that can elicit oxidative damage, disease, and aging, as well as deterioration of functional performance of the organs, and some of the damages may be permanent. However, research literature has shown a remarkable resilience of the organism to deal with ROS, mainly due to the upregulation of cellular antioxidant defense system (Ji 1995). Organisms may also make certain subcellular remodeling to reduce ROS generation through more efficient use of oxygen (i.e., reducing ROS while maintaining the same metabolic rate). A good example might be mitochondrial biogenesis and remodel (Chan 2006; Handschin and Spiegelman 2008). These adaptations could occur in response to many physiological, environmental, and pathological stresses, but the current chapter will mainly focus on a unique metabolic stress animals encounter to maintain mobility to seek food, reproduction, and survival, namely physical activity through muscle contraction.

There is now an abundance of data from both animal and human studies indicating that ROS production is increased in skeletal muscle and myocardium during strenuous exercise (Powers and Jackson 2008). It is also becoming clear that animals and humans engaged in long-term heavy exercise become more resistant to oxidative stress, mainly due to the adaptation of their antioxidant defense systems (Sen 1995; Reid 2001; Ji 2008). Some exercise physiologists pioneered the idea that exercise might stimulate the body's antioxidant defense and thus reduce either ROS production or increase their removal, or both (Jenkins 1993). However, this speculation has taken a long time to gain sufficient scientific support. The discovery of several key redox-sensitive signaling pathways, such as nuclear factor κB (NF-κB) (Baeuerle and Baltimore 1988), mitogen-activated protein kinases (MAPKs) (Allen and Tresini 2000), and peroxisome proliferator–activated receptor γ coactivator 1α (PGC-1α) (Wu et al. 1999) had a tremendous impact on the field and were the milestones for most research occurring during this time period. It has become clear that ROS generated at physiological concentrations are not merely damaging agents inflicting random destruction to the cell, but useful signaling molecules to regulate a wide range of physiological adaptations. An abundant volume of reviews is available on this topic (Hawley and Zierath 2004; D'Autréaux and Toledano 2007; Hamanaka and Chandel 2010; Pourova et al. 2010; Finkel 2011; Collins et al. 2012).

Although cellular antioxidant *reserve* can handle low-level oxidative challenge, the majority of antioxidant adaptation requires de novo protein synthesis through transcription, translation, and protein transport. These cellular events have been termed *signal transduction* or simply *signaling* (Ji 2007). It is noteworthy that signaling pathways do not operate separately but often interact with each other to process and transfer signals, termed *cross talk*, which involves multiple organelles and cellular compartments. This chapter will briefly describe the mechanism, gene targets, and functions of several most important cell signaling pathways that impact on the ability of the organisms to adapt to and withstand oxidative stress. Understanding how the cell controls the level of ROS production and regulates the signal transduction process is essential for understanding the molecular mechanism of hormesis.

## 14.2 GENERAL MECHANISMS OF REDOX-SENSITIVE SIGNALING PATHWAYS

When cells are exposed to oxidative stress such as ultraviolet (UV) irradiation, phorboesters, toxins (such as lipopolysaccharide [LPS]), redox-disturbing agents (such as paraquat, menadione, dithiotheitol [DTT], thioredoxin [Trx], N-acetylcysteine [NAC]), certain growth factors, and anoxia/hypoxia/hyperoxia, intracellular levels of antioxidant defense are increased. It has been known for some time that increased generation of ROS in the cell associated with the exposure of the above physical and chemical agents may stimulate the gene expression of antioxidant enzymes, immunoreactive proteins such as cytokines and chemokines, and transcription factors (Oberley et al. 1987). It is now clear that several redox-sensitive signal transduction pathways are responsible for the observed adaptations, among which NF-κB, the MAPK family, the phosphoinositide 3-kinase ($PI_3K$)/Akt pathway, p53 activation, and the heat shock proteins (HSPs) are most recognized and studied (Allen and Tresini 2000). Increased metabolism (such as mitochondrial respiration), inflammation, ion-channel opening, activation of oxidative enzymes, or simply chemical reactions can promote the production of a number of chemical species such as hydrogen peroxide ($H_2O_2$), nitric oxide (NO), $Ca^{2+}$, and cytokines, which may serve as messengers to transfer signals from the cytoplasm to the nucleus to stimulate gene expression.

A review of the list of potential signaling molecules provided by Allen and Tresini (2000) indicates that $H_2O_2$ by far is the most common messenger, contributing more than 50% of the cases. The reason for $H_2O_2$ to serve this role is several fold: (1) $H_2O_2$ is constantly produced in the mitochondria during normal metabolism; (2) $H_2O_2$ is a relatively stable molecule; (3) it is a strong oxidant capable of oxidizing a variety of moieties (such as sulfhydryl, hydroxyl, and sulfoxide), yet not highly destructive; and (4) $H_2O_2$ is a small enough molecule to diffuse across most, but not all, biomembrane barriers (Chance et al. 1979). It is worthy to note that the intracellular level of $H_2O_2$ is often elevated in response to other signaling molecules such as inflammatory cytokines (Meyer et al. 1994). Figure 14.1 illustrates the various cellular sources of $H_2O_2$ production.

## 14.3 ANTIOXIDANT GENE EXPRESSION CONTROLLED BY REDOX SIGNALING

One of the most important functions of redox signaling is to regulate the cellular antioxidant defense system, composed of antioxidant enzymes, superoxide dismutase (SOD), glutathione peroxidase (GPX), catalase, and low-molecular antioxidants such as glutathione (GSH), Trx, and uncoupling proteins (UCP). Weight Adequate antioxidant capacity not only minimizes potential oxidative damage caused by ROS but also controls the formation and release of $H_2O_2$, a major ROS produced in the

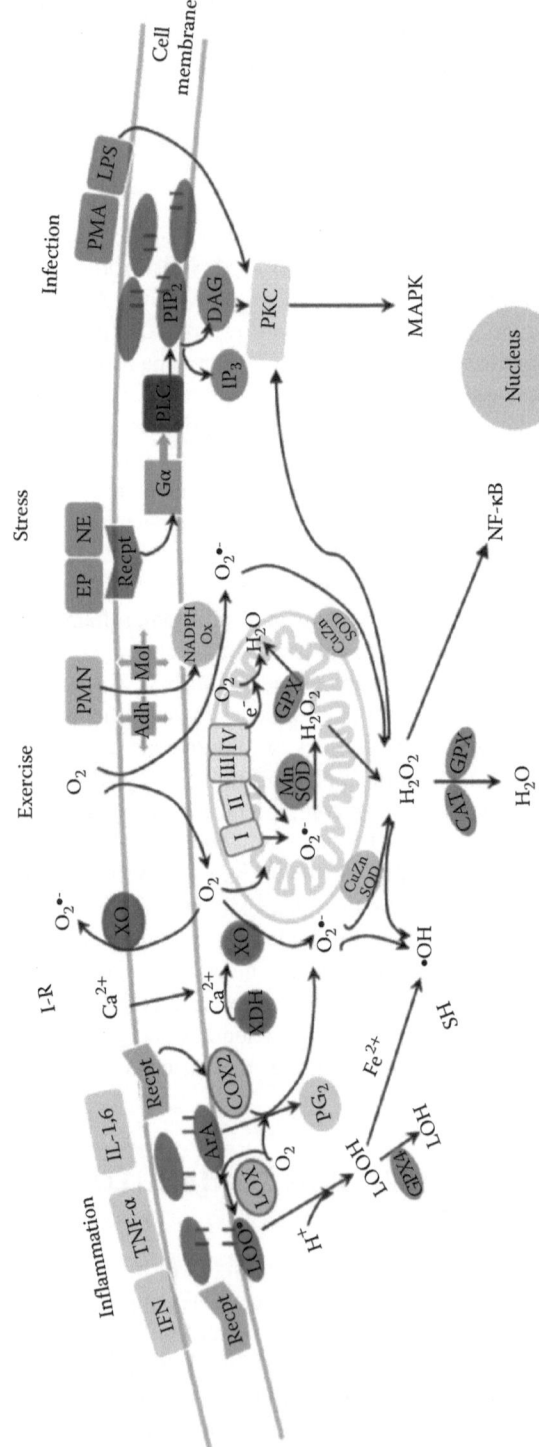

**FIGURE 14.1** Various sources of cellular production of hydrogen peroxide. Abbreviations not specified in the text: Adh Mol, adhesion molecule; ArA, aracidonic acid; DAG, diacylglycerol; EP, epinephrine; Gα, G-protein α subunit; IFN, interferon; $IP_3$, inositol 3-phosphate; I-R, ischemia-reperfusion; LOOH, lipoperoxide; LOH, hydroxylipid; LOX, lipooxygenase; NE, norepinephrine; $PG_2$, prostaglandin 2; $PIP_2$, phosphotidylinositol; PLC, phospholipase C; PMN, polymorphoneutrophil; PTP, tyrosine phosphatases; XDH, xanthine dehydrogenase. (From Ji, L.L., et al., *Infect. Disord. Drug Targets*, 9(4), 428–444, 2009.)

mitochondria that serves as an important redox signaling molecule. Expression of antioxidants at the optimal time and concentration and in the optimal cellular compartment is the key to the maintenance of oxidant–antioxidant homeostasis.

### 14.3.1 Major Signaling Pathways

Redox-sensitive signaling pathways use ROS to transfer signals from membrane to nucleus to stimulate growth, differentiation, proliferation, and apoptosis. Most antioxidant enzymes contain redox-sensitive gene regulatory sequences in their promoter and/or intron regions that can interact with transcription factors to trigger upregulation of gene expression (Allen and Tresini 2000). Although all of these pathways are important, NF-κB and MAPK are considered the most critical for the cells to cope with oxidative stress.

#### 14.3.1.1 NF-κB

NF-κB is a dimeric transcription factor composed of members of the Rel family (Baeuerle and Baltimore 1988; Meyer et al. 1994; Flohé et al. 1997). NF-κB is activated by a variety of external stimulants, such as $H_2O_2$, proinflammatory cytokines (TNF-α, IL-1, IL-6), LPS, phorbol myristate acetate, ionizing irradiation, and viral infection. These signals result in the phosphorylation and activation of IκB kinase (IKK), which phosphorylates two critical serine residues and primes IκB for ubiquitination and proteolytic degradation by the 26S proteasome. IκB dissociation unleashes p50/p65 to dimerize and translocate into the nucleus and bind the κB consensus sequence of the target genes. UV irradiation has also been reported to initiate IκB degradation without prior phosphorylation (Li and Karin 1999).

As the key enzyme to phosphorylate IκB, IKK is known to exist as a complex made of two catalytic subunits, IKKα and IKKβ, and a regulatory subunit, IKKγ (Tegethoff et al. 2003). IKK activation involves a Ser/Thr kinase domain at its N-terminal that can be phosphorylated by several kinases, including NF kappa B inducing kinase, NF-κB-activating kinase, transforming growth factor β-activated kinase (TAK-1), and MAPK/extracellular signal-regulated kinases (MEKK1), among which NIK activation seems to be the most important step (Li and Engelhardt 2006). Interestingly, activation of NIK can only be conferred within a narrow range of $H_2O_2$ concentration of 1–10 μM close to physiological levels within the cell. Since NIK is a member of the MAPK family, this study demonstrates the importance of cross talk between the two redox-sensitive signaling pathways. Other MAPKs may also be selectively involved in NF-κB activation. For example, Jiang et al. (2004) showed that activation of ERK was required to upregulate the IL-1β induced expression of cyclooxygenase-2 (COX-2) and inducible NO synthetase (iNOS), but not Mn-containing SOD (SOD2) in cultured rat vascular smooth cells.

There is a general belief that NF-κB can be activated by ROS and oxidative stress. This belief is based on the following research evidence: (1) exposure of certain types of cells, including T cells, L6 skeletal muscle myotubes, and 70Z/3 pre-B cells, to $H_2O_2$ can lead to activation of NF-κB cascades; (2) the best-known NF-κB activators, such as tumor necrosis factor-α (TNF-α), interleukin-1 (IL-1), LPS, and PMA, all can lead to increased intracellular levels of $H_2O_2$; and (3) treatment of cells with antioxidants, such as GSH and NAC, abolishes NF-κB activation. However, some research evidence argues against the notion that ROS are directly involved in the process. First, $H_2O_2$ activation of NF-κB is known to be highly cell-type specific; many cells, such as HeLa, fibroblast, and Jurkat T cells, are not responsive to $H_2O_2$ treatment. Second, some widely known antioxidants that inhibit NF-κB may exert their effects independent of ROS. Hayakawa et al. (2003) demonstrated that the inhibitory effect of NAC on TNF-α-induced NF-κB activity was due to its ability to attenuate receptor binding of TNF-α instead of reducing ROS. Further, the authors showed that pyrrolidine dithiocarbamate (PDTC) could prevent NF-κB activation due to its inhibition on ubiquitin ligase toward phosphorylated IκB rather than its antioxidant potential.

The best-known proteins and enzymes that require NF-κB binding to activate transcription are SOD2, COX-2, iNOS, γ-glutamylcysteinyl synthetase (GCS), vascular cell adhesion molecule-1 (VCAM-1), and several cytokines. These genes are involved in a wide variety of biological functions, such as antioxidant, inflammation, immunity, and antiapoptosis. Ghosh et al. (1998), Ghosh and Karin (2002), and Zhou et al. (2001) showed that $H_2O_2$ treatment of cultured C2C12 muscle cells increased DNA binding of NF-κB as well as luciferase report gene activity. Transfection with either a transdominant inhibitor of IκBα or a dominant-negative inhibitor of NF-κB blocked the induction of GPX and catalase (CAT) expression by more than 50%. These results suggest that NF-κB is an important mediator of oxidative stress–induced antioxidant gene expression in skeletal muscle. Because NF-κB activation is linked to expression of proinflammatory cytokines and inflammatory response, hyperactivation of NF-κB could lead to oxidative stress. This aspect will be discussed in the following section 14.4.2.3.

### 14.3.1.2 MAPK

MAPK has a complicated hierarchy including ERK, c-Jun N-terminal kinases, and p38$^{MAPK}$, which are regulated by their respective upstream kinases (MAPK kinase/ MAPK kinase kinase) (Allen and Tresini 2000). The primary extracellular stimulators of the MAPK pathway are growth factors (GFs), inflammatory cytokines (TNF-α, IL-1, 6), and phorbol esters. In the ERK and JNK pathways, Ras plays an important role in the initial phase of MAPK activation. After receptor binding by GFs, member-associated proteins Sos, Grb-2, and SHC undergo conformational changes leading to the activation of Ras. Ras stimulates the translocation and phosphorylation of Raf-1, the commander of MEK/MKKs. TNF-α and IL-1 bypass the Ras pathway by increasing cytosolic concentration of $H_2O_2$, which activates several isoforms of protein kinase C (PKC) (Carroll and May 1994). PKC appears to serve as a pivot enzyme in activating MAPK pathways by stimulating multiple MEK/MKKs, as well as the NF-κB pathway by activating NIK (Figure 14.2).

The primary function of the MAPK pathway is to modulate growth, metabolism, differentiation, transcription, translation, and remodeling. This broad MAPK function is outside the scope of this chapter and readers are referred to several excellent recent reviews (Sakamoto and Goodyear 2002; Hawley and Zierath 2004; Long et al. 2004). Besides GF, stress and cytokines can also activate MAPK to trigger inflammation (by inducing inflammatory cytokines in an autocrinary fashion, as well as VCAM-1), degradation (via phospholipase $A_2$), and apoptosis (Allen and

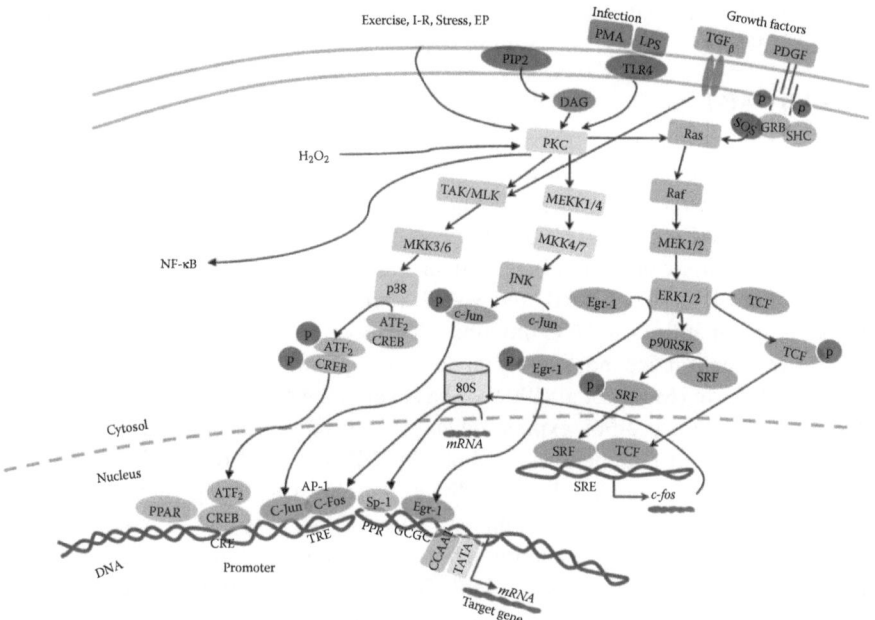

**FIGURE 14.2** Schematic illustration of mitogen-activated protein kinase (MAPK) signaling in the cell. Abbreviations not specified in the text: 80S, ribosome 80S; $ATF_2$, activating transcription factor-2; CRE, cAMP-response element; DAG, diacylglycerol; Egr-1, early growth-responsive-1 protein; EP, epinephrine; Gα, G-protein α subunit; I-R, ischemia-reperfusion; MLK, mixed lineage kinases; $PIP_2$, phosphotidylinositol; p90RSK, p90 ribosomal S6 kinase; PPR, proximal promoter region; SRE, serum response element; SRF, serum response factor; TCF, ternary complex factor; $TGF_β$, transforming growth factor β; TLR4, Toll-like receptor 4; TRE, tetradecanoylphorbolacetate response element. (From Ji, L.L., et al., *Infect. Disord. Drug Targets*, 9(4), 428–444, 2009.)

Tresini 2000). In this regard, it is noteworthy that MAPK participates in the regulation of gene expression controlled primarily by NF-κB signaling, such as antioxidant enzymes. For example, although the established role of ERK1/2 is to phosphorylate transcription factor related to growth and development, ERK has recently been shown to play an important role in the sustained activation of NF-κB in rat vascular smooth muscle, leading to upregulation of iNOS (Jiang et al. 2004). It is noteworthy that several important adaptations in skeletal muscle, such as mitogenesis, hypertrophy, and fiber transformation, are regulated primarily by MAPK signaling and play an important role in determining overall cellular oxidant–antioxidant homeostasis (Lin et al 2002; Sen and Roy 2005).

NF-κB and MAPK are distinct signaling pathways in the cell. NF-κB is primarily responsive to stress, toxins, and cytokines, leading to inflammation, apoptosis, and adaptation, whereas the primary consequence of MAPK activation is growth, development, transcription, translation, and remodeling. However, there are considerable functional overlaps and cross talks between the two pathways. For example, ERK and p38 have been shown to play an important role in the temporal regulation of NF-κB activation by IL-1β and $H_2O_2$ (Catani et al. 2004). Also, NIK and IKK are members of the MEK/MKK family, upstream kinases of MAPK (Allen and Tresini 2000).

### 14.3.1.3 AP-1

AP-1 is another important transcription factor that regulates expression of numerous genes in a redox-sensitive manner (Allen and Tresini 2000). AP-1 is a dimer composed of activating (c-Fos and c-Jun) and inhibitory (Fos-related antigen [Fra]-1 and 2) subunits (Catani et al. 2004). Depending on the cellular redox milieu and cell type, Fos and Jun can form a heterodimer or interact with other transcription factors, such as SRF, ATF, CREB, and Maf, leading to activation or inhibition of gene transcription of antioxidant and immunoactive proteins. One unique characteristic of AP-1 activation is it depends largely on de novo expression of transcription factors as mentioned above (Meyer et al. 1994). It has been known for a long time that TNF-α and IL-1 can induce c-Fos expression in rapid fashion. Recent research suggests that AP-1 function is dependent on both MAPK and NF-κB signaling pathways. IL-1 was shown to induce c-Fos and Fra-1, thereby stimulating IL-8 (also known as cytokine-induced neutrophil chemoattractant, CINC) expression, whereas blockade of MEKK1 by PD98059 suppressed the expression of almost all AP-1 subunits, indicating ERK activation was required (Hoffmann et al. 2005). Furthermore, IL-8 promoter has a p65 NF-κB binding site occupancy that plays an important role in the synergistic action of c-Fos. In contrast, Fra-1 strongly inhibits inducible IL-8 transcription. The role of JNK, which phosphorylates c-Jun and allows for its dimerization with c-Fos to form AP-1, appears less important, at least for IL-1-induced IL-8 expression (Holzberg et al. 2003).

### 14.3.2 Exercise Activation of Redox-Sensitive Signaling Pathway

Goodyear et al. (1996) first reported that the key enzymes in the MAPK pathway, such as JNK, ERK1/2, and p38$^{MAPK}$, were activated in rat skeletal muscle after an acute bout of treadmill running. This research group soon after showed that p42- and p44$^{MAPK}$ (i.e., ERK1/2) were phosphorylated after bicycle exercise in human muscle, along with activation of MEK1 and Raf-1 (MEKK), as well as downstream substrate p90 ribosomal S6 kinase (RSK) (Aronson et al. 1997). Since then, numerous studies have shown that MAPK pathways can be activated by contractile activity in skeletal muscle (Baar and Esser 1999; Ryder et al. 2000; Nader and Esser 2001; Coffey et al. 2006; van Ginneken 2006). The signals triggering MAPK activation have been attributed to a variety of physiological stimuli associated with exercise, including hormones, calcium ion, neural activity, and mechanical force. Biological implications of MAPK activation are widespread, including important functions such as glucose transport, muscle and heart hypertrophy, angiogenesis, and vascular adaptation (Sakamoto and Goodyear 2002; Hawley and Zierath 2004).

The study regarding exercise influence on NF-κB signaling came several years after the study on the MAPK pathway, although the role of NF-κB in antioxidant signaling in vitro was then already well-established (Flohé et al. 1997; Allen and Tresini 2000). Hollander et al. (2001) first reported that NF κB and AP-1 binding was significantly elevated in rat skeletal muscle after an acute bout of prolonged exercise in a fiber-specific manner. The increased NF-κB and AP-1 binding was accompanied by increased manganese-containing SOD (MnSOD) mRNA abundance and protein content in exercised muscle. Ji et al. (2004) showed that NF-κB binding was elevated in rat vastus lateralis muscle after 1 hour of treadmill running, along with increased IKK activity, IκBα phosphorylation and degradation, and p50 nuclear translocation. The exercise-induced activation of NF-κB cascade was abolished in rats injected with PDTC prior to exercise. A time course study revealed that robust IκBα phosphorylation took place shortly after the exercise stimulation, whereas 2–4 hours were required to detect significant p50/65 level in muscle nuclear extracts. Consistent with the above study, Ho et al. (2005) reported that NF-κB activation was elevated twofold in the soleus and red gastrocnemius muscles in response to 1 hour of treadmill exercise in rats. Peak IKKα/β activation was found early during exercise whereas maximal NF-κB binding was at 1–3 hours. Importantly, the authors found that application of p38 and ERK inhibitors reduced IKK activation, suggesting MAPK and NF-κB might work synergistically during exercise.

There has been evidence that ROS are the required chemical signals in the exercise-induced redox signaling in whole-animal studies. Gomez-Cabrera et al. (2005) investigated the role of ROS in exercise-activated MAPK and NF-κB signaling by injecting rats with a xanthine oxidase (XO) inhibitor (allopurinol) prior to a progressive treadmill running. NF-κB binding and ERK1/2 and p38 activities were elevated in rat gastrocnemius muscle after exercise, whereas these effects were severely attenuated in the allopurinol-treated rats. Since a substantial portion of ROS was generated via nonmitochondrial sources in the rat-running protocol, the authors suggested that ROS generated via XO played a crucial role in activating the signaling pathways. Kumar and Boriek (2003) reported that passive mechanical stretch in mouse diaphragm muscle activated the NF-κB pathway, which could be blocked by NAC. Unloading dramatically decreases NF-κB activity, suggesting contractile activity and/or nerve stimulation are required for the basal activity of this signaling pathway (Hunter et al. 2002; Durham et al. 2006).

### 14.3.3 REDOX REGULATION OF ANTIOXIDANT ENZYMES BY MUSCLE CONTRACTION

There is an abundance of literature reporting muscle antioxidant adaptation to chronic exercise training. SOD activity has consistently been shown to increase with exercise training in an intensity-dependent manner, and MnSOD is primarily responsible for the observed increase in SOD activity (Hollander et al. 2001). GPX activity has also been shown to increase after endurance training. The observed training adaptation is due to altered gene expression, with both mRNA and enzyme protein levels being upregulated. Training adaptation of antioxidant enzymes is influenced heavily by a number of physiological and environmental factors, such as gender, age, diet, and medication state. For a thorough review on this topic, the readers are referred to several previous reviews (Ji 1995, 2008; Reid 2001; Powers and Jackson 2008).

Besides SOD and GPX, the rate-limiting enzyme for GSH synthesis, GCS, can be induced by endurance training in rat skeletal muscle and liver (Sen 1995; Ramires and

Ji 2001). Training also increases muscle GSH content, though the adaptation is fiber-specific. In mammalian cells, GCS is a heterodimer consisting of the catalytic heavy-chair subunit (GCS-HS) and regulatory light-chair subunit (GCS-LC). GCS-HS expression is known to be regulated by redox-sensitive mechanism via a variety of oxidants, phenolic antioxidants, and proinflammatory cytokines. Both GCS-HC and GCS-LC promoters contain antioxidant response element (ARE) and nuclear respiratory factor (NRF)-2 binding sites that seem to play a critical role in oxidative stress–induced GCS upregulation. GCS-HC also has NF-κB binding sites that are essential for GCS expression in some, but not all cell types (Chan and Kwong 2000).

NO at low concentration exerts an antioxidant function by neutralizing $O_2^{-\bullet}$ (Reid 2001; Collins et al. 2012). Its vasodilative effect increases blood flow to the working muscle, thereby improving the availability of bloodborne energy substrates and antioxidants. Thus, an increase in NO production via the regulation of NOS may be viewed as indirectly enhancing muscle antioxidant defense during exercise. iNOS is responsive primarily to ROS and inflammatory cytokines through activation of NF-κB and MAPK (Adams et al. 2002; Gomez-Cabrera et al. 2005). In rat skeletal muscle, iNOS mRNA level has been reported to increase after an acute bout of exercise (Balon 1999). Although modest levels of iNOS induction may be viewed as improving muscle antioxidant defense, high levels of NO production would lead to the formation of peroxynitrite, a highly reactive ROS contributing to muscle oxidative damage. Thus, iNOS upregulation may favor either antioxidant or pro-oxidant functions depending on muscle physiological condition.

It is noteworthy that although activation of NF-κB and MAPK is a necessary event in response to oxidative stress and induces antioxidant adaptation, prolonged hyperactivity also promotes the gene expression of proinflammatory cytokines and chemokines, rendering to potential ROS generation, inhibition of PGC-1 function (see 14.4.2.3), and degradative processes such as proteolysis and apoptosis. This could occur during rigorous muscle contraction, especially eccentric exercise and subsequent muscle injury. Chronic NF-κB activation could lead to systemic inflammation, which is a major etiological mechanism for many chronic diseases, such as rheumatoid, atherosclerosis, obesity, and certain types of cancer.

## 14.4 ROLE OF PGC-1α IN MITOCHONDRIAL BIOGENESIS AND REDOX SIGNALING

### 14.4.1 Mitochondrial Adaptation to Functional Needs

When animals are stressed with low oxygen environment (hypoxia), treated with thyroid hormones, under hypothermia, or engage in long-term physical work with high oxygen consumption, there is a substantial increase in mitochondrial volume, density, and oxidative enzyme activity in the skeletal muscle, and to some extent in the myocardium (Holloszy and Coyle 1984). Increased mitochondrial populations are not only capable of utilizing additional oxygen to metabolize fuels to provide ATP as energy for muscle contraction, but also shift fuels from carbohydrate to fat as a more efficient energy source. In addition, proliferation of mitochondria helps distribute oxygen among increased electron transport chains and thus reduces the production of ROS (Davies et al. 1982). Thus, working muscle is alleviated of both metabolic and oxidative stress resulting from heavy workload.

Mitochondrial biogenesis is regulated by complex signaling pathways in response to various stimuli. It is a complex process that requires the synthesis, import, and incorporation of proteins and lipids to the existing mitochondrial reticulum, as well as replication of the mitochondrial DNA (mtDNA) (Mootha et al. 2003; Hood et al. 2006; Manoli et al. 2007; Liesa et al. 2009). Although muscle mitochondrial adaptation to aerobic exercise has been confirmed with a wide range of experimental models and in numerous animal and human studies, the underlying mechanism was not elucidated until the late twentieth century. In the past decade, PGC-1α emerged as a master transcriptional coactivator, which has provided a mechanistic insight for understanding how nuclear regulatory pathways are coupled to the biogenesis of mitochondria, antioxidant defense, and inflammatory response in skeletal muscle. These findings have profound effects on our understanding of signal transduction pathways related to muscle function.

PGC-1α was first identified as a functional activator of the peroxisome proliferator-activated receptor (PPAR) γ receptor in brown adipose tissue (Puigserver et al. 1998) and is known to influence numerous aspects of metabolism (Wu et al. 1999; Lin et al. 2002). PGC-1α has been identified in other mitochondria-rich tissues, including red skeletal muscle and heart, as well as in kidney, liver, and brain (Finck and Kelly 2006). PGC-1α interacts with nuclear receptors and transcription factors to activate transcription of their target genes and its activity is responsive to multiple stimuli, including calcium ion, ROS, insulin, thyroid and estrogen hormone, hypoxia, ATP demand, and cytokines (Puigserver and Spiegelman 2003).

### 14.4.2 Roles of PGC-1α in Mitochondrial Adaptations

#### 14.4.2.1 PGC-1α: A Master Regulator for Mitochondrial Biogenesis

PGC-1α has been thought to be a master regulator of mitochondrial biogenesis by coactivating several transcription factors that in turn bind to the promoters of distinct sets of nuclear-encoded mitochondrial genes (Handschin and Spiegelman 2008; Scarpulla 2008). PGC-1α interacts with several nuclear transcription factors, including PPAR family members, NRF-1 and -2, and estrogen-related receptor-α (ERRα), as well as myocyte enhancer factor-2 (MEF2), forkhead box protein O (FOXO) 1, and sterol regulatory element-binding proteins (SREBP) (Knutti and Kralli 2001; Russell 2005). PGC-1α coactivation of NRF-1, 2 promotes the expression of nuclear-encoded mitochondrial proteins (NEMP) and mitochondrial transcription factor A (Tfam), which directly stimulates mtDNA replication and transcription (Puigserver and Spiegelman 2003; Kelly and Scarpulla 2004; Lin et al. 2005). In skeletal muscle, PGC-1α regulates skeletal muscle fiber type switch, glucose transport, and lipid utilization (Jagoe and Goldberg 2001; Michael et al. 2001; Lin et al. 2002), as well as mitochondrial fusion (Puigserver et al. 1998; Schreiber et al. 2004).

### 14.4.2.2 PGC-1α Is Required for Antioxidant Enzyme Expression

Recent studies have shown that PGC-1α also has a regulatory mechanism for the expression of endogenous antioxidant proteins. Reduced mRNA levels of SOD1 (CuZnSOD), SOD2 (MnSOD), and/or GPx1 (Leick et al. 2008), as well as SOD2 protein content (Geng et al. 2010; Leick et al. 2010) were observed in skeletal muscle from PGC-1α knockout (KO) mice compared to wild type (WT), whereas PGC-1α overexpression mice showed an upregulation of SOD2 protein content in skeletal muscle (Wenz et al. 2009). PGC-1α KO fibroblasts exhibit a decrease in SOD2, catalase, and GPx1 mRNA content relative to WT fibroblasts, and PGC-1α KO mice were more vulnerable to oxidative stress (St-Pierre et al. 2006). In addition, PGC-1α has been shown to regulate the mRNA expression of UCP-2 and-3 in cell culture (St-Pierre et al. 2003), suggesting that PGC-1α may also increase the uncoupling capacity and concomitantly reduce mitochondrial ROS production (see 14.5.3). Furthermore, it has also been shown that PGC-1α promotes SIRT3 gene expression, which is mediated by an ERR-α binding element mapped to the SIRT3 promoter region (Kong et al. 2010). SIRT3 deacetylates and activates mitochondrial enzymes, including SOD2, through a posttranslational mechanism (Bellizzi et al. 2005; Shi et al. 2005). Taken together, PGC-1α seems to have a role in reducing ROS damage by upregulating antioxidant gene expression and activity.

### 14.4.2.3 PGC-1α as a Suppressor of Inflammation

Recent observations further indicate that PGC-1α may also play a role in antiinflammatory effects. Studies in PGC-1α KO animals indicated that PGC-1α modulates local or systemic inflammation and might regulate the expression of inflammatory cytokines and inflammatory markers such as TNF-α and IL-6 (Handschin 2009; Arnold et al. 2010). PGC-1α KO mice showed higher basal mRNA expression of TNF-α, IL-6 in skeletal muscle, as well as higher serum IL-6 level than WT (Handschin and Spiegelman 2008). In addition, PGC-1α overexpressed mice had lower expression of TNF-α and IL-6 mRNA in skeletal muscle, and reduced serum TNF-α and IL-6 levels (Wenz et al. 2009). These data suggest that PGC-1α has a protective role in inflammatory response by reducing proinflammatory cytokine production. Moreover, a single exercise bout elicited a significant increase in skeletal muscle TNF-α mRNA and serum TNF-α content in PGC-1α KO mice but not in WT mice, indicating that PGC-1α normally protects against exercise-induced increases in TNF-α (Handschin and Spiegelman 2008). Previous research has shown that ROS induce inflammatory cytokine production in skeletal muscle and the expression of mitochondrial ROS-detoxifying enzymes was increased by PGC-1α (St-Pierre et al. 2003; Valle et al. 2005). Figure 14.3 is a schematic illustration of the potential cross talks between NF-κB, MAPK, and PGC-1α pathways that may influence antioxidant gene expression, mitochondrial biogenesis, and inflammation in skeletal muscle.

### 14.4.2.4 Mechanism of Exercise Activation of PGC-1α Signaling

PGC-1α is known to be a major regulator of exercise-induced phenotypic adaptation and fiber transformation from type 2 to type 1 (Lin et al. 2002). PGC-1α expression is linked to muscle contraction through $Ca^{2+}$/calmodulin-dependent protein kinase IV

# Oxidative Stress Response Pathways

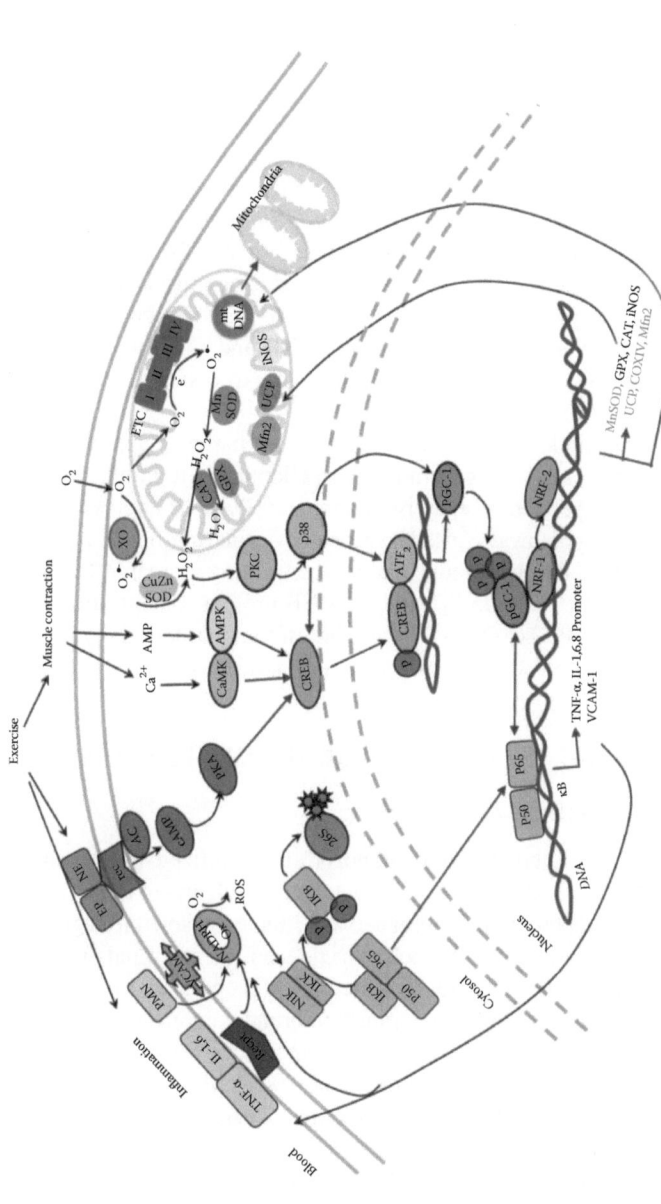

**FIGURE 14.3** Schematic illustration of the cross talks between NF-κB, MAPK, and PGC-1α pathways based on current understanding. Also shown are the major sources of reactive oxygen species (ROS) generation in the cell in response to physiological stimuli, and cellular antioxidant defense systems. The illustration did not intend to include all components and pathways of redox signaling, but those mentioned in the text. There may be substantial differences between different cell types. Abbreviations not specified in the text: 26S, proteasome 26S; AC, adenylate cyclase; AMPK, AMP activated kinase; EP, epinephrine; NE, norepinephrine; p38, p38 mitogen activated protein kinase; p50 and p65, subunit of NF-κB; PKA, protein kinase A; PMN, polymorphoneutrophil. (From Ji, L.L. and Zhang, Y. Antioxidant and anti-inflammatory effects of exercise: Role of redox signaling. *Free Radic. Res.*, in press. With permission.)

(CaMKIV) and it is known CaMKIV and calcineurin A are activated through calcium ion dynamics within the muscle in response to exercise (Baar et al. 2002). The increased calcium signaling during muscle contraction activates several important transcription factors, such as cAMP-response element binding protein (CREB), a target of CaMKIV, and MEF2 (Handschin 2010). Another factor that regulates PGC-1α expression upon exercise involves p38$^{MAPK}$, which activates MEF2 and activating transcription factor 2 (ATF2). p38$^{MAPK}$ in conjunction with ATF2 results in increased expression of PGC-1α (Cao et al. 2004). ATF-2 and subsequent interactions of ATF2–CREB appeared to be an early event in PGC-1α-mediated signaling processes (Vercauteren et al. 2006). p38$^{MAPK}$ also stimulates PGC-1α by phosphorylation in response to cytokine stimulation in muscle cells (Puigserver et al. 2001). Finally, as a metabolic energy deprivation sensor, AMPK is activated by exercise due to increased AMP/ATP ratio and $Ca^{2+}$ flux during muscle contraction, enhancing PGC-1α transcription as well as activity. It was demonstrated that activation of p38$^{MAPK}$-mediated phosphorylation of CREB and subsequent binding to PGC-1α promoter plays a key role in activating PGC-1α expression in response to increased muscle activity (Akimoto et al. 2005).

Exercise-induced upregulation of the PGC-1α signaling pathway appears to be redox sensitive. In rats, reducing ROS generation with allopurinol to inhibit XO, the main ROS source of this type of rats, exercise, attenuated PGC-1α expression and PGC-1α-controlled signaling pathway caused by repeated sprint running (Kang et al. 2009). It was recently reported that although endurance training increased PGC-1α, cytochrome *c* oxidase subunit IV (COXIV) and p-CREB protein contents and in rat skeletal muscle, injection of PDTC, NF-κB inhibitor, and antioxidant daily during training attenuated the observed adaptations (Feng et al. 2013).

## 14.5 MITOCHONDRIAL REMODELING BY REDOX SIGNALING

### 14.5.1 Mitochondrial Fusion and Fission

Morphological data demonstrated that mitochondria are organized into dynamic tubular structures or networks, extending throughout the cytosol, and in close contact with the nucleus, endoplasmic reticulum, Golgi network, and cytoskeleton (Jagoe and Goldberg 2001; Hom and Sheu 2009). Mitochondrial fusion in mammals requires mitofusin (Mfn1, Mfn2), a GTPase family enzyme which is anchored on the mitochondrial outer membrane and contains two transmembrane domains connected by a small intermembrane-space loop. Mfn1 seems to act in concert with Optic atrophy protein 1 (OPA1), whereas Mfn2 acts alone (Song et al. 2009). Mitochondrial fission is controlled by the interaction of two proteins: dynamin-related protein 1 (Drp1) and human fission protein 1 (hFis1).

Several views point to a role of balanced mitochondrial fusion and fission events in the maintenance of mitochondrial integrity. Mitochondrial fusion and related protein (Mfn1/2 and OPA1) are beneficial for oxidative phosphorylation and optimal metabolic output in muscle. In addition, fusion provides a chance for mitochondria to mix their contents, thus enabling rapid transmission of $Ca^{2+}$ signals and electrogenic events in large muscle fibers, protein complementation, and mtDNA repair (Grandemange et al. 2009). Conversely, fragmented mitochondria are thought to be easily transportable and allow for rapid mitochondrial trafficking to energy-demanding regions of the cell. Recent studies showed that fission of the mitochondrial network into individual units is necessary for efficient mitophagy to eliminate damaged mitochondria (Breckenridge et al. 2008). This could occur via proteolytic degradation of the fusion protein OPA1 in energetically compromised mitochondria or by an increased activity of the fission proteins, such as Fis1 (Gomes and Scorrano 2008). This in turn stimulates mitochondrial biogenesis to ensure a stable pool of functional mitochondria within each cell.

The exact mechanisms controlling mitochondrial network remodeling are still elusive. However, ROS, PGC-1α, and NF-κB have been identified as potential regulators in some recent studies. Exercise-induced mitochondrial ROS may contribute to the rapid alteration in mitochondrial fusion/fission protein expression. Koopman et al. (2008) reported that vitamin E supplementation reduced ROS production and restored aberrant mitochondrial morphology in fibroblasts with mitochondrial complex I deficiency, suggesting that ROS are involved in controlling mitochondrial shape in these cells. Another report demonstrated that mitochondria became fragmented and rounded with the treatment of rotenone or antimycin, which induced ROS production (Yu et al. 2006). It has been shown that rats subjected to chronic training had lower Mfn2 protein levels in muscle mitochondria compared to controls, whereas no change was observed in the cytosolic Mfn2. Furthermore, this effect was not seen in trained rats with daily injection of PDTC, an antioxidant, suggesting oxidative process may control Mfn2 expression (Feng et al. 2013). In addition, with a nonapoptotic concentration of $H_2O_2$ treatment, originally long and interconnected mitochondrial tubules became transiently shortened and weakly fragmented (Jendrach et al. 2008). On the basis of the above evidence, it seems that ROS could induce fission of the mitochondrial network to fragmented units, depending on the concentration of ROS and duration of the treatment.

PGC-1 has been shown to be involved in mitochondrial network remodeling through controlled fusion and fission. $H_2O_2$-stimulated Mfn1/2 expression was dramatically attenuated in PGC-1α KO C2C12 muscle cells (St-Pierre et al. 2006), whereas PGC-1α overexpression markedly enhanced Mfn2 mRNA and protein levels in cultured muscle cells (Liesa et al. 2008). PGC-1α also stimulates the transcriptional activity of the human Mfn2 gene promoter, which requires the integrity of the nuclear hormone ERRα-binding element (Soriano et al. 2006). PGC-1β induces almost all fission protein expression in C2C12 myoblasts and myotubes, but increases the length of mitochondrial tubules (Liesa et al. 2008). This means PGC-1β may predominantly influence the rate of mitochondrial fusion. Furthermore, mitochondrial ubiquitin ligase activator of NF-κB (MULAN), which is located in the mitochondria, was identified as regulators of mitochondrial dynamics (Li et al. 2008). This may be a link between mitochondrial dynamics and mitochondria-to-nucleus signaling.

### 14.5.2 Effect of Exercise on Mitochondrial Remodeling

There is strong evidence that alterations of certain mitochondrial dynamics proteins could affect cellular energy metabolism. For example, Mfn2 repression in L6E9 muscle cells led to decreased rates of pyruvate or glucose oxidation, a reduction in mitochondrial membrane potential ($\Delta\psi_m$), and a dramatic discontinuity of the mitochondrial network. These changes occur under conditions in which mitochondrial mass is unaltered (Bach et al. 2003). Accordingly, cells with low Mfn2 activity rely mainly on the use of anaerobic glycolysis to generate energy (Santel and Fuller 2001). Inversely, change of demand in mitochondrial energy metabolism can alter fusion and fission protein expression. During physical exercise, muscle ATP production increases dramatically and so does ROS generation. These changes have been shown to link to profound changes in both the morphology and gene expression of mitochondrial fusion and fission proteins. However, the effect of exercise on mitochondrial fusion and fission has been studied only sparsely and the available data suggest that an acute bout of exercise and chronic exercise may have differential effects. Bo et al. (2010) reported that during a bout of prolonged exercise with incremental durations, increased Fis1 expression but decreased Mfn1/2 expression occurred, and the magnitudes of these alterations depended on exercise duration. These alterations were also associated with increased ROS generation and state 4 respiration, but with decreased state 3 respiration and attenuated ATP synthase. These findings show that heavy exercise may induce a tendency toward more fragmented mitochondrial network, and impairment in OXPHOS.

On the other hand, endurance exercise leads to an expansion of the mitochondrial reticulum through a fusion process. Garnier et al. (2005) observed a coordinated increase in maximal mitochondrial respiratory capacity, with increased mRNA levels of Mnf2 and Drp1 in the trained subjects. Increased gene expression appears to occur in the recovery period following an acute exercise bout. It was reported that mRNA levels of Mfn1/2 and Fis1 were all elevated significantly above the resting levels 24 hours postexercise (Ding et al. 2009). Similar findings were also obtained by Cartoni et al. (2005), who found that Mfn1 and Mfn2 mRNA were increased in muscle biopsies obtained from cyclists at 24 hours postexercise. However, the effects of endurance training on mitochondrial fusion and fission protein expression are still controversial and conflicting reports can be found in the literature. For example, Feng et al. (2013) recently showed that Mfn2 protein content was decreased in trained rats compared to controls. These results suggest that establishment of higher level balance of mitochondrial fusion and fission may be an important process behind the functional adaption to endurance training.

### 14.5.3 Role of UCP in Mitochondrial Antioxidant Defense

Mitochondrial production of ROS is partly dependent on the cross-membrane proton motive force ($\Delta\psi_m$), thus increasing mitochondrial membrane permeability with uncouplers (such as DNP and FCCP) to reduce $\Delta\psi_m$ is viewed as a classic way to reduce ROS production. Located in the mitochondrial inner membrane, UCPs are a heterogeneous family of proteins that play an important role in partially dissipating the proton electrochemical gradient $\Delta\psi_m$ (Klingenberg and Huang 1999). The best characterized UCP1 is expressed exclusively in the brown adipose tissue of rodents with a key function of adaptive thermogenesis, so that when environmental temperature drops part of the electrons are shunted back to the matrix, bypassing ATP synthase ($F_0$–$F_1$ complex, or complex V) to generate heat. UCP2 is expressed ubiquitously with a physiological role yet to be understood (Krauss et al. 2005). UCP3, expressed primarily in the skeletal muscle, shares 59% homology to that of UCP1 and is regarded as a plausible regulator of transmembrane proton potential and hence efficiency of oxidative phosphorylation (Ljubicic et al. 2004). Several previous studies have shown that UCP3 expression was increased in response to an acute bout of exercise or contractile activity in mammalian skeletal muscle (Cortright et al. 1999; Zhou et al. 2000; Ljubicic et al. 2004). Furthermore, Goglia and Skulachev (2003) postulated that by translocating fatty acid peroxides from inner to the outer membrane leaflet, UCP may fulfill a role in antioxidant defense of the mitochondria. It is possible that UCP3 may be induced in skeletal muscle during exercise in part to shunt protons back to the matrix and maintain a modest $\Delta\psi$, thus reducing superoxide radical production and regulating the balance between ATP production and antioxidant function.

Jiang et al. (2009) reported that in response to a prolonged bout of exercise UCP3 mRNA and protein expression in rat skeletal muscle were remarkably elevated. ROS production increased progressively during exercise, but showed a dramatic drop at 150 minutes when UCP3 protein reached the highest level. State 4 respiration rate increased up until 120 minutes, but returned to resting rate at 150 minutes. These data demonstrate that UCP3 expression in rat skeletal muscle can be rapidly upregulated during prolonged exercise and this induction can reduce ROS production, thus serving as a means of antioxidant defense to protect mitochondria from oxidative stress. Similar findings were observed in rat heart, wherein exercise-induced UCP2 coincided with a reduction of ROS production in a time-coordinated manner (Bo et al. 2008).

The cellular mechanism by which muscle contraction increases UCP expression in the mitochondria is still unclear. Anderson et al. (2007) demonstrated that UCP3 gene expression was increased by an acute bout of exercise in mice gastrocnemius muscle and this upregulation was dependent on mitochondrial $H_2O_2$ production. In the UCP3$^{-/-}$ mice, exercise failed to increase mitochondrial uncoupling respiration, whereas $H_2O_2$ production was greater compared to that in the WT mice. St-Pierre et al. (2006) showed that while $H_2O_2$ stimulated UCP3 and UCP2 expression in WT muscle cells, PGC-1α KO abolished these effects. Thus, PGC-1α could be an important mediator in the upregulation of UCP3.

## 14.6 CONCLUSION

Evolution has turned ROS into an essential component of cellular life through redox signaling. In skeletal muscle, contraction-induced ROS production underlies most of the major adaptations that have been observed during the past half-century. Muscle contraction can dramatically change the balance between ROS production and antioxidant defense, resulting in a shift in cellular redox status, which activates several signal transduction pathways, including but not limited to NF-κB, MAPK, and PGC-1α. Due to a broad range of cellular functions controlled by these important pathways, redox signaling can lead to proper adaptations related to metabolism, thermogenesis, antioxidant defense, mitochondrial remodeling, and apoptosis that are vital to cell life. It is thus proposed that redox signaling is one of the primary cellular mechanisms underlining hormesis that has been observed in many physiological and pathological conditions covered in this book.

# REFERENCES

Adams V, Nehrhoff B, Spate U, Linke A, Schulze PC, Baur A, Gielen S, Hambrecht R, and Schuler G. 2002. Induction of iNOS expression in skeletal muscle by IL-1beta and NFkappaB activation: An in vitro and in vivo study. *Cardiovasc. Res.* 54(1): 95–104.

Akimoto T, Pohnert SC, Li P, Zhang M, Gumbs C, Rosenberg PB, Williams RS, and Yan Z. 2005. Exercise stimulates Pgc-1α transcription in skeletal muscle through activation of the p38 MAPK pathway. *J. Biol. Chem.* 280: 19587–19593.

Allen RG and Tresini M. 2000. Oxidative stress and gene regulation. *Free Rad. Biol. Med.* 28: 463–499.

Anderson EJ, Yamazaki H, and Neufer PD. 2007. Induction of endogenous uncoupling protein 3 suppresses mitochondrial oxidant emission during fatty acid-supported respiration. *J. Biol. Chem.* 282: 31257–31266.

Arnold AS, Egger A, and Handschin C. 2010. PGC-1α and myokines in the aging muscle—A mini-review. *Gerontology.* 57: 37–43.

Aronson D, Violan MA, Dufresne SD, Zangen D, Fielding RA, and Goodyear LJ. 1997. Exercise stimulates the mitogen-activated protein kinase pathway in human skeletal muscle. *J. Clin. Invest.* 99: 1251–1257.

Baar K and Esser K. 1999. Phosphorylation of p70(S6K) correlates with increased skeletal muscle mass following resistance exercise. *Am. J. Physiol. Cell. Physiol.* 276: C120–C127.

Baar K, Wende AR, Jones TE, Marison M, Nolte LA, Chen M, Kelly DP, and Holloszy JO. 2002. Adaptations of skeletal muscle to exercise: Rapid increase in the transcriptional coactivator PGC-1. *FASEB J.* 16: 1879–1886.

Bach D, Pich S, Soriano FX, Vega N, Baumgartner B, Oriola J, Daugaard JR, et al. 2003. Mitofusin-2 determines mitochondrial network architecture and mitochondrial metabolism. A novel regulatory mechanism altered in obesity. *J. Biol. Chem.* 278: 17190–17197.

Baeuerle PA and Baltimore D. 1988. Activation of DNA-binding activity in an apparently cytoplasmic precursor of the NFκB transcription factor. *Cell.* 53: 211–217.

Balon TW. 1999. Integrative biology of nitric oxide and exercise. *Exerc. Sport Sci. Rev.* 27: 219–253.

Bellizzi D, Rose G, Cavalcante P, Covello G, Dato S, De Rango F, Greco V, et al. 2005. A novel VNTR enhancer within the SIRT3 gene, a human homologue of SIR2, is associated with survival at oldest ages. *Genomics.* 85: 258–263.

Bo H, Jiang N, Ma G, Qu J, Zhang G, Cao D, Wen L, Liu S, Ji LL, and Zhang Y. 2008. Regulation of mitochondrial uncoupling respiration during exercise in rat heart: Role of reactive oxygen species (ROS) and uncoupling protein 2. *Free Rad. Biol. Med.* 44(7): 1373–1381.

Bo H, Zhang Y, and Ji LL. 2010. Redefining the role of mitochondria in exercise: A dynamic remodeling. *Ann. N. Y. Acad. Sci.* 1201: 121–128.

Breckenridge DG, Kang BH, Kokel D, Mitani S, Staehelin LA, and Xue D. 2008. *Caenorhabditis elegans* drp-1 and fis-2 regulate distinct cell-death execution pathways downstream of ced-3 and independent of ced-9. *Mol. Cell.* 31: 586–597.

Cao W, Daniel KW, Robidoux J, Puigserver P, Medvedev AV, Bai X, Floering LM, Spiegelman BM, and Collins S. 2004. p38 mitogen-activated protein kinase is the central regulator of cyclic AMP-dependent transcription of the brown fat uncoupling protein 1 gene. *Mol. Cell. Biol.* 24: 3057–3067.

Carroll MP and May WS. 1994. Protein kinase C-mediated serine phosphorylation directly activates Raf-1 in murine hematopoietic cells. *J. Biol. Chem.* 269: 1249–1256.

Cartoni R, Léger B, Hock MB, Praz M, Crettenand A, Pich S, Ziltener JL, et al. 2005. Mitofusins 1/2 and ERR alpha expression are increased in human skeletal muscle after physical exercise. *J. Physiol.* 567. 349–358.

Catani MV, Savini I, Duranti G, Caporossi D, Ceci R, Sabatini S, and Avigliano L. 2004. Nuclear factor kappaB and activating protein 1 are involved in differentiation-related resistance to oxidative stress in skeletal muscle cells. *Free Rad. Biol. Med.* 37: 1024–1036.

Chan DC. 2006. Dissecting mitochondrial fusion. *Dev. Cell.* 11: 592–594.

Chan JY and Kwong M. 2000. Impaired expression of glutathione synthetic enzyme genes in mice with targeted deletion of the Nrf2 basic-leucine zipper protein. *Biochim. Biophys. Acta.* 1517: 19–26.

Chance B, Sies H, and Boveris A. 1979. Hydroperoxide metabolism in mammalian organs. *Physiol. Rev.* 59: 527–605.

Cheshire JL, Willuans BRG, and Baldwin AS. 1999. Involvement of double-stranded RNA-activated protein kinase in the synergistic activation of nuclear factor κB by tumor necrosis factor-α and γ-interferon in preneuronal cells. *J. Biol. Chem.* 274: 4801–4806.

Coffey VG, Zhong Z, Shield A, Canny BJ, Chibalin AV, Zierath JR, and Hawley JA. 2006. Early signaling responses to divergent exercise stimuli in skeletal muscle from well-trained humans. *FASEB J.* 20: 190–192.

Collins Y, Chouchani ET, James AM, Menger KE, Cochemé HM, and Murphy MP. 2012. Mitochondrial redox signalling at a glance. *J. Cell. Sci.* 125(Pt 4): 801–806.

Cortright RN, Zheng D, Jones JP, Fluckey JD, DiCarlo SE, Grujic D, Lowell BB, and Dohm GL. 1999. Regulation of skeletal muscle UCP-2 and UCP-3 gene expression by exercise and denervation. *Am. J. Physiol. Endocrinol. Metab.* 276: E217–E221.

D'Autréaux B and Toledano MB. 2007. ROS as signalling molecules: Mechanisms that generate specificity in ROS homeostasis. *Nature Rev. Mol. Cell. Bio.* 8: 813–824.

Davies KJA, Quintanilha AT, Brooks GA, and Packer L. 1982. Free radicals and tissue damage produced by exercise. *Biochem. Biophys. Res. Comm.* 107: 1198–1205.

Ding H, Jiang N, Liu H, Liu X, Liu D, Zhao F, Wen L, Liu S, Ji LL, and Zhang Y. 2009. Response of mitochondrial fusion and fission protein gene expression to exercise in rat skeletal muscle. *Biochim. Biophys. Acta.* 1800: 250–256.

Durham WJ, Arbogast S, Gerken E, Li YP, and Reid MB. 2006. Progressive nuclear factor-kappaB activation resistant to inhibition by contraction and curcumin in mdx mice. *Muscle Nerve.* 34: 298–303.

Feng H, Kang C, Dickman JR, Koenig R, Awoyinka I, Zhang Y, and Ji LL. 2013. Training-induced mitochondrial adaptation: Role of PGC-1α, NFκB, and β-Blockade. *Exp. Physiol.* 98: 784–795.

Finck BN and Kelly DP. 2006. PGC-1 coactivators: Inducible regulators of energy metabolism in health and disease. *J. Clin. Invest.* 116: 615–622.

Finkel T. 2011. Signal transduction by reactive oxygen species. *J. Cell. Biol.* 194: 7–15.

Flohé L, Brigelius-Flohé R, Saliou C, Traber M, and Packer L. 1997. Redox regulation of NF-kappa B activation. *Free Rad. Biol. Med.* 22: 1115–1126.

Garnier A, Fortin D, Zoll J, N'Guessan B, Mettauer B, Lampert E, Veksler V, and Ventura-Clapier R. 2005. Coordinated changes in mitochondrial function and biogenesis in healthy and diseased human skeletal muscle. *FASEB J.* 19: 43–52.

Geng T, Li P, Okutsu M, Yin X, Kwek J, Zhang M, and Yan Z. 2010. PGC-1α plays a functional role in exercise-induced mitochondrial biogenesis and angiogenesis but not fiber-type transformation in mouse skeletal muscle. *Am. J. Physiol. Cell Physiol.* 298: 572–579.

Ghosh S and Karin M. 2002. Missing pieces in the NF-kappaB puzzle. *Cell.* 109(Suppl): S81–S96.

Ghosh S, May MJ, and Kopp EB. 1998. NF-kappa B and Rel proteins: Evolutionarily conserved mediators of immune responses. *Annu. Rev. Immunol.* 16: 225–260.

Goglia F and Skulachev VP. 2003. A function for novel uncoupling proteins: Antioxidant defense of mitochondrial matrix by translocating fatty acid peroxides from the inner to the outer membrane leaflet. *FASEB J.* 17: 1585–1591.

Gomes LC and Scorrano L. 2008. High levels of Fis1, a pro-fission mitochondrial protein, trigger autophagy. *Biochim. Biophys. Acta.* 1777: 860–866.

Gomez-Cabrera M-C, Borras C, Pallardó FV, Sastre J, Ji LL, and Vina J. 2005. Decreasing xanthine oxidase-mediated oxidative stress prevents useful cellular adaptations to exercise in rats. *J. Physiol. London.* 567: 113–120.

Gomez-Lazaro M, Bonekamp NA, Galindo MF, Jordan J, and Schrader M. 2008. 6-Hydroxydopamine (6-OHDA) induces Drp1-dependent mitochondrial fragmentation in SH-SY5Y cells. *Free Rad. Biol. Med.* 44: 1960–1969.

Goodyear L, Chang P, Sherwood D, Dufresne S, and Moller D. 1996. Effects of exercise and insulin on mitogen-activated protein kinase signaling pathways in rat skeletal muscle. *Am. J. Physiol.* 271: E403–E408.

Grandemange S, Herzig S, and Martinou JC. 2009. Mitochondrial dynamics and cancer. *Semin. Cancer Biol.* 19: 50–56.

Hamanaka RB and Chandel NS. 2010. Mitochondrial reactive oxygen species regulate cellular signaling and dictate biological outcomes. *Trends. Biochem. Sci.* 35(9): 505–513.

Handschin C. 2009. Peroxisome proliferator-activated receptor-γ coactivator-1α in muscle links metabolism to inflammation. *Clin. Exp. Pharmacol. Physiol.* 36: 1139–1143.

Handschin C. 2010. Regulation of skeletal muscle cell plasticity by the peroxisome proliferator-activated receptor γ coactivator 1α. *J. Recep. Signal. Trans.* 30: 376–384.

Handschin C and Spiegelman BM. 2008. The role of exercise and PGC1α in inflammation and chronic disease. *Nature.* 454: 463–469.

Hawley JA and Zierath JR. 2004. Integration of metabolic and mitogenic signal transduction in skeletal muscle. *Exerc. Sport Sci. Rev.* 32: 4–8.

Hayakawa M, Miyashita H, Sakamoto I, Kitagawa M, Tanaka H, Yasuda H, Karin M, and Kikugawa K. 2003. Evidence that reactive oxygen species do not mediate NF-kappaB activation. *EMBO J.* 22: 3356–3366.

Ho, R. C.; Hirshman, M. F.; Li, Y.; Cai, D.; Farmer, J. R.; Aschenbach, W. G.; Witczak, C. A.; Shoelson, S. E.; Goodyear, L. J. *Am J Physiol Cell Physiol* 2005, 289, C794–801.

Ho YS, Howard AJ, and Crapo JD. 1991. Molecular structure of a functional rat gene for manganese-containing superoxide dismutase. *Am. J. Respir. Cell Mol. Biol.* 4: 278–286.

Hoffmann E, Thiefes A, Buhrow D, Dittrich-Breiholz O, Schneider H, Resch K, and Kracht M. 2005. MEK1-dependent delayed expression of Fos-related antigen-1 counteracts c-Fos and p65 NF-kappaB-mediated interleukin-8 transcription in response to cytokines or growth factors. *J. Biol. Chem.* 280: 9706–9718.

Hollander J, Fiebig R, Gore M, Ookawara T, Ohno H, and Ji LL. 2001. Superoxide dismutase gene expression is activated by a single bout of exercise in rat skeletal muscle. *Pflug. Arch. Eur. J. Physiol.* 442: 426–434.

Holloszy JO and Coyle EF. 1984. Adaptations of skeletal muscle to endurance exercise and their metabolic consequences. *J. Appl. Physiol.* 56: 831–838.

Holzberg D, Knight CG, Dittrich-Breiholz O, Schneider H, Dorrie A, Hoffmann E, Resch K, and Kracht M. 2003. Disruption of the c-JUN-JNK complex by a cell-permeable peptide containing the c-JUN delta domain induces apoptosis and affects a distinct set of interleukin-1-induced inflammatory genes. *J. Biol. Chem.* 278: 40213–40223.

Hom J and Sheu SS. 2009. Morphological dynamics of mitochondria—A special emphasis on cardiac muscle cells. *J. Mol. Cell. Cardiol.* 46: 811–820.

Hood DA, Irrcher I, Ljubicic V, and Joseph AM. 2006. Coordination of metabolic plasticity in skeletal muscle. *J. Exp. Biol.* 209: 2265–2275.

Hunter RB, Stevenson E, Koncarevic A, Mitchell-Felton H, Essig DA, and Kandarian SC. 2002. Activation of an alternative NF-kappaB pathway in skeletal muscle during disuse atrophy. *FASEB J.* 16: 529–538.

Jagoe RT and Goldberg AL. 2001. What do we really know about the ubiquitin–proteasome pathway in muscle atrophy? *Curr. Opin. Clin. Nutr. Metab. Care.* 4: 183–190.

Jendrach M, Mai S, Pohl S, Voth M, and Bereiter-Hahn J. 2008. Short- and long-term alterations of mitochondrial morphology, dynamics, and mtDNA after transient oxidative stress. *Mitochondrion*. 8: 293–304.
Jenkins RR. Exercise, Oxidative stress and antioxidant: A review. *Intl J Sports Nutr.* 1993; 3: 356-375.
Jezek P and Plecitá-Hlavatá L. 2009. Mitochondrial reticulum network dynamics in relation to oxidative stress, redox regulation, and hypoxia. *Int. J. Biochem. Cell Biol.* 41(10): 1790–1804.
Ji, L. L. and Y. Zhang. Antioxidant and anti-inflammatory effects of exercise: role of redox signaling. *Free Rad. Res.* 2013, ISSN 1071-5762 print/USSB 1029–2470 online.
Ji LL. 1995. Exercise and oxidative stress: Role of the cellular antioxidant systems. *Exerc. Sport Sci. Rev.* 23: 135–166.
Ji LL. 2007. Antioxidant signaling in skeletal muscle: A brief review. *Exp. Gerontol.* 42: 582–593.
Ji LL. 2008. Modulation of skeletal muscle antioxidant defense by exercise: Role of redox signaling. *Free Rad. Biol. Med.* 44: 142–152.
Ji LL, Gomez-Cabrera M-C, Steinhafel N, and Vina J. 2004. Acute exercise activates nuclear factor (NF)κB signaling pathway in rat skeletal muscle. *FASEB J.* 18: 1499–1506.
Ji LL, Gomez-Cabrera M-C, and Vina J. 2009. Role of antioxidants in muscle health and pathology. Infectious disorders special issue. *Infect. Disord. Drug Targets.* 9(4): 428–444.
Jiang B, Xu S, Hou X, Pimentel DR, Brecher P, and Cohen RA. 2004. Temporal control of NF-kappaB activation by ERK differentially regulates interleukin-1beta-induced gene expression. *J. Biol. Chem.* 279: 1323–1329.
Jiang N, Zhang G, Bo H, Qu J, Ma G, Cao D, Wen L, Liu S, Ji LL, and Zhang Y. 2009. Upregulation of uncoupling protein-3 in skeletal muscle during exercise: A potential antioxidant function. *Free Rad. Biol. Med.* 46(2): 138–145.
Kang C, O'Moore KM, Dickman JR, and Ji LL. 2009. Exercise activation of muscle peroxisome proliferator-activated receptor-γ coactivator-1α signaling is redox sensitive. *Free Rad. Biol. Med.* 47: 1394–1400.
Kelly DP and Scarpulla RC. 2004. Transcriptional regulatory circuits controlling mitochondrial biogenesis and function. *Genes. Dev.* 18: 357–368.
Klingenberg M and Huang SG. 1999. Structure and function of the uncoupling protein from brown adipose tissue. *Biochim. Biophys. Acta*. 1415: 271–296.
Knutti D and Kralli A. 2001. PGC-1, a versatile coactivator. *Trends Endocrinol. Metab.* 12: 360–365.
Kong X, Wang R, Xue Y, Liu X, Zhang H, Chen Y, Fang F, and Chang Y. 2010. Sirtuin 3, a new target of PGC-1α, plays an important role in the suppression of ROS and mitochondrial biogenesis. *PLoS One.* 5: e11707.
Koopman WJ, Verkaart S, van Emst-de Vries SE, Grefte S, Smeitink JA, Nijtmans LG, and Willems PH. 2008. Mitigation of NADH: Ubiquinone oxidoreductase deficiency by chronic Trolox treatment. *Biochim. Biophys. Acta*. 1777: 853–859.
Krauss S, Zhang CY, and Lowell BB. 2005. The mitochondrial uncoupling-protein homologues. *Nat. Rev. Mol. Cell. Biol.* 6: 248–261.
Kumar A and Boriek AM. 2003. Mechanical stress activates the nuclear factor-kappaB pathway in skeletal muscle fibers: A possible role in Duchenne muscular dystrophy. *FASEB J.* 17: 386–396.
Leick L, Lyngby SS, Wojtaszewski JF, and Pilegaard H. 2010. PGC-1α is required for training-induced prevention of age-associated decline in mitochondrial enzymes in mouse skeletal muscle. *Exp. Gerontol.* 45: 336–342.
Leick L, Wojtaszewski JF, Johansen ST, Kiilerich K, Comes G, Hellsten Y, Hidalgo J, and Pilegaard H. 2008. PGC-1α is not mandatory for exercise- and training-induced adaptive gene responses in mouse skeletal muscle. *Am. J. Physiol. Endocrinol. Metab.* 294: 463–474.
Li N and Karin M. 1999. Is NFκB the sensor of oxidative stress? *FASEB J.* 13: 1137–1143.

Li Q and Engelhardt JF. 2006. Interleukin-1β induction of NFκB is partially regulated by $H_2O_2$-mediated activation of NFκB-inducing kinase. *J. Biol. Chem.* 281: 1495–1505.

Li W, Bengtson MH, Ulbrich A, Matsuda A, Reddy VA, Orth A, Chanda SK, Batalov S, Joazeiro CA. Genome-wide and functional annotation of human E3 ubiquitin ligases identifies MULAN, a mitochondrial E3 that regulates the organelle's dynamics and signaling. PLoS One 2008; 3:e1487.

Liesa M, Borda-d'Agua B, Medina-Gomez G, Lelliott CJ, Paz JC, Rojo M, Palacin M, Vidal-Puig A, and Zorzano A. 2008. Mitochondrial fusion is increased by the nuclear coactivator PGC-1beta. *PLoS One.* 3: e3613.

Liesa M, Palacin M, and Zorzano A. 2009. Mitochondrial dynamics in mammalian health and disease. *Physiol. Rev.* 89: 799–845.

Lin J, Handschin C, and Spiegelman BM. 2005. Metabolic control through the PGC-1 family of transcription coactivators. *Cell Metab.* 1: 361–370.

Lin J, Wu H, Tarr PT, Zhang CY, Wu Z, Boss O, Michael LF, et al. 2002. Transcriptional co-activator PGC-1α drives the formation of slow-twitch muscle fibres. *Nature.* 418: 797–801.

Ljubicic V, Adhihetty PJ, and Hood DA. 2004. Role of UCP3 in state 4 respiration during contractile activity-induced mitochondrial biogenesis. *J. Appl. Physiol.* 97: 976–983.

Long YC, Widegren U, and Zierath JR. 2004. Exercise-induced mitogen-activated protein kinase signalling in skeletal muscle. *Proc. Nutr. Soc.* 63: 227–232.

Manoli I, Alesci S, Blackman MR, Su YA, Rennert OM, and Chrousos GP. 2007. Mitochondria as key components of the stress response. *Trends Endocrinol. Metab.* 18: 190–198.

Meyer M, Pahl HL, and Baeuerle PA. 1994. Regulation of the transcription factors NF-kB and AP-1 by redox changes. *Chem. Biol. Interact.* 91: 91–100.

Michael LF, Wu Z, Cheatham RB, Puigserver P, Adelmant G, Lehman JJ, Kelly DP, and Spiegelman BM. 2001. Restoration of insulin-sensitive glucose transporter (GLUT4) gene expression in muscle cells by the transcriptional coactivator PGC-1. *Proc. Natl. Acad. Sci. USA.* 98: 3820–3825.

Mootha VK, Lindgren CM, Eriksson KF, Subramanian A, Sihag S, Lehar J, Puigserver P, et al. 2003. PGC-1α-responsive genes involved in oxidative phosphorylation are coordinately downregulated in human diabetes. *Nature Genet.* 34: 267–273.

Nader G and Esser K. 2001. Intracellular signaling specificity in skeletal muscle in response to different modes of exercise. *J. Appl. Physiol.* 90: 1936–1942.

Oberley LW, St Clair DK, Autor AP, and Oberley TD. 1987. Increase in manganese superoxide dismutase activity in the mouse heart after X-irradiation. *Arch. Biochem. Biophys.* 254: 69–80.

Pourova J, Kottova M, Voprsalova M, and Pour M. 2010. Reactive oxygen and nitrogen species in normal physiological processes. *Acta. Physiol. Oxf.* 198(1): 15–35.

Powers SK and Jackson MJ. 2008. Exercise-induced oxidative stress: Cellular mechanisms and impact on muscle force production. *Physiol. Rev.* 88: 1243–1276.

Puigserver P, Rhee J, Lin J, Wu Z, Yoon JC, Zhang CY, Krauss S, Mootha VK, Lowell BB, and Spiegelman BM. 2001. Cytokine stimulation of energy expenditure through p38 MAP kinase activation of PPARgamma coactivator-1. *Mol. Cell.* 8: 971–982.

Puigserver P and Spiegelman BM. 2003. Peroxisome proliferator-activated receptor-gamma coactivator 1α (PGC-1α): Transcriptional coactivator and metabolic regulator. *Endocr. Rev.* 24: 78–90.

Puigserver P, Wu Z, Park CW, Graves R, Wright M, and Spiegelman BM. 1998. A cold-inducible coactivator of nuclear receptors linked to adaptive thermogenesis. *Cell.* 92: 829–839.

Ramires P and Ji LL. 2001. Glutathione supplementation and training increases myocardial resistance to ischemia-reperfusion in vivo. *Am. J. Physiol.* 281: H679–H688.

Reid MB. 2001. Redox modulation of skeletal muscle contraction: What we know and what we don't. *J. Appl. Physiol.* 90: 724–731.

Russell AP. 2005. PGC-1α and exercise: Important partners in combating insulin resistance. *Curr. Diabetes Rev.* 1: 175–181.

Ryder J, Fahlman R, Wallberg-Henriksson H, Alessi D, Krook A, and Zierath J. 2000. Effect of contraction on mitogen-activated protein kinase signal transduction in skeletal muscle. Involvement of the mitogen- and stress-activated protein kinase 1. *J. Biol. Chem.* 275: 1457–1462.

Santel A and Fuller MT. 2001. Control of mitochondrial morphology by a human mitofusin. *J. Cell Sci.* 114: 867–874.

Sakamoto K and Goodyear LJ. 2002. Invited review: Intracellular signaling in contracting skeletal muscle. *J. Appl. Physiol.* 93: 369–383.

Scarpulla RC. 2008. Transcriptional paradigms in mammalian mitochondrial biogenesis and function. *Physiol. Rev.* 88: 611–638.

Schreiber SN, Emter R, Hock MB, Knutti D, Cardenas J, Podvinec M, Oakelev EJ, and Kralli A. 2004. The estrogen-related receptor α (ERRα) functions in PPARgamma coactivator 1α (PGC-1α)-induced mitochondrial biogenesis. *Proc. Natl. Acad. Sci. USA.* 101: 6472–6477.

Sen CK. 1995. Oxidants and antioxidants in exercise. *J. Appl. Physiol.* 79: 675–686.

Sen CK and Roy S. 2005. Relief from a heavy heart: Redox-sensitive NF-kappaB as a therapeutic target in managing cardiac hypertrophy. *Am. J. Physiol. Heart Circ. Physiol.* 289: H17–H19.

Shi T, Wang F, Stieren E, and Tong Q. 2005. SIRT3, a mitochondrial sirtuin deacetylase, regulates mitochondrial function and thermogenesis in brown adipocytes. *J. Biol. Chem.* 280: 13560–13567.

Song Z, Ghochani M, McCaffery JM, Frey TG, and Chan DC. 2009. Mitofusins and OPA1 mediate sequential steps in mitochondrial membrane fusion. *Mol. Biol. Cell.* 20: 3525–3532.

Soriano FX, Liesa M, Bach D, Chan DC, Palacin M, and Zorzano A. 2006. Evidence for a mitochondrial regulatory pathway defined by peroxisome proliferator-activated receptor-gamma coactivator-1 alpha, estrogen-related receptor-alpha, and mitofusin 2. *Diabetes.* 55: 1783–1791.

St-Pierre J, Drori S, Uldry M, Silvaggi JM, Rhee J, Jager S, Handschin C, et al. 2006. Suppression of reactive oxygen species and neurodegeneration by the PGC-1 transcriptional coactivators. *Cell.* 127: 397–408.

St-Pierre J, Lin J, Krauss S, Tarr PT, Yang R, Newgard CB, and Spiegelman BM. 2003. Bioenergetic analysis of peroxisome proliferator-activated receptor gamma coactivators 1alpha and 1beta (PGC-1alpha and PGC-1beta) in muscle cells. *J. Biol. Chem.* 278: 26597–26603.

Tegethoff S, Behlke J, and Scheidereit C. 2003. Tetrameric oligomerization of IkappaB kinase gamma (IKKgamma) is obligatory for IKK complex activity and NF-kappaB activation. *Mol. Cell. Biol.* 23: 2029–2041.

Valle I, Alvarez-Barrientos A, Arza E, Lamas S, and Monsalve M. 2005. PGC-1α regulates the mitochondrial antioxidant defense system in vascular endothelial cells. *Cardiovasc. Res.* 66: 562–573.

van Ginneken MM, de Graaf-Roelfsema E, Keizer HA, van Dam KG, Wijnberg ID, van der Kolk JH, and van Breda E. 2006. Effect of exercise on activation of the p38 mitogen-activated protein kinase pathway, c-Jun NH2 terminal kinase, and heat shock protein 27 in equine skeletal muscle. *Am. J. Vet. Res.* 67: 837–844.

Vercauteren K, Pasko RA, Gleyzer N, Marino VM, and Scarpulla RC. 2006. PGC-1-related coactivator: Immediate early expression and characterization of a CREB/NRF-1 binding domain associated with cytochrome c promoter occupancy and respiratory growth. *Mol. Cell. Biol.* 26: 7409–7419.

Wenz T, Rossi S, Rotundo RL, Spiegelman BM, and Moraes CT. 2009. Increased muscle PGC-1α expression protects from sarcopenia and metabolic disease during aging. *Proc. Natl. Acad. Sci. USA*. 106: 20405–20410.

Wu Z, Puigserver P, Andersson U, Zhang C, Adelmant G, Mootha V, Troy A, et al. 1999. Mechanisms controlling mitochondrial biogenesis and respiration through the thermogenic coactivator PGC-1. *Cell*. 98: 115–124.

Yu T, Robotham JL, and Yoon Y. 2006. Increased production of reactive oxygen species in hyperglycemic conditions requires dynamic change of mitochondrial morphology. *Proc. Natl. Acad. Sci. USA*. 103: 2653–2658.

Zhou LZ, Johnson AP, and Rando TA. 2001. NF kappa B and AP-1 mediate transcriptional responses to oxidative stress in skeletal muscle cells. *Free Rad. Biol. Med.* 31: 1405–1416.

Zhou M, Lin BZ, Coughlin S, Vallega G, and Pilch PF. 2000. UCP-3 expression in skeletal muscle: Effects of exercise, hypoxia, and AMP-activated protein kinase. *Am. J. Physiol. Endocrinol. Metab.* 279: 622–629.

# Section IV

## Hormesis in Risk Assessment

# 15 Relating Hormesis to Ethics and Policy
## Conceptual Issues and Scientific Uncertainty

*George R. Hoffmann*

**CONTENTS**

15.1 Models for the Effects of Chemicals and Radiation at Low Doses..............308
    15.1.1 Thresholds and Linear Nonthreshold as Standard Models ..............308
    15.1.2 Hormesis: A Biphasic Dose-Response Relationship ......................309
    15.1.3 Hormesis as a Challenge to Threshold and Linear Nonthreshold Models ..............................................................................................310
15.2 Nature of Hormesis............................................................................................311
    15.2.1 Difficulty of Detecting and Measuring Hormesis ..........................311
    15.2.2 Evidence Supports Hormesis as Real ..............................................312
    15.2.3 Apparent Hormesis Arising as an Artifact......................................314
15.3 Biological Stress Responses .............................................................................314
    15.3.1 Adaptive Responses and Preconditioning ......................................314
    15.3.2 Parallels between Adaptive Responses and Hormesis....................316
15.4 Mechanisms of Hormesis and Stress Responses .............................................317
15.5 Prospects for Hormesis in Risk Assessment....................................................318
    15.5.1 Controversy over Assimilating Hormesis into Policy .....................318
    15.5.2 Precautionary Principle and Scientific Reality................................319
    15.5.3 Hormesis and Biomedical Ethics......................................................321
15.6 Challenges of Assimilating Hormesis into Risk Assessment.......................321
    15.6.1 Default Assumptions for Low-Dose Effects ....................................321
    15.6.2 Disagreement about the Generalizability of Hormesis ...................322
    15.6.3 Disagreement about the Prevalence of Hormesis...........................323
    15.6.4 Uncertainties in the Quantification of Hormetic Effects.................323
    15.6.5 Accounting for Heterogeneity in Susceptibility .............................324
    15.6.6 Interactions among Agents ..............................................................326
    15.6.7 Hormesis and Concomitant Toxicity ...............................................326
    15.6.8 Feasibility of Hormesis-Based Risk Assessment.............................327
15.7 Why an Understanding of Hormesis Is Essential..........................................327
    15.7.1 Concerns about Hormesis and Risks of Ignoring It .......................327
    15.7.2 Optimizing the Benefits of Mild Stress Responses ........................327

15.7.3 Avoiding Unforeseen Risks to Public Health .................................. 328
15.7.4 Agricultural Productivity and Environmental Quality .................... 329
15.8 Conclusions ................................................................................................ 330
Acknowledgments ............................................................................................... 330
References ........................................................................................................... 331

## 15.1 MODELS FOR THE EFFECTS OF CHEMICALS AND RADIATION AT LOW DOSES

### 15.1.1 Threshold and Linear Nonthreshold as Standard Models

Humans are routinely exposed to low doses of toxicants and radiation, but it is extremely difficult or impossible to measure biological effects at the low doses that are of interest. Dose-response models are therefore useful not only in assessing risks associated with measured effects but also in shaping our expectations for effects at dosages below which accurate measurements are impossible or impractical. Figure 15.1 shows threshold and linear nonthreshold (LNT) models that have dominated thought about low doses in toxicology and radiation biology. The curves show the frequency of an adverse effect plotted against dosage. In the threshold model (Figure 15.1a), there is a dosage below which the frequency does not differ from the unexposed control population. This dosage represents a biological threshold, often represented in toxicological studies as a *no observed adverse effect level* (NOAEL). In contrast, the linear nonthreshold model (Figure 15.1b), often referred to as LNT, extrapolates to the spontaneous frequency on the ordinate.

The dominant dose-response model in toxicology is a threshold model. A threshold dose-response relationship (Figure 15.1a) has a slope of zero at low doses, followed by an increasing response, which may be nonlinear, in the zone of toxicity above the threshold. It is often represented in toxicology textbooks as a sigmoid curve that describes the proportion of a defined population showing a quantal characteristic, such as death, with increasing dosage (Eaton and Gilbert 2008). For risk assessment purposes, the aim under the assumption of a threshold model is to ensure

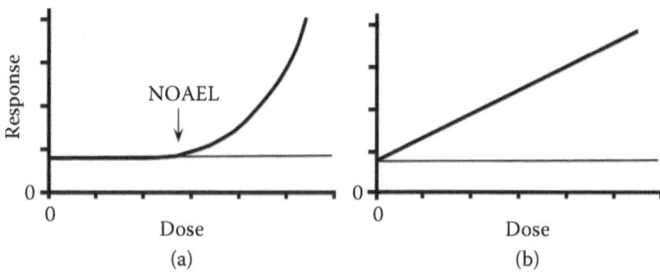

**FIGURE 15.1** Dose-response relationships: Threshold (a) and linear nonthreshold (LNT) (b) dose-response models. The response is the frequency of an adverse effect, and the thin horizontal line is its spontaneous frequency in an unexposed control population. A threshold dose is indicated by the NOAEL (no observed adverse effect level). (Adapted from Hoffmann, G. R., *Dose-Response* 7, 1–51, 2009.)

that exposures are below the threshold or NOAEL. Although the threshold model is the prevailing assumption for biological endpoints other than mutation and cancer, its acceptance is not universal, and the debate has intensified with arguments for (White et al. 2009) and against (Rhomberg et al. 2011; Bukowski et al. 2013) the use of linear extrapolation for noncancer effects.

In contrast to the threshold model, all doses in the LNT model (Figure 15.1b) are considered to have an effect, and one strives to ensure that exposures are either zero or small enough that the risk is negligible. LNT has been the prevailing assumption in genetic toxicology (Doak et al. 2007; Gocke and Müller 2009; Lutz and Lutz 2009; Bryce et al. 2010) and carcinogenesis (EPA 2005; Brenner and Sachs 2006; Preston and Hoffmann 2013). The adoption of LNT for genotoxic agents was based on conceptual, historical, and experimental reasons (Hoffmann 2009). Hit theory supporting an assumption of linearity for the induction of genetic damage was central to radiation biology, and radiation biology provided the historical foundation for interpreting chemical mutagenesis, which was discovered 15 years after the classical demonstration of x-ray mutagenesis by H.J. Muller in 1927. Although the early studies concentrated on radiation doses that would be moderate or high by today's standards, linearity appeared to extend to the lowest doses (Brenner et al. 2003). However, resolution of the shape of dose-response curves for mutagenesis in the low-dose zone is extremely difficult, owing to the fact that the low spontaneous frequencies of genetic alterations make it hard to measure small changes.

Evidence that accumulated over decades made it clear that mutagenesis is not a unitary interaction between agent and target as envisioned in hit theory. Biological systems show genetic and physiological responses to damage inflicted by radiation and chemicals, and they cannot be thought of as inert elements to which mutagenesis simply happens. Rather, such factors as mutagen uptake and metabolism, complex interactions with DNA, cellular processing of damage through repair and recombination, regulation of cellular proliferation, and factors in mutant expression can all lead to deviations from linearity (Hoffmann 2009). Mechanistic reasons for nonlinearity are also supported by experimental evidence that dose-responses are often sublinear (Lutz and Lutz 2009; Dobo et al. 2011). Although LNT remains the best model for describing some genotoxic effects (Bryce et al. 2010; Spassova et al. 2013), thresholds are now well documented in many other cases (Doak et al. 2007; Gocke and Müller 2009; Lutz and Lutz 2009; Bryce et al. 2010; Hoffmann et al. 2013; Thomas et al. 2013).

### 15.1.2 Hormesis: A Biphasic Dose-Response Relationship

Hormesis differs fundamentally from the threshold and LNT models in that the hormesis model proposes biphasic responses to toxicants or radiation (Calabrese and Baldwin 2001a; Calabrese 2008a; Hoffmann 2009). The threshold and LNT curves can be described as monotonic, in that they show either an increase or a decrease in response over the full range of doses that have an effect. In contrast, the hormetic curve is nonmonotonic, meaning that the response changes in more than one direction with a unidirectional change in dose (Davis and Svendsgaard 1994). Using the term *hormesis* does not refer to a specific mechanism or pathway but, rather, to the shape of the dose-response curve.

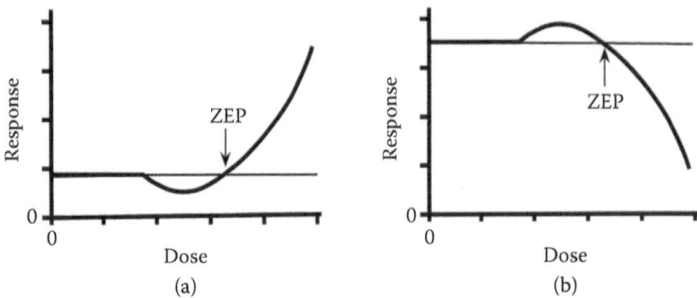

**FIGURE 15.2** Hormetic curves. The *J-shaped* curve (a) shows hormesis at low doses and an adverse effect at high doses. The *inverted U* curve (b) shows hormesis for a biological function that is stimulated by low doses and inhibited by high doses. At the zero equivalent point (ZEP) the curve crosses the level of response of the unexposed control. (Adapted from Hoffmann, G. R., *Dose-Response* 7, 1–51, 2009.)

Figure 15.2 shows two manifestations of hormesis. In Figure 15.2a, sometimes called a J-shaped curve, the response is the induction of an adverse effect at high doses, while low doses lead to a reduction of its frequency below the level in the unexposed control population. The response in Figure 15.2b is a biological function that is inhibited by high doses but stimulated by low doses. For example, one would expect such a curve, called an *inverted U*, if a growth-inhibiting chemical actually stimulated growth in the low-dose zone. A concept befitting the hormesis model is the zero equivalent point (ZEP), defined as the point at which the biphasic curve crosses the level of response of the unexposed control (Calabrese 2005a). The ZEP is a variation on the NOAEL that is independent of whether effects are adverse or beneficial, and it differs from a no-effect level, in that there is a biological effect at dosages above and below this point (Hoffmann 2009). Thus, the defining feature of hormesis is the biphasic nature of the response, not whether that response happens to be detrimental or beneficial. If hormesis is defined on the basis of opposite responses at high and low doses, it encompasses a broad array of phenomena for which low doses stimulate a process while high doses inhibit it, and vice versa. Although one may be inclined to think of the hormetic zone as beneficial and the high-dose zone as harmful, there are circumstances where the reverse is true.

### 15.1.3 Hormesis as a Challenge to Threshold and Linear Nonthreshold Models

Hormesis presents a challenge for toxicology at two levels: first, it implies that the dose-response relationships that are a cornerstone of the discipline may often be wrong (Calabrese 2005a, 2009); and second, its biphasic nature implies that low doses of toxicants and radiation may be beneficial. The most heated disagreement about hormesis concerns the latter, especially as it relates to regulatory policy and protection of public health. A perspective will be offered in this chapter on the merits of the proposal that hormesis should be incorporated into risk assessment.

Aspects of hormesis apart from toxicological and radiological risk are less controversial, but they are not yet widely recognized. It would be unfortunate if arguments against the application of hormesis in risk assessment impeded the elucidation of hormesis as a biological phenomenon. There is growing evidence that hormesis-like mechanisms contribute to neurological, cardiovascular, skeletal, and muscular well-being (Arumugam et al. 2006; Radak et al. 2008), and that mild stress can contribute to healthy aging (Rattan 2008; Le Bourg 2009). Hormesis is also relevant to environmental issues, because hormetic mechanisms may figure into how organisms respond to ecosystem disturbance and agricultural practices (Hoffmann 2009).

## 15.2 NATURE OF HORMESIS

### 15.2.1 Difficulty of Detecting and Measuring Hormesis

Much evidence supports the view that hormesis is a real biological phenomenon, not merely an artifact of data selection or a consequence of random variation (Davis and Svendsgaard 1994; Calabrese and Baldwin 2001a; Calabrese et al. 2006). Yet, it is not readily detected. One must be able to detect change at low doses both above and below the background level of damage, and this may not be possible for effects that are rare or absent in an unexposed population (Calabrese and Baldwin 2001b). Similarly, hormesis may not be observable when the background level of exposure is already above a toxicity threshold, as has been suggested for lead, or if the agent mimics endogenous substances, such as estrogens, that are themselves risk factors for adverse effects (Welshons et al. 2003; Hoffmann 2009).

The design of many studies works against detecting hormesis because toxicology and radiation biology emphasize adverse effects. Doses that have little effect in preliminary studies are often not used in follow-up studies. The dosages commonly used are those that elicit measurable responses, that is, the high-dose range. Hormetic effects tend to have small deviations from the control, and there are typically too few doses below a NOAEL to evaluate the shape of the curve in the low-dose zone (Calabrese and Baldwin 2001b). In addition, there are often too few replicates to generate the statistical power required to measure small changes in the low frequency of events in that zone. These factors can make observations of possible hormesis uncertain, in that the effects can also be attributed to random variation, and the evidence of hormesis is rarely robust enough to exclude other dose-response models (Hoffmann 2009).

Even studies conducted on a large scale can be consistent with several models. For example, the $ED_{01}$ study of the National Center for Toxicological Research evaluated the carcinogenicity of 2-acetylaminofluorene specifically to evaluate the low-dose zone. Despite using more than 24,000 mice to have large sample sizes, the responses were considered to show a threshold for bladder cancer but no threshold for liver cancer (Eaton and Gilbert 2008). Moreover, there was a debatable claim of hormesis for bladder cancer (Bruce et al. 1983; Kodell et al. 1983). Speaking on behalf of a Society of Toxicology task force, Bruce et al. (1983) pointed out that "statistical uncertainty makes it impossible to establish the true shape of the dose-response curve." Scaling up the size of experiments may therefore fail to resolve responses to low doses, and mechanistic evidence on the mode of action is more apt to clarify which carcinogens and carcinogenic effects show thresholds (Andersen et al. 2003; EPA 2005). Extending from thresholds to hormesis will require expanded efforts to understand hormesis mechanistically.

### 15.2.2 Evidence Supports Hormesis as Real

Early claims of hormesis were often based on the observation of curves that appeared to be biphasic. For example, Townsend and Luckey (1960) cited more than 100 examples of what they called *hormoligosis* a half-century ago. The identification was based on a biphasic response, which they called a β-*pattern*. Many early reports were essentially anecdotal, in that they relied largely on an accumulation of examples and ad hoc criteria for hormesis (Calabrese et al. 1999). Nonetheless, they implied that hormesis is a common phenomenon. Stronger evidence for hormesis has come from quantification of the frequency of hormetic curves in scientific literature (Davis and Svendsgaard 1994; Calabrese and Baldwin 2001a, 2003) and analysis of data from the high-throughput screening of chemicals for reasons other than measuring hormesis (Calabrese et al. 2006, 2010). The reliance on literature surveys and preexisting databases rather than experiments specifically designed to measure hormesis has been criticized (Kitchin and Drane 2005; Mushak 2009; Elliott 2011), but it is nevertheless true that both scientific and historical factors have hindered a more direct approach.

Crump (2001) succinctly described a problem in many claims of hormesis, pointing out the lack of a valid statistical test for hormesis and the lack of objective criteria for measuring its prevalence. Studies suggesting that hormesis is common typically lacked an appropriate denominator, and the inclusion of studies because they appeared to be hormetic was a source of bias in some evaluations (Crump 2001). Yet, the fact that some low dose-responses differed significantly from the control in the opposite direction from high dose-responses, while not definitive, gave credibility to the claim of hormesis. Moreover, an analysis of the literature by Davis and Svendsgaard (1994) provided limited but objective evidence of hormesis.

Davis and Svendsgaard (1994) estimated a frequency of biphasic curves from the Environmental Protection Agency Integrated Risk Information System database of reference doses (RfDs). An RfD is an estimate of a daily exposure that is likely to be without an appreciable risk of adverse effects over a lifetime. It is typically derived from a NOAEL by dividing by an uncertainty factor. Criteria for excluding published toxicology papers were delineated, and 147 papers were evaluated by at least two people. They acknowledged that an appropriate statistical test for frequencies of U-shaped curves was lacking, so they applied such ad hoc criteria as a change of 5% or more compared to the control in an initial evaluation and a difference of two standard errors in a follow-up evaluation. These papers contained 780 dose-responses, and they judged 12% of them to meet their criteria for U-shaped biphasic responses. An unidentified *independent academic toxicologist* that they asked to review the same articles judged the frequency of U-shaped curves to be 24%. Thus, there was an estimate of 12%–24% prevalence of hormesis.

A survey of peer-reviewed toxicological literature by Calabrese and Baldwin (2001a) led to the interpretation that only a small fraction of papers (195/20,285) permitted an evaluation of hormesis. The majority were excluded because of the lack of at least two sub-NOAEL doses on which to make the evaluation, the lack of proper controls, or the lack of a toxic response at high dose (Calabrese and Baldwin 2001a,b). Hormesis was indicated by either of two criteria: statistical significance or a 10% difference from the control in three or more doses below a NOAEL. Of 668 dose-response relationships in the qualifying papers, 245 were judged to provide evidence of hormesis (Calabrese and Baldwin 2001b). Slightly more than 20% of responses below the NOAEL differed significantly from the control and, among these, 19.5% differed in the direction expected for hormesis, whereas only 0.6% differed in the direction of toxicity (Calabrese and Baldwin 2001a). The data correspond to a prevalence of 19%–37% hormesis, depending on the stringency of the criteria used for classifying a response as hormetic. In a follow-up study, Calabrese and Baldwin (2003) compared the hormesis model with a threshold model, using 664 dose-response relationships that contained 1800 sub-NOAEL doses. They evaluated the assumption that the sub-NOAEL doses should fall above and below the NOAEL with equal frequency if the data fit a threshold model. The hypothetical 1:1 ratio was not observed. Rather, the ratio was 2.5:1 in the direction predicted by hormesis.

The analysis of databases from the high-throughput screening of chemicals for toxicity complements the analyses of published literature. Both support the reality of hormesis. Calabrese et al. (2006) analyzed data on yeast growth in a National Cancer Institute antitumor drug-screening database for roughly 57,000 dose-responses representing more than 2,000 chemicals. There was a wild-type strain and 12 mutants whose genotypes are relevant to toxicant responses, such as having altered DNA repair genes. A benchmark dose (BMD), roughly equivalent to a NOAEL, was calculated as the minimum dose causing toxicity. If the threshold model were correct, one should expect responses below the BMD to be randomly distributed above and below the control. Growth at doses below the BMD was significantly greater than the control in all strains, both for highly toxic compounds and relatively nontoxic compounds. The mean growth for chemicals grouped by toxicity was 103.6%–106.6% of the control when the yeast were exposed to sub-NOAEL doses, and hormetic response patterns were four times more common than would be expected by chance. Several different modes of analysis supported the occurrence of hormesis in the yeast database (Calabrese et al. 2006, 2008).

A bacterial database provided qualitatively similar evidence of hormetic effects but the magnitude of the low-dose stimulation of growth was smaller than that in yeast (Calabrese et al. 2010). The data were measurements of growth of *Escherichia coli* after exposure to 11 concentrations of 1888 chemicals. Growth at concentrations below the threshold was significantly greater than that in the controls. The determination of the stimulatory effect was not straightforward because of differences between chemicals of different toxicity and between edge rows and interior rows of the 96-well plates used in high-throughput screening. An estimate of the hormetic effect is 1%–4% above the controls (Calabrese et al. 2010).

### 15.2.3 APPARENT HORMESIS ARISING AS AN ARTIFACT

A difficulty in the detection of hormesis is that there are artifacts that can resemble hormetic responses. Controls that happen to be atypical can cause low doses that are not different from the true control value to seem hormetic (Thayer et al. 2005). For example, a higher-than-normal frequency of tumors in a control can make low doses that are consistent with the true or historical control value appear to show a hormetic response for carcinogenesis. Pooling biological endpoints can also give an erroneous appearance of hormesis (Thayer et al. 2005). For example, if a chemical causes a modest decrease in the incidence of tumors at one site but a large increase in tumors at another site at a higher dose, the composite response for total tumors may appear as a biphasic curve. Rather than being hormesis, which is by definition biphasic, such a case is actually the summing of two monotonic curves.

Essentiality is another possible cause of an illusion of hormesis (Kefford et al. 2008). If a compound is physiologically required or can substitute for an essential nutrient, it may stimulate growth at low doses and then cause a decline in growth in the toxic zone. Although the distinction between a xenobiotic and an essential nutrient is usually obvious, it may not always be so. For example, if plants are treated with a chemical mixture containing a mineral nutrient required for growth, the nutrient may stimulate growth at doses below which the toxicant exerts any effect. The resultant inverted-U curve actually represents the summing of the two curves, rather than a hormetic effect of the toxicant at low dose. Vigilance is needed to ensure that noncritical interpretation of biphasic curves does not lead to the misidentification of responses as hormetic.

## 15.3 BIOLOGICAL STRESS RESPONSES

### 15.3.1 ADAPTIVE RESPONSES AND PRECONDITIONING

Sequential exposures to a toxicant or radiation show that the outcome is dependent not only on the agent causing damage but also on the organism's biological response in processing the damage. A first exposure, typically at a low dose, often causes a reduction in susceptibility that manifests itself as a smaller effect of a subsequent, larger exposure. This phenomenon has been described under different names in different disciplines (Calabrese et al. 2007), but it is most commonly called an adaptive response, preconditioning, or a stress response that leads to tolerance.

The first reported adaptive response entailed diminished bacterial mutagenicity of the potent mutagen $N$-methyl-$N'$-nitro-$N$-nitrosoguanidine (MNNG) caused by a small prior exposure of the bacteria to the same agent (Samson and Cairns 1977). A few years later, studies in human lymphocytes provided the first evidence of an adaptive response to ionizing radiation. Cells chronically given a small exposure to β-particles from tritiated thymidine experienced fewer chromatid aberrations after irradiation with 1.5 Gy x-rays than did cells without the β-particle exposure (Olivieri et al. 1984).

Small priming doses of x-rays similarly induced an adaptive response that conferred reduced susceptibility to the induction of chromosomal aberrations by a subsequent higher dose (Shadley and Wolff 1987). Such responses are sometimes called stress responses, and they are known in organisms throughout the phylogenetic hierarchy. They confer resistance to diverse stressors, including hypoxia, high osmotic strength, oxidants, and various toxicants and metabolites (Samson and Cairns 1977; Olivieri et al. 1984; Davies et al. 1995; Wiese et al. 1995; Wolff 1998; Miura 2004; Calabrese et al. 2007; Hoffmann 2009; Guan et al. 2012; Morano et al. 2012).

There is overlap between adaptive responses, in that exposure to one agent may confer resistance to others. For example, a small exposure to $H_2O_2$ can confer resistance not only to $H_2O_2$ but also to other inducers of oxidative stress, such as menadione or paraquat (Temple et al. 2005). Conversely, resistance to $H_2O_2$ can be induced by exposures to NaCl (Guan et al. 2012), heat, or lipid peroxidation products (Temple et al. 2005). In some instances, an adaptive response to one agent may be accompanied by enhanced susceptibility to another, as in the case of γ-radiation reducing the induction of chromosomal damage in human lymphocytes by bleomycin or mitomycin C but increasing the damage caused by methyl methanesulfonate (Wolff et al. 1988). Although adaptive responses often show cross-resistances (Wolff 1996; Wheeler and Wong 2007), these are not always reciprocal, and the patterns suggest that there are several adaptive pathways with overlapping components (Temple et al. 2005; Morano et al. 2012).

The dose dependence of the induction of adaptive responses is not well-understood. In some cases it occurs within a window of effective dosage, such that smaller doses are insufficient to trigger the response, and higher doses are ineffective in inducing the response or contribute to damage to an extent that swamps the adaptive response. Thus, the induction is biphasic and resembles a hormetic curve. For example, in the cytogenetic studies of Shadley and Wolff (1987), an adaptive response to x-rays was observed after priming doses from 0.5 to 20 cGy, but not after exposure at high doses. Similarly, a priming dose of 13 cGy x-rays was optimal for the induction of an adaptive response that made the growth of HE22 human embryonic fibroblasts more tolerant of a challenge dose of 2 Gy (Ishii and Watanabe 1996). Unlike these studies, which suggest a biphasic induction, others found similar adaptive responses over a broad range of priming doses. For example, doses of 1–500 mGy γ-rays at a low dose rate conferred roughly the same extent of reduced susceptibility to the induction of micronuclei by 4 Gy γ-rays in human AG1522 fibroblasts (Broome et al. 2002). Similarly, an adaptive response to the induction of chromosomal inversions by x-rays in pKZ1 mice was induced to a comparable extent by a 1000-fold range of priming doses (Day et al. 2007). The explanation for such discrepancies may lie in factors other than dose itself, such as other stressors, dose rate, physiological conditions, and genetic susceptibility (Mitchel 2010).

The term *adaptive response* is common in genetics, but in other fields similar phenomena are sometimes called *preconditioning* (Murry et al. 1986; Arumugam et al. 2006; Lin et al. 2008). The unifying feature that defines these responses is that cells or organisms that are exposed to a mild stress become tolerant of more severe stress (Arumugam et al. 2006). For example, the severity of myocardial infarction caused by coronary occlusion in dogs was reduced if the dogs were preconditioned by exposing them to brief ischemia from mild coronary occlusion (Murry et al. 1986). A similar phenomenon has been reported in humans, in that survival after cardiogenic shock was higher in patients who had preinfarction angina than in those without angina (Le Bourg 2009). Taken as a whole, adaptive responses, preconditioning, and the overlapping qualities of the various biological responses suggest the existence of a broad family of conserved responses to environmental stressors.

### 15.3.2 Parallels between Adaptive Responses and Hormesis

A possible relationship of hormesis to stress responses was noted by an early advocate of hormesis, T.D. Luckey (1968), who suggested that levels of a stressing agent that are too small to be detrimental will be stimulatory to the organism. About 30 years later, a paper coauthored by a large group of scientists proposed that adaptive responses are manifestations of hormesis (Calabrese et al. 2007). A striking parallel between them is that both entail a reduction in damage conferred by a small exposure to a stressor. Although the proposed relationship is conceivable, it is not trivial that the conditions under which the two phenomena are observed differ sharply. Hormesis, as traditionally defined, involves a biphasic response to a single exposure. The hormetic response reduces the damage to a level below that of the control. In contrast, adaptive responses depend on sequential exposures, in which the first exposure modifies susceptibility to the second.

Hormesis and adaptive responses share a temporal component that affects their detection and quantitative characteristics. In both cases, one should expect a lag after exposure for the induction of the response, a period of protection, and then a return to the ground state (Calabrese and Baldwin 2001b; Calabrese et al. 2007; Hoffmann 2009). This pattern has been observed for adaptive responses in various experimental systems (Shadley and Wolff 1987; Ishii and Watanabe 1996; Stecca and Gerber 1998; Wolff 1998; Zhang et al. 2009; Guan et al. 2012; Hoffmann et al. 2013). Therefore, measurements made too early or too late may fail to detect an adaptive response or hormesis.

Some studies showing an adaptive response also suggest the occurrence of hormesis, per se. For example, Davies et al. (1995) reported that yeast exposed to low doses of $H_2O_2$ survived a subsequent higher dose of $H_2O_2$ and continued to divide at normal rates, whereas yeast that had not been pretreated were arrested by the challenge. The response for single exposures to $H_2O_2$ suggested hormesis, in that viability increased to 125% of the control value at 0.4 mM and then declined sharply at higher doses (Davies et al. 1995). Adaptive responses are more easily detected and measured than hormesis because an adaptive response entails a substantial reduction in a high level of damage caused by a high dosage. In contrast, the detection of hormesis requires measuring a modest reduction in a low level of spontaneous damage (Figure 15.2a) or a slightly enhanced biological function (Figure 15.2b). It is therefore not surprising that most reports of adaptive responses do not provide evidence of whether the adaptive response is associated with hormesis per se.

It is uncertain whether the parallels between hormesis and stress responses are sufficient to consider them manifestations of the same phenomenon. Hormesis may be separable from adaptive responses for mechanistic reasons. One should not expect to see hormesis if the adaptive response is based on an inducible response that is specific to a lesion that does not contribute significantly to spontaneous damage (Hoffmann et al. 2013). It has been suggested that the proposed linkage is an excessively broad application of the hormesis concept (Jonas 2010; Elliott 2011). The finding that the same conditions that induce an adaptive response to $H_2O_2$ in a yeast genetic assay show no evidence of hormesis in its original sense is consistent with this view, but it is difficult to exclude other explanations, such as subtle differences in the time course for expression of the responses (Hoffmann et al. 2013). Given the fact that adaptive responses and hormesis are measured under different circumstances and with different levels of difficulty, the hypothesis of a fundamental linkage is difficult to test experimentally. In lieu of further evidence, it seems preferable to maintain a clear distinction between hormesis as originally defined and stress responses that depend on sequential treatments. Although it is possible that both are part of a broad family of evolutionarily conserved responses to stress (Calabrese et al. 2007), pooling the two kinds of phenomena may obfuscate the elucidation of the basic properties of hormesis itself.

## 15.4 MECHANISMS OF HORMESIS AND STRESS RESPONSES

There is a growing understanding of mechanisms of how adaptive responses prevent damage and enhance repair (Stecca and Gerber 1998; Wolff 1998; Miura 2004; Hoffmann 2009; Morano et al. 2012). For example, responses to oxidative stress involve cell-cycle alterations and such antioxidant defenses as endogenous scavengers and detoxication enzymes that inactivate reactive oxygen species, reduce their production, and repair the damage that they cause (Benzie 2000; Miura 2004; Arumugam et al. 2006; Morano et al. 2012). Such responses involve reorganization of gene expression and metabolism that occurs by the regulation of transcription, translation, and posttranslational processes (Temple et al. 2005; Shenton et al. 2006; Morano et al. 2012).

The relative ease of detection makes adaptive responses more amenable to mechanistic exploration than in hormesis. Some adaptive responses are well characterized at the mechanistic level, such as inducible DNA repair, making an organism less susceptible to a second challenge. Such mechanisms could, in principle, lead to hormesis in the traditional sense if a small exposure activates a repair process that also removes spontaneously occurring damage. It has been speculated that hormesis occurs when there is a disruption of homeostasis, and the biological system responds to the stress with overcompensation (Calabrese 2002; Conolly and Lutz 2004) as balance is reestablished (Rattan 2008). Such speculation draws support from parallels with adaptive responses and from those laboratory systems that permit an experimental analysis of hormesis, such as a cell transformation system in which radiation hormesis has been ascribed to inducible DNA repair (Redpath and Elmore 2007).

Hormesis and adaptive responses can arise by upregulation of genes encoding protective proteins, growth factors, cytokines, and enzymes of signaling pathways (Stecca and Gerber 1998; Mattson et al. 2004; Miura 2004; Arumugam et al. 2006; Hoffmann 2009). Other hormetic effects have been ascribed to substances interacting with stimulatory and inhibitory receptor subtypes of regulatory systems (Calabrese 2002; Conolly and Lutz 2004) and to enhanced immune responses (Conolly and Lutz 2004). Selective death may also contribute to hormesis if direct killing or apoptosis occurs preferentially in abnormal cells (Bauer 2007; Portess et al. 2007; Redpath and Elmore 2007). Other mechanisms include inducible repair processes; interactions among cell proliferation, cell-cycle delay, and apoptosis after DNA damage; and enhanced intercellular communication at low doses (Stecca and Gerber 1998; Rouse and Jackson 2002; Conolly and Lutz 2004; Miura 2004; Fukushima et al. 2005; Arumugam et al. 2006; Bauer 2007; Calabrese et al. 2007; Portess et al. 2007; Redpath and Elmore 2007; Rattan 2008; Hoffmann 2009). Thus, there is a multiplicity of mechanisms that may contribute to hormesis. However, their distribution among cell types, organisms, agents, and biological endpoints is not yet well understood.

## 15.5 PROSPECTS FOR HORMESIS IN RISK ASSESSMENT

### 15.5.1 Controversy over Assimilating Hormesis into Policy

It has been proposed that considering hormesis in risk assessment for toxic substances and radiation can confer health benefits that would be lost by adhering to threshold or LNT models (Cook and Calabrese 2006a,b; Calabrese 2011). A related argument is that economic resources are being wasted by regulating to levels of exposure that cause no harm and may even be beneficial (Calabrese 2004a, 2011). These views have been strongly challenged (Axelrod et al. 2004; Thayer et al. 2005, 2006; Mushak 2007, 2009; Elliott 2011). Critics of hormesis commonly acknowledge that the phenomenon of hormesis occurs (Thayer et al. 2005), but they question the assertion that it is highly prevalent (Mushak 2007). A major concern relates to public policy—the fear that acceptance of the viewpoint that beneficial effects of hormesis are likely at low doses can lead to weaker standards for environmental policies and public health protection (Axelrod et al. 2004).

The heated debate over hormesis and policy has led to the suggestion that those who may benefit from broader recognition of hormesis are influenced by conflicts of interest (Axelrod et al. 2004; Shrader-Frechette 2008; Elliott 2011). Not surprisingly, contrary arguments have suggested political motivations and bias favoring an exaggeration of risks that can hinder the broader acceptance of hormesis (Calabrese 2005b, 2008a). Although one cannot cleanly separate scientific analysis from the social and political judgments that enter into scientific policy, the heated debate over hormesis calls for skepticism about unequivocal opinions that overlook or minimize the uncertainty that surrounds complex issues.

Monotonic dose-response models lend themselves to one principal objective—avoiding harm. Thus, the aim in public health policies based on threshold and LNT models is prevention of disease and disability. Proponents of the assimilation of hormesis into policy argue that hormesis offers the prospect of improving public health by harvesting the hormetic benefit (Cook and Calabrese 2006a,b). In areas such as preconditioning and physiological stress, this may be an achievable goal. For example, mild stress through exercise may improve health through hormetic mechanisms while conferring negligible risk (Radak et al. 2008). In the case of exposure to toxicants, however, the risk in the toxic zone is appreciable, and attempts to acquire a hormetic benefit would entail exposures much closer to the NOAEL than in current practice (Hoffmann 2009). Great certainty about the ability to target the hormetic zones would be required, so as not to fall into the toxic zone by error.

The hormetic zone is a relatively small target. A large database of hormetic responses indicates that the hormetic zone begins immediately below the ZEP, and its width in dosage is less than 10-fold in roughly half the examples and less than 100-fold in the great majority. The database also suggests that hormetic effects are modest, typically amounting to 30%–60% differences from a control value (Calabrese and Blain 2005). Potential risks and benefits are influenced by asymmetry around the ZEP (Figure 15.2), in that toxic effects to the right can be large, while hormetic effects to the left would be relatively small (Hoffmann 2009).

### 15.5.2 Precautionary Principle and Scientific Reality

A precautionary principle often underlies regulatory policy, and it may be formulated in various ways. The basic idea is that plausible evidence of likely and significant harm warrants public action to protect individuals, society, or the environment even in the absence of scientific certainty about the risk (Vineis 2005; Elliott 2011). This view effectively shifts the burden of proof toward demonstrating the absence of risk rather than unequivocally showing its presence (Vineis 2005). The aim of toxicological risk assessment under a threshold model is to ensure that exposures are in the no-effect zone. The imposition of a safety factor, also called an uncertainty factor, below the NOAEL is a standard means of keeping exposures substantially to the left of an NOAEL (Figure 15.1) and thereby minimizing risk. In the case of LNT, all doses are treated as though they confer risk, and the aim is to keep exposures low enough that the risk is negligible or acceptable. The precautionary principle in this instance would hold that it is better to overestimate risk than to underestimate it. Excessive caution is widely viewed as preferable to too little caution, and this preference guides conservative risk assessment. Risk estimation may therefore reflect a tension between basing policy on the best science available and basing it on a blend of science and conservative risk-assessment philosophy (Hoffmann 2009).

If LNT is used in radiation risk assessment, we may be choosing a model that is incorrect in light of evidence of nonlinearity at low doses and mechanisms that can explain this nonlinearity. Nevertheless, the LNT model may be judged prudent for its tendency to overestimate risks. In supporting LNT, the U.S. National Academy of Sciences' Committee on the Biological Effects of Ionizing Radiation directly acknowledged this fact, justifying its use of LNT on the grounds of prudent conservatism (NRC 1980, 2006), while also acknowledging that current data make it impossible to be confident about actual biological responses at very low doses. The Committee thereby attempted to avoid blurring the boundary between scientific interpretation and policy judgment. Thus, LNT may be the best model for radiation protection, even though it may not offer the most accurate scientific description of low-dose effects (Breckow 2006).

As the gap between a policy position and scientific evidence becomes larger, it provides impetus for finding an alternative risk assessment strategy. On the basis of cancer epidemiology and laboratory studies of carcinogenesis in animals, the French Academy of Sciences and French Academy of Medicine issued a joint report rejecting LNT as a realistic model for low doses of ionizing radiation (Aurengo et al. 2005; Tubiana et al. 2006). The report contends that carcinogenic effects at doses less than roughly 100 mSv are substantially overestimated by LNT. It argues for the existence of thresholds for radiation carcinogenesis and is receptive to hormesis (Aurengo et al. 2005; Tubiana et al. 2006). It has been suggested that about 40% of animal carcinogenicity studies with low linear energy transfer radiation show evidence of hormesis (Duport 2003; Tubiana et al. 2006).

It has been argued that carcinogenic risk is apt to be more strongly overestimated at very low doses (e.g., below 10 mSv) than at doses whose effects are readily measured (Aurengo et al. 2005; Tubiana et al. 2006). This view has been countered by the contention that low-dose risks may actually be supralinear rather than sublinear owing to phenomena not typically included in risk assessment (Brenner and Sachs 2006). Bystander effects are processes whereby unirradiated cells experience such effects as chromosome aberrations, mutations, morphological transformation, cytotoxicity, and apoptosis if they receive signals from irradiated cells by diffusible messengers or cellular contact at gap junctions (Morgan 2003a,b; Zhou et al. 2003; Kadhim et al. 2013). Genomic instability refers to delayed biological effects, including mutations and chromosome aberrations, that continue to occur in the progeny of irradiated cells (Morgan 2003a,b; Kadhim et al. 2013). Such nontargeted and delayed effects may enlarge the effective target size for ionizing radiation beyond the cells that received the radiation damage (Brenner et al. 2001, 2003; Morgan 2003a,b, 2006; Zhou et al. 2003; Kadhim et al. 2013). If so, low-dose risks may be greater than previously anticipated (Kadhim et al. 2013). A much better understanding of the balance between the protective and adverse elements of low-dose phenomena will be required to resolve the controversy. Such uncertainties may place LNT in the range of a reasonable middle ground, rather than a gross overestimation (Mossman 2001).

Conservative risk assessment is sometimes criticized because tighter regulatory practices cause an economic burden. Many would argue, however, that public health risks should take priority over economic interests and that overestimation of risk is benign. The latter point has been challenged, in that a large overestimation of risk may not support public health. Avoidance of valuable diagnostic or therapeutic procedures can be an unintended adverse effect of the exaggeration of radiation risks (Aurengo et al. 2005; Scott and Di Palma 2006; Tubiana et al. 2006), and proponents of radiation hormesis have argued that the diagnostic procedures themselves (e.g., dental x-rays, chest x-rays, mammograms, and thyroid scans) may confer a hormetic benefit. The argument calls for a balanced perspective on the possibility of harm stemming from the underestimation or overestimation of risks.

Growing evidence of hormesis calls for a better understanding of the phenomenon. It does not, however, necessitate changing risk assessment policies because of the possibility of hormetic effects immediately below the NOAEL. Rather, one might decide not to factor hormesis into risk assessment on the basis of uncertainty (Hoffmann and Stempsey 2008). Although it may be in the public interest to disregard the possibility of hormesis in risk assessment, this should not lead us to deny the existence of hormesis or to assume that it is unimportant in other circumstances. For example, hormetic effects benefiting a bacterial pathogen would be detrimental to public health, and assimilating knowledge of hormesis into policies would, in that case, be in the public interest (Hoffmann 2009). Finding a proper way to consider hormesis with respect to public health relates to the relative weights placed on avoidance of harm versus conferring of benefits.

### 15.5.3 Hormesis and Biomedical Ethics

Policies related to toxic substances and radiation are typically based on protecting against harm, rather than accruing benefit. This emphasis is related to the ethical principles of nonmaleficence and beneficence (Beauchamp and Childress 2001; Hoffmann and Stempsey 2008). Nonmaleficence is based on avoiding the causation of harm, which coincides with the medical tenet *above all, do no harm*. In contrast, beneficence entails conferring a benefit that is properly balanced against risks and costs (Beauchamp and Childress 2001). Although the Hippocratic origins of medical ethics actually encompass doing good as well as avoiding harm, the latter usually takes precedence (Ross 1988; Beauchamp and Childress 2001).

These principles are relevant to hormesis because its biphasic responses (Figure 15.2) open the prospect of conferring benefit as well as avoiding harm, whereas the threshold and LNT curves (Figure 15.1) lend themselves to the latter only. On the basis of this difference, it has been proposed that hormesis can be used to improve public health (Cook and Calabrese 2006a,b). If doing so were to entail either giving or allowing toxicant exposures to accrue the hormetic benefit, it would represent a shift from the principle of nonmaleficence to that of beneficence (Hoffmann and Stempsey 2008). Public health policies based on beneficence, such as fluoridation of public drinking water and mandatory vaccinations, tend to generate controversy, and they typically have to meet a high standard with respect to efficacy and safety before they can be accepted (Hoffmann and Stempsey 2008). A lack of precision in defining and targeting a hormetic zone would make this unlikely with respect to toxicant exposures (Thayer et al. 2005, 2006). The hurdle to acceptance would undoubtedly be smaller for hormetic stressors that pose little risk of harm, such as using mild exercise to encourage healthy aging or to reduce the likelihood or slow the onset of degenerative diseases (Arumugam et al. 2006; Radak et al. 2008; Rattan 2008; Le Bourg 2009). However, caution is essential, as even mild stresses are not necessarily beneficial and may sometimes have negative effects (Le Bourg 2009).

## 15.6 CHALLENGES OF ASSIMILATING HORMESIS INTO RISK ASSESSMENT

### 15.6.1 Default Assumptions for Low-Dose Effects

Risk assessment often relies on default assumptions about what one might expect at low doses unless there is compelling evidence to the contrary. Such assumptions are based on general scientific knowledge and policy judgments when specific scientific information is lacking (NRC 1994; Preston and Hoffmann 2013). Default assumptions have been used because of the impracticality or impossibility of determining what actually occurs in the low-dose zone. Although it is hoped that mechanistic understanding of the mode of action of toxicants can replace default assumptions (EPA 2005), current approaches to risk assessment rely on a combination of the two (Bolt and Huici-Montagud 2008; Preston and Hoffmann 2013).

An LNT model has been the default assumption for genotoxic carcinogens (EPA 2005; Brenner and Sachs 2006; Preston and Hoffmann 2013), while thresholds are widely recognized for nongenotoxic carcinogens (Bolt and Huici-Montagud 2008) and other toxicological effects (Eaton and Gilbert 2008; Hoffmann 2009). The assertion that hormesis should be the default assumption for risk assessment (Calabrese 2004b) is a challenge to these interpretations. The question of whether hormesis qualifies as a default assumption is complicated by the difficulty of detecting hormesis in individual cases and the diversity of mechanisms that may contribute to or diminish hormetic effects. It has been argued that hormesis would only qualify as a default assumption if it were so prevalent as to be a nearly universal phenomenon (Crump 2001). However, there is disagreement about reported frequencies of hormetic curves in toxicological studies and claims about the generalizability of the phenomenon.

### 15.6.2 Disagreement about the Generalizability of Hormesis

Although the reality of hormesis has received growing acceptance, its generalizability continues to generate debate. It has been asserted that hormesis is broadly generalizable without regard to the specific agent, organism, biological endpoint, or genetic susceptibility (Calabrese and Baldwin 1998, 2001b; Calabrese et al. 1999; Calabrese 2004a, 2008, 2010; Calabrese and Mattson 2011). This view has been contested on the basis of ambiguity in the criteria for generalizability, mechanistic and statistical considerations, and uncertainties about the frequency and reproducibility of hormetic responses (Crump 2001; Axelrod et al. 2004; Kitchin and Drane 2005; Mushak 2007; Elliott 2011).

The evidence for hormesis is stronger for some biological endpoints than for others, and carcinogenesis has often been the focus of debate. It has been claimed that hormesis can be generalized to carcinogens (Calabrese and Baldwin 1998; Calabrese 2008a), but this view has been challenged (Mushak 2007). A problem in obtaining persuasive evidence of hormesis for carcinogenicity is the difficulty of measuring decreases in the spontaneous incidence of tumors in studies of limited size. Even the massive $ED_{01}$ study was equivocal in this respect (Bruce et al. 1983; Kodell et al. 1983; Mushak 2007). The same is true of epidemiological studies, where the arguments often lie in confounding variables. It might be hoped that data from short-term tests for carcinogens could prove fruitful in the detection of hormesis because they offer controlled conditions and larger sample sizes. Mutagenicity and carcinogenicity have been linked by the historic development of the fields, mechanistic considerations, many agents that exhibit both properties, and the use of mutagenicity testing as an indicator or surrogate for likely carcinogenicity. Although it is clear that there are nongenotoxic carcinogens and that correlations have been overstated, there remain many overlaps between these endpoints. Convincing examples of genotoxic effects that exhibit biphasic curves in the low-dose zone can be found in the scientific literature (Hooker et al. 2004; Gocke and Müller 2009; Thomas et al. 2013), but there is little evidence of widespread hormesis supporting the notion of generalizability.

The analysis of databases on inhibition of growth in microorganisms has provided evidence of hormesis (Calabrese et al. 2006, 2010), so databases from mutagenicity testing may seem to offer a promising way to get comparable information for genotoxicity. The most widely used of all mutagenicity tests is the Salmonella/microsome assay, commonly known as the Ames test. It has been claimed that Ames test data from systematic chemical screening show a high incidence of hormetic curves (Calabrese et al. 2011), but the methods used in this analysis have been challenged (Zeiger and Hoffmann 2012). The Ames test, while an excellent assay for bacterial mutagenicity, should be considered an inappropriate model for studying hormesis because there is a complex interaction between mutagenicity and toxicity, such that a reduction in colony count is apt to reflect toxicity rather than a reduction in mutation frequency below the spontaneous level (Zeiger and Hoffmann 2012). The generalizability of the hormesis model to genetic toxicology remains in doubt.

### 15.6.3 Disagreement about the Prevalence of Hormesis

There is no agreement on the frequency of hormesis that would be needed to support the view that hormesis is a highly prevalent phenomenon. Studies using clearly defined methods and criteria to evaluate published papers for hormesis are few, but they are reasonably consistent in estimating frequencies of nonmonotonic curves. Davis and Svendsgaard (1994) reported a prevalence of 12%–24%, whereas Calabrese and Baldwin (2001a,b) reported 19.5%–37%. Such values have been interpreted as evidence both for (Calabrese and Baldwin 2001a) and against (Mushak 2007) the prevalence of hormesis. These numbers may themselves be challenged but, even if accepted, they do not make a compelling case for hormesis being a default assumption for risk assessment, given that two-thirds or more of dose-responses do not show hormesis. On the other hand, the fact that up to one-third of responses show hormesis argues that hormetic responses are common. This constitutes good reason to be cognizant of hormesis, while showing restraint in asserting that it is a highly prevalent or broadly generalizable phenomenon.

### 15.6.4 Uncertainties in the Quantification of Hormetic Effects

To qualify as a default assumption for risk assessment, hormesis would need to be a reliable, quantitative substitute for actual low-dose data (Mushak 2007). It is not clear that the evidence for hormesis has gone far enough beyond an accumulation of examples (Calabrese and Blain 2005) into the realm of systematic quantitative analysis to support such an assumption (Crump 2001; Kitchin and Drane 2005; Mushak 2007). It has often been claimed on the basis of examples of hormesis in peer-reviewed literature that a deviation of 30%–60% from the control is typical of hormesis (Calabrese and Baldwin 1998; Calabrese et al. 1999; Calabrese 2002, 2004b, 2008a, 2010; Calabrese and Blain 2005; Calabrese and Mattson 2011). The magnitude of the response has been called "the most consistent quantitative feature of the hormetic dose response" (Calabrese 2010). By comparison, the hormetic responses in analyses of databases from chemical screening were roughly 3%–7% in one study (Calabrese et al. 2006) and 1%–4% in another (Calabrese et al. 2010). The reason for the difference is unclear, but it may lie in differences among species, agents, and endpoints or proportions of responses that are actually hormetic. In any case, it suggests uncertainty with respect to the generalizability of hormesis and its quantitative consistency.

Other studies suggest that adaptive responses are also highly variable. For example, in a study of ten human lymphoblastoid cell lines in a leading cytogenetics laboratory, six showed an adaptive response for the induction of micronuclei by γ-rays, three showed no adaptive response, and one showed an amplified response (Sorensen et al. 2002). The responses varied in magnitude, and only five of the six lines that showed an adaptive response responded consistently in repeat tests (Sorensen et al. 2002). Variability among donors is also reported for an adaptive response to the alkylating agent MNNG in human lymphocytes (Morimoto et al. 1986).

### 15.6.5 Accounting for Heterogeneity in Susceptibility

Heterogeneity in susceptibility to toxicants raises difficult questions about hormesis, notably whether hormetic responses are equally likely for different genotypes, and how that might relate to ethical considerations in risk assessment. Failure to consider heterogeneity among individuals has been raised as a weakness of the hormesis model for risk assessment (Axelrod et al. 2004). Although this problem applies to all models, it presents some unique challenges under the hormesis model.

It has been suggested on the basis of published dose-response curves that sensitive genotypes do show hormesis, but it occurs at lower doses (Calabrese and Baldwin 2002a; Cook and Calabrese 2006a; Calabrese 2008a). Thus, the dose-response curve is shifted to the left, as shown in Figure 15.3. Evidence consistent with this interpretation includes the finding that a radiation-sensitive, cancer-prone mouse strain showed a longer latency period for spontaneous lymphomas and osteosarcomas when treated with a low dose of γ-rays administered at a low dose rate (Mitchel et al. 2003). The fact that 12 mutant yeast strains, altered in genes that can affect toxicant responses, all showed hormesis comparable to that in the wild-type strain supports the same view (Calabrese et al. 2006). It would be premature, however, to assume that sensitive genotypes will consistently show hormesis. Adaptive responses to alkylating agents and ionizing radiation can be blocked by mutations that alter repair functions (Kleibl 2002; Hoffmann 2009) and by inhibition of poly(ADP-ribose)polymerase (PARP), an enzyme involved in the repair of DNA strand breaks (Shadley and Wolff 1987; Stecca and Gerber 1998; Miura 2004), respectively. Hormesis in its original sense also appears to be blocked in some genotypes. For example, radiation hormesis in a mammalian cell assay for neoplastic transformation is inhibited by inhibition of PARP (Pant et al. 2003), and mutations in genes of the insulin-like signaling pathway in the worm *Caenorhabditis elegans* block a hormetic response to heat stress (Cypser and Johnson 2003). The variation in these responses leaves uncertainty about how one should expect sensitive subpopulations to respond to low doses.

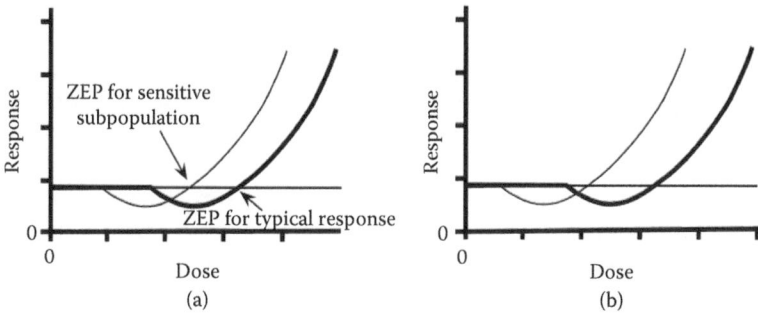

**FIGURE 15.3** Hormesis and sensitive subgroups. (a) A hormetic dose-response curve for a typical population (bold line) and a sensitive subpopulation (fine line). The zero equivalent point (ZEP) corresponds to a no observed adverse effect level (NOAEL), in that the response above this dose is an adverse effect. Even if both populations show a hormetic response, part of the hormetic zone for the general population is in the toxic zone for the sensitive population. (b) A greater difference between the populations, such that there would be little overlap between their hormetic zones. (Adapted from Hoffmann, G. R. and Stempsey, W.E., *BELLE Newsl.* 14, 3, 11–17, 2008.)

If the hypothesis is correct that sensitive subpopulations show hormesis but that it occurs at lower doses (Calabrese and Baldwin 2002a), it may present a unique challenge for the assimilation of hormesis into risk assessment (Thayer et al. 2005, 2006). Differences in susceptibility may be associated with genetic constitution, age, sex, and health status. Figure 15.3 shows a hormetic curve for a sensitive subpopulation in comparison to the general population. If an attempt were made to allow exposures in the hormetic zone, it may lead to part of the population benefiting while another part is harmed. This dilemma does not exist in the threshold and LNT models because they are monotonic. Defining an ethical policy for targeting the hormetic zone under these circumstances would be problematic. In the spectrum of likely viewpoints, one pole is to stay far below the NOAEL and forego the hormetic benefit to the majority to protect the minority. The opposite pole is to seek the hormetic zone and provide the greatest good for the greatest number. The principle of nonmaleficence would call for a course like the former, rather than the utilitarian ethics of the latter (Hoffmann and Stempsey 2008; Hoffmann 2009). Of course, intermediate courses of action might also be formulated.

If the curves are like those in Figure 15.3a, it may be possible to have exposures below the NOAEL that benefit both populations, as suggested by Cook and Calabrese (2006a). However, if there is a larger difference between the two groups, as in Figure 15.3b, that goal may be unattainable. If the difference between populations were larger than those in Figure 15.3b, there would be no overlap in the hormetic zones. In actual situations, unlike these idealized curves, it would be unlikely that the hormetic zone and the degree of sensitivity would be so well defined that proper levels of exposure could be identified. The prudent course of action is not to use a hormesis-based risk assessment strategy.

### 15.6.6 INTERACTIONS AMONG AGENTS

Unlike controlled laboratory exposures, human chemical exposures occur in mixtures of agents, some of which are identified and others not. Effects of toxicants may be additive or they can exhibit such interactions as potentiation, synergism, or antagonism (Eaton and Gilbert 2008). Little is known about possible interactions when several agents are simultaneously in the hormetic zone. It has been argued that if hormesis is maximal at a dose equivalent to one-fifth of the NOAEL, then concurrent exposure to more than five such agents would move the total effect into the toxic zone (Axelrod et al. 2004; Shrader-Frechette 2008). Without a clear justification for the assumption of additivity, this assertion remains in the realm of speculation. An alternative proposal is that hormetic maxima are in the range of 30%–60% and that simultaneous hormetic exposures are apt to drive the total effect to the higher part of that range (Calabrese 2008a). Without a clear justification for the assumption of cumulative hormetic effect, whether additive or synergistic, this assertion is also speculation.

In sum, both the possibility that simultaneous exposures to agents that have similar toxic effects might move the response from the hormetic zone to the toxic zone (Thayer et al. 2005) and the possibility of cumulative hormetic effects (Calabrese 2008a) warrant further consideration. Although interactions among agents with respect to hormesis remain largely a mystery, promising studies in plants suggest that the dimensions of hormetic stimulation of mixed treatments may be predicted, at least roughly, from that of the individual responses (Belz et al. 2008). Studies in *Drosophila* of combined treatments with agents that induce mild stress (cold and hypergravity) support the same interpretation (Le Bourg 2012). However, much more needs to be learned before generalizations are reached about the likelihoods of additivity, synergy, or antagonisms with respect to hormesis.

### 15.6.7 HORMESIS AND CONCOMITANT TOXICITY

The problems of heterogeneity in susceptibility apply not only to individuals, but also to differences among endpoints and differences among tissues and organs. The hormetic zone is near to the classical NOAEL, so it would not be unusual for a given exposure that is hormetic for one endpoint or tissue to be in the toxic zone for others. For example, a lower incidence of testicular tumors in rats treated with cadmium chloride (Waalkes et al. 1988) has been ascribed to hormesis (Calabrese and Baldwin 1998), but it is accompanied by an increase in prostate tumors at the same doses (Waalkes et al. 1988; Thayer et al. 2006).

An analysis of carcinogenicity data from the National Toxicology Program (NTP) bioassays indicated that more than 90% of NTP-tested chemicals in the survey showed at least one statistically significant decrease in site-specific tumor frequency. Random variability can account for some of the decreases because of the many comparisons made, but others probably reflect anticarcinogenic effects (Haseman and Johnson 1996). If combined with carcinogenic effects at a higher dose, such decreases might be ascribed to hormesis. If associated with carcinogenic effects at another site at higher dose, the pooling of the two monotonic curves may give an artifactual appearance of hormesis, as has been argued about a claim of hormesis for the carcinogenicity of 2,3,7,8-tetrachlorodibenzodioxin (Thayer et al. 2005). When hormesis is associated with carcinogenicity, the carcinogenic effect would typically override a small hormetic benefit in importance. If hormesis were factored into risk assessment, one would need to be confident about whether a hormetic effect for one endpoint is accompanied by detriment elsewhere.

### 15.6.8 Feasibility of Hormesis-Based Risk Assessment

More than 5 years ago, a critic of the prospect of hormesis-based risk assessment for toxicants, including chemical carcinogens, extended a challenge to its proponents. Mushak (2007) stated that "proponents have not yet laid out a convincing methodological schematic that actually walks the reader or risk assessor through a hormesis-based quantitative risk assessment." To my knowledge, this remains true today. The pitfalls that must be overcome suggest that hormesis-based risk assessment would certainly be premature and possibly not even feasible.

## 15.7 WHY AN UNDERSTANDING OF HORMESIS IS ESSENTIAL

### 15.7.1 Concerns about Hormesis and Risks of Ignoring It

Concerns about possible misapplications of hormesis in risk assessment have led some critics of hormesis to emphasize gaps in the evidence for its existence, to deny its common occurrence, or to minimize its potential importance (Shrader-Frechette 2008; Mushak 2009). However, critics have also correctly identified substantive questions that must be resolved with respect to risk assessment (Axelrod et al. 2004; Thayer et al. 2005; Mushak 2007, 2009; Elliott 2011). An inadvertent consequence of this debate is that it may encourage a disregard for hormesis more globally. If so, this could be disadvantageous for public health and environmental quality, because potential benefits of mild hormetic stress may go unrealized, detrimental consequences of hormesis may be overlooked, and opportunities for better environmental policies could be lost.

### 15.7.2 Optimizing the Benefits of Mild Stress Responses

There is accumulating evidence that stress responses play a role in the aging process, such that mild stress confers hormetic benefits, whereas severe or chronic stress can exacerbate the degradative aspects of aging (Rattan 2008). An understanding of hormetic mechanisms may therefore promote a healthier aging process. The hormesis model is also relevant to understanding how responses to mild stress can benefit cardiovascular, skeletal, muscular, and neurological health (Arumugam et al. 2006; Radak et al. 2008). The biphasic nature of the responses is reflected in the common observation that regular exercise, unlike inactivity, is beneficial but that the benefit can be offset by excessive exercise and overtraining (Radak et al. 2008). Conditioning can also confer benefits for the avoidance or slowing of neurodegenerative disorders (Mattson et al. 2004; Mattson and Cheng 2006), and biphasic responses are reported from many areas of neurobiology (Calabrese 2008b). Diverse mechanisms can contribute to these phenomena, notably including modulation of the formation of reactive oxygen species (Arumugam et al. 2006; Radak et al. 2008). The beneficial effects of a vegetable-rich diet are ascribable, at least in part, to antioxidant effects, but it may also include hormetic effects of small exposures to toxic phytochemicals (Mattson and Cheng 2006).

There is growing evidence that exercise, cognitive stimulation, and calorie restriction can improve longevity and lower the risk of Alzheimer's disease, Parkinson's disease, stroke, and other age-related disorders through hormetic mechanisms (Mattson et al. 2004; Arumugam et al. 2006). Generalizations must be reached cautiously, however, as some effects may be specific to particular organisms or genotypes. For example, calorie restriction, which is known to increase longevity in rodent studies, may not do so consistently in primates (Mattison et al. 2012). It has been hypothesized that life-history strategies are important in predicting whether dietary restriction will improve longevity, such that short-lived species that spend their lives in a small geographic area are more likely to benefit than longer-lived species that can migrate over great distances (Le Bourg 2010).

Current understanding of the extent to which hormesis influences longevity and disease resistance is at a formative stage. A lack of understanding of hormesis or failure to recognize its occurrence can impede the optimization of practices that take advantage of natural adaptive responses. A more complete understanding of biological stress responses offers the promise of improved therapies for age-related disorders and better dietary and behavioral approaches for the improvement of public health (Mattson et al. 2004; Arumugam et al. 2006; Rattan 2008).

### 15.7.3 Avoiding Unforeseen Risks to Public Health

In the context of toxicological risk assessment, it is often assumed that hormesis refers to a beneficial effect. Although this may be correct at some level, it can be misleading. Calabrese has argued persuasively that hormesis is better defined on the basis of the characteristics of its biphasic dose-response relationship and not on the basis of benefit and harm (Calabrese and Baldwin 2002b; Calabrese 2008a). In some instances hormetic effects can be detrimental to health, and it is important that they be recognized and avoided. For example, the hormetic effect would be deleterious if a low dose of an inhibitory drug were to stimulate a harmful hyperplasia. Thus, specific instances can differ with respect to whether the high-dose range or the low-dose range is beneficial (Calabrese 2008a; Hoffmann 2009).

There is widespread awareness that antibiotic use selects for antibiotic resistance, but it is less generally appreciated that antibiotics in insufficient dosages can stimulate bacterial growth through a hormetic mechanism. Davies et al. (2006) advanced the argument that low doses of antibiotics commonly serve as signaling molecules and have stimulatory effects in bacteria. If unrecognized, the public health consequences can be substantial. Linares et al. (2006) found that three classes of antibiotics can stimulate the opportunistic pathogen *Pseudomonas aeruginosa*, which colonizes the lungs of cystic fibrosis patients, causing serious deterioration. The need to understand hormetic processes is obvious if antibiotics at low doses can confer hormetic benefits on bacterial pathogens to the detriment of their human hosts.

The hormetic stimulation of surviving tumor cells by low levels of residual chemotherapy drugs after cytotoxic therapy also deserves consideration, as hormesis is reported to be common in human tumor cells (Calabrese 2005a). The temporary stimulation of metastatic breast cancer by tamoxifen in some patients before the drug reaches sufficient dosages to be inhibitory may also have an underlying hormetic mechanism (Brandes 2005). A lack of understanding of hormesis or failure to recognize its occurrence or its consequences can contribute to ineffectiveness in combating detrimental hormetic effects.

Although the hormesis model (Figure 15.2) implies that benefit and harm lie on opposite sides of the ZEP, it should not be accepted uncritically that this is necessarily true of nonmonotonic dose-response curves. Some nonmonotonic responses, such as those for endocrine disruptors, are reported to entail harmful effects both at high and low doses (Timms et al. 2005; Vandenberg et al. 2012). These responses and hormetic responses both suggest that effects at low doses are not readily predicted by effects at high doses, but they have different implications with respect to risk. Although the hormetic J-shaped curve (Figure 15.2a) suggests that a threshold model (Figure 15.1a) would overestimate risk at low doses, adverse effects of endocrine disruptors at low doses have been interpreted as reason to believe that a threshold model can underestimate low-dose risk (Weltje et al. 2005).

### 15.7.4 Agricultural Productivity and Environmental Quality

An appreciation for hormesis can offer useful insight for environmental policies (Hoffmann 2009). The hormetic stimulation of bacteria by low doses of antibiotics, which is an obvious concern for public health, can also have detrimental environmental consequences through the formation of bacterial biofilms (Linares et al. 2006). One may speculate on consequences of hormesis in disturbed ecosystems, but our understanding of hormetic effects in microorganisms in natural environments is rudimentary.

Differences among plant and animal species in susceptibility to toxicants are often not well enough known to predict low-dose responses in natural environments and agricultural settings. Many herbicides, including such widely used compounds as glyphosate, show hormetic effects in plants at low doses (Duke et al. 2006; Cedergreen et al. 2007). It is difficult to detect hormetic effects in communities of mixed species because stimulatory effects associated with exposure to a chemical may also be ascribable to altered interspecific competition (Duke et al. 2006). Nevertheless, hormesis was evident in plants even when controlling for this confounding factor (Cedergreen et al. 2007). Hormetic effects of glyphosate have been measured in several plant species, and hormesis may be the explanation underlying the use of the herbicide to stimulate the accumulation of sucrose in sugarcane (Velini et al. 2008).

Although pesticides are typically applied under field conditions at doses sufficient to be effective, pests peripheral to the treatment zone may experience hormetic effects of low doses, and this may contribute to subsequent outbreaks of the pest (Morse 1998). Different classes of insecticides, including organochlorine, organophosphate, carbamate, and pyrethroid, have all been reported to cause hormetic stimulation in insects (Morse and Zareh 1991). For example, insecticide doses that cause high mortality when fed to citrus thrips also suppressed their reproductive rate, but doses that caused less than 1% mortality led to increased fecundity (Morse and Zareh 1991). It is likely that such phenomena can contribute to the resurgence of pests. Conclusions, however, are not always straightforward, as ecological manifestations of hormesis, like those in public health, may involve complex patterns where a hormetic benefit in one area is offset by detriment elsewhere. For example, a hormetic increase in numbers of offspring may be offset by their lower survival (Duke et al. 2006; Hoffmann 2009). A recent review by Cutler (2013) identifies many possible cases of hormesis in insects and discusses other factors that can mimic hormesis in insect populations, including reduced competition from other herbivores, changes in pest behavior, altered host-plant nutrition, and increased attractiveness of the host plant. Such observations argue that an understanding of toxic effects in pests, their hosts, predators, and competitors should be accompanied by a better understanding of their responses to low doses (Morse 1998; Kefford et al. 2008; Hoffmann 2009; Cutler 2013).

## 15.8 CONCLUSIONS

Hormesis describes a dose-response relationship in which effects at low doses are opposite to those at high doses. Substantial evidence supports the reality of hormesis, but much disagreement remains about its prevalence and broader implications. Much of the controversy stems from the proposal that the hormetic response should be incorporated into risk assessment and public policy with respect to toxicants. Such applications would depend on hormesis being a highly prevalent and consistent response to toxicants and on the feasibility of acquiring a hormetic benefit without unduly risking toxicity. There is insufficient evidence on these points, and heterogeneity among species, genotypes, and tissues substantially complicates extrapolations. Much evidence suggests that basing toxicological risk assessment on the principle of hormesis would be premature and is probably not feasible. At the same time, mild stress is known to stimulate adaptive responses that can be beneficial, and there is reason to think that factors such as exercise and cognitive stimulation contribute to good health through hormetic mechanisms. On the other hand, the possibility of adverse effects occurring through hormetic stimulation of bacteria, parasites, and tumors also deserves consideration. An understanding of hormesis and stress responses is therefore important for public health and medicine, and it has become increasingly clear that understanding how these phenomena function in microorganisms, plants, and animals can be important for environmental policies and agriculture.

## ACKNOWLEDGMENTS

The author thanks Edward Calabrese for many interesting discussions of hormesis, Justin McAlister and Errol Zeiger for helpful suggestions on the manuscript, and Darlene Colonna for excellent secretarial assistance.

# REFERENCES

Andersen, M. E., R. B. Conolly, and D. W. Gaylor. 2003. Letter to the editor. *Toxicol. Sci.* 74: 486–487.
Arumugam, T. V., M. Gleichmann, S.-C. Tang, and M. P. Mattson. 2006. Hormesis/preconditioning mechanisms, the nervous system and aging. *Ageing Res. Rev.* 5: 165–178.
Aurengo, A., D. Averbeck, A. Bonnin, B. Le Guen, R. Masse, R. Monier, M. Tubiana, A.-J. Valleron, and F. de Vathaire. 2005. Dose–effect relationships and estimation of the carcinogenic effects of low doses of ionizing radiation. Joint Report of the Académie des Sciences (Paris) and of the Academie Nationale de Médecine. March 30, 2005. http://www.ecolo.org/documents/documents_in_english/low_dose-acad-05-complete.doc
Axelrod, D., K. Burns, D. Davis, and N. von Larebeke. 2004. "Hormesis"—An inappropriate extrapolation from the specific to the universal. *Int. J. Occup. Environ. Health* 10: 335–339.
Bauer, G. 2007. Low dose radiation and intercellular induction of apoptosis: Potential implications for the control of oncogenesis. *Int. J. Radiat. Biol.* 83: 873–888.
Beauchamp, T. L., and J. F. Childress. 2001. *Principles of Biomedical Ethics*, 5th Edition, pp. 114–116. New York: Oxford University Press.
Belz, R. G., N. Cedergreen, and H. Sørensen. 2008. Hormesis in mixtures—Can it be predicted? *Sci. Tot. Environ.* 404: 77–87.
Benzie, I. F. F. 2000. Evolution of antioxidant defence mechanisms. *Eur. J. Nutr.* 39: 53–61.
Bolt, H. M., and A. Huici-Montagud. 2008. Strategy of the Scientific Committee on Occupational Exposure Limits (SCOEL) in the derivation of occupational exposure limits for carcinogens and mutagens. *Arch. Toxicol.* 82: 61–64.
Brandes, L. J. 2005. Hormetic effects of hormones, antihormones, and antidepressants on cancer cell growth in culture: In vivo correlates. *Crit. Rev. Toxicol.* 35: 587–592.
Breckow, J. 2006. Linear-no-threshold is a radiation-protection standard rather than a mechanistic effect model. *Radiat. Environ. Biophys.* 44: 257–260.
Brenner, D. J., and R. K. Sachs. 2006. Estimating radiation-induced cancer risks at very low doses: Rationale for using a linear no-threshold approach. *Radiat. Environ. Biophys.* 44: 253–256.
Brenner, D. J., J. B. Little, and R. K. Sachs. 2001. The bystander effect in radiation oncogenesis: II. A quantitative model. *Radiat. Res.* 155: 402–408.
Brenner, D. J., R. Doll, D. T. Goodhead, E. J. Hall, C. E. Land, J. B. Little, J. H. Lubin et al. 2003. Cancer risks attributable to low doses of ionizing radiation: Assessing what we really know. *Proc. Natl. Acad. Sci. USA* 100: 13761–13766.
Broome, E. J., D. L. Brown, and R. E. J. Mitchel. 2002. Dose responses for adaptation to low doses of $^{60}$Co γ rays and $^{3}$H β particles in normal human fibroblasts. *Rad. Res.* 158: 181–186.
Bruce, R. D., W. W. Carlton, D. Clayson, K. H. Ferber, D. H. Hughes, J. F. Quast, D. S. Salsburg et al. 1983. The SOT Task Force's response to the NCTR Letter. *Fundam. Appl. Toxicol.* 3: 9/a–12/a.
Bryce, S. M., S. L. Avlasevich, J. C. Bemis, S. Phonethepswath, and S. D. Dertinger. 2010. Miniaturized flow cytometric in vitro micronucleus assay represents an efficient tool for comprehensively characterizing genotoxicity dose–response relationships. *Mutat. Res.* 703: 191–199.
Bukowski, J., M. Nicolich, and R. J. Lewis. 2013. Extreme sensitivity and the practical implications of risk assessment thresholds. *Dose-Response* 11: 130–153.
Calabrese, E. J. 2002. Hormesis: Changing view of the dose–response, a personal account of the history and current status. *Mutat. Res.* 511: 181–189.
Calabrese, E. J. 2004a. Hormesis—Basic, generalizable, central to toxicology and a method to improve the risk-assessment process. *Int. J. Occup. Environ. Health* 10: 466–467.

Calabrese, E. J. 2004b. Hormesis: From marginalization to mainstream: A case for hormesis as the default dose–response model in risk assessment. *Toxicol. Appl. Pharmacol.* 197: 125–136.

Calabrese, E. J. 2005a. Cancer biology and hormesis: Human tumor cell lines commonly display hormetic (biphasic) dose responses. *Crit. Rev. Toxicol.* 35: 463–582.

Calabrese, E. J. 2005b. Historical blunders: How toxicology got the dose–response relationship half right. *Cell. Mol. Biol.* 51: 643–654.

Calabrese, E. J. 2008a. Hormesis: Why it is important to toxicology and toxicologists. *Environ. Toxicol. Chem.* 27: 1451–1474.

Calabrese, E. J. 2008b. Dose–response features of neuroprotective agents: An integrative summary. *Crit. Rev. Toxicol.* 38: 253–348.

Calabrese, E. J. 2009. Getting the dose–response wrong: Why hormesis became marginalized and the threshold model accepted. *Arch. Toxicol.* 83(3): 227–247.

Calabrese, E. J. 2010. Hormesis is central to toxicology, pharmacology and risk assessment. *Hum. Exp. Toxicol.* 29: 249–261.

Calabrese, E. J. 2011. Toxicology rewrites its history and rethinks its future: Giving equal focus to both harmful and beneficial effects. *Environ. Toxicol. Chem.* 30(12): 2658–2673.

Calabrese, E. J., and L. A. Baldwin. 1998. Can the concept of hormesis be generalized to carcinogenesis? *Regul. Toxicol. Pharmacol.* 28: 230–241.

Calabrese, E. J., and L. A. Baldwin. 2001a. The frequency of U-shaped dose responses in the toxicological literature. *Toxicol. Sci.* 62: 330–338.

Calabrese, E. J., and L. A. Baldwin. 2001b. Hormesis: A generalizable and unifying hypothesis. *Crit. Rev. Toxicol.* 31(4–5): 353–424.

Calabrese, E. J., and L. A. Baldwin. 2002a. Hormesis and high-risk groups. *Regul. Toxicol. Pharmacol.* 35: 414–428.

Calabrese, E. J., and L. A. Baldwin. 2002b. Defining hormesis. *Hum. Exp. Toxicol.* 21: 91–97.

Calabrese, E. J., and L. A. Baldwin. 2003. The hormetic dose–response model is more common than the threshold model in toxicology. *Toxicol. Sci.* 71: 246–250.

Calabrese, E. J., and R. Blain. 2005. The occurrence of hormetic dose responses in the toxicological literature, the hormesis database: An overview. *Toxicol. Appl. Pharmacol.* 202: 289–301.

Calabrese, E. J., and M. P. Mattson. 2011. Hormesis provides a generalized quantitative estimate of biological plasticity. *J. Cell. Commun. Signal.* 5: 25–38.

Calabrese, E. J., L. A. Baldwin, and C. D. Holland. 1999. Hormesis: A highly generalizable and reproducible phenomenon with important implications for risk assessment. *Risk Anal.* 19: 261–281.

Calabrese, E. J., J. W. Staudenmayer, E. J. Stanek, and G. R. Hoffmann. 2006. Hormesis outperforms threshold model in National Cancer Institute antitumor drug screening database. *Toxicol. Sci.* 94: 368–378.

Calabrese, E. J., K. A. Bachmann, A. J. Bailer, P. M. Bolger, J. Borak, L. Cai, N. Cedergreen et al. 2007. Biological stress response terminology: Integrating the concepts of adaptive response and preconditioning stress within a hormetic dose–response framework. *Toxicol. Appl. Pharmacol.* 222: 122–128.

Calabrese, E. J., E. J. Stanek, M. A. Nascarella, and G. R. Hoffmann. 2008. Hormesis predicts low-dose responses better than threshold models. *Int. J. Toxicol.* 27: 369–378.

Calabrese, E. J., G. R. Hoffmann, E. J. Stanek, and M. A. Nascarella. 2010. Hormesis in high-throughput screening of antibacterial compounds in *E. coli*. *Hum. Exp. Toxicol.* 29: 667–677.

Calabrese, E. J., E. J. Stanek III, and M. A. Nascarella. 2011. Evidence for hormesis in mutagenicity dose–response relationships. *Mutat. Res.* 726: 91–97.

Cedergreen, N., J. C. Streibig, P. Kudsk, S. K. Mathiassen, and S. O. Duke. 2007. The occurrence of hormesis in plants and algae. *Dose-Response* 5: 150–162.

Conolly, R. B., and W. K. Lutz. 2004. Nonmonotonic dose–response relationships: Mechanistic basis, kinetic modeling, and implications for risk assessment. *Toxicol. Sci.* 77: 151–157.

Cook, R., and E. J. Calabrese. 2006a. The importance of hormesis to public health. *Environ. Health Perspect.* 114: 1631–1635.

Cook, R. R., and E. J. Calabrese. 2006b. Hormesis is biology, not religion. *Environ. Health Perspect.* 114: A688.

Crump, K. 2001. Evaluating the evidence for hormesis: A statistical perspective. *Crit. Rev. Toxicol.* 31(4–5): 669–679.

Cutler, G. C. 2013. Insects, insecticides and hormesis: Evidence and considerations for study. *Dose-Response* 11: 154–177.

Cypser, J. R., and T. E. Johnson. 2003. Hormesis in *Caenorhabditis elegans* dauer-defective mutants. *Biogerontology* 4: 203–214.

Davies, J., G. B. Spiegelman, and G. Yim. 2006. The world of subinhibitory antibiotic concentrations. *Curr. Opin. Microbiol.* 9: 445–453.

Davies, J. M. S., C. V. Lowry, and K. J. A. Davies. 1995. Transient adaptation to oxidative stress in yeast. *Arch. Biochem. Biophys.* 317: 1–6.

Davis, J. M., and D. J. Svendsgaard. 1994. Nonmonotonic dose–response relationships in toxicological studies. In *Biological Effects of Low Level Exposures: Dose–Response Relationships*, E. J. Calabrese (ed.). Chapter 5, pp. 67–85. Boca Raton, FL: Lewis Publishers/CRC Press.

Day, T. K., G. Zeng, A. M. Hooker, M. Bhat, D. R. Turner, and P. J. Sykes. 2007. Extremely low doses of X-radiation can induce adaptive responses in mouse prostate. *Dose-Response* 5: 315–322.

Doak, S. H., G. J. S. Jenkins, G. E. Johnson, E. Quick, E. M. Parry, and J. M. Parry. 2007. Mechanistic influences for mutation induction curves after exposure to DNA-reactive carcinogens. *Cancer Res.* 67: 3904–3911.

Dobo, K. L., R. D. Fiedler, W. C. Gunther, C. J. Thiffeault, Z. Cammerer, S. L. Coffing, T. Shutsky, and M. Schuler. 2011. Defining EMS and ENU dose–response relationships using the *Pig-a* mutation assay in rats. *Mutat. Res.* 725: 13–21.

Duke, S. O., N. Cedergreen, E. D. Velini, and R. G. Belz. 2006. Hormesis: Is it an important factor in herbicide use and allelopathy? *Outlooks Pest Manag.* 17: 29–33.

Duport, P. 2003. A database of cancer induction by low-dose radiation in mammals: Overview and initial observations. *Int. J. Low Radiat.* 1: 120–131.

Eaton, D. L., and S. G. Gilbert. 2008. Principles of toxicology. In *Casarett and Doull's Toxicology: The Basic Science of Poisons*, C. D. Klaassen (ed.). 7th Edition, pp. 11–43. New York: McGraw-Hill.

Elliott, K. C. 2011. *Is a Little Pollution Good for You? Incorporating Societal Values in Environmental Research*. New York: Oxford University Press.

EPA (U.S. Environmental Protection Agency). 2005. *Guidelines for Carcinogen Risk Assessment*. EPA/630/P-03/001B. March 2005. http://www.epa.gov/cancerguidelines/.

Fukushima, S., A. Kinoshita, R. Puatanachokchai, M. Kushida, H. Wanibuchi, and K. Morimura. 2005. Hormesis and dose-response-mediated mechanisms in carcinogenesis: Evidence for a threshold in carcinogenicity of non-genotoxic carcinogens. *Carcinogenesis* 26: 1835–1845.

Gocke, E., and L. Müller. 2009. In vivo studies in the mouse to define a threshold for the genotoxicity of EMS and ENU. *Mutat. Res.* 678: 101–107.

Guan, Q., S. Haroon, D. González Bravo, J. L. Will, and A. P. Gasch. 2012. Cellular memory of acquired stress resistance in *Saccharomyces cerevisiae*. *Genetics* 192: 495–505.

Haseman, J. K., and F. M. Johnson. 1996. Analysis of National Toxicology Program rodent bioassay data for anticarcinogenic effects. *Mutat. Res.* 350: 131–141.

Hoffmann, G. R. 2009. A perspective on the scientific, philosophical, and policy dimensions of hormesis. *Dose-Response* 7: 1–51.

Hoffmann, G. R., and W. E. Stempsey. 2008. The hormesis concept and risk assessment: Are there unique ethical and policy considerations? *BELLE Newsl.* 14(3): 11–17. http://www.belleonline.com/newsletters/volume14/vol14-3.pdf.

Hoffmann, G. R., A. V. Moczula, A. M. Laterza, L. K. MacNeil, and J. P. Tartaglione. 2013. Adaptive response to hydrogen peroxide in yeast: Induction, time course, and relationship to dose–response models. *Environ. Mol. Mutagen.* 54: 384–396.

Hooker, A. M., M. Bhat, T. K. Day, J. M. Lane, S. J. Swinburne, A. A. Morley, and P. J. Sykes. 2004. The linear no-threshold model does not hold for low-dose ionizing radiation. *Radiat. Res.* 162: 447–452.

Ishii, K., and M. Watanabe. 1996. Participation of gap-junctional cell communication on the adaptive response in human cells induced by low dose of X-rays. *Int. J. Radiat. Biol.* 69: 291–299.

Jonas, W. B. 2010. What dose metaphor? *Hum. Exp. Toxicol.* 29: 271–273.

Kadhim, M., S. Salomaa, E. Wright, G. Hildebrandt, O. V. Belyakov, K. M. Prise, and M. P. Little. 2013. Non-targeted effects of ionising radiation—Implications for low dose risk. *Mutat. Res.* 752(2): 84–98.

Kefford B. J., L. Zalizniak, J. S. Warne, and D. Nugegoda. 2008. Is the integration of hormesis and essentiality into ecotoxicology now opening Pandora's box? *Environ. Pollut.* 151: 516–523.

Kitchin, K. T., and J. W. Drane. 2005. A critique of the use of hormesis in risk assessment. *Hum. Exp. Toxicol.* 24: 249–253.

Kleibl, K. 2002. Molecular mechanisms of adaptive response to alkylating agents in *Escherichia coli* and some remarks on $O^6$-methylguanine DNA-methyltransferase in other organisms. *Mutat. Res.* 512: 67.

Kodell, R. L., D. W. Gaylor, D. L. Greenman, N. A. Littlefield, and J. H. Farmer. 1983. Response to the Society of Toxicology task force re-examination of the $ED_{01}$ study. *Fundam. Appl. Toxicol.* 3: 3/a–8/a.

Le Bourg, E. 2009. Hormesis, aging and longevity. *Biochim. Biophys. Acta.* 1790(10): 1030–1039.

Le Bourg, É. 2010. Predicting whether dietary restriction would increase longevity in species not tested so far. *Ageing Res. Rev.* 9: 289–297.

Le Bourg, É. 2012. Combined effects of two mild stresses (cold and hypergravity) on longevity, behavioral aging, and resistance to severe stresses in *Drosophila melanogaster*. *Biogerontology* 13: 313–328.

Lin, J. H.-C., N. Lou, N. Kang, T. Takano, F. Hu, X. Han, Q. Xu et al. 2008. A central role of connexin 43 in hypoxic preconditioning. *J. Neurosci.* 28: 681–695.

Linares, J. F., I. Gustafsson, F. Baquero, and J. L. Martinez. 2006. Antibiotics as intermicrobial signaling agents instead of weapons. *Proc. Natl. Acad. Sci. USA* 103: 19484–19489.

Luckey, T. D. 1968. Insecticide hormoligosis. *J. Econ. Entomol.* 61: 7–12.

Lutz, W. K., and R. W. Lutz. 2009. Statistical model to estimate a threshold dose and its confidence limits for the analysis of sublinear dose–response relationships, exemplified for mutagenicity data. *Mutat. Res.* 678: 118–122.

Mattison, J. A., G. S. Roth, T. M. Beasley, E. M. Tilmont, A. M. Handy, R. L. Herbert, D. L. Longo et al. 2012. Impact of caloric restriction on health and survival in rhesus monkeys from the NIA study. *Nature* 489: 318–321.

Mattson, M. P., and A. Cheng. 2006. Neurohormetic phytochemicals: Low-dose toxins that induce adaptive neuronal stress responses. *Trends Neurosci.* 29: 632–639.

Mattson, M. P., W. Duan, R. Wan, and Z. Guo. 2004. Prophylactic activation of neuroprotective stress response pathways by dietary and behavioral manipulations. *NeuroRx* 1: 111–116.

Mitchel, R. E. 2010. The dose window for radiation-induced protective adaptive responses. *Dose-Response* 8: 192–208.

Mitchel, R. E. J., J. S. Jackson, D. P. Morrison, and S. M. Carlisle. 2003. Low doses of radiation increase the latency of spontaneous lymphomas and spinal osteosarcomas in cancer-prone, radiation-sensitive *Trp53* heterozygous mice. *Radiat. Res.* 159: 320–327.

Miura, Y. 2004. Oxidative stress, radiation-adaptive responses, and aging. *J. Radiat. Res.* 45: 357–372.

Morano, K. A., C. M. Grant, and W. S. Moye-Rowley. 2012. The response to heat shock and oxidative stress in *Saccharomyces cerevisiae*. *Genetics* 190: 1157–1195.

Morgan, W. F. 2003a. Non-targeted and delayed effects of exposure to ionizing radiation: I. Radiation-induced genomic instability and bystander effects in vitro. *Radiat. Res.* 159: 567–580.

Morgan, W. F. 2003b. Non-targeted and delayed effects of exposure to ionizing radiation: II. Radiation-induced genomic instability and bystander effects in vivo, clastogenic factors and transgenerational effects. *Radiat. Res.* 159: 581–596.

Morgan, W. F. 2006. Will radiation-induced bystander effects or adaptive responses impact on the shape of the dose response relationships at low doses of ionizing radiation? *Dose-Response* 4: 257–262.

Morimoto, K., M. Sato-Mizuno, and A. Koizumi 1986. Adaptation-like response to the chemical induction of sister chromatid exchanges in human lymphocytes. *Hum. Genet.* 73: 81–85.

Morse, J. G. 1998. Agricultural implications of pesticide-induced hormesis of insects and mites. *Hum. Exp. Toxicol.* 17: 266–269.

Morse, J. G., and N. Zareh. 1991. Pesticide-induced hormoligosis of citrus thrips (Thysanoptera: Thripidae) fecundity. *J. Econ. Entomol.* 84: 1169–1174.

Mossman, K. L. 2001. Deconstructing radiation hormesis. *Health Phys.* 80: 263–269.

Murry, C. E., R. B. Jennings, and K. A. Reimer. 1986 Preconditioning with ischemia: A delay of lethal cell injury in ischemic myocardium. *Circulation* 74(5): 1124–1136.

Mushak, P. 2007. Hormesis and its place in nonmonotonic dose–response relationships: Some scientific reality checks. *Environ. Health Perspect.* 115: 500–506.

Mushak, P. 2009. Ad hoc and fast forward: The science of hormesis growth and development. *Environ. Health Perspect.* 117: 1333–1338.

NRC (National Research Council Committee on the Biological Effects of Ionizing Radiations). 1980. *The Effects on Populations of Exposure to Low Levels of Ionizing Radiation* (BEIR III). Washington, DC: National Academy Press.

NRC (National Research Council). 1994. *Science and Judgment in Risk Assessment*. Washington, DC: National Academy Press.

NRC (National Research Council Committee to Assess Health Risks from Exposure to Low Levels of Ionizing Radiations). 2006. *Health Risk from Exposure to Low Levels of Ionizing Radiation* (BEIR VII, Phase 2). Washington, DC: National Academy Press.

Olivieri, G., J. Bodycote, and S. Wolff. 1984. Adaptive response of human lymphocytes to low concentrations of radioactive thymidine. *Science* 223: 594–597.

Pant, M. C., X.-Y. Liao, Q. Lu, S. Molloi, E. Elmore, and J. L. Redpath. 2003. Mechanisms of suppression of neoplastic transformation in vitro by low doses of low LET radiation. *Carcinogenesis* 24: 1961–1965.

Portess, D. I., G. Bauer, M. A. Hill, and P. O'Neill. 2007. Low-dose irradiation of nontransformed cells stimulates the selective removal of precancerous cells via intercellular induction of apoptosis. *Cancer Res.* 67: 1246–1253.

Preston, R. J., and G. R. Hoffmann. 2013. Genetic toxicology. In *Casarett and Doull's Toxicology: The Basic Science of Poisons*, C. D. Klaassen (ed.). 8th Edition, Chapter 9. New York: McGraw-Hill.

Radak, Z., H. Y. Chung, E. Koltai, A. W. Taylor, and S. Goto. 2008. Exercise, oxidative stress and hormesis. *Ageing Res. Rev.* 7: 34–42.

Rattan, S. I. S. 2008. Hormesis in aging. *Ageing Res. Rev.* 7: 63–78.

Redpath, J. L., and E. Elmore. 2007 Radiation-induced neoplastic transformation in vitro, hormesis and risk assessment. *Dose-Response* 5: 123–130.

Rhomberg, L. R., J. E. Goodman, L. T. Haber, M. Dourson, M. E. Andersen, J. E. Klaunig, B. Meek, P. S. Price, R. O. McClellan, and S. M. Cohen. 2011. Linear low-dose extrapolation for noncancer health effects is the exception, not the rule. *Crit. Rev. Toxicol.* 41: 1–19.

Ross, W. D. 1988. *The Right and the Good*, pp. 21–22. Indianapolis, IN: Hackett Publishing.

Rouse, J., and S. P. Jackson. 2002. Interfaces between the detection, signaling, and repair of DNA damage. *Science* 297: 547–551.

Samson, L., and J. Cairns. 1977. A new pathway for DNA repair in *Escherichia coli*. *Nature* 267: 281–283.

Scott, B. R., and J. Di Palma. 2006. Sparsely ionizing diagnostic and natural background radiations are likely preventing cancer and other genomic-instability-associated diseases. *Dose-Response* 5: 230–255.

Shadley, J. D., and S. Wolff. 1987. Very low doses of X-rays can cause human lymphocytes to become less susceptible to ionizing radiation. *Mutagenesis* 2: 95–96.

Shenton, D., J. B. Smirnova, J. N. Selley, K. Carroll, S. J. Hubbard, G. D. Pavitt, M. P. Ashe, and C. M. Grant. 2006. Global translational responses to oxidative stress impact upon multiple levels of protein synthesis. *J. Biol. Chem.* 281: 29011–29021.

Shrader-Frechette, K. 2008. Ideological toxicology: Invalid logic, science, ethics about low-dose pollution. *Hum. Exp. Toxicol.* 27: 647–657.

Sorensen, K. J., C. M. Attix, A. T. Christian, A. J. Wyrobek, and J. D. Tucker. 2002. Adaptive response induction and variation in human lymphoblastoid cell lines. *Mutat. Res.* 519: 15–24.

Spassova, M. A., D. J. Miller, D. A. Eastmond, N. S. Nikolova, S. V. Vulimiri, J. Caldwell, C. Chen, and P. D. White. 2013. Dose–response analysis of bromate-induced DNA damage and mutagenicity is consistent with low-dose linear, nonthreshold processes. *Environ. Mol. Mutagen.* 54: 19–35.

Stecca, C., and G. B. Gerber. 1998. Adaptive response to DNA-damaging agents. *Biochem. Pharmacol.* 55: 941–951.

Temple, M. D., G. G. Perrone, and I. W. Dawes. 2005. Complex cellular responses to reactive oxygen species. *Trends Cell Biol.* 15: 319–326.

Thayer, K. A., R. Melnick, K. Burns, D. Davis, and J. Huff. 2005. Fundamental flaws of hormesis for public health decisions. *Environ. Health Perspect.* 113: 1271–1276.

Thayer, K. A., R. Melnick, J. Huff, K. Burns, and D. Davis. 2006. Hormesis: A new religion? *Environ. Health Perspect.* 114: A632–A633.

Thomas, A. D., G. J. Jenkins, B. Kaina, O. G. Bodger, K. H. Tomaszowski, P. D. Lewis, S. H. Doak, and G. E. Johnson. 2013. Influence of DNA repair on nonlinear dose–responses for mutation. *Toxicol. Sci.* 132: 87–95.

Timms, B. G., K. L. Howdeshell, L. Barton, S. Bradley, C. A. Richter, and F. S. vom Saal. 2005. Estrogenic chemicals in plastic and oral contraceptives disrupt development of the fetal mouse prostate and urethra. *Proc. Natl. Acad. Sci. USA* 102: 7014–7019.

Townsend, J. F., and T. D. Luckey. 1960. Hormoligosis in pharmacology. *J. Amer. Med. Assoc.* 173: 44–48.

Tubiana, M., A. Aurengo, D. Averbeck, and R. Masse. 2006. Recent reports on the effects of low doses of ionizing radiation and its dose–effect relationship. *Radiat. Environ. Biophys.* 44: 245–251.

Vandenberg, L. N., T. Colborn, T. B. Hayes, J. J. Heindel, D. R. Jacobs Jr., D. H. Lee, T. Shioda et al. 2012. Hormones and endocrine-disrupting chemicals: Low-dose effects and nonmonotonic dose responses. *Endocr. Rev.* 33(3): 378–455.

Velini, E. D., E. Alves, M. C. Godoy, D. K. Meschede, R. T. Souza, and S. O. Duke. 2008. Glyphosate applied at low doses can stimulate plant growth. *Pest Manag. Sci.* 64(4): 489–496.

Vineis, P. 2005. Scientific basis for the precautionary principle. *Toxicol. Appl. Pharmacol.* 207: S658–S662.

Waalkes, M. P., S. Rehm, C. W. Riggs, R. M. Bare, D. E. Devor, L. A. Poirier, M. L. Wenk, and J. R. Henneman. 1988. Cadmium carcinogenesis in male Wistar [Crl:(WI)BR] rats: Dose–response analysis of tumor induction in the prostate and testes and at the injection site. *Cancer Res.* 48: 4656–4663.

Welshons, W. V., K. A. Thayer, B. M. Judy, J. A. Taylor, E. M. Curran, and F. S. vom Saal. 2003. Large effects from small exposures. I. Mechanisms for endocrine-disrupting chemicals with estrogenic activity. *Environ. Health Perspect.* 111: 994–1006.

Weltje, L., F. S. vom Saal, and J. Oehlmann. 2005. Reproductive stimulation by low doses of xenoestrogens contrasts with the view of hormesis as an adaptive response. *Hum. Exp. Toxicol.* 24: 431–437.

Wheeler, D. S., and H. R. Wong. 2007. The heat shock response and acute lung injury. *Free Radic. Biol. Med.* 42: 1–14.

White, R. H., I. Cote, L. Zeise, M. Fox, F. Dominici, T. A. Burke, P. D. White, D. B. Hattis, and J. M. Samet. 2009. State-of-the-science workshop report: Issues and approaches in low dose–response extrapolation for environmental health risk assessment. *Environ. Health Perspect.* 117: 283–287.

Wiese, A. G., R. E. Pacifici, and K. J. A. Davies. 1995. Transient adaptation of oxidative stress in mammalian cells. *Arch. Biochem. Biophys.* 318: 231–240.

Wolff, S. 1996. Aspects of the adaptive response to very low doses of radiation and other agents. *Mutat. Res.* 358: 135–142.

Wolff, S. 1998. The adaptive response in radiobiology: Evolving insights and implications. *Environ. Health Perspect.* 106: 277–283.

Wolff, S., V. Afzal, J. K. Wiencke, G. Olivieri, and A. Michaeli. 1988. Human lymphocytes exposed to low doses of ionizing radiations become refractory to high doses of radiation as well as to chemical mutagens that induce double-strand breaks in DNA. *Int. J. Radiat. Biol.* 53: 39–48.

Zeiger, E., and G. R. Hoffmann. 2012. An illusion of hormesis in the Ames test: Statistical significance is not equivalent to biological significance. *Mutat. Res.* 746: 89–93.

Zhang, Y., J. Zhou, J. Baldwin, K. D. Held, K. M. Prise, R. W. Redmond, and H. L. Liber. 2009. Ionizing radiation-induced bystander mutagenesis and adaptation: Quantitative and temporal aspects. *Mutat. Res.* 671: 20–25.

Zhou, H., G. Randers-Pehrson, C. R. Geard, D. J. Brenner, E. J. Hall, and T. K. Hei. 2003. Interaction between radiation-induced adaptive response and bystander mutagenesis in mammalian cells. *Radiat. Res.* 160: 512–516.

# 16 Hormesis and Risk Assessment

*Edward J. Calabrese*

## CONTENTS

16.1 Introduction ........................................................................................... 339
16.2 Historical Overview ............................................................................... 339
16.3 Risk Assessment and Dose-Response Model Validation ...................... 345
16.4 Improving Harmonization in Risk Assessment ..................................... 348
16.5 How Hormesis Could Be Applied to Risk Assessment ........................ 348
16.6 Hormesis as the Default Assumption in Risk Assessment ................... 350
16.7 Hormesis and the Definition of Risk Assessment ................................ 351
16.8 Hormetic Risk Assessment Concerns .................................................... 352
References ....................................................................................................... 354

## 16.1 INTRODUCTION

Risk assessment is a four-step process, involving hazard assessment, exposure assessment, dose-response estimation/assessment, and an integrative risk characterization. Many factors may therefore affect the estimation of risk within the population (National Research Council [NRC] 1983). The risk assessment process has become codified and integrated into environmental, food safety, and occupational health practices throughout the world. This chapter explores how the concept of hormesis could affect process of risk assessment.

## 16.2 HISTORICAL OVERVIEW

From a historical perspective, the threshold dose-response was the first dose-response model that was adopted to guide various governmental agencies and their advisory bodies. This was the case even before the concept of risk assessment had been more formerly assessed and codified (NRC 1983). The threshold dose-response model became established during the early decades of the twentieth century (Calabrese 2008a), being used to assess risks and to establish acceptable exposures for toxic chemicals, ionizing radiation, and pharmaceuticals. The threshold model, however, soon came to be challenged following the report of Muller (1927) indicating that x-rays could induce mutations in the germ cells of male and female fruit flies. Even though Muller's research was conducted at extremely high levels of exposure, he proposed the concept of a *proportionality rule* to account for the nature of the dose-response for ionizing radiation–induced mutation (Muller 1930). The proportionality rule was the operational expression for the linear dose-response model, referred to here as the linear nothreshold (LNT) model, a term first used by Olson and Lewis (1928). These authors offered the hypothesis that cosmic/terrestrial ionizing radiation–induced germ cell mutation was the mechanism of evolutionary change, based on the mutational findings of Muller (1927) (see Muller and Mott-Smith [1930] for a critique of this proposal).

The proportionality rule for x-ray-induced mutations was provided a mechanistic framework in 1935 when Timoféeff-Ressovsky et al. integrated the single-hit mutational hypothesis into the LNT model based on the concept of target theory. This integrated framework provided the basis for the belief that a single ionization or a single-molecule-induced genetic mutation could enhance cancer risks. The LNT model found an application when it became part of the occupational health standard setting debate for ionizing radiation during the 1930s concerning the emerging concept of a *permissible dose*. Permissible dose is a concept that was able to make a distinction between a safe (i.e., no risk) exposure and an acceptable exposure to the same agent (Calabrese 2009a). Permissible dose was viewed as an exposure that was acceptable but one that was not necessarily believed to be without risk. This was a dose-response concept that was derived directly from an LNT framework.

Debate was intensified on the nature of the dose-response for ionizing radiation during the years leading up to World War II, based on Muller's findings. The Manhattan Project, the secret research project during World War II in the United States to build an atomic bomb, funded research at the University of Rochester (Rochester, New York) under the direction of professor and geneticist Curt Stern to provide a definite answer to the following question: Was the nature of the dose-response in the low-dose zone that of a threshold or was it linear, as argued by Muller (1930) within the context of his proportionality rule? There were very significant clinical, public health, scientific, economic, and policy implications depending on the answer.

The Stern research within the Manhattan Project started in 1943 but the results of these investigations were not published in the open literature until after World War II in 1948, as these scientific reports became declassified in 1947 (Calabrese 2011a). Three key articles emerged from the Stern research, one dealing with acute exposures (Spencer and Stern 1948), one with chronic (i.e., 1/13, 200th the dose rate of the acute study) exposures (Caspari and Stern 1948), and a third that integrated the previously published work and three additional significant studies using the low-dose rate (Uphoff and Stern 1949). The research that was undertaken by Stern was strongly influenced by Muller, who was a paid consultant to the project, providing the Muller-5 *Drosophila* fruit fly strain, and advising on all aspects of the study design and endpoint selection (i.e., sex-linked recessive mutations). Muller also helped resolve uncertainties over control group background mutation rate and provided key guidance on manuscripts that would be submitted for publication (Calabrese 2011a).

The findings of Stern and his colleagues would be interpreted as demonstrating that the nature of the dose-response in the low-dose zone for the effects of ionizing radiation on germ cell mutation supported and confirmed the linear dose response hypothesis. The Stern studies were extremely influential as they provided the basis for a key recommendation of the U.S. National Academy of Sciences (NAS) Biological Effects of Atomic Radiation (BEAR), Genetics Panel in 1956 (of which Muller was a member) to a switch from a threshold to a linear dose-response model for assessing the risks of ionizing radiation for germ cell mutation. The LNT concept was then generalized to the concept of somatic effects the following year, creating a linear dose-response modeling framework for assessing cancer risks from ionizing radiation and chemical carcinogens (Calabrese 2009a).

This recommendation of the NAS BEAR I Committee, Genetics Panel would serve as the most significant risk assessment guidance document of the twentieth century, providing the theoretical foundation for cancer risk assessment for suspect human chemical carcinogens and ionizing radiation. As a result of the recommendations of this 1956 report, risk assessment would make use of two types of dose-response models for the remainder of the twentieth century to the present: the threshold dose-response model for the assessment of noncarcinogenic endpoint responses and the LNT model for carcinogens.

The threshold dose-response model incorporated the use of multiple uncertainty factors (i.e., uncertainty factors for interspecies variation, interindividual variation, less than lifetime exposure, differing exposure routes in test animals versus human exposure and other possibilities). In contrast to how the threshold dose-response model would be used in the risk assessment process, the cancer risk assessment methodology using the LNT model would assume that humans and animal models would display the same susceptibility to each assessed carcinogen. The assumption of interindividual variability would also be ignored in the cancer risk assessment process. Thus, there are several basic differences in the process of noncarcinogen and carcinogen risk assessment that were adopted by most governmental agencies worldwide. Furthermore, the Environmental Protection Agency (EPA) introduced another fundamental difference in the risk assessment of noncarcinogens versus carcinogens. This involves the concept of dose normalization. In the case of noncarcinogens, the EPA has based dosing on normalization to body weight, whereas in the case of carcinogens the dosing was normalized to body surface area. The difference in normalization strategies can have important practical implications as the use of surface area for dose normalization for carcinogens would yield a 13-fold lower dose if the data were based on mice and a 6-fold lower dose if the findings were based on rats. The risk assessment process was therefore such that different dose-response assumptions were made for noncarcinogens and carcinogens as well as for the dose normalization process (Calabrese 1978; Calabrese and Kenyon 1991).

Although the above dose-response evolution was unfolding with respect to the acceptance and integration of the threshold and linear dose-response models within the public health and regulatory communities, the hormetic dose-response was excluded from this discussion, being treated as a marginalized concept by the scientific community, not included in university textbooks, omitted from the toxicological and biomedical general research agenda, and not included in federal agency regulatory directives or the U.S. NAS evaluation processes at any level of detail or consideration (Calabrese, 2011b). It simply was ignored, and not given any scientific standing. To make matters more difficult for this concept, it was commonly associated with the high-dilution wing of the practice of homeopathy, an association that would not only reinforce in the marginalization of hormesis but also subject it to ridicule within the scientific and especially the biomedical communities (Calabrese 2011b).

Despite the exclusion of the hormetic concept from the risk assessment discussion throughout the twentieth century (NRC 1983, 1993, 1994; Risk Commission 1997a,b), Calabrese and Baldwin (2000a–e) reported that scientific studies on hormetic-like biphasic dose-responses were consistently reported since the initial reports of Schulz in 1887 and 1888 (see Chapter 1). The findings from the 1880s to the 1940s were based principally on research with plants, micro-organisms (e.g., bacteria, yeast), and to a lesser extent, insects. This is the case for both the chemical and radiation domains. In fact, there was a scientific journal published from 1924 to 1930 in the German language on the hormetic concept, entitled "Cell Stimulation Research" (Popoff and Gleisberg 1924–1930). Secondly, during the 1970s there was another journal publication, called the *Stimulation Newsletter* (1970–1976), that was devoted to the publication of hormetic dose-responses (Calabrese and Baldwin, 2001).

Although there were a substantial number of publications in leading journals and by some notable investigators during the early decades of the twentieth century (Calabrese 2009b), the area of biphasic dose-response did not mature and develop into an integrative scientific area; that is, one that could be compared, contrasted, and possibly integrated with the threshold dose-response model. Why was this the case? Through what appears to be an unusual series of professional developments, no single researcher out of a grouping of several dozen possibilities continued to develop the biphasic dose-response concept. The careers of each of the leading or emerging hormetic researchers during the first half of the twentieth century tended to transition into academic administration or into other research areas. This contributed to the failure of the hormetic concept to develop the necessary research continuity to properly develop and to be vetted in the research domain (Calabrese 2009b).

The term *hormesis*, from the Greek meaning "to excite," would not enter into the scientific literature until 1943 following the publication of Southam and Ehrlich (1943) concerning the effects of extracts of the red cedar tree on the growth of fungi in a research project designed to address issues relating to factors affecting the rotting of wood. Prior to this time the biphasic dose-response was often referred to as the Arndt–Schulz Law or Hueppe's Rule after Ferdinand Hueppe, a protégé of Robert Koch. The term hormesis was not assimilated into the contemporary scientific literature very quickly and in fact was rarely cited in the leading indices over the next four decades, despite the fact that biphasic dose-responses were commonly reported. It might have been more effectively promoted soon after its introduction into the scientific literature had Southam and Ehrlich remained interested in their research topic. However, both would soon change research directions, having significant subsequent scientific careers but ones that left the dose-response concept and hormesis behind.

This leaderless activity on hormesis would begin to change in the 1940s with the initial research of Thomas Luckey concerning the effects of antibiotics on animal models that were evaluated within gnotobiotic experimental protocols. Luckey would continue to provide a focus on the concept of hormesis up to the present, a duration of nearly seven decades. Despite his long interest in the hormetic concept, it would take Luckey some 30 years to provide the first linkage of the hormetic concept to the issue of environmental regulation. This delay was not a failure of Luckey to make the connection of hormesis to risk assessment but due to the fact it would take regulatory agencies until the mid-1970s to come to grips with the nature of the dose-response in the low-dose zone. In a prophetic statement in the preface of his 1975 book entitled *Heavy Metal Toxicity Safety and Hormology* (Luckey et al. 1975), Luckey suggested to the toxicology and regulatory communities the need to consider the hormetic dose-response and the concept of response benefit in the risk assessment process, a suggestion that would go unheeded.

> Understanding the extent of this phenomenon [hormesis] is essential before worldwide committees and legislative bodies make recommendations which consider only toxic actions of heavy metals.

In 1980, Luckey published a book entitled *Ionizing Radiation and Hormesis* that was to have an impact. This book would be a catalyst for generating a regulatory focus on hormesis. In this case, Luckey's book stimulated key leaders of the Japanese electric power industry to join with the U.S. Electric Power Research Institute to conduct a conference on radiation hormesis, which was held in August 1985. It was the first such conference and provided an important organizing stimulus on hormesis, leading to the publication of a conference proceedings in the journal *Health Physics* in 1987.

The hormetic concept had become potentially important to the risk assessment process for practical considerations. In the early 1980s, the use of quantitative risk assessment for chemical carcinogens and ionizing radiation using LNT-based dose-response modeling was seen as having an enormous impact on the cost of environmental regulations, environmental remediation, and other occupational and environmental compliance activities. The switch from the threshold dose-response model to the LNT model had proven to have formidable economic implications for the electric power industry, for the chemical industry, as well as for public utility companies such as those providing community drinking water and other companies.

One of the challenges was that the chronic bioassay used by the U.S. government was very limited, only being designed to answer the question of whether an agent could cause cancer or not. It could not properly address the question of whether the test agent was carcinogenic at doses that would be normally encountered by people at their jobs or within the nonoccupational normal living activities. The original chronic bioassay only used two doses: the maximum tolerated dose (MTD), that is, a daily dose of an agent that could be tolerated for its entire adult life (i.e., 2 years for mice and rats) without noticeable signs of illness or a loss of body weight by more than 10%; and one half of the MTD. Particularly problematic were observations that organ-specific toxicity may often occur at the MTD even with a <10% decrease in body weight. Damage to an organ would result in reparative processes that could act as tumor-promotional stimuli. Under such situations the risk observed at the higher dose would misrepresent the risk assessment process, assuming low-dose linearity. The chronic bioassay was therefore not designed to address the issue of low-dose extrapolation, yet that was how it came to be applied, thereby creating higher uncertainty with estimations and much debate within the toxicology and risk assessment communities.

Within the framework of the chronic bioassay, it was not possible to confidently determine whether the experimental data better fit that of a threshold or linear dose-response model due to the lack of sufficient doses and their very limited dose spacing. In situations where it was not possible to differentiate between the threshold or LNT models, the regulatory agencies in the United States would default to the most conservative estimate of risk, following the concept of the precautionary principle. This practice frustrated the regulated community as regulatory decisions would invariably be based on unverified predictions of the LNT model, the most conservative and, from the industry's perspective, the most expensive option. At some point, a decision was made to explore the hormetic option based on the inspiration and credibility provided by the 1980 book of Luckey. The hormetic dose-response concept provided an alternative theoretical vehicle by which to challenge the linear dose-response model for carcinogen risk assessment, because the threshold model set within a chronic bioassay framework had failed to do so (Figure 16.1).

Although the above historical reconstruction reflects a potential economic incentive for a hormetic revitalization, at about the same time the concept of biphasic dose-responses began to be reported with heightened frequency from a broad range of biological disciplines, using different terms, such as U-shaped, J-shaped, bimodal, biphasic, dual effects, hormetic, adaptive responses, nonmonotonic, paradoxical responses, opposite effects, overshoot, rebound effect, and others. Despite the diversity of terms there was a strong dose-response conceptual convergence that led to considerable interest in biphasic dose-responses (Calabrese 2008b).

During this period of conceptual convergence, there emerged the issue of language and terminology for biphasic dose-response relationships. The lack of leadership on the concept of biphasic dose-responses and the operational independence of specific subfields with in the biological sciences led to the use of a plethora of names that describe the same dose-response phenomenon, adding confusion to the field while also impeding its biological integration. For example, epidemiology typically

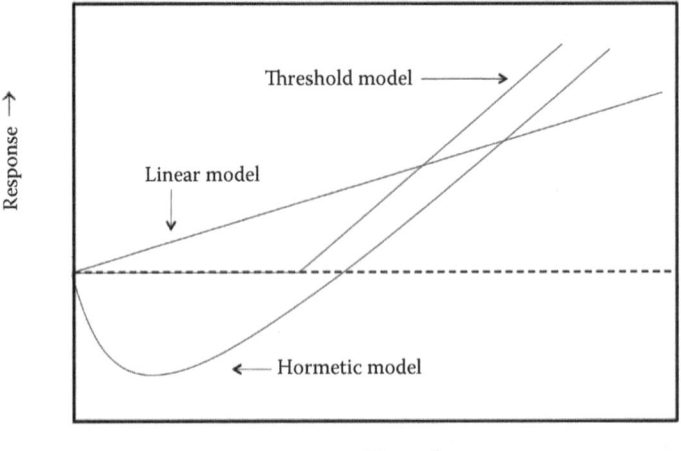

**FIGURE 16.1** Dose-response models: hormetic, linear, and threshold.

used the term *U-shaped*, the stress biology field often uses the term *Yerkes–Dodson Law*, the endocrine disruption area often uses the term *nonmonotonic*, laser biology has often used the term *Arndt–Schulz Law* and the field of radiation biology uses the term *adaptive response* while a range of medical disciplines use the term *preconditioning* for phenomena that also follow a biphasic dose-response. Of interest is that the quantitative features of these diverse biphasic dose-responses are indistinguishable and appear to represent a common adaptive dose-response strategy.

Although the issue of language is dynamic, with progressive modifications and refinements over time, the term hormesis is unique among the above descriptive terms in that it has been defined to represent biphasic dose-responses with very specific features. These include having specific quantitative features relating to the amplitude of the stimulatory response, the width of the stimulatory response, and the relationship of the stimulatory response to the threshold (Calabrese and Blain 2005, 2009, 2011). In fact, it is the consistency of the quantitative features of the hormetic dose-response and its relationship to the threshold which make it extremely valuable to risk assessors. The high predictability of the quantitative features of the dose-response would be an important attribute for a default risk assessment model.

The term hormesis is still relatively new within the biomedical and toxicological literature, having only made it into major textbooks within the past decade. Nonetheless, a sign of its growth can be seen in the citation frequency of the Web of knowledge database. In the 1980s, the terms hormesis or hormetic were cited only about 10–15 times per year. In 2012, the number of citations exceeded 4500, showing strong growth over the past several decades. Such growth in the scientific evaluation of the hormetic concept is potentially important for regulatory applications as scientific developments transformed the hormetic concept into one that would be science driven.

As suggested by the increase in citations in scientific databases over the past two decades, the hormetic dose-response has been shown to be reproducible, common (Calabrese and Blain 2005, 2011), and more frequent than the threshold and linear dose-response models in head-to-head comparisons (Calabrese et al. 2006, 2008, 2010). Hormetic mechanisms have also been documented for several hundred dose-response relationships at the level of receptor and cell signaling pathway (Calabrese 2013). These findings suggest that the hormetic dose-response should be considered a dose-response model that would have application to the risk assessment process. Table 16.1 provides a listing of biological features of hormetic dose-responses, whereas Table 16.2 integrates these features into a set of hormetic concepts and principles.

## 16.3 RISK ASSESSMENT AND DOSE-RESPONSE MODEL VALIDATION

The threshold dose-response model was never validated during the entire twentieth century. It was accepted/assumed to provide accurate estimates of responses both above and below the threshold response. This historical failing may seem hard to believe because entire federal governmental regulatory programs in numerous countries for chemical safety were based on the threshold dose-response model. This is the model that was incorporated into leading toxicology, pharmacology, and risk assessment textbooks. The realization that the threshold dose-response model was never validated

**TABLE 16.1**
**Hormetic Concepts and Principles**

Low/modest stress induces pro-survival responses

The quantitative features of the hormetic dose-response are similar across species and individuals and independent of differential susceptibility, agent potency, and mechanism

The magnitude of the stimulatory response is constrained by and defines the plasticity of the biological model

Hormetic responses occur at multiple levels of biological organization, such as the cellular, organ, individual, and population levels

Downstream processes integrate responses from multiple independent stressor agents/excitatory stimuli and affect an integrated dose-response (i.e., molecular vector) reflecting the hormetic dose-response

Hormetic responses represent both a general response to environmentally induced stress/damage and some elements of chemical structure specificity for endpoint induction

*Source:* Calabrese, E.J., *Environ. Toxicol. Chem.*, 27, 7, 2008.

---

**TABLE 16.2**
**Principal Hormetic Dose-Response Research Findings**

Most commonly observed dose-response relationship

Distinctive quantitative features, making it a unique biphasic dose-response relationship

Most unique feature is the modest magnitude of the stimulatory response, typically less than twice the control values

The low-dose stimulation occurs via a direct stimulation or by an overcompensation to a disruption of homeostasis/modest toxicity

Hormetic dose-responses are an adaptive response that ensures tissue repair in an efficient manner and protects against damage from subsequent and more massive exposures

Hormetic dose-responses are very generalizable. They are independent of biological model, endpoint measured, chemical class, and mechanism

Numerous specific mechanisms have been reported to account for hormetic dose-responses, without affecting the quantitative feature of the dose-response

*Source:* Calabrese, E.J., *Environ. Toxicol. Chem.*, 27, 7, 2008.

---

occurred unexpectedly. In my evaluation of the hormetic concept, I explored methods that would permit a means to test the validity of the hormetic dose-response model to make accurate predictions in the low-dose zone. This challenge suggested that a similar question had been raised about the threshold dose-response in past decades and that the threshold dose-response model must have been validated by the toxicological community or by various research-oriented regulatory agencies. However, a prolonged and detailed search for such attempts at validation of the threshold dose-response model never revealed evidence that such a validation had been attempted and/or completed. Although it is not possible to prove a negative, it appeared that no individual or group had ever made the attempt to validate the capacity of the threshold dose model to make accurate predictions for responses below the threshold response.

On the basis of my failure to find such documentation, I obtained three complementary, large, and independent databases to test the capacity of the threshold, linearity, and hormetic models to be validated based on their capacity to make accurate predictions in the low-dose zone (Calabrese et al. 2006, 2008, 2010). The results of these three validation tests revealed that the threshold and linear models poorly predicted responses in the low-dose zone. However, the hormetic dose-response made uniformly accurate predictions in the low-dose zone. The only dose-response model, therefore, that was excluded from serious scientific consideration, and the model not given any standing for use in the risk assessment process, was the only one that performed well in the validation test procedures.

Although no formal attempt was made to validate the threshold dose-response during the twentieth century as noted above, the U.S. Food and Drug Administration attempted to validate the linear dose-response model during the late 1970s by testing a single chemical called 2-acetylaminofluorene (2-AAF) using about 24,000 mice, the so-called mega-mouse study. The data, which were reviewed in considerable detail by a 14-member expert panel of the U.S. Society of Toxicology (Bruce et al. 1981) showed that AAF induced a hormetic dose-response for bladder cancer, a principal a priori endpoint of the study. The findings were surprising, being at odds with both the linear and threshold dose-response model predictions. Of importance is that quasi replication validations of the experiment were carried out in six different testing rooms and in each setting the observed dose-response relationship was that of a j-shaped dose-response relationship indicative of the hormetic dose-response (Figure 16.2). The Society of Toxicology expert panel noted on page 77:

The $ED_{01}$ study provides more than evidence of a "threshold". It provides statistically significant evidence that low doses of a carcinogen are beneficial.

The most striking aspect of figure 10 is the reduction in probability of bladder tumor from controls to doses 30, 35, and 45 PPM. This reduction occurs in all six rooms and is statistically significant.

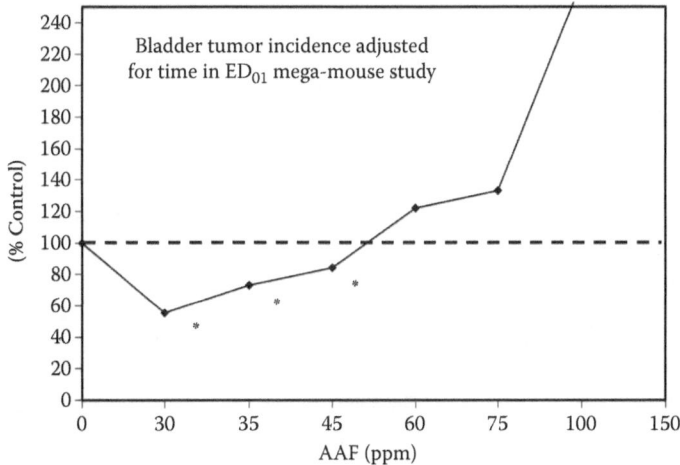

**FIGURE 16.2** Bladder tumor incidence adjusted for time in $ED_{01}$ mega-mouse study. (From Bruce, R.D. et al., *Fundam. Appl. Toxicol.*, 1, 67–80, 1981.)

## 16.4 IMPROVING HARMONIZATION IN RISK ASSESSMENT

The hormetic dose-response and its quantitative features are similar across all biological models, endpoints, inducing agents, and mechanisms. This would mean that the dose-response features for noncarcinogens and carcinogens are similar. This is contrary to what is the common practice of risk assessment throughout the world. The generalizability of the hormetic dose-response across endpoints argues that the cancer and noncancer regulatory dichotomy is incorrect. If the hormetic dose-response was the default model, it could provide a scientifically valid framework for the harmonization of noncarcinogen and carcinogen dose-response relationships. According to the concept of hormesis, there is no justification for the use of two different dose-response models to assess risks from carcinogenic and noncarcinogenic agents.

## 16.5 HOW HORMESIS COULD BE APPLIED TO RISK ASSESSMENT

For dose-responses displaying the biphasic hormetic dose-response, there are two thresholds for all responses, one occurring at the traditional threshold (i.e., threshold 1) and the second (i.e., threshold 2) occurring at a lower dose, once the response has regressed to the control group response (Figure 16.3). As doses proceed below the traditional toxicological or pharmacological threshold, there is a response that is opposite to that observed at higher doses. This may be due to either an overcompensation response to a disruption in homeostasis or a direct stimulation. The toxicological and pharmacological response immediately below threshold response 1 may be seen in risk assessment terms as desirable/beneficial, undesirable/harmful or, it could be of unknown biological–biomedical significance. In the case of a beneficial response, this may be seen if lifespan were prolonged, if memory were increased, or if disease incidence were reduced. In the case of an undesirable/harmful response that may be seen if there was a significant increase in prostate size sufficient to affect urination, a hypersensitivity response to immune system active agents, or stimulation of tumor cell proliferation.

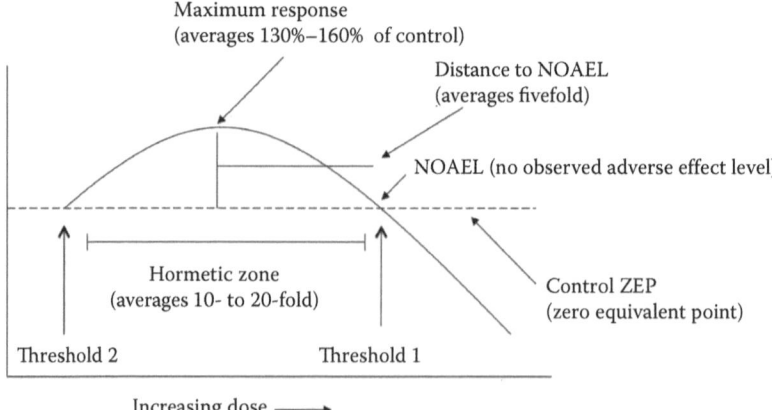

**FIGURE 16.3** Dose-response curve depicting the quantitative features of hormesis and its application to the concept of enhanced biological performance.

# Hormesis and Risk Assessment

These two options have different clinical and public health implications for environmental and clinical/therapeutic risk assessment. In the case of a beneficial response at low dose, there are two likely exposure scenarios. Dose-response scenario 1 represents the optimal benefit but also the exposure level closest to the toxicity threshold. However, if the width of the low-dose stimulatory zone were wider, as shown in Figure 16.4, it would result in a wider dose difference between an optimized benefit and the toxicity threshold. Knowledge of the quantitative features of the hormetic dose-response provides information that could lead to exposure standards that optimize public health and therapeutic benefits. Dose-response 2 offers the risk assessor an improved optimal benefit to harm ratio than provided in dose-response 1. This would also be the case in clinical medicine, where it is common to seek out optimized exposure strategies that maximize health benefits while minimizing potential for adverse responses. Ignoring the benefit potential by the use of large uncertainty factors is a strategy based on avoidance of harm without consideration to benefit.

Regulatory agencies should be required to provide detailed evaluations of potential benefits and harm, including the number of people affected (benefited/harmed) at lower doses, to permit risk managers to properly weigh policy options. The case of a harmful effect both above and below the threshold provides no option for making use of a low-dose beneficial response. For this type of hormetic dose-response, any change from the control value would be considered undesirable. Exposure in this case is driven only by an avoidance of the possibility of harm.

Of relevance to both examples are the widths of the respective stimulatory zones. Although the height of the stimulatory zone is not very variable, the width of the stimulatory zone often times can markedly differ between hormetic dose-response relationships. Although most hormetic dose-responses have a response width in the

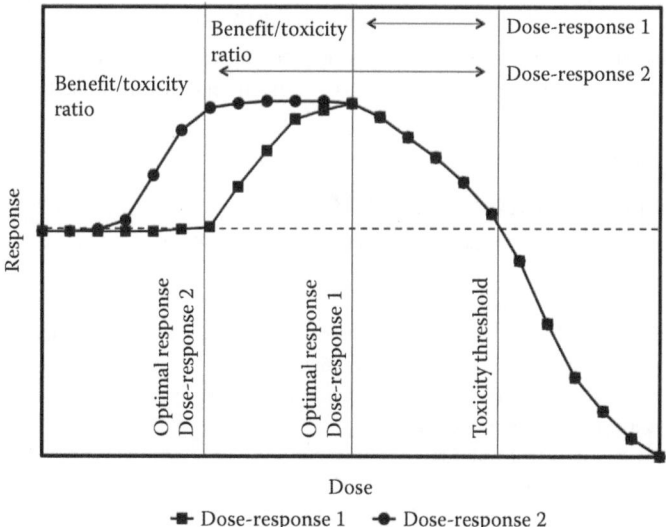

**FIGURE 16.4** Comparative benefit to toxicity ratio of dose-responses with differing widths of the beneficial component of the hormetic dose-response.

stimulatory zone of less than 20-fold, about 15% of cases exceed a 100-fold range and about 5%–10% exceed a 1000-fold range. This type of width variation could affect the decision-making process for risk assessment. Differing widths of the stimulatory zone alter the benefit-to-harm ratio. The reason is that the optimal response location could be moved further from the onset of toxicity, making the benefit-to-harm ratio higher. The wider the stimulatory zone for a beneficial effect, the more this situation could favor an increase in the benefit-to-harm ratio. The reverse would be the case if the width of the stimulatory response were decreased. The only implication for the second case would be to further decrease exposure as this would keep a constant margin of safety.

## 16.6 HORMESIS AS THE DEFAULT ASSUMPTION IN RISK ASSESSMENT

The default dose-response model in risk assessment should be the dose-response model that can be demonstrated to make the most accurate predictions of treatment effects in the low-dose zone, independent of biological model, endpoint, inducing agent, and mechanism. Default dose-response models established by regulatory agencies were established based on assumptions, without validation testing. In the only broad-based validated testing yet conducted, the hormetic dose-response model far outperformed the threshold and linear dose-response models for making accurate predictions of response in the low-dose zone. This suggests that the hormetic dose-response model should replace the threshold and linear dose-response models and become the default dose-response model for all endpoints. Table 16.3 provides criteria to consider in the selection of a default dose-response model. The hormetic dose-response can satisfy each of these criteria. However, this is not the case for the threshold and LNT models, which performed poorly with criteria 1, 2, and 3. The LNT model would also have considerable difficulty with criteria 8. In fact, the mega-mouse study was called the $ED_{01}$ study because cancer risks could not be estimated below $10^{-2}$ despite the use of massive numbers of animals. If the hormetic

### TABLE 16.3
### Default Dose-Response Model Criteria

Generalizability by biological model, endpoint measured, chemical class/physical agent, and mechanism
Frequency in the toxicological literature
Application of dose-response model for endpoints of relevance to risk assessment
Capacity for false positive and negative estimates
Impact of model on hazard assessment study requirements
Capacity to estimate risk quantitatively
Ability to validate risk estimates
Capacity to assess public health implications
Capacity to assess continuous and dichotomous endpoints of risk assessment relevance

*Source:* Calabrese, E.J., *Environ. Toxicol. Chem.*, 27, 1, 2008.

dose-response model were the default dose-response, there would still be a range of options that could be used in the derivation of exposure estimates as suggested in the examples as given immediately above.

Because the NTP (National Toxicology Program) bioassay currently uses only three doses, the MTD, 1/2 MTD, and 1/4 MTD, it is not likely to detect hormetic dose-responses in this testing framework. If hormesis were the default dose-response model, it would be necessary to estimate the threshold and then an optimized hormetic response could be derived based on data within the Hormetic Database (Calabrese and Blain 2005, 2011). The optimized hormetic response could be based on the entire database or it could be tailored to the specific model, endpoint, or other feature or combination of parameters.

It is also important to note that the shape of the hormetic dose-response is independent of the mechanism of action. Not only is the shape of the dose-response consistent for the above-mentioned parameters but the magnitude of the stimulation in the hormetic zone is also similar across biological model, endpoint, inducing agent, and mechanism.

## 16.7 HORMESIS AND THE DEFINITION OF RISK ASSESSMENT

Although there are numerous ways in which the hormetic concept might influence the risk assessment process (Table 16.4), there is no possibility that it can affect it until regulatory agencies incorporate this concept explicitly to the regulatory

### TABLE 16.4
### How Hormesis Affects Toxicology/Risk Assessment and Clinical Practices/Pharmaceutical Companies

**Toxicology/Risk Assessment**

Changes strategy for hazard assessment, guiding animal model and endpoint selection, and study design, including number of doses, dose range, and number of subject per dose

Employs biostatistical modeling to predict estimates of response below control background disease incidence

Differentiates dose optima (i.e., benefits) for normal- and high-risk segments of the population

Provides evaluative framework to assess benefits or harm below traditional toxicological threshold

Provides novel framework for quantitatively altering the magnitude of uncertainty factors in the risk assessment process

**Clinical Practices/Pharmaceutical Companies**

Drug performance expectation will be constrained and described by the quantitative features of the hormetic dose-response

Drugs that act at high doses may have hormetic effects at low doses, with possible undesirable effects (e.g., tumor cell proliferation)

Modification of biological set points will affect the quantitative features of the hormetic dose-response

Clinical trials need to recognize interindividual variation in the hormetic dose-response

Clinical trials need to plan for the quantitative features of the hormetic dose-response in study design

*Source:* Calabrese, E.J., *Environ. Toxicol. Chem.*, 27, 7, 2008.

definition of a risk assessment. In the case of the U.S. Environmental Protection Agency (U.S. EPA), their definition of a risk assessment purposefully excludes any consideration of the concept of adaptive response and benefit (EPA 2004). The EPA risk assessment process is designed only to avoid damage. If an exposure could affect a benefit, even one relating to enhancing lifespan, it would not be addressed in the risk assessment and standard setting process. Thus, the EPA actions would ignore the possibility of extending life to have an exposure standard that would exclude the possibility for harm, even if the probability for a significant life extension far exceeded the possibility of even a de minimus risk. It seems clear, at least as far as the EPA is concerned, that hormesis cannot be considered in the risk assessment process if the dose-response suggested a benefit. However, hormesis could be considered within the risk assessment if the low-dose stimulation affected adverse health effects. The EPA, therefore, is not opposed to the concept of hormesis, only against the concept of the risk assessment process incorporating benefit, which is however usually based on a hormetic dose-response.

Although the case for switching to a hormetic model has been proposed in this chapter, what happens if society changes to this model, and the hormetic model is incorrect for some agents? Assuming the hormetic model is completely incorrect and that the assumption of a linear at low-dose-response were correct, how much harm would take place? In general, it would be expected that the risk of developing cancer in a lifetime would increase by about 100-fold from a risk of $10^{-6}$ to $10^{-4}$. It should be noted that given the capacity of epidemiology to detect harmful effects in the population, it will not have the capacity to ever detect an increased risk of $10^{-4}$. Thus, even if the hormetic model were wrong, it would not be able to be discerned, except if this involved a rare form of cancer.

## 16.8 HORMETIC RISK ASSESSMENT CONCERNS

Although this chapter explores a possible role of hormesis in the risk assessment process, it is important to consider possible arguments against such an application of the hormetic concept. It is often difficult to demonstrate hormesis in standard hazard assessment bioassays because of the typically modest size of the stimulatory response, especially within a context of an endpoint with considerable variability. Most standard bioassays would need to be strengthened in a number of significant ways.

When there is a negligible or very low control group response incidence, there is little or no opportunity to assess a possible hormetic dose-response. Even when the control group has an appropriate disease frequency incidence, the bioassay would still need to add more doses, especially in the low-dose zone. The sample size would also need to be increased in the low-dose zone to enhance statistical power. Furthermore, it is likely that an initial hazard assessment bioassay would have the limited focused goal of threshold estimation. Follow-up experimentation would be needed to assess the hormetic hypothesis once the threshold response had been reasonably estimated. Experiments demonstrating apparent hormetic effects would need to be replicated in a manner that shows the reliability of the hormetic response. In addition, it may not be clear what the biological significance of a modest increase in a parameter is.

The above experimental changes, if implemented, would have a profound impact on the hazard assessment and risk assessment process. In practical terms, they would result in a marked prolongation of the hazard assessment process; the cost of the hazard assessment would also markedly increase whereas the number of chemicals that could be assessed per year would decrease. These factors highlight important implications and impediments for the implementation of the hormetic concept within the hazard assessment process.

Although the above assessment represents an argument to maintain the status quo for risk assessment and to not implement the hormetic concept, there are other considerations that need to be addressed. It may be argued that the hormetic concept should not have to be demonstrated in each hazard assessment for each chemical and endpoint of concern. This position is supported by reports in the literature that the hormetic dose-response has a very high frequency and is more common than either the threshold or linear dose-response in head-to-head comparisons (Calabrese et al. 2006, 2008, 2010). It would be possible to simply use the hormetic concept as the so-called *default* model. If this was the case, it would be relatively easy to provide estimates of the expected quantitative features of the hormetic dose-response for risk assessment purposes based on thousands of examples of hormetic dose-responses. These estimates have the potential to be tailored to the specific endpoints of concern. By using the hormetic concept as the default risk assessment model, it would not be necessary to change the current hazard assessment process. This would permit the hazard assessment to proceed as usual with no time delay, no increase in cost, and no reduction in the amount of agents evaluated per year. In fact, this has been how regulatory and public health agencies have dealt with the LNT model for risk assessment.

Although the LNT model cannot be assessed experimentally in the low-dose areas of interest, this is not the case with the concept of hormesis. If it were of value to experimentally test a hormetic hypothesis, this could be readily done. With the use of cell culture testing it is now possible to evaluate hormetic dose-response hypotheses in a manner consistent with recent recommendations by the US NRC concerning toxicity testing for the twenty-first century and to do so in a cost-effective manner (2007). The use of the hormetic dose-response model offers those in risk management the opportunity to assess the concept of risks and benefits in a more comprehensive manner than is currently done.

Changing the risk assessment paradigm after nearly half century is not something that should be lightly considered. However, a strong case can be made that the scientific and regulatory communities got the nature of the dose-response in the low-dose zone wrong and implemented this view within the risk assessment process. Such failings need to be addressed in an objective manner with a broad consideration given to dose-response model validation, generalizability for application to the risk assessment process, and to enhance the capacity for comprehensive cost-benefit assessments.

# REFERENCES

BEAR I. 1956. Genetic effects of atomic radiation. *Science* 123:1157–1164.

Bruce, R.D., Carlton, W.W., Ferber, K.H., Hughes, D.H., Quast, J.F., Salsburg, D.S., and Smith, J.M. et al. 1981. Re-examination of the ED01 study—Adjusting for time on study. *Fundamental and Applied Toxicology* 1: 67–80.

Calabrese, E.J. 1978. *Methodological Approaches to the Development of Environmental and Occupational Health Standards.* New York: John Wiley and Sons.

Calabrese, E.J. 2008a. Hormesis: Principles and applications for pharmacology and toxicology. *American Journal of Pharmacology and Toxicology* 3: 59–71.

Calabrese, E.J. 2008b. Hormesis: Why it is important to toxicology and toxicologists. *Environmental Toxicology and Chemistry* 27(7): 1451–1474.

Calabrese, E.J. 2009a. The road to linearity: Why linearity at low doses became the basis for carcinogen risk assessment. *Archives of Toxicology* 83: 203–225.

Calabrese, E.J. 2009b. Getting the dose-response wrong: Why hormesis became marginalized and the threshold model accepted. *Archives of Toxicology* 83: 227–247.

Calabrese, E.J. 2011a. Key studies used to support cancer risk assessment questioned. *Environmental and MolecularMutagenesis* 52: 595–606.

Calabrese, E.J. 2011b. Toxicology rewrites its history and rethinks its future: Giving equal focus to both harmful and beneficial effects. *Environmental Toxicology and Chemistry* 30(12): 2658–2673.

Calabrese, E.J. 2013. Hormetic mechanisms. *Critical Reviews in Toxicology* 43(7): 580–606.

Calabrese, E.J., and Baldwin, L.A. 2000a. Chemical hormesis: Its historical foundations as a biological hypothesis. *Human and Experimental Toxicology* 19(1): 2–31.

Calabrese, E.J., and Baldwin, L.A. 2000b. The marginalization of hormesis. *Human and Experimental Toxicology* 19(1): 32–40.

Calabrese, E.J., and Baldwin, L.A. 2000c. Radiation hormesis: Its historical foundations as a biological hypothesis. *Human and Experimental Toxicology* 19(1): 41–75.

Calabrese, E.J., and Baldwin, L.A. 2000d. Radiation hormesis: The demise of a legitimate hypothesis. *Human and Experimental Toxicology* 19(1): 76–84.

Calabrese, E.J., and Baldwin, L.A. 2000e. Tales of two similar hypotheses: The risk and fall of chemical and radiation hormesis. *Human and Experimental Toxicology* 19(1): 85–97.

Calabrese, E.J., and Baldwin, L.A. 2001. Hormesis: U-shaped dose responses and their centrality in toxicology. *Trends in Pharmacological Sciences* 22(6): 285–291.

Calabrese, E.J., and Blain, R.B. 2005. The occurrence of hormetic dose responses in the toxicological literature, the hormesis database: An overview. *Toxicology and Applied Pharmacology* 202(3): 289–301.

Calabrese, E.J., and Blain, R.B. 2009. Hormesis and plant biology. *Environmental Pollution* 157(1): 42–48.

Calabrese, E.J., and Blain, R.B. 2011. The hormesis database: The occurrence of hormetic dose responses in the toxicological literature. *Regulatory Toxicology and Pharmacology* 61(1): 73–81.

Calabrese, E.J., and Kenyon, E. 1991. *Air Toxics and Risk Assessment.* Lewis Publishers, Boca Raton, FL.

Calabrese, E.J., Hoffmann, G.R., Stanek III, E.J., and Nascarella, M.A. 2010. Hormesis in high-throughput screening of antibacterial compounds in *E. coli. Human and Experimental Toxicology* 29: 667–677.

Calabrese, E.J., Stanek III, E.J., Nascarella, M.A., and Hoffmann, G.R. 2008. Hormesis predicts low-dose responses better than threshold model. *International Journal of Toxicology* 27: 369–378.

Calabrese, E.J., Staudenmayer, J.W., Stanek, E.J., and Hoffmann, G.R. 2006. Hormesis outperforms threshold model in NCI anti-tumor drug screening data. *Toxicological Sciences* 94: 368–378.
Caspari, E., and Stern, C. 1948. The influence of chronic irradiation with gamma rays at low dosages on the mutation rate in *Drosophila melanogaster*. *Genetics* 33(1948): 75–95.
Luckey, T.D. 1980. *Ionizing Radiation and Hormesis*. CRC Press, Boca Raton, FL.
Luckey, T.D., Venugopal, B., and Hutcheson, D. 1975. *Heavy Metal Toxicity, Safety and Hormology*. Academic Press, New York.
Muller, H.J. 1927. Artificial transmutation of the gene. *Science* 66(1699): 84–87.
Muller, H.J. 1930. Radiation and genetics. *American Naturalist* 64(692): 220–251.
Muller, H.J., and Mott-Smith, L.M. 1930. Evidence that natural radioactivity is inadequate to explain the frequency of "natural" mutations. *Proceedings of the National Academy of Sciences of the United States of America* 16: 277–285.
National Academy of Sciences (NAS)/National Research Council (NRC). 1956. *The Biological Effects of Atomic Radiation (BEAR): A Report to the Public*. NAS/NRC, Washington, DC.
National Research Council (NRC). 1983. *Risk assessment in the Federal Government: Managing the Progress*. National Research Council, National Academy Press, Washington, DC.
National Research Council (NRC). 1993. *Issues in Risk Assessment. I. Use of the Maximum Tolerated Dose in Animal Bioassays for Carcinogenicity*. National Research Council, National Academy Press, Washington, DC.
National Research Council (NRC). 1994. *Science and Judgment in Risk Assessment*. National Research Council, National Academies Press, Washington, DC.
National Research Council (NRC). 2007. *Toxicity Testing in the 21st Century: A Vision and a Strategy*. National Academy Press, Washington, DC.
Olson, A.R., and Lewis, G.N. 1928. Natural reactivity and the origin of species. *Nature* 121(3052): 673–674.
Popoff, M., and Gleisberg, W. (eds). 1924–1930. *Zell-Stimulation Forschungen*, Volumes 1–3. Paul Parey.
Risk Commission. 1997a. *Presidential/Congressional Commission on Risk Assessment and Risk Management*, Volume 1. Government Printing Office, Washington, DC.
Risk Commission. 1997b. *Presidential/Congressional Commission on Risk Assessment and Risk Management*, Volume 2. Government Printing Office, Washington, DC.
Schulz, H. 1887. Zur Lehre von der Arzneiwirdung. *Virchows Archiv Pathology Anatomy Physiologiel Klinische Medizin* 108: 423–445.
Schulz, H. 1888. Uber Hefegifte. *Pflugers Archiv Physiol Menschen Tiere* 42: 517–541.
Southam, C., and Ehrlich, J. 1943. Effects of extract of western red-cedar heartwood on certain wood-decaying fungi in culture. *Phytophatology* 33: 517–524.
Spencer, W.P., and Stern, C. 1948. Experiments to test the validity of the linear R-dose/mutation at low dosage. *Genetics* 33(1948): 43–74.
Stimulation Newsletter. 1970–1976. (Nos 1–10). Institut für Strahlenbotanik der Gesellschaft für Strahlen-und Umweltforschung MBH München.
Timoféeff-Ressovsky, N.W., Zimmer, K.G., and Delbruck, M. 1935. Uber die Natur der Genmutation und der Genstruktur. Nachrichten von der Gesellschaft der Wissenschaften zu Gottingen:Mathematische-Physikalische Klass, Fachgruppe VI *Biologie* 1: 189–245.
EPA 2004. *An Examination of EPA Risk Assessment Principles and Practices*. EPA/100/B/001 March, Washington, DC.
Uphoff, D.E., and C. Stern. 1949. The genetic effects of low intensity in irradiation. *Science* 109: 609–610.

# Index

## A

Absorbed dose, 109
2-Acetylaminofluorene (2-AAF), 347
Activating transcription factor-6 (ATF6), 235–236
Acute myocardial infarction (AMI), 168
Adaptive recovery, from stressful input, 208–209
Adaptive response, 345
    and hormesis, parallels between, 316–317
    and preconditioning, 314–316
ADF, see Alternate-day fasting
Adjusted odds ratios (AORs), 123
ADMF, see Alternate-day modified fast
Aging, 83–84
Agricultural productivity, hormesis and, 329–330
Akt, see Protein kinase B (PKB)
α–Radiation-induced lung cancers, dose-response relationship for, 114
Alternate-day fasting (ADF), 81–84
Alternate-day modified fast (ADMF), 81
Alzheimer's disease, 9
    transgenic mouse model, 83
Ames test, 323
AMI, see Acute myocardial infarction
Amlexanox, 263
AMP-activated protein kinase (AMPK), 233–234
Anaerobic glycolysis, 171
Angina pectoris, 177
Antioxidant DNA response element (ARE), 134
Antioxidant gene expression control, redox-sensitive signaling pathways, 283, 285
    AP-1, 288
    MAPK, 286–288
    NF-κB, 285–286
Antioxidant response, 231
Antitumor immune response, activation of, 132
AORs, see Adjusted odds ratios
AP, see Autophagy
Apoptosis, 135–136
AP-1 pathway, 288
ARE, see Antioxidant DNA response element
Arndt–Schulz Law, 342, 345
Arsenic, 260
Arsenite/cadmium, homologous postconditioning hormesis, 21
Ataxia-telangiectasia mutated (ATM), 234
ATF6, see Activating transcription factor-6
ATM, see Ataxia-telangiectasia mutated
Atomic bomb survivors, health status in, 128–130
Autophagy (AP), 135–136, 232

## B

Balneotherapy, 156
    hyperthermia to, 153–155
Basal-like breast cancer, 253
BDNF, see Brain-derived neurotrophic factor
BEAR, see Biological Effects of Atomic Radiation
BEIR, see Biological Effects of Ionizing Radiation
Benchmark dose (BMD), 313
β-pattern, 312
Biological Effects of Atomic Radiation (BEAR), 341
Biological Effects of Ionizing Radiation (BEIR), 110–111, 120, 139
Biological stress responses, hormesis and, 314–317
Biophotons, 205
Biphasic dose-response relationship, 3, 309–310
BMD, see Benchmark dose
Brain-derived neurotrophic factor (BDNF), 219

## C

Canadian Breast Fluoroscopy Study, 117
    breast cancer mortality in, 118
Canadian Nuclear Safety Commission (CNSC), 114
Cancer, 252–254
Cardiac protection, ischemia/reperfusion injury and mechanisms of, 171–176
Cardiovascular diseases (CVDs), 168
    iron deficiency
        ferric carboxymaltose therapy, 98
        risk factor, 99
    polypill, 46
Cell protection
    concept of, 176–177
    model of, 172–173
Cellular quality control, 27
Cerebral ischemia
    ischemic tolerance, see Ischemic tolerance
    transient ischemic attack, 186
        neuroprotection conferred by, 193–195
CHD, see Coronary heart disease
Chernobyl accident, health effects of, 124
Childhood leukemia, 124
Chronic heart failure, iron deficiency, 99
Chronic iron deficiency, 98

357

Chronic obstructive pulmonary disease (COPD), 158
Cigarette smoking, lung cancer induction by, 114–115
c-Jun-N-terminal kinase (JNK) pathway, 254–257
Climatotherapy, 156
CNSC, see Canadian Nuclear Safety Commission
Cognitive dissonance
  creation of, 211–212
  pain and gain, 213
Cognitive health, 83–84
Cold baths, 158
Cold water, 156
Complexity theory, 216
  MindBodyMatrix, 207–208
Computed tomography (CT), 117
Concomitant toxicity, hormesis and, 326–327
Conductive heating, 153
Convective heating, 153
COPD, see Chronic obstructive pulmonary disease
Coronary heart disease (CHD), 168
Cross-conditioning, ischemic tolerance, 194
Cullin 3, 47
CVDs, see Cardiovascular diseases

### D

DDR, see DNA damage response
Default model, hormetic concept, 353
Defensive collapse, stress and, 208–209
Deferoxamine (DFO), 95
Deoxyribonucleic acid repair, 133–134
DFO, see Deferoxamine
Diagnostic medical exposures, therapeutic and, 116–120
Diagnostic radiation procedures, 117–118
Dietary phytochemicals, 258
Dis-ease, 220
Dis-order, 220
DNA damage response (DDR), 234–235
DNA-dependent protein kinase catalytic subunit (DNA-PKcs), 251
Dose equivalent, 109
Dose-response model, 308–309, 344
  criteria, 350
  validation, risk assessment and, 345–347
Dose-response relationship, 206–207, 210
Double-strand breaks (DSBs), 133
  DNA damage response, 234
*Drosophila melanogaster*, 135, 137
DSBs, see Double-strand breaks

### E

Ecological methodology, 110
$ED_{01}$ mega-mouse study, 347

EGCG, see Epigallocatechin gallate
Eicosapentaenoic acid (EPA), 206
Electric Power Research Institute (EPRI), 115, 139
Endoplasmic SR, 235
Endothelial cells, LPS stimulation of, 259
Energy quanta, 205
Energy stress response (ESR), 233–234
Environmental background radiation, 120–121
Environmental Protection Agency (EPA), 341, 352
Environmental quality, hormesis and, 329–330
Environmental stressors, 203–204
EPA, see Eicosapentaenoic acid; Environmental Protection Agency
Epigallocatechin gallate (EGCG), 49, 252
Epigenetic alterations, 136–138
Epigenetic mechanisms, 64–66
EPRI, see Electric Power Research Institute
ERK pathway, see Extracellular signal–regulated kinase pathway
ERR, see Excess relative risk
ESR, see Energy stress response
Eustress, 211
Excess relative risk (ERR), 130
Exercise
  activation by
    PGC-1α signaling, 292, 294
    redox-sensitive signaling, 288–289
  mitochondrial remodeling, effect on, 295–296
  oxidative stress response, 232
  resilience and adaptability, 205–206
Exercise hormesis
  antioxidants and ROS, 40–41
  bell-shaped hormetic curve, 41
  exercise-associated adaptive response, 38–39
  regular exercise acts, 39–40
Extracellular signal–regulated kinase (ERK) pathway, 254–257

### F

Fangotherapy, 157
Fasting
  chemotherapy, reduce side effects of, 80
  periodic, see Periodic fasting
Fenton and Haber–Weiss reactions, 94
Ferdinand Hueppe, see Arndt–Schulz Law
Ferric Iron Sucrose in Heart Failure (FERRIC-HF), 98
Flow optimization, information and energy, 205–206
Fluid matrix, 208
Forkhead box O transcription factors, 61–62
Free radical, 169
Free radical scavenging, 134–135
Frustration, 245
  optimal, 202–203

# Index

## G

Geographical methodology, 110
Glucose-regulated proteins (GRPs), 23
Glutathione-$S$-transferases, 63
Grieving process, 214–216
Ground regulation system, 204–205
Growth-promoting opportunity, 204
GRPs, see Glucose-regulated proteins

## H

Handicap, transient ischemic attack and, 192
Harmonization, in risk assessment, 348
HAT, see Histone acetyltransferase
Health effects, professionally exposed groups, 110–111
Health status
  in atomic bomb survivors, 128–130
  in medical professional groups, 111–113
Heart failure (HF), 167
Heat-shock element (HSE), 27
Heat shock factors (HSFs), 27, 160, 228, 230
Heat shock, homologous postconditioning hormesis, 20–21
Heat-shock proteins (HSPs), 17, 154, 228
  augmented basal levels of, 26
  induction of, 84
Heat shock response (HSR), 135, 228, 230
Heat shock transcription factors, 51–53
Heme iron intakes, 100
Heterologous postconditioning hormesis
  molecular aspects of, 25
  similarity, issue of, 24
HIFs, see Hypoxia-inducible factors
High-intensity interval training, 205–206
Histone acetyltransferase (HAT), 57
Hit theory, 309
$H_2O_2$
  redox-sensitive signaling pathways, 283
  sources of cellular production, 284
Homeothermal, 156
Homologous postconditioning hormesis, 20–21
Hormesis, 136, 342, 345
  APK/ERK signaling in toxicology, inflammation, and, 257–260
  appliying to risk assessment, 348–350
  biological fields, 7
  and biomedical ethics, 321
  biomedical sciences, 9–10
  biphasic dose-responses, 3–4
  cell protection concept, 176–177
  chaotic perspective of inflammatory pathways in, 265–269
  concepts and principles, 346
  default assumption in risk assessment, 350–351
  definition, 206, 207
  essential understanding, 327–330
  example of, 161
  generalizability, 322–323
  heterogeneity in susceptibility, 324–325
  history of, 7–8
  interactions among agents, 326
  low-dose stimulatory responses, 5
  nature of
    artifact, 314
    detection and measurement, difficulty with, 311–312
    evidence supports, 312–313
    NF-κB in inflammation, role of, 262–265
    periodic fasting in, 84
    prevalence of, disagreement about, 323
    radiation and chemical hormesis perspectives, 5, 6
    risk assessment, 318, 351–352
    threshold dose-response model, 4–5
    in Web of Knowledge database, 8
Hormetic cytoprotectant, in thermal waters, 159–160
Hormetic dose-response relationship, 206–207, 341, 348
  research findings, 346
Hormetic effects, uncertainties in quantification, 323–324
Hormetic stress response, 158
Hormetins, 237
Hormoligosis, 312
HPA axis function, see Hypothalamic–pituitary-adrenal axis function
$H_2S$, 159
HSE, see Heat-shock element
HSFs, see Heat shock factors
HSPs, see Heat-shock proteins
HSR, see Heat shock response
Hueppe's Rule, see Arndt–Schulz Law
Human body, normal temperature of, 154
Human vascular endothelial cells, 62
Hydrogen sulfide, as hormetic cytoprotectant in thermal waters, 159–160
Hydrotherapy, 156
Hydrothermal hormesis, therapeutic applications of, 155–159
Hydrothermal therapy, 161
Hyperbaric oxygen preconditioning, 196
Hyperphosphorylation of PERK, 260
Hyperthermal, 156
Hyperthermia, to balneotherapy, 153–155
Hypothalamic–pituitary–adrenal (HPA) axis function, 158–159
Hypothermal, 156
Hypoxia-inducible factors (HIFs), 231–232
Hypoxia, oxidative stress response, 231

## I

IAE, *see* International Atomic Energy
IκB kinase (IKK), NF-κB pathway, 285
Immersion, in water, 157
Immune system, 244
Immune system response, 131–133
Inflammation, 170, 243, 246
   and hormesis
      PI3K/Akt signaling in, 250–252
      role of NF-κB in, 262–265
      toxicology, APK/ER K signaling in, 257–260
   in network and chaotic perspective, 243–246
   PGC-1α as suppressor, 292, 294
Inflammatory pathways, 243
   hormesis, chaotic perspective of, 265–269
Inflammatory response, 236–237
Inhibitor of κB kinase (IKK), 262
Inositol-requiring enzyme-1 (IRE1), 235
Insulin resistance, 99–101
Intermittent fasting, *see* Periodic fasting
International Atomic Energy (IAE), 124
International Commission on Radiological Protection, 111
Interventional cardiologists, 112
In vitro model, ischemic tolerance, 194
In vivo model, ischemic tolerance, 193–194
IPC, *see* Ischemic preconditioning
Iron
   deficiency, 96–99
   iron overload and insulin resistance/type 2 diabetes, 99–101
   osteogenesis, 96
   ROS and RNS, 94–95
Iron overload, 99–101
Ischemia/reperfusion
   homologous postconditioning hormesis, 21
   injury and mechanisms of cardiac protection, 171–176, 178
Ischemic preconditioning (IPC), 172
   short- and long-term protection of, 176
Ischemic stroke, 186
Ischemic tolerance
   clinical application of, 195–197
   definition, 193
   endogenous mechanisms of, 195
   remote preconditioning and cross-conditioning, 194
   time windows for, 194–195
   in vitro and in vivo models, 193–194
Isoflavonoids, 66

## K

Kelch-like ECH-associated protein 1 (Keap1), 47

## L

Lacunar strokes, 192
LC3, *see* Light chain 3
LD ionizing radiation, 108, 110, 136
LD radiation, 120, 130, 132, 134, 139
LD radioimmunotherapy, 119
LD x-ray radiation exposure, effects of, 112
LET radiation, *see* Linear energy transfer radiation
Leukemia, 128
Lifespan, 83–84
Light chain 3 (LC3), nutritional stress response, 233
Linear energy transfer (LET) radiation, 127
Linear no-threshold (LNT) model, 108, 139, 308–309, 340, 344, 350, 353
   hormesis as a challenge, 310–311
   and radiation protection, 319
Liposarcoma cells, pretreatment of, 57
LNT-based dose response modeling, 343
LNT model, *see* Linear no-threshold model
Long-term follow-ups, in nuclear test participants, 116
Long-term health consequences of accidents, nuclear power plants, 123–128
Low-dose effects, assumptions for, 321–322
Low-dose radiation effects in human studies, 109–110
   diagnostic and therapeutic medical exposures, 116–120
   environmental background radiation, 120–121
   health status in atomic bomb survivors, 128–130
   long-term health consequences, nuclear power plants, 123–128
   professionally exposed groups, health effects in, 110–116
   residential radon and lung cancer risk, 121–123
LPS-induced endothelial cells, ICAM-1 expression of, 259
Luciferase, 28
Lung cancer risk, residential radon and, 121–123

## M

Macroautophagy, 232
Magnetic resonance imaging (MRI), transient ischemic attack, 186
Mammalian target of rapamycin (mTOR), autophagy, 232
MAPK kinase kinases (MAP3Ks), 256
MAPKs, *see* Mitogen-activated protein kinases
*Matrix and Matrix Regulation: Basis for a Holistic Theory in Medicine*, 205
Maximum tolerated dose (MTD), 343

# Index

Medical professional groups, health status in, 111–113
Medication, poison *vs.*, 217–218
Mega-mouse study, 347
Mental potential optimization, 219–220
Metabolic processes
    phase I metabolism, 63
    phase II metabolism, 63–64
    phase III transporters and cross talk, 64
microRNAs (miRNAs), 60
Mild radium therapy, 107
Mild stress, 19
    responses, optimization of benefits, 327–328
MindBodyMatrix, 204, 206–208
miRNAs, *see* microRNAs
Mitochondrial biogenesis, 290–292, 294
Mitochondrial permeability transition pores (mPTP), 173
Mitochondrial remodeling
    exercise, effect of, 295–296
    fusion and fission, 294–295
    UCP role in antioxidant defense, 296–297
Mitogen-activated protein kinases (MAPKs), 174, 254–257, 286–288
    cascade, 265–267
    in toxicology, inflammation, and hormesis, 257–260
Molecular SR pathways
    DNA damage response, 234–235
    energy stress response, 233–234
    heat shock response, 228, 230
    inflammatory response, 236–237
    nutritional stress response, 232–233
    overview, 227–228
    oxidative stress response, 231–232
    stressors and effectors, 229
    unfolded protein response, 235–236
Mortality, 129
    Canadian cohort study of, 113
mPTP, *see* Mitochondrial permeability transition pores
MRI, *see* Magnetic resonance imaging
MTD, *see* Maximum tolerated dose
mTOR, *see* Mammalian target of rapamycin
Mudpacks, 157
Multiple sclerosis pathogenesis, 170
Muscle contraction, redox regulation of antioxidant enzymes, 289–290

## N

$Na^+/H^+$ exchanger (NHE), 171
NAM mononucleotide (NMN), 58
National Academy of Sciences (NAS), 341
National Center for Toxicological Research, 311–312
National Health Interview Survey, 80
National Toxicology Program (NTP) bioassay, 351
Network paradigm, 244
Neurodegenerative autoimmune diseases, reactive oxygen species and, 170
New York Heart Association (NYHA), 98
NFE2L2, *see* Nuclear factor (erythroid-derived 2)-like 2
NF-κB, *see* Nuclear factor-kappa B
NHE, *see* $Na^+/H^+$ exchanger
NMN, *see* NAM mononucleotide
NOAEL, *see* No observed adverse effect level
Nonalcoholic fatty liver disease, 100
Nonmonotonic, endocrine disruption, 345
No observed adverse effect level (NOAEL), 308
Normoxia, 232
NSR, *see* Nutritional stress response
NTP bioassay, *see* National Toxicology Program bioassay
Nuclear factor-kappa B (NF-κB)
    inflammatory response, 236
    inhibition of, 54
    PARP-1, 55–57
    pathway, 285–286
    RelA–p65 protein complex of, 60
    signaling pathway, 260–265
Nuclear factor (erythroid-derived 2)-like 2 (NFE2L2), 134, 159, 231
    EGCG, 49
    Keap1–Nrf2–ARE pathway, 51
    oxidative activation of, 48
Nuclear power plants, long-term health consequences of accidents, 123–128
Nuclear test participants, long-term follow-ups in, 116
Nuclear worker cohort studies, 113–115
Nucleophilic tone, 51
Nutritional components
    epigenetic mechanisms, 64–67
    forkhead box O transcription factors, 61–62
    heat shock transcription factors, 51–53
    metabolic processes, 62–64
    NFE2L2, 47–51
    NF-κB, 53–57
    polypill, 46
    sirtuin 1, 57–61
Nutritional stress response (NSR), 232–233
NYHA, *see* New York Heart Association

## O

Optimal frustration, 202–203
Optimal HSR, 230
Optimal stress, 204
    traumatic stress *vs.*, 210–211
OR, 247–250
Oscillating systems, 246
OSR, *see* Oxidative stress response

Osteogenesis, 96
Oxidative damage markers, 42
Oxidative-hormetic model, 264
Oxidative stress, 265
Oxidative stress response (OSR), 231–232
Oxidative stress theory, 42
Oxygen-dependent radicals, 169

## P

Pathogen-associated molecular pattern molecules (PAMPs), 257
Patient's defensive structures, challenge *vs.* support, 209–210
PDEs, *see* Phosphodiesterases
PDK1, *see* Phosphoinositide-dependent kinase-1
Periodic fasting
    cardioprotective benefits, 82
    cognitive health, aging, and lifespan, 83–84
    energy balance and cardiometabolic health, 81–82
    health improvement strategy, 86–87
    hormes in, 84–86
    overview of, 80–81
PERK, *see* Protein kinase RNA-like ER kinase
Permissible dose, 340
Perturbation, self-organizing systems resist, 209
PGC-1α, mitochondrial adaptations role, 291–294
PG/GAG, *see* Proteoglycan/glycosaminoglycan
Phlebotomy, 100
Phosphatase and tensin homolog (PTEN), 175, 249
Phosphodiesterases (PDEs), 60
Phosphoinositide-dependent kinase-1 (PDK1), 247
Phosphoinositide 3-kinase (PI3K), 175, 247
Phosphorylation, MAPKs, 256
Physical inactivity, 37
PI3K/Akt pathway
    in inflammation and hormesis, 250–252
    OR, 247–250
Poison *vs.* medication, 217–218
Polyphenols, 46
Polypill, 46
Poly(ADP-ribose)-polymerase (PARP)-1, 55
Postconditioning hormesis
    action, terminology and mechanism of, 16–17
    arsenite/cadmium, 21
    aspects, 15
    case of, 14
    cross-sensitization, 23–25
    descriptions, 16
    homologous heat postconditioning, 20–21
    recuperative mechanisms, enhancement of, 27–28
    sensitization, issue of, 19–21
    stress conditions, 22

Posttranslational modifications (PTMs), 53
Precautionary principle, hormesis in risk assessment, 319–321
Preconditioning, for phenomena, 345
    adaptive responses and, 314–316
Preconditioning hormesis
    action, terminology and mechanism of, 16–17
    aspects and clinical settings, 15
    case of, 14
    cross-tolerance, 23
    desensitization, issue of, 18–19
    experimental settings, 15–16
    homologous and heterologous strategies, 22
    protective mechanisms, induction of, 26–27
Professionally exposed groups, health effects in, 110–111
    medical professional groups, health status in, 111–113
    nuclear test participants, long-term follow-ups in, 116
    nuclear worker cohort studies, 113–115
Proportionality rule, 339, 340
Protein denaturation, 28
Protein kinase B (PKB), 247, 249
Protein kinase RNA-like ER kinase (PERK), 236
Proteoglycan/glycosaminoglycan (PG/GAG), 205
Psychological resilience, 218
Psychopharmacology, hormetic effect in, 206
Psychotherapy, 216
PTEN, *see* Phosphatase and tensin homolog
PTMs, *see* Posttranslational modifications
Public health, hormetic effects and, 328–329

## Q

Quality control systems, 27

## R

Radiant heat, 153
Radiation carcinogenesis, LNT model of, 108
Radiation dose, 116
    units of, 109
Radiation exposure, 107–109
    dose, units of, 109
    low-dose radiation effects in human studies, *see* Low-dose radiation effects
    radiation hormesis, mechanisms of, 130–131
        autophagy and apoptosis, 135–136
        deoxyribonucleic acid repair, 133–134
        epigenetic alterations, 136–138
        free radical scavenging, 134–135
        heat-shock response, 135
        immune system response, 131–133
Radiation-induced foci (RIFs), 133–134
Radioactivity, thermal and mineral waters, 161
Radiotherapy, 118–120

# Index

Radon gas, 121, 161
Radon therapy, 161
Radon-thermal water, health effects of, 161
Reactive nitrogen species (RNS), 94
Reactive oxygen species (ROS), 38, 47, 84, 85, 94, 169, 282
    dose-response, bell-shape curve type of, 42
    maximal/optimal levels of, 41
    and neurodegenerative autoimmune diseases, 170
    oxidative stress response, 232
Redox factor-1 (Ref-1), 175
Redox-sensitive signaling pathways
    antioxidant gene expression control, 283, 285
        AP-1, 288
        MAPK, 286–288
        NF-κB, 285–286
    exercise activation of, 288–289
    general mechanisms of, 283
    hormesis and, 282–283
    mitochondrial remodeling by, 294–297
    muscle contraction, 289–290
    PGC-1α in mitochondrial biogenesis and, *see* Mitochondrial biogenesis
Ref-1, *see* Redox factor-1
Reference doses (RfDs), 312
Reference group, 123
Regular exercise, benefit, 37–38
Relentless hope, 213–214
Remote preconditioning, ischemic tolerance, 194, 196
Renaturation process, 28
Repeated mild heat-shock (RMHS), 26
Reperfusion injury, 171
Residential radon, 121–123
Resilience, *see* Psychological resilience
RfDs, *see* Reference doses
RIFs, *see* Radiation-induced foci
Risk assessment, 339
    and dose-response model validation, 345–347
    historical overview, 339–345
    hormesis applying to, 348–351
    hormetic in concerns, 352–353
    improving harmonization in, 348
RMHS, *see* Repeated mild heat-shock
RNS, *see* Reactive nitrogen species
Rodent models, SIRT1 protein expression, 85
ROS, *see* Reactive oxygen species

## S

Sandpile model, 216–217, 245
Sauna therapy, 157
*Schisandra chinensis*, 258
Schulz's biphasic dose-response, 4
Self-organizing systems
    examples of, 208

    resist perturbation, 209
Sigmoid curve, threshold model, 308
Signaling pathways, 160
Sirtuin 1
    cAMP–PKA pathway, 59–60
    JNK1 activation, 59
    miRNAs, 60
    NAD+ levels, 58
    physiological and pathophysiological functions of, 57
Sirtuins (SIRTs), energy stress response, 233–234
SMR, *see* Standardized mortality ratio
Sober acceptance, relentless hope and, 213–214
Solid cancer risk, Chernobyl emergency workers, 126
SR, *see* Stress response
Standardized mortality ratio (SMR), 112
*Staphylococcus epidermidis*, 9
Stress, *see also* Optimal stress
    condition, 20
    defensive collapse and adaptive recovery, 208–209
    definition, 227
    dose-response relationship, 206–207
    environmental factors, 203–204
    MindBodyMatrix, paradoxical response, 206
    optimal frustration and challenge, 202–203
    paradoxical impact of, 216–217
    psychological factors, 203
    traumatic, 210–211
Stressors, 257
    moderate induction of, 264
Stress proteins, 228
Stress response (SR), 227, 314–317
    hormesis and, mechanisms of, 317–318
    molecular pathways, *see* Molecular SR pathways
    optimization of benefits, 327–328
Stroke, *see also* Cerebral ischemia
    ischemic, *see* Ischemic stroke
    lacunar, 192
    transient ischemic attack, 186
Survival stimulation factor, 25

## T

Tacrolimus, intra-coronary administration of, 254
Therapeutic medical exposures, diagnostic and, 116–120
Therapeutic targets, molecular and, 159–160
Thermal baths, 155
Thermal therapies
    hormetic cytoprotectant in thermal waters, 159–160

hydrothermal hormesis, therapeutic
    applications of, 155–159
  from hyperthermia to balneotherapy, 153–155
  radon-thermal water, health effects of, 161
Thermal waters, hydrogen sulfide as hormetic
    cytoprotectant, 159–160
Three Mile Island accident, 124
Threshold dose-response model, 4–5, 339, 341
Threshold model, 308–311
TIA, *see* Transient ischemic attack
Time window, cerebral ischemic tolerance, 194–195
Tissue injury, process of, 254
TLR, *see* Toll-like receptor
TNF, *see* Tumor necrosis factor
TNF-α, *see* Tumor necrosis factor-α
Toll-like receptor (TLR), 259
Transient ischemic attack (TIA)
  outcome after subsequent ischemic stroke, 192–193
  prestroke, effect of, 192
  stroke, risk factor of, 186
  World Health Organization (WHO), TIA, 186
Traumatic stress, 204
  *vs.* optimal stress, 210–211
Trigger phase, 172
Tumor necrosis factor (TNF), 236–237
Tumor necrosis factor-α (TNF-α), 54
Type 2 diabetes, 100–101

**U**

UCP, *see* Uncoupling proteins
Ultrasensitivity, 266
Uncertainty factor, 319
Uncoupling proteins (UCP), mitochondrial
    antioxidant defense role, 296–297
UNDP, *see* United Nations Development
    Program

Unfolded protein response (UPR), 235–236
United Nations Development Program (UNDP), 124
United Nations Scientific Committee on
    the Effects of Atomic Radiation
    (UNSCEAR), 124
UPR, *see* Unfolded protein response
U.S. Department of Agriculture (USDA), 6
U.S. Food and Drug Administration, 9
U-shaped, dose-responses, 345
U.S. National Academy of Sciences, *see* National
    Academy of Sciences

**V**

Veratrine, 3

**W**

Web of life, 204–205

**X**

Xenobiotic response elements (XREs), 64
X-ray-induced mutations, proportionality rule
    for, 340
X-ray repair cross-complementing group 1
    protein (XRCC1) expression, 251
XREs, *see* Xenobiotic response elements

**Y**

Yerkes–Dodson Law, 14, 345
Youth Risk Behavior Survey, 80

**Z**

Zero equivalent point (ZEP), 310